Polymer Blends

VOLUME 1

Contributors

R. A. Dickie

J. R. Fried

H. B. Hopfenberg

F. E. Karasz

Sonja Krause

T. K. Kwei

W. J. MacKnight

D. R. Paul

Isaac C. Sanchez

R. S. Stein

H. Van Oene

T. T. Wang

Souheng Wu

POLYMER BLENDS

Edited by
D. R. PAUL
The University of Texas, Department of Chemical Engineering
Austin, Texas

SEYMOUR NEWMAN
Ford Motor Company, Plastics, Paint, and Vinyl Division
Detroit, Michigan

VOLUME 1

ACADEMIC PRESS, INC.
Harcourt Brace Jovanovich, Publishers
San Diego New York Berkeley Boston
London Sydney Tokyo Toronto

ACADEMIC PRESS, INC.
San Diego, California 92101

United Kingdom Edition published by
ACADEMIC PRESS LIMITED
24-28 Oval Road, London NW1 7DX

Library of Congress Cataloging in Publication Data

Main entry under title:

Polymer blends.

Bibliography: p.
1. Polymers and polymerization. I. Paul, Donald R.
II. Newman, Seymour, Date
TP1087.P64   668.4'1   77–6606
ISBN 0–12–546801–6 (v. 1)

Contents

List of Contributors   ix
Preface   xi
Contents of Volume 2   xv

Chapter 1 Background and Perspective

D. R. Paul

Terminology	2
Effect of Molecular Weight	4
Miscibility through Specific Interactions	7
Phase Equilibria and Transitions	9
Phase Morphology	11
Applications	13
References	14

Chapter 2 Polymer–Polymer Compatibility

Sonja Krause

I.	Introduction	16
II.	Theoretical Aspects	20
III.	Experimental Data	32
IV.	Prediction Schemes for Polymer–Polymer Compatibility	41
	References	106

Chapter 3 Statistical Thermodynamics of Polymer Blends

Isaac C. Sanchez

I.	Introduction	115
II.	Flory Theory	117
III.	Lattice Fluid	121
IV.	Predicting Polymer Blend Compatibility	132
	References	139

Chapter 4 Phase Separation Behavior of Polymer–Polymer Mixtures

T. K. Kwei and T. T. Wang

I.	Introduction	141
II.	Upper and Lower Critical Solution Temperatures	146
III.	Phase Separation	152
IV.	Experiment	159
V.	Results	165
VI.	Conclusion	183
	References	183

Chapter 5 Solid State Transition Behavior of Blends

W. J. MacKnight, F. E. Karasz, and J. R. Fried

I.	Criteria for Miscibility	186
II.	Measurements of the Glass Transition Temperature in Polyblends	193
III.	Limitations on the Use of Glass Transition Temperatures as a Criterion for Blend Compatibility	197
IV.	Solid State Transition Behavior of Poly(2,6-Dimethylphenylene Oxide)-Polystyrene Blends	201
V.	Solid State Transition Behavior of PPO-Polychlorostyrene Blends	218
VI.	Unsolved Problems	229
VII.	Commercial Blends of PPO-Polystyrene: Noryl	234
	References	238

Chapter 6 Interfacial Energy, Structure, and Adhesion between Polymers

Souheng Wu

I.	Introduction	244
II.	Surface Tension	245
III.	Interfacial Tension	263
IV.	Adhesion between Polymers	278
V.	Summary	288
	References	288

Chapter 7 Rheology of Polymer Blends and Dispersions

H. Van Oene

I.	Introduction	296
II.	Flow Behavior of Viscoelastic Fluids	298

Contents

	III.	Cocurrent Flow of Two Fluids	304
	IV.	Rheological Behavior of Suspensions of Rigid Axisymmetric Particles in Viscoelastic Fluids	310
	V.	Rheological Behavior of Suspensions of Deformable Droplets	318
	VI.	Rheology of Filled Polymers	325
	VII.	Rheology of Polymer Blends	330
	VIII.	Conclusion	347
		References	349

Chapter 8 Mechanical Properties (Small Deformations) of Multiphase Polymer Blends

R. A. Dickie

	I.	Introduction	353
	II.	Background	354
	III.	Modulus–Composition Dependence	356
	IV.	Temperature Dependence of Composite Modulus	369
	V.	Time–Temperature Dependence	381
		References	388

Chapter 9 Optical Behavior of Polymer Blends

R. S. Stein

	I.	Introduction	393
	II.	Interaction of Radiation with Polymers	394
	III.	The Transparency and Scattering of Blends	401
	IV.	Other Optical Properties of Blends	433
	V.	Conclusions	442
		References	442

Chapter 10 Transport Phenomena in Polymer Blends

H. B. Hopfenberg and D. R. Paul

	I.	Introduction	445
	II.	Permeation through Multiphase (Heterogeneous) Polymeric Systems	454
	III.	Sorption of Vapors and Liquids in Polymer Blends	468
	IV.	Membranes Prepared from Polymer Blends	479
	V.	Summary	485
		References	487

Appendix Conversion Factors to SI Units

491

Index   493

List of Contributors

Numbers in parentheses indicate the pages on which the authors' contributions begin.

R. A. *Dickie* (353), Engineering and Research Staff, Ford Motor Company, Dearborn, Michigan 48121

J. R. *Fried* (185), Corporate Research Department, Monsanto Company, St. Louis, Missouri 63166

H. B. *Hopfenberg* (445), Department of Chemical Engineering, North Carolina State University, Raleigh, North Carolina

F. E. *Karasz* (185), Department of Polymer Science and Engineering, University of Massachusetts, Amherst, Massachusetts 01003

Sonja Krause (15), Department of Chemistry, Rensselaer Polytechnic Institute, Troy, New York

T. K. *Kwei* (141), Bell Laboratories, Murray Hill, New Jersey 07974

W. J. *MacKnight* (185), Department of Polymer Science and Engineering, University of Massachusetts, Amherst, Massachusetts 01003

D. R. *Paul* (1, 445), Department of Chemical Engineering, The University of Texas, Austin, Texas 78712

*Isaac C. Sanchez** (115), Materials Research Laboratory and Department of Polymer Science and Engineering, University of Massachusetts, Amherst, Massachusetts 01003

R. S. *Stein* (393), Polymer Research Institute and Department of Chemistry, University of Massachusetts, Amherst, Massachusetts 01003

H. *Van Oene* (295), Engineering and Research Staff, Ford Motor Company, Dearborn, Michigan 48121

T. T. *Wang* (141), Bell Laboratories, Murray Hill, New Jersey 07974

Souheng Wu (243), Central Research and Development Department, Experimental Station, E. I. du Pont de Nemours & Company, Wilmington, Delaware 19898

* Present address: National Bureau of Standards, Polymer Division 311.00, Washington, D.C. 20237.

ix

Preface

At different times in the history of polymer science, specific subjects have come to "center stage" for intense investigation because they represented new and important intellectual challenges as well as technological opportunities. Dilute solution behavior, chain statistics, rubber elasticity, tacticity, single crystal formation, and viscoelastic behavior have all had their day of peak interest and then taken their place for continuing investigation by the community of polymer scientists. These periods of concentrated effort have served to carve out major new areas of macromolecular science to add to and build on the efforts of previous workers.

Polymer blends have now come to the fore as such a major endeavor. Their current and potential technological importance is remarkable and their ubiquitous presence in consumer products is testimony to their commercial importance.

Furthermore, pursuit of our understanding of the physical and mechanical properties of blends has uncovered new principles, refined earlier fundamental concepts, and revealed further opportunities for research and practical problem solving. In this last respect, polymer blends or "polyblends" offer a strong analogy to the previously established role of copolymerization as a means of combining the useful properties of different molecular species, but blends allow this to be done through physical rather than chemical means.

Our purpose, therefore, in organizing this work has been to underscore the importance of mixed polymer systems as a major new branch of macromolecular science as well as to provide academic and industrial research scientists and technologists with a broad background in current principles and practice. A wide range of subjects must be covered to meet these objectives, and no individual could be expected to have the knowledge and experience required to write an authoritative treatment for any significant fraction of these subjects. Consequently, the book was written by a group of authors—selected by the editors because of their particular expertise and contributions. However, a strong effort was made to have the outcome represent a cohesive treatment rather than a collection of separate contributions. The final outcome is the result of numerous decisions, and clearly different decisions could have been made. Considerable

xi

thought and consultation were devoted to a selection of topics that covered the main principles within the constraints of time and space.

Once the authors were selected, the final chapter outlines evolved through continuing consultation among the editors and authors. Each chapter has been reviewed and altered as necessary. In some cases, editorial footnotes have been added for clarification. There is extensive cross-referencing among chapters to emphasize the relations among chapters and to reduce duplication of content. A loose system of common nomenclature was developed, but exceptions were permitted to allow each subject to be developed in terms most commonly used in the literature and best understood by the reader. A mixture of unit systems appears here because various disciplines of science and technology are in different stages of conversion to the metric and SI systems at this time; hence, we have included in the Appendix selected conversion factors to help minimize the problems this creates.

We have elected to define a polymer blend as any combination of two or more polymers resulting from common processing steps. In keeping with this broad definition, we have elected to include, for example, the chapter on multilaminate films by Schrenk and Alfrey and the treatment of bicomponent fibers by Paul, both in Volume 2. However, since block and graft copolymers have been treated extensively in other recent publications, we have included only two chapters specifically devoted to these systems, i.e., Block Copolymers in Blends with Other Polymers by Kraus (Volume 2) and Interfacial Agents ("Compatibilizers") for Polymer Blends by Paul (Volume 2). However, additional references to such systems are made in other chapters.

For thermodynamic reasons, most polymer pairs are immiscible. Nevertheless, the degree of compatibility may vary widely, and since this aspect of polymer blends is of such underlying importance to morphology and properties, we have set aside a considerable portion of the book for this purpose. Krause's chapter, Polymer–Polymer Compatibility (Volume 1), provides an authoritative, comprehensive summary and interpretation of available information on a great number of systems reported in the literature. A review of more recent statistical mechanical theories for mixtures with particular attention given to lower critical solution temperatures is presented by Sanchez in Volume 1, whereas an interpretation of experimental work on phase separation and boundaries using NMR and other techniques is found in the contribution by Kwei and Wang, also in Volume 1.

Compatible polymers represent such unique cases in the realm of blends that the editors have sought to single out some of these for special discussion. Aside from relevant discussions by Krause and Kwei and Wang, the reader is referred to the following chapters: Blends Containing

Poly(ϵ-caprolactone) and Related Polymers, by Koleske (Volume 2); Solid State Transition Behavior of Blends, by MacKnight, Karasz, and Fried; Polymeric Plasticizers, by Hammer; and Transport Phenomena in Polymer Blends, by Hopfenberg and Paul.

A second cornerstone along with thermodynamics in the building of the often complex structure of incompatible disperse phases is the interaction of melt flow with interfacial behavior and the viscoelastic properties of multicomponent systems. We have sought to lay the groundwork for this fundamental aspect of polyblends in Wu's chapter, Interfacial Energy, Structure, and Adhesion between Polymers (Volume 1), and Van Oene's chapter, Rheology of Polymer Blends and Dispersions (Volume 1). More technologically oriented extensions of this subject will be found in Polyolefin Blends: Rheology, Melt Mixing and Applications, by Plochocki (Volume 2), and Rubbery Thermoplastic Blends, by Kresge (Volume 2).

Overall, mixed polymers provide an incredible range of morphological states from coarse to fine. One need only glance at Interpenetrating Networks (Thomas and Sperling, Volume 1) and Rubbery Thermoplastic Blends (Kresge) to sense the range of possibilities. However, aside from unusual possibilities of disperse phase, size, shape, and geometric arrangement such as described in Fibers from Polymer Blends (Paul, Volume 2) even more subtle complexities are possible with crystalline polymers wherein intricate arrangements of the crystalline and amorphous phases are possible as is shown by Stein in Optical Behavior of Blends (Volume 1). Here special investigatory techniques have been brought to bear on the disposition of molecules in the submicroscopic range. This chapter also presents the basic principles for understanding the optical properties of blends important in some applications.

Properties, along with composition and morphology, represent the third major area of interest. Historically, the modification of glassy polymers with a disperse rubber phase was one of the first major synthetic efforts directed to useful polymer blends. It is of particular importance because the mechanical properties of these synergistic mixtures are not related simply to the sum of the properties of the components. We have attempted to trace the growth of the concept of rubber modification in Newman's chapter, Rubber Modification of Plastics (Volume 2). A more detailed study of toughening theory for a restricted range of systems is presented in Fracture Phenomena in Polymer Blends, by Bucknall (Volume 2). These chapters deal largely with ultimate properties. Small strain behavior is dealt with in Mechanical Properties (Small Deformations) of Multiphase Polymer Blends, by Dickie.

Another major area of commercial importance of blending is the rubber industry, which capitalizes on the combination of properties each component contributes to the blend. The chapter by McDonel, Baranwal, and

Andries in Volume 2 illustrates the advantages and compromises in the very large area of application, Elastomer Blends in Tires. Somewhat related are the contrasting combinations of rubbery polymers and plastics treated by Hammer in Polymeric Plasticizers and Kresge in Rubbery Thermoplastic Blends.

The connections between polymer blends and the field of reinforced composites and its techniques of analysis are clearly evident in the chapters by Dickie on mechanical properties and by Hopfenberg and Paul on sorption and permeation. This is also seen in the chapters on films by Schrenk and Alfrey and fibers by Paul and to a lesser extent in the chapters by Thomas and Sperling and Newman.

A remarkable balance of diverse properties is achievable with blends as is evidenced by the chapters mentioned in the previous paragraphs. One unusual case of great commercial importance heretofore largely unnoticed in fundamental research is represented in Low Profile Behavior (Atkins, Volume 2) where polyblends of thermoplastics in cross-linked styrenated polyesters are used to control the volume of molded objects. The list of useful performance characteristics that can be controlled by polyblending is as long as the list of properties themselves.

An introductory chapter, written after all of the other chapters were completed and reviewed, has been included to fill some gaps and to provide a certain degree of perspective.

We have selected the above topics in order to achieve our goal of a comprehensive treatment of the science and technology of polymer blends. However, the growth of the field has made it necessary to publish this treatise in two volumes. We have attempted to divide the chapters equally between the two and to concentrate the fundamental and general topics in Volume 1 and the more specific and commercially oriented subjects in Volume 2.

In Volume 2, the topics dealt with are more specific and have a greater orientation toward commercial interests. It has been our purpose to make each chapter self-contained without making it necessary for the reader to be referred continually to other chapters. Nevertheless, some overlap is unavoidable in a cooperative undertaking. This does, however, permit the presentation of different points of view and thereby contributes to a balanced presentation. On the other hand, we have also sought to tie together the various chapters, and this has been accomplished in part by the addition of cross-references.

Many people have contributed to the development and completion of these two volumes through their advice and encouragement. However, we wish to mention specifically the help of F. P. Baldwin, J. W. Barlow, R. E. Bernstein, C. A. Cruz, S. Davison, P. H. Hobson, and J. H. Saunders.

Contents of Volume 2

11. Interpenetrating Polymer Networks
    D. A. Thomas and L. H. Sperling

12. Interfacial Agents ("Compatibilizers") for Polymer Blends
    D. R. Paul

13. Rubber Modification of Plastics
    Seymour Newman

14. Fracture Phenomena in Polymer Blends
    C. B. Bucknall

15. Coextruded Multilayer Polymer Films and Sheets
    W. J. Schrenk and T. Alfrey, Jr.

16. Fibers from Polymer Blends
    D. R. Paul

17. Polymeric Plasticizers
    C. F. Hammer

18. Block Copolymers in Blends with Other Polymers
    Gerard Kraus

19. Elastomer Blends in Tires
    E. T. McDonel, K. C. Baranwal, and J. C. Andries

20. Rubbery Thermoplastic Blends
    E. N. Kresge

21. Polyolefin Blends: Rheology, Melt Mixing, and Applications
    A. P. Plochocki

22. Blends Containing Poly(ϵ-caprolactone) and Related Polymers
    J. V. Koleske

23. Low-Profile Behavior
    K. E. Atkins

Appendix   Conversion Factors to SI Units

Index

Polymer Blends

VOLUME 1

Chapter 1
Background and Perspective

D. R. Paul

Department of Chemical Engineering
University of Texas
Austin, Texas

The field of polymer science and technology has undergone an enormous expansion over the last several decades primarily through chemical diversity. First, there was the development of new polymers from a seemingly endless variety of monomers. Next, random copolymerization was used as an effective technique for tailoring or modifying polymers. Later, more controlled block-and-graft copolymerization was introduced. The list of new concepts in polymer synthesis has not been exhausted. However, it has become clear that new chemical structures or organizations are not always needed to meet new needs or to solve old problems.

The concept of physically blending two or more existing polymers to obtain new products or for problem solving has not been developed as fully as the chemical approach but is now attracting widespread interest and commercial utilization. The successful implementation of this concept requires different knowledge and techniques than that used to develop new polymers. It is the purpose of this book to bring together the pertinent principles needed to implement and advance this more physical approach to polymer products. In the first part of this book, these principles are discussed in general or scientific terms; in the latter part they are illustrated by particular systems or with reference to specific applications.

The purpose of this introductory chapter is to develop some fundamental *background* pertinent to the early chapters and to give a *perspective* that to some degree will tie together the diverse subjects and viewpoints presented by the various authors throughout this book.

1

TERMINOLOGY

Polymer blends are often referred to by the contraction "polyblends" and sometimes as "alloys" to borrow a term from metallurgy. Various restricted definitions might be offered for any of these or other terms; however, the boundaries of what is intended are invariably imprecise, and the terms are not used with the same meaning by everyone. No attempt will be made here to adopt any rigid terminology, and the concept of blending will be discussed in the broadest possible manner. This book covers materials or products made by combining two or more polymers through processing steps into random or structured arrangements and includes geometries that might be regarded as polymer–polymer composites (see, e.g., Volume 2, Chapter 15 and parts of Chapter 16). We do not include, obviously, separately processed polymer items that are subsequently assembled into finished products. Block-and-graft copolymers share many common features and purposes as blends, but these materials, which generally differ from blends by only a few chemical bonds, are not included here except when they are components of blends (see Volume 2, Chapters 12 and 18).

The thermodynamics of polymer–polymer mixtures or composites is one of the most important fundamental elements since it plays a major role in the molecular state of dispersion, the morphology of two phase mixtures, the adhesion between phases, and consequently influences most properties and applications. Because of this, the early chapters including this one are devoted mainly to this subject. Like many aspects of polymer blends, an impediment to understanding the thermodynamics of blends has been a lack of suitable experimental techniques and theories; however, recent activities have made a start toward removing these deficiencies as will be seen here and in subsequent chapters. One of the first, but not the only, thermodynamic questions concerns the equilibrium miscibility or solubility of two polymeric components in a blend.

Very often the term "compatibility" is used synonymously with miscibility. However, in materials technology compatibility is a more general term with a wider diversity of meanings and implications, which in the extreme might result in two materials being classified as *incompatible* because they are *miscible*. In a strict technological sense, compatibility is often used to describe whether a desired or beneficial result occurs when two materials are combined together. If one is concerned with identifying polymeric plasticizers, then complete miscibility is desired, whereas, for rubbery impact modifiers of glassy plastics complete miscibility is not desirable. Generally speaking, most polymer pairs are not miscible (but more are miscible than was recognized only a few years ago) and by this terminology are incompatible. It is very likely that the use of the word *incompatible* as part of this general rule has been an unfortunate psychological impediment to commercial development of polymer blends because of the accompanying implication that poor results are inevitable when *incompatible* materials are combined. For many purposes, miscibility in polymer blends is neither a requirement nor desirable; however, adhesion between the components frequently is. In a fundamental sense, however, adhesion, interfacial energies, and miscibility are all interrelated thermodynamically in a complex way to the interaction forces between the two polymers.

Miscibility in polymer–polymer mixtures has been the subject of considerable discussion and debate in the literature. Frequently, the concern is over the size of the phases or domains implied by a particular observation; or, is mixing on a molecular or segmental scale (see e.g., [1, 2])? Interestingly, these questions are almost never raised about solutions of low molecular weight compounds, but apparently they arise naturally for *macro*molecules. Similar concerns existed many years ago about solutions of polymers in low molecular weight solvents and only disappeared when appropriate thermodynamic theories and experimental data appeared [3] which demonstrated that these solutions were not unusual or unique once the conformations and large size of the polymer chains were correctly considered.

Miscibility in every case is best understood, and ultimately can only be defined, in thermodynamic terms rather than through attempts that place overdue emphasis on details at the molecular or segmental level. Until recently, techniques for examining the thermodynamics of miscible polymer–polymer mixtures critically and unambiguously were extremely limited. The neutron scattering results that are now beginning to appear [4–6] seem to fill this need for conceptual clarification and quantitative results. In the experiments of interest here, one polymer is "dissolved" in a different polymer (the two may be regarded as solute and solvent), and the "solutions" are studied via neutron scattering in a manner analogous to classical light scattering of polymer in a solvent. Preferably, one of the polymers is deuterated, but this is of no fundamental concern here. Scattered intensities have been measured as a function of "solute" concentration and scattering angle and subsequently analyzed by the familiar Zimm plot used in treating *light* scattering from dilute polymer solutions. Interestingly, this analysis gives the correct molecular weight for the solute polymer, which is confirmation that these "solutions" are classical ones having miscibility in a true thermodynamic sense. Furthermore, the radius of gyration of the "solute" polymer was found to be of the size one expects in bulk or dilute solution and has a similar molecular weight dependence.

This is the first conclusive evidence that conformations in polymer–polymer solutions are substantially the same as those in other better-understood polymer states and implies that the segments of the two polymers are more or less in random contact just as solvent molecules and polymer segments mix in solution [3]. This kind of behavior was observed only for blends that showed a single glass transition temperature (T_g) (e.g., poly(methyl methacrylate)/styrene–acrylonitrile copolymer—see Stein *et al.* [7]). Zimm plots readily indicative of immiscibility were obtained for blends that showed two glass transitions [e.g., poly(methyl methacrylate)–poly(α-methylstyrene)]. This adds additional fundamental justification for use of the common and useful criteria of a single glass transition to indicate blend miscibility (see Chapter 5). In addition to the above important results, meaningful second virial coefficients consistent with other observations were obtained for the miscible blends. These results show that miscibility can be detected fundamentally in polymer–polymer solutions and has the same meaning as always. Furthermore, detailed knowledge about the scale of mixing is not required to decide whether miscibility exists or not; however, such information is essential when the miscibility or mixing is not complete in the thermodynamic sense. The technique of neutron scattering may be expected to be an important source of detailed, fundamental thermodynamic information on polymer–polymer systems in the future.

EFFECT OF MOLECULAR WEIGHT

The unique factor affecting the thermodynamics of polymer blends compared with other systems is the large molecular weight of *both* components. Generally, the qualitative thermodynamic argument, which limits miscibility to a rare occurrence in polymer blends, recognizes that the entropy of mixing ΔS_{mix} in the free energy of mixing expression

$$\Delta G_{mix} = \Delta H_{mix} - T \Delta S_{mix} \tag{1}$$

will be very small owing to the small number of moles of each polymer in the blend as a result of their large molecular weights. While the sign of the combinatorial entropy favors mixing, it is usually too small to result in the necessary negative free energy because the heat of mixing ΔH_{mix} is generally thought to be positive, at least for relatively nonpolar systems. It will be useful background to examine this situation in more detail here.

The Flory–Huggins solution theory [3], although inadequate for some purposes, provides a useful first approximation for the terms in Eq. (1), and its applications to polymer blends is treated in detail in Chapter 2. This theory [see Eq. (1) of Chapter 2] gives for the free energy of mixing polymers A and B

$$\Delta G_{mix} = RTV \left\{ \frac{\phi_A \ln \phi_A}{\tilde{V}_A} + \frac{(1 - \phi_A) \ln(1 - \phi_A)}{\tilde{V}_B} + \tilde{\chi}_{AB} \phi_A (1 - \phi_A) \right\} \tag{2}$$

where ϕ_A is the volume fraction of A, V_i the molar volume of i, and $\tilde{\chi}_{AB}$ (χ_{AB}/V_r in the notation of Chapter 2) is an interaction parameter related to the heat of mixing, which is positive for endothermic systems. The first two terms arise from the combinatorial entropy of mixing, and each is inversely related to the size or molecular weight of that component. It will simplify this discussion to assume that both polymers have the same molecular weight M and density ρ and to replace $\tilde{\chi}_{AB}$ with an equivalent parameter $2\rho/M_{cr}$, where M_{cr} will be a critical molecular weight. For this case, Eq. (2) can be rewritten as

$$\Delta G_{mix} = \frac{\rho VRT}{M_{cr}} \left\{ \frac{M_{cr}}{M} [\phi_A \ln \phi_A + (1 - \phi_A) \ln(1 - \phi_A)] + 2\phi_A(1 - \phi_A) \right\} \quad (3)$$

Figure 1 shows various cases calculated from Eq. (3) when the factor in front has any arbitrary value. The extreme upper and lower curves are the heat of mixing (for this model, $\Delta H_{mix} = \Delta G_{mix}$ when $M \to \infty$) and the entropy term in the free energy when $M = M_{cr}$, respectively. The

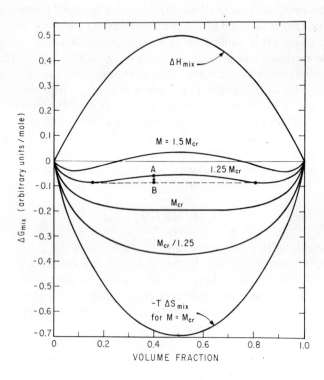

Fig. 1 Free energy of mixing and component terms of polymers A and B with the same molecular weight M. The curves were computed from Eq. (3).

intermediate curves are the free energy of mixing for various polymer molecular weights M, expressed relative to M_{cr}. Clearly, the free energy tends to more positive values as the molecular weight is increased as expected from the discussion of Eq. (1). It is, in fact, positive for some composition regions, but not all, for the case in which M is 50 % larger than M_{cr}, but it is never positive when M is only 25 % larger. For the latter case, a mixture whose composition locates it at Point A will have a negative free energy of mixing, and, therefore, simplistically one might expect the two polymers to mix since this is one of the thermodynamic requirements for processes to occur spontaneously. However, a mixture represented by Point A, if it did occur, would be unstable since it can lower its free energy even further to Point B by separating into two phases with compositions given by the end points of the dotted line. Thermodynamic stability of a one phase mixture exists only when

$$(\partial^2 \Delta G_{mix}/\partial \phi^2)_{T,p} > 0 \qquad (4)$$

as discussed further in Chapters 2–4 (or see [8, 9]). As it turns out, M_{cr} is the critical molecular weight [see Eq. (3a) of Chapter 2] where this condition is no longer satisfied for all compositions, whereas, for $M < M_{cr}$, the condition of Eq. (4) is fulfilled for all compositions. When $M > M_{cr}$, stable one-phase mixtures can exist at the extremities of the composition range, that is, partial miscibility; however, these composition zones become smaller as the ratio M/M_{cr} increases. For cases where $M_A \neq M_B$, the free energy curves in Fig. 1 would be skewed toward the side of the lower molecular weight component rather than be symmetrical as shown there.

As discussed in Chapter 2, the $\bar{\chi}_{AB}$ parameter, or the heat of mixing, can be estimated from the solubility parameters δ_A and δ_B if the components are relatively nonpolar. For the above example, rearrangement of Eq. (12) from Chapter 2 permits an estimate of M_{cr} for the nonpolar case

$$M_{cr} \simeq 2\rho RT/(\delta_A - \delta_B)^2 \qquad (5)$$

When the solubility parameters differ by 1.0 $(cal/cm^3)^{1/2}$, M_{cr} is less than 1200 at 25°C; however, when the two polymers are matched more closely and this difference is 0.1 $(cal/cm^3)^{1/2}$, M_{cr} rises to about 120,000 (see Table VII of Chapter 2). Thus, polymer molecular weight has a strong influence on miscibility for systems with endothermic heats of mixing. The situation is different for exothermic systems.

Interestingly, Flory's equation of state analysis has shown that the actual entropy of mixing for some systems is less than that given by the combinatorial value employed in Eq. (2) and may even be negative when the two polymers have very large molecular weights [10].

MISCIBILITY THROUGH SPECIFIC INTERACTIONS

As shown in the last section, miscibility of nonpolar polymers of substantial molecular weights occurs only when the solubility parameters are precisely matched. Attempts to find miscible polymer pairs by matching similar structures, or solubility parameters, is doomed to produce very few cases, unless the molecular weights are low, since the enthalpy of mixing at the very best can only approach zero. However, the enthalpy of mixing can be negative, or exothermic, if certain specific interactions between polar groups are involved, and consequently ΔG_{mix} will be negative in spite of the small entropy. These interactions may arise from a variety of mechanisms, such as dipole–dipole forces, but it is often useful to think in terms of donor and acceptor groups in analogy to hydrogen bonding. As a result, it may be possible to select two polymers having chemical moieties within or attached to the chains which have the proper *complementary dissimilarity* to yield an exothermic heat of mixing, although, it should be recognized that there will still be an endothermic contribution from the dispersive interactions, or van der Waals forces, between the remaining parts of the structure that do not interact specifically. This is illustrated nicely by some calorimetry data on model compounds designed to understand the solvating power of cyclic ethers and ketones for poly(vinyl chloride) [11]. Heats of mixing were measured for various singly and doubly chlorinated hydrocarbons (4–6 carbons) and poly(vinyl chloride) (PVC) with tetrahydrofuran (THF) and for hexane with cyclohexane. The heats of mixing ΔH_{mix} were all essentially parabolic with composition such that the ratio shown on the ordinate in Fig. 2 was nearly constant for a particular system. Two monomer units of PVC were used in computing the mole fractions x_A and x_B. These data are plotted versus the extent of chlorination in Fig. 2 where the cyclohexane–hexane system is taken as a nonpolar model for estimating the endothermic contribution of the dispersive bonding to ΔH_{mix} for the chlorinated hydrocarbon–THF system. It is interesting that all of the data for singly chlorinated hydrocarbons with THF fall in a very narrow band, while the doubly chlorinated ones fall in another similar band. The position of the chlorines has essentially no effect including replacing the α hydrogen with a methyl group. The latter appears to rule out the hydrogen bonding argument frequently invoked for such systems [11]. Instead, there seems to be some direct interaction between the chlorines and the oxygen, and all that matters is the number of interactions. The data in Fig. 2 are nicely connected by a straight line suggesting that the contributions from specific and dispersive interactions are simply additive as one might expect, that is,

$$\Delta H_{mix} = \Delta H_{mix}^{(dis)} + N \, \Delta \tilde{H}_{mix}^{(sp)} \tag{6}$$

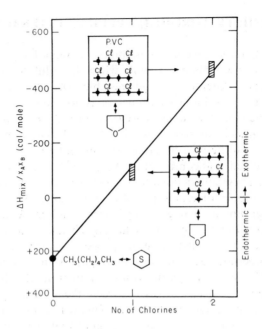

Fig. 2   Calorimetrically determined heats of mixing showing the effect of a specific inter-action between chlorine atoms on hydrocarbons with the oxygen in THF.

where N is the number of such interactions (related to number of chlorines in Fig. 2) and $\Delta \tilde{H}_{mix}^{(sp)}$ is their intensity, which is negative whereas $\Delta H_{mix}^{(dis)}$ is positive.

The specific interaction involved in Fig. 2 perhaps explains the many examples of miscibility or partial miscibility among halogenated polymers and those containing oxygen (e.g., ester groups) (see the tables in Chapter 2). However, as Eq. (6) and Fig. 2 show, there must be sufficient interactions to outweigh the dispersive contribution to make ΔH_{mix} exothermic or a sufficiently small positive value. The striking structural example provided by the much greater miscibility with PVC of the syndiotactic compared to the isotactic isomer of poly(methyl methacrylate) [12] teaches that, in addition to having the proper type and number of interacting groups, they must also be properly articulated spatially to produce the interaction.

Quantitative knowledge about specific interactions is limited; however, such data would be very useful for understanding the miscibility in certain blend systems and to design or select components for miscibility. Such data can be obtained by a well-planned calorimetry program using model compounds.

It is evident from examining the known cases of miscibility in blends that there is a greater opportunity for discovering new miscible systems by trying to select complementing dissimilar structures rather than by matching similar structures as most general rules teach. The number of examples of polymer–polymer miscibility is expanding rapidly and may be expected to continue as this relatively new viewpoint is exploited. Careful research that quantifies the specific interaction and its origin will be most useful.

PHASE EQUILIBRIA AND TRANSITIONS

Above their melting points, metals are usually quite miscible with one another and some other elements [13]. However, mixing of different metals in the crystal lattice below the melting point is more restricted and only occurs when certain size and valency requirements, summarized in the Hume–Rothery rules [9], are met. Solid-metal alloys may consist of one or two phases and each offers certain advantages. For polymer blends, there seems to be no established cases in which the two cocrystallize into the same lattice. In fact, isomorphism of comonomers randomly or regularly distributed in the chain is not common but does occur in some instances [14]. Evidently, miscibility in polymer blends is restricted to amorphous phases.

It is well recognized that many solutions of low molecular weight compounds have limits of solubility, and the same is true for polymer blends.

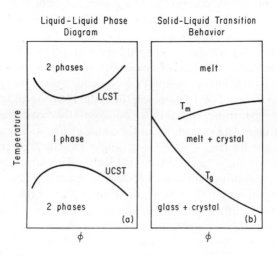

Fig. 3   Possible phase and transition behavior in polymer blends.

There are various mechanisms for these limits. The phase separation discussed in connection with Fig. 1 results in two "liquid" phases and is very much affected by temperature. Figure 3a shows one general pattern of liquid–liquid phase equilibrium, which includes upper (UCST) and lower (LCST) critical solution temperature behavior. These and other phase diagram forms are discussed in Chapters 2–4.

Until recently, there was little mention of such phase boundaries in the literature on polymer blends. Lower critical solution temperature behavior, however, is apparently common. The few systems in which this behavior has been reported are reviewed in Chapters 2–4; however, since the time when these chapters were written, LCST behavior has also been found in the following systems [15]: poly(vinylidene fluoride) (PVF_2)–poly(ethyl methacrylate) (PEMA); PVF_2–poly(methyl methacrylate) (PMMA); PVF_2–poly(methyl acrylate); PVF_2–poly(ethyl acrylate); and poly(ε-caprolactone)–polycarbonate.

The literature on polymer blends has been more concerned in the past with solid–liquid-type transitions (T_g or T_m, as discussed in Chapter 5) than the liquid–liquid type. Figure 3b shows the expected behavior of these transitions for a mixture of two miscible polymers free of any phase boundaries of the type on the left in the region of interest. If both polymers are amorphous, the glass transition T_g varies monotonically with composition as shown. If one polymer is crystallizable, its melting point T_m will be depressed slightly [16] as it is diluted with the other polymer. Presumably, it crystallizes by itself, and the remaining amorphous phase is a homogeneous mixture whose composition is somewhat altered by the removal of some but not all of the crystallizable component. This amorphous phase will be rubbery above T_g and glassy below. While the crystallization does represent a phase separation, it is of a different origin than the liquid–liquid type in Fig. 3a. Crystallization of one component offers the system *another* way to lower its free energy below that of a single phase in analogy to the change from A to B in Fig. 1. However, the system may be quite stable to any liquid–liquid separation, and thus it possesses the basic thermodynamic attributes for complete miscibility in the sense of Fig. 1. The propensity to crystallize might be removed by some slight structural alteration of this component or it may be avoided kinetically. The latter case does not represent an equilibrium state with respect to crystallization; however, the one phase system may still be in equilibrium with respect to liquid–liquid phase separation. Until recently [16, 17], there was relatively little information in the literature on potential miscibility for blend systems with crystallizable components apparently because it was felt that the crystallization itself indicated immiscibility and thus such systems were of no interest. However, it seems more pertinent to focus on the state of miscibility in the remaining amorphous phase since the crystallization per se should not rule out interest in potential usefulness—many important polymers crystallize.

For some systems, the two types of phase behavior shown in Fig. 3 no doubt have regions of overlap (see, e.g., Bernstein *et al.* [15] and Kwei *et al.* [18]). This may result in a rather complex situation with regard to a detailed fundamental study; however, it may not alter practical interest.

A central point here is that historically polymer blends have often been regarded as miscible or compatible only when they have one phase for all component proportions and are stable with respect to all types of phase changes (see Chapter 2); however, there is a growing recognition that the many types of partial miscibility [17], largely ignored in the past, offer interesting areas for fundamental study and opportunities for commercial utilization.

PHASE MORPHOLOGY

Blends with two phases can be organized into a variety of morphologies as the subsequent chapters demonstrate. Many properties, and subsequently uses, of a blend depend critically on the nature of this arrangement of the two phases. One phase may be dispersed in a matrix of the other, and in this case the matrix phase dominates the properties. A parallel arrangement allows both phases to contribute to many properties in direct proportion to their composition in the blend, but this is a nonisotropic structure since perpendicular to this direction the system may represent a series arrangement with properties that disproportionately favor one phase. For mechanical properties, adhesion between phases is an issue that is more critical for some morphologies than others.

An intriguing morphological concept, which avoids some of the dilemmas pointed out above, is one in which both polymers are continuous simultaneously and thereby form interpenetrating networks (IPN) of *phases*. A related but different concept is the idealized interpenetration of two *molecular* networks (see Volume 2, Chapter 11). The IPN structure can (a) be spatially isotropic, (b) allow each phase to contribute to properties more nearly in proportion to their concentration in the blend, and (c) remove somewhat the stringent necessity for adhesion inherent to other morphologies. There are at least two routes to forming IPN type morphologies that might be used: one is by judicious control of rheological factors during processing as described in Volume 2, Chapter 20, and the other is phase separation from a homogeneous phase via the spinodal decomposition mechanism described in Chapter 4. The latter has resulted in practical applications in other materials technology [19] but apparently not for polymer blends.

It would be of interest to examine more closely how such properties as modulus (Chapter 8) or permeability (Chapter 10) depend on composition and component properties for an IPN type structure. This is a very complex geometry for exact analysis; however, an approximate treatment for the modulus is shown in Fig. 4. Kraus and Rollman [20] proposed an isotropic extension of a Takayanagi type model (see Chapter 8) that in one limit (their parameter $b \to 1$) reduces to a very special IPN. This IPN is a regular lattice of cubes, like that shown in Fig. 4, each containing a regular arrangement of Phases 1 and 2 (light and shaded respectively). The volume fraction of Phase 2 is related to the linear fraction a it contributes to the cube edge: $\phi_2 = a^2(3 - 2a)$. Via the Takayanagi approach, approximate upper and lower bounds for the IPN modulus can be deduced and these have been used to define the shaded zone in Fig. 4. Within this zone should lie a reasonable estimate for actual IPN structures. Also shown for comparison in Fig. 4 are the upper and lower bounds for all phase arrangements, viz., parallel and series respectively, plus a dispersion of Component 2 in a matrix of 1. It is interesting to note that the IPN structure gives a modulus in all directions only slightly below the parallel arrangement, that is, just below perfect

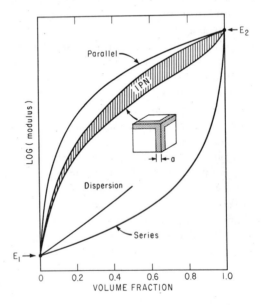

Fig. 4   Mechanical modulus of various polymer–polymer phase arrangements. The cube shows unit cell of an idealized interpenetrating network (IPN) structure. The shaded area is determined by upper- and lower-bound estimates for this model.

additivity—note that on arithmetic coordinates the parallel arrangement would yield a straight line.

The IPN structure is a way to achieve the maximum contribution from each component simultaneously and offers opportunities for combining unique properties into a blend.

APPLICATIONS

The ultimate goal of polymer blending is a practical one of achieving commercially viable products through either unique properties or lower cost than some other means might provide. Commercial products have been based on miscible blends, for example, polystyrene–poly(phenylene oxide), PVF_2–PMMA, PVF_2–PEMA, and PVC–nitrile rubber, and immiscible blends, for example, rubber blends in tires (Volume 2, Chapters 19 and 20), impact-modified plastics (Volume 2, Chapters 12–14), and coextruded film and fibers (Volume 2, Chapters 15 and 16). What one can accomplish with miscible versus immiscible blends is fundamentally different and in some respects resembles the differences between random versus block copolymers.

The latter chapters in this book deal with a wide spectrum of examples of present and future applications of blends, and point out their advantages and disadvantages. A careful study of these should result in a general understanding of what is required of the components thermodynamically to achieve the desired objective. This book does not include, obviously, all present examples of blend usage or future possibilities as illustrated by the following. Frequently, a small amount of a second polymer may be blended with PVC to serve as a "processing aid" or as a "lubricant." Similar advantages have been observed recently for a wider range of polymers (see, e.g., Kraus and Rollmann [21]). These effects are complex and poorly understood at the present but offer potentially significant benefits that apparently derive from rheological processes dependent on the immiscibility, or at most partial miscibility, of the two polymers. Another attractive possibility is to use polymeric stabilizers for polymers, for example, antioxidants and UV absorbers, since they would offer the advantage of very low migration [22] similar to the permanence of polymeric plasticizers (Volume 2, Chapter 17). At least some miscibility is needed in this application.

Acknowledgment

The author gratefully acknowledges the helpful criticism and suggestions of S. Newman in preparing this chapter.

REFERENCES

1. D. S. Kaplan, *J. Appl. Polym. Sci.* **20**, 2615 (1976).
2. R. L. Jalbert, "Modern Plastics Encyclopedia," Vol. 52, p. 107. McGraw-Hill, New York, 1975–1976.
3. P. J. Flory, "Principles of Polymer Chemistry." Cornell Univ. Press, Ithaca, New York, 1953.
4. R. G. Kirste and B. R. Lehnen, *Makromol. Chem.* **177**, 1137 (1976).
5. W. A. Krause, R. G. Kirste, J. Haas, B. J. Schmitt, and D. J. Stein, *Makromol. Chem.* **177**, 1145 (1976).
6. D. G. H. Ballard, M. G. Rayner, and J. Schelten, *Polymer* **17**, 640 (1976).
7. D. J. Stein, R. H. Jung, K. H. Illers, and H. Hendus, *Angew. Makromol. Chem.* **36**, 89 (1974).
8. J. M. Prausnitz, "Molecular Thermodynamics of Fluid-Phase Equilibria," Chapter 7. Prentice-Hall, Englewood Cliffs, New Jersey, 1969.
9. A. T. Di Benedetto, "The Structure and Properties of Materials," Chapter 7. McGraw-Hill, New York, 1967.
10. P. J. Flory, B. E. Eichinger, and R. A. Orwoll, *Macromolecules* **1**, 287 (1968).
11. J. Pouchly and J. Biros, *Polym. Lett.* **7**, 463 (1969).
12. J. W. Schurer, A. de Boer, and G. Challa, *Polymer* **16**, 201 (1975).
13. K. M. Ralls, T. H. Courtney, and J. Wulff, "Introduction to Materials Science and Engineering," Chapter 8. Wiley, New York, 1976.
14. G. Natta, G.. Allegra, I. W. Bassi, D. Sianesi, G. Caporiccio, and E. Torti, *J. Polym. Sci. Part A*, 4263 (1965).
15. R. E. Bernstein, C. A. Cruz, D. R. Paul, and J. W. Barlow, *Macromolecules* **10**, 681 (1977).
16. R. L. Imken, D. R. Paul, and J. W. Barlow, *Polym. Eng. Sci.* **16**, 593 (1976).
17. D. R. Paul, J. W. Barlow, C. A. Cruz, R. N. Mohn, T. R. Nassar, and D. C. Wahrmund, *Amer. Chem. Soc. Div. Org. Coatings Plast. Chem. Prepr.* **37** (1), 130 (1977).
18. T. K. Kwei, G. D. Patterson, and T. T. Wang, *Macromolecules* **9**, 780 (1976).
19. J. W. Cahn, *Trans. Metall. Soc. AIME* **242**, 166 (1968).
20. G. Kraus and K. W. Rollmann, *in* "Multicomponent Polymer Systems" (N. A. J. Platzer, ed.), Vol. 99, Adv. Chem. Ser., p. 189. Amer. Chem. Soc., Washington, D.C., 1971.
21. C. K. Shih, *Polym. Eng. Sci.* **16**, 742 (1976); *J. Polym. Sci. Polym. Phys. Ed.* **14**, 1729 (1976).
22. Y. Mizutani and K. Kusumoto, *J. Appl. Polym. Sci* **19**, 713 (1975).

Chapter 2
Polymer–Polymer Compatibility

Sonja Krause

Department of Chemistry
Rensselaer Polytechnic Institute
Troy, New York

I.	Introduction	16
	A. Scope	16
	B. Literature Search and Referencing	16
	C. Definitions of Compatibility	17
	D. Experimental Determination of Compatibility	18
II.	Theoretical Aspects	20
	A. Types of Phase Diagrams	20
	B. Flory–Huggins Theory of Polymer Solutions	25
	C. Equation of State Theories of Polymer Solutions	31
III.	Experimental Data	32
	A. Arrangement of Data	32
	B. Tables of Polymers That May Be Compatible at Room Temperature	34
IV.	Prediction Schemes for Polymer–Polymer Compatibility	41
	A. Review of the Literature	41
	B. A Simple Scheme Based on Flory–Huggins Theory	45
	C. A Critique of the Simple Scheme and Comparison with Data	50
	Appendix: Review of Data on Polymer Mixtures in the Literature	52
	A. Cellulose Derivatives and Other Cellulose Derivatives	53
	B. Cellulose Derivatives and Other Polymers	54
	C. Polyisoprene and Other Polymers	63
	D. Poly(vinyl chloride) (PVC) and Other Polymers	65
	E. Polyethylene (PE) and Other Polymers	69
	F. Polybutadiene (PBD) and Other Polymers	70
	G. Poly(vinyl acetate) (PVA) and Other Polymers	71
	H. Polystyrene (PS) and Other Polymers	74
	I. Acrylic Polymers and Other Acrylic Polymers	79
	J. Acrylic Polymers and Other Polymers	82
	K. Miscellaneous Homopolymers	86
	L. Miscellaneous Homopolymers and Miscellaneous Copolymers	91
	M. Copolymers and Other Copolymers	98
	N. Mixtures of Three Polymers	102

O. Same Copolymer but Different Compositions   103
Acknowledgments   106
References   106

I. INTRODUCTION

A. Scope

In this chapter, the words "polymer compatibility" refer to the total miscibility, on a molecular scale, of homopolymers and of random copolymers with each other in various combinations. In this connection, it is understood that miscibility on a molecular scale is not necessarily random; interactions between similar or different molecules may lead to a small amount of clustering or other nonrandom arrangements of the polymer segments.

In this chapter, no distinction is made between polymers in the rubbery or glassy state. Crystallizable polymers are dealt with only in those cases in which the polymer remains at least partially amorphous and in which the miscibility with other polymers was investigated in the amorphous state. Mixed crystal formation is not considered in this chapter; microphase separation or other phase separation in block copolymers and in mixtures of block copolymers with other polymers also is not considered. Within these limits, this discussion covers the miscibility of homopolymers and copolymers, in bulk and in solution, from the experimental and from the theoretical viewpoints. Methods for predicting compatibility are reviewed briefly, and one simple method for predicting compatibility is presented at length. Limitations of this simple method are discussed.

B. Literature Search and Referencing

This chapter is an extension of a recent review on polymer compatibility by the same author (reprinted from Ref. [1], pp. 251–314, by courtesy of Marcel Dekker, Inc.). Early work in this field was reviewed by Flory [2] and Tompa [3]; more recent reviews of various aspects of polymer compatibility have been published by Bohn [4], Fettes and Maclay [5], Voorn [6], Friese [7, 8], Thinius [9], Gerrens [10], Schmieder [11], Corish and Powell [12], Tlusta and Zelinger [13], Pazonyi and Dimitrov [14], and Kuleznev and Krokhina [15].

The literature survey made for this chapter ended in June, 1976, and is probably not complete. Work on polymer–polymer compatibility does not necessarily appear under that title, thus making a literature survey on this subject somewhat uncertain. After my first review on this subject, several authors, specifically S. Akiyama and V. N. Kuleznev, graciously sent me some of their papers that I had overlooked.

Some of the papers reviewed here have appeared in more than one journal; in these cases, both references have been given the same reference number, especially since the two references are usually to the same paper printed in two different languages.

In most cases in which experimenters worked with trademarked polymers, I was able to determine the chemical compositions of the samples. There was only one reference [16] in which I was unable to decide the composition of many of the samples unequivocally. Those samples have been left out of the tabulations in this chapter.

C. Definitions of Compatibility

The definition of compatibility as miscibility on a molecular scale to be used in this chapter is only one of many definitions of polymer compatibility. To some workers, compatible polymers are polymer mixtures that have desirable physical properties when blended. To other workers, compatible polymers are polymer mixtures that do not exhibit gross symptoms of phase separation when blended. These three definitions of polymer compatibility are somewhat interrelated since it is only reasonable to suppose that polymer blends exhibiting no gross symptoms of phase separation on blending and having desirable properties show at least some mixing of polymer segments on a microscopic scale. This implies either a certain amount of thermodynamic compatibility or a physical constraint that prevents demixing such as grafting, cross-linking, the presence of block copolymers, interpenetrating network (IPN) formation, or the quenching of a mixed system to a temperature at which demixing is thermodynamically but not kinetically favored.

Many methods used in the literature to investigate polymer compatibility do not actually probe thermodynamic mixing on a molecular scale and are covered only briefly here. Such methods include studies of the viscosities of polymer mixtures in solution [17–23] and in bulk [24], tensile strength, elongation at break, and other mechanical properties of polymer mixtures [25–27], thermal oxidative degradation of polymer mixtures [28, 29], and densities of polymer mixtures [30, 31]. An estimate of compatibility from mutual adhesion of two polymers may, however, be possible [32]. Some of the other properties mentioned above may be used to correlate with other properties of polymer mixtures and some, on occasion, correlate well with polymer compatibility or with the composition at which phase separation occurs [27]. Some authors [33] have postulated various kinds of "pseudo-compatibility" to explain the above data.

D. Experimental Determination of Compatibility

Unambiguous methods of studying polymer–polymer compatibility are hard to find. In bulk, compatible polymers form transparent films and fibers that exhibit no heterogeneity under considerable magnification in the phase contrast microscope or in the electron microscope, no matter what staining methods are employed. In addition, mixtures of compatible polymers should have only a single glass transition temperature no matter how this is measured. These criteria appear to be unambiguous in principle, but a number of ambiguities may appear in practice. First, incompatible polymers form transparent films when both polymers have the same refractive index or, if the refractive indexes are different, they may form two-layered films that appear transparent when a solution of polymers is evaporated [34]. These situations are not too difficult to work with, since transparent films formed from incompatible polymers will exhibit the two glass transition temperatures characteristic of the components of the mixture if the two glass transition temperatures are far enough apart to be resolvable by the measuring technique employed for their detection. A number of workers have obtained two glass transition temperatures on clear, transparent films [35–37].

In connection with the use of multiple glass transition temperatures as a criterion of incompatibility, there is a small amount of evidence indicating that the sensitivity of different measurement techniques toward the glass transitions in mixtures depends on the physical dimensions of the phases [38, 39]. Dynamic mechanical measurements can apparently detect smaller phases than differential calorimetry. At least one measuring technique, pulsed nuclear magnetic resonance (NMR) determination of T_2 of a polymer mixture (see Chapter 4), appears to be sensitive to microheterogeneities so small that they should be called clusters rather than phases in a mixture that is compatible by other standards,[†] in this case, polystyrene and poly-(vinyl methyl ether) [40].

A more difficult situation has been observed experimentally for several polymer–polymer systems that are on the "edge" of compatibility, that is,

---

† A similar conclusion could be drawn for the small-angle x-ray scattering results on the poly(vinyl chloride)–poly(ε-caprolactone) system described in Chapter 9. [—Editors]

the free energy change or driving force leading to phase separation is very small, not always enough to overcome the obstacle to phase separation presented by the high viscosity of the bulk mixture of polymers. For example, some mixtures of polystyrene with poly(vinyl methyl ether) form transparent films with only a single glass transition temperature when cast from toluene, whereas they form cloudy films with two glass transition temperatures when cast from trichloroethylene [38]. It became known later that these mixtures, at room temperature, are close to both an upper and to a lower critical solution temperature so that small changes in temperature or other conditions could lead to phase separation [41]. Heat treatment caused phase separation in a system that originally showed no phase separation, poly-(vinyl acetate)–poly(methyl methacrylate) [42]. On at least one occasion during measurements of T_g in mixtures of polystyrene with poly(2,6-dimethyl-1,4-phenylene ether), dynamic mechanical measurements indicated two loss peaks[†] in the glass transition temperature range, while differential calorimetry showed only one transition [43]. Mixtures of chlorinated rubber with an ethylene–vinyl acetate copolymer form transparent films when cast from hydrocarbon solvents or CCl_4, but they form opalescent films when cast from partially chlorinated hydrocarbons [44]. In all these systems, the free energies of both the equilibrium state and the nonequilibrium state are probably very close together, and much work may be necessary to discover whether the polymers are miscible at room temperature.

These considerations lead to the suspicion that any observation made on polymer mixtures in bulk should be somewhat suspect because it is very difficult to ascertain whether a bulk polymer sample is in its equilibrium state. It is usually assumed without proof that a mixed polymer film deposited from solution is more likely to be in its equilibrium state than a mixed polymer sample prepared by mastication of the two polymers in bulk. Careful workers generally assume that they are working with mixtures at equilibrium if the same results are obtained using solution-deposited films and using masticated or milled samples, and if the results do not change on temperature cycling of the sample.

Because of the problems associated with bulk mixtures of polymers, a number of workers have studied polymer compatibility in solution in mutual solvents. Compatible polymers form a single transparent phase with the mutual solvent, whereas incompatible polymers exhibit phase separation, at least in concentrated solution. Equilibrium is not difficult to achieve if the solutions are very dilute, that is, nonviscous, but becomes problematical in concentrated, viscous solutions even after a waiting period of weeks or

---

† The two peaks in this miscible system very likely resulted from incomplete mixing in the preparation procedure. [—Editors]

months. As it happens, many polymers are compatible in the easily studied very dilute solution. If two polymers, however, show phase separation in dilute solution in a mutual solvent, they are usually also incompatible in bulk. Incompatibility is therefore much easier to demonstrate than compatibility for a particular set of polymers.

In some cases, details of the physics or chemistry of the polymer mixture may affect the measurement of compatibility. For example, two glass transitions, signifying incompatibility, may turn into a single intermediate glass transition, signifying compatibility, when a mixed rubber such as polybutadiene–poly(butadiene-*co*-styrene), is vulcanized [45]. Another example involves some work on polystyrene–ethylcellulose mixtures in dilute solution, which at first indicated that phase separation may be reversed at a high enough shear rate [46, 47]. Later work, however, indicated that the two phases were still present during shear, but formed long strings [48]. In other words, phase separation may be very difficult to detect in solutions that are being sheared, that is, stirred or agitated.

In the Appendix to this chapter, the methods used for obtaining glass transition temperatures in various polymer mixtures usually is mentioned. Most methods are well known, and include dilatometry, various dynamic mechanical methods, dielectric measurements, refractive-index temperature measurements, differential scanning calorimetry (DSC), and differential thermal analysis (DTA). Good data apparently may also be obtained from thermooptical analysis (TOA) [49], essentially a light-transmission measurement, and from radiothermoluminescence, a method in which a polymer is irradiated at low temperature with electrons or γ rays and the luminescence emitted by the sample during heating is recorded—peaks in the luminescence are seen in the vicinity of any glass transition temperature [50].

II. THEORETICAL ASPECTS

A. Types of Phase Diagrams

The thermodynamic theory of phase separation in mixtures of polymers is, in principle, exactly the same as that used to describe phase separation in mixtures of small molecules. If the free energy of mixing can be calculated for all possible compositions of the mixture of interest, then it becomes possible to calculate the values of temperature, pressure, and composition at which the mixture will form a stable single phase, that is, will be miscible. At the same time, it becomes possible to calculate not only the compositions at which the mixture will always separate into more than one phase (unstable compositions), but also those compositions at which the mixture may either form a single phase if clean and relatively undisturbed or, at equilibrium, will separate into several phases (metastable compositions). The calculated phase diagram for a strictly two-component system, that is, a mixture of two monodisperse polymers, could resemble any of those in Fig. 1 in which systems with either an upper critical solution temperature (UCST) or a lower critical solution temperature (LCST) or both or a tendency toward both are shown. Real polymeric mixtures, even though synthetic polymers are never monodisperse, exhibit phase diagrams that resemble these with some variations, with the exception of the diagram shown in Fig. 1c. In addition, two-peaked coexistence curves have been observed for certain polymer mixtures such as polystyrene–polybutadiene [51] in the presence of solvent and for mixture of styrene and isoprene oligomers in the absence of solvent (Fig. 2) [52]. Mixtures that have positive (endothermic) heats and entropies of mixing usually tend to exhibit UCST, whereas mixtures that have negative (exothermic) heats and entropies of mixing usually exhibit LCST.

In general, metastable compositions separate rapidly into two phases in mixtures of low molecular weight molecules, and it is extremely difficult to observe these compositions before phase separation begins [53]. However, it seems possible that metastable compositions could remain as a single phase almost indefinitely in very viscous polymer mixtures. One might therefore expect that such compositions, when observed experimentally, would sometimes appear as a single phase, while at other times they would separate into several phases. This is one possible explanation for the conflicting data on almost compatible systems mentioned in Section I.D above.

Figures 1 and 2 can be used to demonstrate a dilemma that affects any discussion of miscibility or compatibility. At temperatures above a UCST or below an LCST, the binary mixtures are completely miscible at all compositions. Below the UCST and above the LCST, there are still compositions, generally those in which one of the components is present in small percentages, in which only a single phase is observed. At intermediate compositions, phase separation occurs. But shall we call such a mixture as a whole compatible or incompatible? In this chapter, and in general, such mixtures are usually called incompatible, although I have called some such mixtures in which the compatible compositions span a reasonable range "almost compatible."

In any mixture the boundary between stable and metastable compositions is called the binodal, and the boundary between metastable and unstable compositions is called the spinodal. Compositions along the binodal are the ones that define the phases into which unstable and metastable compositions will separate, that is, at temperature T_1, compositions A, B, and C

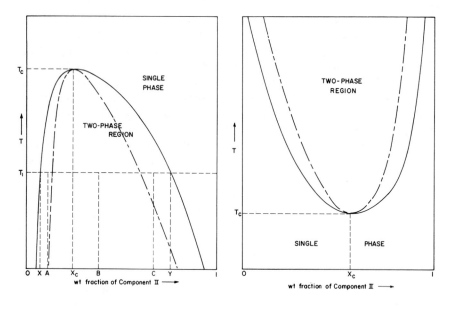

Fig. 1a   Fig. 1b

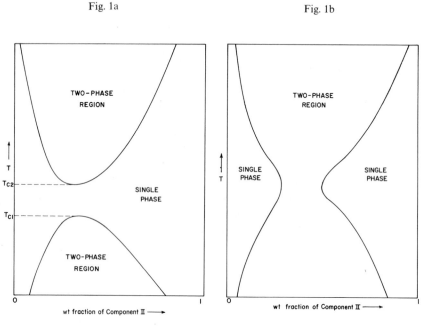

Fig. 1d   Fig. 1e

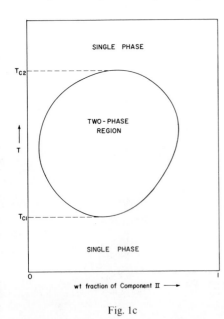

SINGLE PHASE

T_{c2}

TWO - PHASE
REGION

T

T_{cl}

SINGLE PHASE

0   1

wt fraction of Component Ⅱ ⟶

Fig. 1c

Fig. 1 Phase diagrams for various mixtures as a function of temperature. (a) A mixture with a UCST: (——) binodal; (— ·—) spinodal. (b) A mixture with an LCST: (——) binodal; (— ·—) spinodal. (c) A mixture with a UCST above an LCST (closed phase diagram). (d) A mixture with an LCST above a UCST. (e) A mixture with a tendency toward greater solubility at intermediate temperatures.

in Fig. 1a will separate into compositions X and Y. The mixture forms a single phase whenever its composition is outside the range $X–Y$. Recently, it has become possible to determine experimentally not only the binodal in polymeric systems, but also the spinodal; this was done specifically for polystyrene–poly(vinyl methyl ether) mixtures [54].

The binodal, although very important, is extremely difficult to calculate, even for a binary system. The spinodal, on the other hand, can usually be calculated fairly easily, at least with the aid of a computer. Easiest of all to calculate are the critical points on any phase diagram, such as T_c and the composition X_c associated with it. For a binary system, a critical point indicates the circumstances under which the system will just begin to separate into two phases, that is, the limits of compatibility. For this reason, I give equations for the binary critical point below, and I use those equations for the most part to show how compatibility may be predicted. Three-component systems and their phase diagrams are more complex, and it usually is not possible to discuss them in terms of a single parameter such as a critical point. Simplifications can sometimes be made, and these are discussed below.

The type of three-component system of major interest in this Chapter is the polymer 1–polymer 2–solvent system, usually shown at a single temperature on an equilateral triangle phase diagram as shown in Fig. 3. Each corner of the triangle represents one of the pure components, each edge

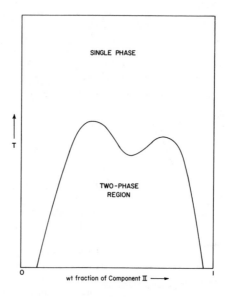

Fig. 2 Two-peaked coexistence curve as seen in the case of polystyrene–polybutadiene–solvent mixtures [51] and for mixtures of styrene and isoprene oligomers [52].

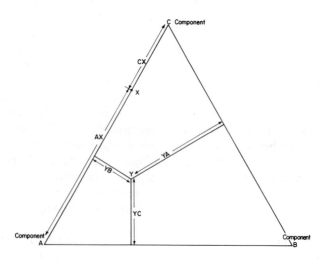

Fig. 3 Equilaterial triangle as used for three-component phase diagrams at a given T and p.

of the triangle represents the mixtures of the two components shown on adjacent corners, and the interior of the triangle represents mixtures of all three components. For mixtures of A and B, distance AX represents the fraction of B in the mixture, while distance BX represents the fraction of A. For mixtures of A, B, and C, for point Y on Fig. 3, y_A, y_B, and y_C represent the fractions of A, B, and C in the mixture. An enormous variety of phase diagrams can be calculated for three-component mixtures, complete with binodals, spinodals, and critical points. The types of data that have appeared in the literature for real mixtures are shown in Fig. 4a and b; those for calculated mixtures are given in Fig. 4c. Figure 4a shows data for a mixture of polymers that is incompatible in the absence of solvent [55]; Fig. 4b shows data for a mixture of polymers that is compatible in the absence of solvent, and in which each polymer is separately miscible with the solvent in all proportions, but where phase separation occurs in some compositions when all three components are mixed [52]; Fig. 4c shows a hypothetical phase diagram [56] that is possible but has not yet been observed, at least to my knowledge.

Phase diagrams in three-component systems vary with temperature, adding a third dimension to the phase diagram. The results are useful but very complex; such diagrams have been published by Koningsveld [57] for mixtures of polyethylene, polypropylene, and diphenyl ether.

B. Flory–Huggins Theory of Polymer Solutions

1. Homopolymers

It is best to begin any discussion of free energy of mixing by demolishing a misconception that pervades the polymer literature. Koningsveld [52] has made a previous attempt at this, hitherto without results. In order to have a compatible mixture, it is a necessary condition that the free energy of mixing $\Delta G_{mix} < 0$. This is not a sufficient condition for compatibility, how-ever; ΔG_{mix} of most incompatible[†] systems fulfills the same condition. The equations below show the complete thermodynamic criteria for compat-ibility, which are somewhat harder to visualize than the necessary, but insufficient criterion that $\Delta G_{mix} < 0$. Tompa[3] discusses the proper criteria at some length with diagrams that aid visualization.

---

† By the author's definition, compatibility implies complete miscibility across the entire composition range, whereas a system that shows any region of incomplete miscibility is regarded as incompatible or "almost compatible." In such regions, ΔG_{mix} may be negative for the one-phase mixture; however, the system is able to achieve an even lower free energy state by splitting into two phases [3] as discussed in Chapter 1. A stability analysis is required to determine whether a particular state is stable. [—EDITORS]

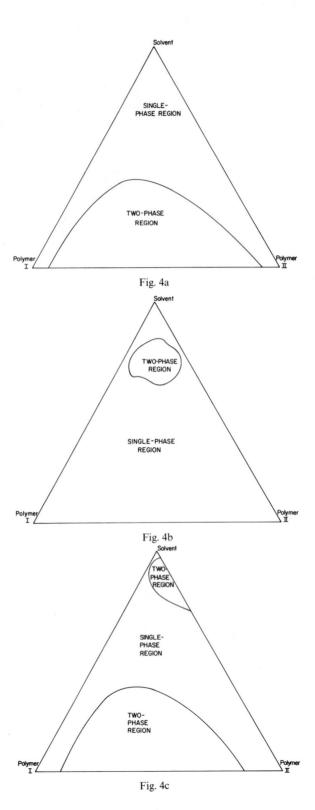

Fig. 4a

Fig. 4b

Fig. 4c

Scott [58] and Tompa [59] were the first to apply the Flory [60, 61]–Huggins [62, 63] theory of polymer solutions to mixtures of polymers, with or without added solvent. Scott [58] obtained essentially the following expression for the Gibbs free energy of mixing a total volume V of two polymers:

$$\Delta G_{\mathrm{mix}} = (RTV/V_r)[(\phi_A/x_A)\ln\phi_A + (\phi_B/x_B)\ln\phi_B + \chi_{AB}\phi_A\phi_B] \quad (1)$$

where V_r is a reference volume which is taken as close to the molar volume of the smallest polymer repeat unit as possible, ϕ_A and ϕ_B are the volume fractions of polymers A and B, respectively, x_A and x_B are the degrees of polymerization of polymer A and polymer B in terms of the reference volume V_r, respectively, and χ_{AB} is related to the enthalpy of interaction of the polymer repeat units, each of molar volume V_r (see Eq. 12).

Scott found the critical conditions in such a system; these can be found from Eq. (1) by letting [3] (at constant T and p)

$$\partial^2\Delta G_{\mathrm{mix}}/\partial\phi_A{}^2 = \partial^3\Delta G_{\mathrm{mix}}/\partial\phi_A{}^3 = 0 \quad (2)$$

The critical conditions are

$$(\chi_{AB})_{cr} = \tfrac{1}{2}[1/x_A^{1/2} + 1/x_B^{1/2}]^2 \quad (3a)$$

$$(\phi_A)_{cr} = \frac{x_B^{1/2}}{x_A^{1/2} + x_B^{1/2}} \quad (3b)$$

$$(\phi_B)_{cr} = \frac{x_A^{1/2}}{x_A^{1/2} + x_B^{1/2}} \quad (3c)$$

Scott noted, using these equations, that $(\chi_{AB})_{cr}$ would be very small for two polymers having appreciable degrees of polymerization, and that polymers of infinite molecular weight would be incompatible if there were any positive heat of mixing at all.

The equation for the spinodal (at constant T and p) is calculated from

$$\partial^2\Delta G_{\mathrm{mix}}/\partial\phi_A{}^2 = 0 \quad (4)$$

and is equal to

$$(\chi_{AB})_{sp} = \tfrac{1}{2}[1/x_A(\phi_A)_{sp} + 1/x_B(\phi_B)_{sp}] \quad (5)$$

Fig. 4 Phase diagrams for Polymer I–Polymer II–solvent mixtures. (a) Polymers are incompatible in bulk but each is miscible with the solvent, such as polystyrene–polypropylene–toluene [55]. (b) Components are miscible in all proportions when taken two at a time but phase separation occurs at some compositions when all three components are mixed, such as benzene–butyl rubber–EPDM rubber or diphenyl ether–atactic polypropylene–polyethylene [52]. (c) Phase diagram that may be observed when polymers are incompatible in bulk and only one of the polymers is miscible with solvent in all proportions [56].

28   *Sonja Krause*

The equations for the binodal are calculated by setting the chemical potential of both polymer A and polymer B equal in the two coexisting phases.

Using primes to designate one phase, and double primes the other phase, the binodal equations for a mixture of two polymers are

$$\ln \phi_A' + (1 - x_A/x_B)\phi_B' + x_A\chi_{AB}(\phi_B')^2$$
$$= \ln \phi_A'' + (1 - x_A/x_B)\phi_B'' + x_A\chi_{AB}(\phi_B'')^2 \tag{6a}$$

$$\ln \phi_B' + (1 - x_B/x_A)\phi_A' + x_B\chi_{AB}(\phi_A')^2$$
$$= \ln \phi_B'' + (1 - x_B/x_A)\phi_A'' + x_B\chi_{AB}(\phi_A'')^2 \tag{6b}$$

All equations for binodals contain a mix of logarithmic and nonlogarithmic terms and the number of equations to be solved simultaneously equals the number of components times the number of coexisting phases minus one. (The maximum number of coexisting phases that are not vapor or crystalline expected at any temperature and pressure equals the number of components in the mixture.) For this reason, equations for binodals are very difficult to solve even with a computer, and very few such equations are given in this chapter.

Scott [58] also discussed mixtures of two polymers in the presence of a solvent—a three-component mixture. He obtained equations that led to a Gibbs free energy of mixing

$$\Delta G_{mix} = RTV/V_S[\phi_S \ln \phi_S + (\phi_A/x_A) \ln \phi_A + (\phi_B/x_B) \ln \phi_B$$
$$+ \chi_{AB}\phi_A\phi_B + \chi_{AS}\phi_A\phi_S + \chi_{BS}\phi_B\phi_S] \tag{7}$$

where V_S, the reference volume, is equal to the molar volume of the solvent, and x_A, χ_{AB}, χ_{AS}, and χ_{BS} must be considered in terms of reference volume V_S; χ_{AS} and χ_{BS} are the interaction parameters between polymers A and B, respectively, and the solvent. The equation for the spinodal in a three-component system is [3]

$$(\partial^2 \Delta G_{mix}/\partial \phi_A^2)(\partial^2 \Delta G_{mix}/\partial \phi_B^2) = (\partial^2 \Delta G_{mix}/\partial \phi_A \partial \phi_B)^2 \tag{8}$$

leading to a complex equation that has been discussed by Tompa [3].

Scott [58] calculated some binodals for special cases; one example is that in which $\chi_{AS} = \chi_{BS}$ and $x = x_A = x_B$, that is, the case in which polymers of comparable degree of polymerization are dissolved in a solvent that has the same interaction parameter with each of the two polymers. Scott called this the "symmetrical" case. He found an equation for the binodal:

$$\ln \theta_A' + x\chi_{AB}(1 - \phi_S)(\theta_B')^2 = \ln \theta_B' + x\chi_{AB}(1 - \phi_S)(\theta_A')^2 \tag{9}$$

where $\theta_A = \phi_A/(\phi_A + \phi_B)$ and $\theta_B = \phi_B/(\phi_A + \phi_B)$. Equation (9) looks exactly like Eqs. (6a) and (6c) when both polymers have the same degree of polymerization, $x = x_A = x_B$, that is,

$$\ln \phi_A' + x\chi_{AB}(\phi_B')^2 = \ln \phi_B' + x\chi_{AB}(\phi_A')^2 \tag{10}$$

except that in Eq. (9) $\chi_{AB}(1 - \phi_S)$ takes the place of χ_{AB} in Eq. (10); θ_A and θ_B refer to relative volume fractions of polymers A and B in the polymer portion of the three-component mixture, and are therefore exactly analogous to ϕ_A and ϕ_B in the binary mixture.

Scott noted that the binodal Eq. (9) in the symmetrical case, did not depend on χ_{AS} or χ_{BS} and that the presence of solvent served only to diminish the effective interaction parameter between the polymers, that is, $\chi_{AB}(1 - \phi_S)$ can be considered the effective interaction parameter between the polymers in this case. When there is a great deal of solvent present, that is, when $\phi_S \simeq 1$, the effective interaction parameter between the two polymers approaches zero, and the whole system will form a single phase. These conclusions can be restated as follows: *No matter how incompatible two polymers may be, it is always possible to make a very dilute solution containing both polymers, as long as a solvent that dissolves both polymers exists.* For this reason it is possible to obtain light-scattering data on very dilute solutions of polymer mixtures, even for polymers that are very incompatible in the absence of solvent.

Scott used the exact result he calculated for the symmetrical case to reach some approximate general conclusions for all polymer–polymer–solvent systems. He calculated an approximate plait point, which is analogous to the critical point in binary mixtures of polymers:

$$(\chi_{AB})_{pl} = \tfrac{1}{2}\{1/x_A^{1/2} + 1/x_B^{1/2}\}^2 \{1/1 - \phi_S\} \tag{11a}$$

$$(\theta_A)_{pl} = x_A^{1/2}/(x_A^{1/2} + x_B^{1/2}) \tag{11b}$$

$$(\theta_B)_{pl} = x_B^{1/2}/(x_A^{1/2} + x_B^{1/2}) \tag{11c}$$

These equations are exactly like Eqs. (3a)–(3c) if we remember that θ_A and θ_B are volume fractions of polymer A and B in the polymer portion of the mixture, and if we use $\chi_{AB}(1 - \phi_S)$ as the effective interaction parameter between the polymers in the presence of solvent. Equations (11a)–(11c) allow calculation of the minimum volume fraction of solvent necessary to "compatibilize" the two polymers, that is, the minimum volume fraction necessary to form a single-phase solution; this minimum volume fraction of solvent will depend on the degree of polymerization of each polymer and on the interaction parameter between the polymers. Equations (11a)–(11c) indicate that the same minimum volume fraction of any solvent that dissolves each polymer separately will give a single phase in the three-component system. This prediction is not strictly true, as we shall see below, but it serves as a reasonably accurate rule for most systems.

Since the advent of high-speed computers, it is no longer necessary to consider only the symmetrical case $\chi_{AS} = \chi_{BS}$, and several workers have taken advantage of this. Zeman and Patterson [64] calculated spinodals for systems where $\chi_{AS} \neq \chi_{BS}$ and actually predicted phase diagrams like Fig. 4b.

Hsu and Prausnitz [65] then calculated binodals for these and other systems.

The theoretical work discussed up to this point deals with strictly mono-disperse polymers; monodisperse in molecular weight, composition, degree of branching, and all other possible variables. The usual polydisperse (with respect to molecular weight) polymers do not follow the equations given here except approximately, but these equations nevertheless serve as valuable guides for real polymer mixture behavior. Calculations of phase diagrams for mixtures of polydisperse polymers with and without solvent have been published by Koningsveld [57], who also allowed χ_{AB} to vary with composition. He later determined [66] that the critical point in such polymer A–polymer B mixtures depends on $\overline{M}_z$ of each polymer. In connection with the variation of χ_{AB} with composition, Koningsveld [67] was able to predict two-peaked phase diagrams like that shown in Fig. 2.

2. Copolymers

Scott [68] also investigated the theoretical effects of composition distri-bution in random copolymers; he found that most random copolymer samples that have an appreciable composition distribution should be expected to separate into two or more phases. Before showing the results of some of his calculations, let us recall that the interaction parameter between molecules of comparable size can often be written in terms of Hildebrand [69, 70] solubility parameters:

$$\chi_{AB} = (V_r/RT)(\delta_A - \delta_B)^2 \tag{12}$$

where δ_A and δ_B are the Hildebrand solubility parameters for A and B, respectively. These solubility parameters are discussed in Section IV.B.

In general, the solubility parameters of copolymer molecules with different compositions will be different (see Eq. 14 below), but it will be possible to imagine an average solubility parameter $\bar{\delta}$ for the whole sample. Scott [68] gives equations in which the composition, density, and molecular weight distributions of any random copolymer samples can be inserted for calcula-tion of the circumstances under which phase separation will take place. For purposes of illustration, he calculated some results for a copolymer sample in which all fractions have the same density, all chemical species have the same molecular weight distribution, and the distribution of solubility parameters around the sample average is symmetrical—specifically, a gaussian distribution. In this restricted case, if phase separation is not desired, 90% of the material in the sample must lie within a range of solubility parameters,

$$\Delta\delta \leqslant 3.3(RT\rho/2\overline{M}_w)^{1/2} \tag{13}$$

Scott used this expression to calculate $\Delta\delta$ for a typical case at room temperature, $T \simeq 300°$K, $\rho = 1$; when $\overline{M}_w = 10^6$, $\Delta\delta = 0.057$, and when $\overline{M}_w = 10^4$, $\Delta\delta = 0.57$. Comparable values are compared with experimental data below; unfortunately, most investigators neglect the molecular weight dependence of compatibility.

Krause and co-workers [71] showed how to calculate the interaction parameters between random copolymers and homopolymers or other random copolymers. Their rather complex-appearing equation may be re-written in the case of binary mixtures to look exactly the same as Eq. (1), with χ_{AB} given by Eq. (12), but subscripts A and B now refer to random copolymers as well as to homopolymers. The Hildebrand solubility parameter of a random copolymer then turns out to be

$$\delta_C = \sum \delta_i \phi_i^C \qquad (14)$$

where δ_i is the solubility parameter of the homopolymer that corresponds to monomer i in the random copolymer and the summation is taken over all the different repeat units in the copolymer. The solubility parameter calculated from Eq. (14) can then be used to calculate interaction parameters from Eq. (12) and these can then be used to calculate compatibility. An example is given in Section IV.B.

As will be seen below, the combination of solubility parameters and Flory–Huggins theory gives remarkably good results in calculations of polymer compatibility when it is used with some care. Nevertheless, the in-adequacies of this theoretical treatment have led to a new approach to the thermodynamics of polymer solutions.

C. Equation of State Theories of Polymer Solutions

Most of the new approaches to the thermodynamics of polymer solutions are based on a theory of corresponding states for liquids pioneered by Prigogine [72]. This new approach is based on the assumption that a single equation of state involving appropriate reduced values of temperature, volume, and pressure is sufficient to describe all liquids and liquid solutions. Polymers and polymer mixtures are considered to be liquids if they are noncrystalline. Flory and co-workers have modified these ideas and applied them to mixtures of polymers, specifically mixtures of polymethylene with polyisobutylene [73]. Fairly laborious calculations based on precise data on thermal expansion coefficients and compressibility coefficients of the pure polymers led them to predict that the solubility of either polymer in the other would be immeasurably low if the molecular weights were high. Since then, McMaster [74] has made more extensive calculations on polymer–polymer mixtures and has been able to predict both upper and lower critical solution temperatures. He has also found, by calculation [74] that

Flory–Huggins theory gives results comparable to those from equation of state theory when the reduced variables for both polymers are comparable; essentially, this means that *Flory–Huggins theory works quite well when both polymers have similar thermal expansion coefficients.* Since the thermal expansion coefficients of most polymers are more similar to each other than to those of most solvents, this implies that Flory–Huggins theory will be closer to the facts in the case of polymer–polymer mixtures than in the case of polymer–solvent mixtures. Partly for this reason and partly because the mathematics are simpler. Flory–Huggins theory is used in Section IV.B for a scheme to predict polymer–polymer compatibility. Further details of the equation of state theories are given in Chapter 3 in which a new treatment developed by Sanchez and Lacombe is described, which should eventually allow simple predictions of polymer–polymer compatibility. More data for each polymer of interest will be needed to use the Sanchez–Lacombe equation of state; this implies that there will be need for the Flory–Huggins type prediction scheme for some time in the future. It must be emphasized that even though calculations using corresponding states theories should yield much more accurate predictions of compatibility than those based on Eqs. (1)–(14), data on coefficients of thermal expansion and coefficients of compressibility are necessary for these more accurate predictions, and, even then, the calculations are laborious. In addition to the comments made by McMaster [74] about the utility of Flory–Huggins theory, Biros *et al.* [75] have shown that the solubility parameter concept embodies many of the ideas used in the newer theory and have concluded that the great popularity of the solubility parameter approach was entirely justified. Among other things, they noted that

$$(\partial \ln \delta / \partial T)_p \simeq -\alpha \qquad (15)$$

and

$$(\partial \ln \delta / \partial p)_T = \beta \qquad (16)$$

This allows calculation of solubility parameters at temperatures and pressures other than the 25°C and 1 atm at which these parameters are usually known.

III. EXPERIMENTAL DATA

A. Arrangement of Data

The type of data available for different polymer pairs varies considerably in quantity, quality, and in the sort of experimental information given. This makes tabulation of the data extremely difficult. Nevertheless, an attempt has been made to tabulate those polymer pairs and triads for which the data in the literature indicate compatibility.

Four tables have been prepared for Section III.B of this Chapter. Those polymer pairs that appear to be compatible in all proportions at reasonably high molecular weights at room temperature are listed in Table I. A system number, from the Appendix to this chapter, is listed with each polymer pair so that the reader may review the available data for himself. References [76–264] are given in the Appendix and are not repeated in the tables.

Those polymer pairs that appear to be compatible under certain conditions at room temperature are listed in Table II. It may be that compatibility exists (1) only for certain compositions of a copolymer in the mixture, or (2) only in the amorphous phase if one of the components crystallizes, or (3) only at very low molecular weight.

Those polymer pairs for which the fragmentary evidence in the literature indicates either compatibility or conditional compatibility at room temperature are listed in Table III. The data, however, seem insufficient to allow the inclusion of the polymer pair in Table I or II, and many of the pairs included in Table III may turn out to be incompatible in later experiments. Again, the data on these polymer pairs are mentioned under their system numbers in the Appendix, and the reader may reach his or her own conclusions.

Those mixtures of three polymers for which literature data indicate at least conditional compatibility are given in Table IV.

In the Appendix, data on all polymer mixtures that I have found in the literature are summarized. This includes all mixtures listed in the tables plus all those mixtures that seem to be incompatible. Only a very short summary of the data is presented in the case of polymer pairs that have been studied very extensively or that are obviously incompatible. All references, are given, however. When few data exist or when they are contradictory, more information is given.

The data are presented in nontabular form for the polymer mixtures investigated as listed in the chapter contents.

B. Tables of Polymers That May Be Compatible at Room Temperature

Table I

Polymer Pairs That Appear to Be Compatible in All Proportions at Room Temperature

System no.	Polymer I Poly( )	Polymer II Poly( )
b42	nitrocellulose	vinyl acetate
b44	nitrocellulose	methyl methacrylate
b45	nitrocellulose	methyl acrylate
d54	vinyl chloride	α-methylstyrene-*co*-methacrylonitrile-*co*-(ethyl acrylate)/58:40:2 wt %
g12	vinyl acetate	vinyl nitrate
h31	styrene	2,6-dimethyl-1,4-phenylene ether
h33	styrene	2,6-diethyl-1,4-phenylene ether
h33	styrene	2-methyl-6-ethyl-1,4-phenylene ether
h33	styrene	2-methyl-6-propyl-1,4-phenylene ether
h33	styrene	2,6-dipropyl-1,4-phenylene ether
h33	styrene	2-ethyl-6-propyl-1,4-phenylene ether
h36	styrene	oxycarbonyloxy(2,6-dimethyl-1,4-phenylene)isopropylidene(3,5-dimethyl-1,4-phenylene)
i18	isopropyl acrylate	isopropyl methacrylate
k35	α-methylstyrene	2,6-dimethyl-1,4-phenylene ether
k56	2,6-dimethyl-1,4-phenylene ether	2-methyl-6-phenyl-1,4-phenylene ether
k57	2,6-dimethyl-1,4-phenylene ether	2-methyl-6-benzyl-1,4-phenylene ether
l44	vinyl butyral	styrene-*co*-(maleic acid) (certain commercial samples)

Table II

Polymer Pairs That Appear to Be Conditionally Compatible

System no.	Polymer I Poly( )	Polymer II Poly( )	Conditions for compatibility
b49	nitrocellulose	ε-caprolactone	>50% I to prevent crystallizatic of II
b59	nitrocellulose	butadiene-*co*-acrylonitrile	Depends on composition of II; more data needed
b106	cellulose acetate–butyrate	butadiene-*co*-acrylonitrile	Depends on compositions of I ar II; more data needed
c4	isoprene	butadiene	Both polymers should not have high *cis*-1,4 content

(*continue*

Table II (*continued*)

System no.	Polymer I poly ( )	Polymer II poly ( )	Conditions for compatibility
d29	vinyl chloride	ε-caprolactone	>50% I to prevent crystallization of II
d38	vinyl chloride	butadiene-*co*-acrylonitrile	II must contain 30–40% acrylonitrile
d40	vinyl chloride	ethylene-*co*-(vinyl acetate)	High temperature may be necessary; evidence for UCST between room temperature and processing temperature
d42	vinyl chloride	ethylene-*co*-(vinyl acetate)-*co*-(sulfur dioxide) (E–VA–SO$_2$)	E/VA/SO$_2$: 72.7/18.5/8.8 compatible from 60 to 90% II; E/VA/SO$_2$: 89.0/7.8/3.2 incompatible at 30% II
e2	ethylene	propylene	Probably a temperature above the polyethylene melting point in the mixture of interest
f5	butadiene	butadiene-*co*-styrene	Styrene content of II must be <25 or 30%
f9	butadiene	EPDM rubber	Not sure because commercial samples were used
h9	styrene	α-methylstyrene	Molecular weights of I and/or II less than 10^5
h30	styrene	vinyl methyl ether	Near room temperature because of LCST and UCST close to room temperature
j13	acrylic acid	ethylene oxide	>50% I to avoid crystallinity of II; avoid extremes of pH
j23	methyl methacrylate	vinylidene fluoride	>65% I to avoid crystallinity of II
j43	methyl methacrylate	(vinyl chloride)-*co*-(vinyl acetate)	Certain compositions of II; not defined because of commercial samples used
j49	methyl methacrylate	styrene-*co*-acrylonitrile	9–20 or 30% acrylonitrile in II
j57	ethyl methacrylate	vinylidene fluoride	>60% I to avoid crystallinity of II
k34	ethylene oxide	vinyl naphthalene	>54% II to avoid crystallinity of I
k54	ethylene oxide	propylene oxide	Very low molecular weight oligomers only
k63	ε-caprolactone	phenoxy	See system
l6	chlorinated isoprene	ethylene-*co*-(vinyl acetate)	Insufficient data; depends on composition and crystallinity of II
l57	ε-caprolactone	styrene-*co*-acrylonitrile	Insufficient data; 28% acrylonitrile in II gave compatible mixture
l61	2,6-dimethyl-1,4-phenylene ether	styrene-*co*-(*p*-chlorostyrene)	Mole fraction styrene in II ≥ 0.347

(*continued*)

Table II (*continued*)

System no.	Polymer I poly ( )	Polymer II poly ( )	Conditions for compatibility
180	oxy-1,4-phenylenesulfonyl-1,4-phenyleneoxy(2,6-diisopropyl-1,4-phenylene)isopropylidene(3,5-diisopropyl-1,4-phenylene)	styrene-*co*-acrylonitrile	Acrylonitrile content of II between 13 and 16%

Table III

Polymer Pairs That May Be or Must Be Compatible
or Conditionally Compatible at Room Temperature

(Date are insufficient for inclusion of Systems in Table I or Table II)

System no.	Polymer I poly ( )	Polymer II poly ( )
a4	cellulose acetate	cellulose acetate
a9	nitrocellulose	cellulose acetate propionate
b25	cellulose acetate	(*N*-vinylpyrrolidone)-*co*-(vinyl acetate)
b46	nitrocellulose	ethyl acrylate
b46	nitrocellulose	*n*-butyl acrylate
b46	nitrocellulose	ethyl methacrylate
b46	nitrocellulose	*n*-butyl methacrylate
b51	nitrocellulose	some urethanes
b52	nitrocellulose	ester
b61	nitrocellulose	(vinyl chloride)-*co*-(vinyl propionate)
b65	nitrocellulose	ethylene-*co*-(vinyl acetate)
b66	nitrocellulose	styrene-*co*-acrylonitrile
b67	nitrocellulose	styrene-*co*-(methyl methacrylate)
b72	nitrocellulose	Indene resin
b94	cellulose acetate butyrate	vinyl acetate
b107	cellulose acetate butyrate	ethylene-*co*-(vinyl acetate)
b141	ethyl cellulose	butadiene-*co*-acrylonitrile
b151	ethyl cellulose	styrene-*co*-(maleic acid)
b152	ethyl cellulose	styrene-*co*-(maleic acid ester)
b155	ethyl cellulose	Indene resin
b165	benzyl cellulose	ethyl methacrylate
c3	isoprene	isobutene
c14	isoprene	butadiene-*co*-styrene

Table III (*continued*)

System no.	Polymer I poly ( )	Polymer II poly ( )
d2	vinyl chloride	chloroprene
d16	vinyl chloride	ethyl methacrylate
d17	vinyl chloride	*n*-propyl methacrylate
d19	vinyl chloride	*n*-butyl methacrylate
d20	vinyl chloride	isobutyl methacrylate
d30	vinyl chloride	urethane
d32	vinyl chloride	(vinyl chloride)-*co*-(vinyl acetate)
d33	vinyl chloride	(vinyl chloride)-*co*-(vinyl propionate)
d34	vinyl chloride	(vinyl chloride)-*co*-(vinyl acetate)-*co*-(vinyl alcohol)
d35	vinyl chloride	(vinyl chloride)-*co*-(vinylidene chloride)-*co*-(acrylic acid ester)
d44	vinyl chloride	ethylene-*co*-acrylonitrile
d45	vinyl chloride	ethylene-*co*-(N,N-dimethyl acrylamide)
d46	vinyl chloride	ethylene-*co*-(N-methyl-N-vinyl acetamide)
d47	vinyl chloride	ethylene-*co*-(*n*-butyl urethane)
d48	vinyl chloride	ethylene-*co*-(4-vinylpyridine)
d49	vinyl chloride	ethylene-*co*-(N,N-dimethylaminoethyl methacrylate)
d50	vinyl chloride	(vinyl acetate)-*co*-(N-vinylpyrrolidone)
d52	vinyl chloride	(methyl methacrylate)-*co*-(ethyl methacrylate)
d56	vinyl chloride	Indene resin
g6	vinyl acetate	ethyl acrylate
g26	vinyl acetate	(vinyl acetate)-*co*-ethylene
g27	vinyl acetate	(vinyl acetate)-*co*-(vinyl chloride)
g38	vinyl acetate	styrene-*co*-(maleic acid ester)
g39	vinyl acetate	(methyl vinyl ether)-*co*-(maleic anhydride)
h44	styrene	styrene-*co*-butadiene
h45	styrene	styrene-*co*-(ethyl acrylate)
h46	styrene	styrene-*co*-acrylonitrile
h47	styrene	styrene-*co*-(methyl methacrylate)
h48	styrene	styrene-*co*-(maleic acid)
h49	styrene	styrene-*co*-(maleic acid ester)
h60	styrene	Miscellaneous resins
i15	ethyl acrylate	(ethyl acrylate)-*co*-(methyl methacrylate)
i26	butyl acrylate	(butyl acrylate)-*co*-(methyl methacrylate)
i28	acrylic acid	sodium acrylate
i30	acrylic acid	(acrylic acid)-*co*-(methacrylic acid)
i41	methyl methacrylate	(ethyl acrylate)-*co*-(methyl methacrylate)
i42	methyl methacrylate	(butyl acrylate)-*co*-(methyl methacrylate)
i46	propyl methacrylate	butyl methacrylate
i47	propyl methacrylate	hexyl methacrylate
i50	butyl methacrylate	hexyl methacrylate

(*continued*)

Table III *(continued)*

System no.	Polymer I poly ( )	Polymer II poly ( )
i52	hexyl methacrylate	octyl methacrylate
i54	methacrylic acid	sodium methacrylate
i55	methacrylic acid	(acrylic acid)-*co*-(methacrylic aid)
j8	ethyl acrylate	(ethyl acrylate)-*co*-styrene
j11	*n*-butyl acrylate	chlorinated isoprene
j21	methyl methacrylate	chlorinated vinyl chloride
j25	methyl methacrylate	α-methylstyrene
j39	methyl methacrylate	(methyl methacrylate)-*co*-styrene
j45	methyl methacrylate	(vinyl chloride)-*co*-(vinyl acetate) -*co*-(vinyl alcohol)
j48	methyl methacrylate	(*N*-vinylpyrrolidone)-*co*-(vinyl acetate)
j51	methyl methacrylate	styrene-*co*-(maleic acid ester)
j52	methyl methacrylate	(methyl vinyl ether)-*co*-(maleic anhydride)
j54	methyl methacrylate	Indene resin
j56	ethyl methacrylate	chlorinated isoprene
j58	*n*-butyl methacrylate	chlorinated isoprene
k16	vinylidene chloride	urethane
k18	vinylidene chloride	Indene resin
k29	vinylpyrrolidone	vinyl butyral
k52	epichlorohydrin	propylene glycol
k53	ethylene glycol	propylene glycol
k62	ε-caprolactone	epichlorohydrin
k79	Various resins	Various resins
l3	isobutene	butadiene-*co*-styrene
l8	chloroprene	butadiene-*co*-styrene
l9	chloroprene	butadiene-*co*-acrylonitrile
l12	chlorosulfonated ethylene	ethylene-*co*-(vinyl acetate)
l14	chlorosulfonated ethylene	(vinyl chloride)-*co*-(vinyl acetate)
l15	chlorosulfonated ethylene	(vinyl chloride)-*co*-(vinyl propionate)
l18	chlorosulfonated ethylene	(vinyl chloride)-*co*-(vinylidene chloride)-*co*-(acrylic acid ester)
l22	chlorinated vinyl chloride	butadiene-*co*-styrene
l23	chlorinated vinyl chloride	butadiene-*co*-acrylonitrile
l25	vinylidene chloride	ethylene-*co*-(vinyl acetate)
l30	vinylidene chloride	(vinyl chloride)-*co*-(vinylidene chloride)-*co*-(acrylic acid ester)
l34	vinylidene fluoride	(vinylidene fluoride)-*co*-tetrachloroethylene
l43	vinyl butyral	(*N*-vinylpyrrolidone)-*co*-vinyl acetate
l45	vinyl butyral	styrene-*co*-(maleric acid ester)
l50	vinyl pyrrolidone	(vinyl chloride)-*co*-(vinylidene chloride)-*co*-(acrylic acid ester)
l51	vinyl pyrrolidone	(*N*-vinylpyrrolidone)-*co*-(vinyl acetate)

Table III (*continued*)

System no.	Polymer I poly ( )	Polymer II poly ( )
I54	epichlorohydrin	styrene-*co*-(methyl methacrylate)
I55	epichlorohydrin	styrene-*co*-acrylonitrile
I84	urethane	(vinyl chloride)-*co*-(vinyl acetate)
I85	urethane	(vinyl chloride)-*co*-(vinyl propionate)
I86	urethane	(vinyl chloride)-*co*-(vinyl acetate)-*co*-(vinyl alcohol)
I87	urethane	(vinyl chloride)-*co*-(vinylidene chloride)-*co*-(acrylic acid ester)
I88	urethane	(*N*-vinylpyrrolidone)-*co*-(vinyl acetate)
I99	ester	(vinyl chloride)-*co*-(vinyl acetate)-*co*-(vinyl alcohol)
I100	ester	(vinyl chloride)-*co*-(vinylidene chloride)-*co*-(acrylic acid ester)
I101	ester	(*N*-vinylpyrrolidone)-*co*-(vinyl acetate)
I104, I114	Various resins	(vinyl chloride)-*co*-(vinyl acetate)
I105, I107, I111, I113	Various resins	butadiene-*co*-acrylonitrile
I106, I108	Various resins	carboxylated butadiene-*co*-acrylonitrile
I109	Various resins	ethylene-*co*-(vinyl acetate)
I117	Indene resin	(vinyl chloride)-*co*-(vinyl acetate)-*co*-(vinyl alcohol)
I119	Various resins	(vinyl chloride)-*co*-(vinylidene chloride)-*co*-(acrylic acid ester)
m4	ethylene-*co*-(vinyl acetate)	(vinyl chloride)-*co*-(vinyl acetate)
m6	ethylene-*co*-(vinyl acetate)	(vinyl chloride)-*co*-(vinyl acetate)-*co*-(vinyl alcohol
m7	ethylene-*co*-(vinyl acetate)	(vinyl chloride)-*co*-(vinyl acetate)-*co*-(maleic acid)
m8	ethylene-*co*-(vinyl acetate)	(vinyl chloride)-*co*-(vinylidene chloride)-*co*-(acrylic acid ester)
m10	ethylene-*co*-(vinyl acetate)	styrene-*co*-(maleic acid)
m11	ethylene-*co*-(vinyl acetate)	styrene-*co*-(maleic acid ester)
m14	butadiene-*co*-styrene	butadiene-*co*-acrylonitrile
m15	butadiene-*co*-styrene	(vinyl chloride)-*co*-(vinyl acetate)
m21	butadiene-*co*-acrylonitrile	(vinyl chloride)-*co*-(vinyl acetate)
m22	butadiene-*co*-acrylonitrile	(vinyl chloride)-*co*-(vinyl acetate)-*co*-(vinyl alcohol)
m23	butadiene-*co*-acrylonitrile	(vinyl chloride)-*co*-(vinyl acetate)-*co*-(maleic acid)
m24	butadiene-*co*-acrylonitrile	(vinyl chloride)-*co*-(vinyl propionate)
m25	butadiene-*co*-acrylonitrile	(vinyl chloride)-*co*-(vinylidene chloride)-*co*-(acrylic acid ester)
m28	butadiene-*co*-acrylonitrile	styrene-*co*-acrylonitrile
m31	butadiene-*co*-acrylonitrile	carboxylated butadiene-*co*-acrylonitrile
m32	(vinyl chloride)-*co*-(vinyl acetate)	(vinyl chloride)-*co*-(vinyl propionate)

(*continued*)

Table III (*continued*)

System no.	Polymer I poly ( )	Polymer II poly ( )
m33	(vinyl chloride)-*co*-(vinyl acetate)	(vinyl chloride)-*co*-(vinyl acetate)-*co*-(vinyl alcohol)
m34	(vinyl chloride)-*co*-(vinyl acetate)	(vinyl chloride)-*co*-(vinyl acetate)-*co*-(maleic acid)
m35	(vinyl chloride)-*co*-(vinyl acetate)	(vinyl chloride)-*co*-(vinylidene chloride)-*co*-(acrylic acid ester)
m36	(vinyl chloride)-*co*-(vinyl acetate)	(*N*-vinylpyrrolidone)-*co*-(vinyl acetate)
m43	(vinyl chloride)-*co*-(vinyl acetate)	carboxylated butadiene-*co*-acrylonitrile
m44	(vinyl chloride)-*co*-(vinyl propionate)	(vinyl chloride)-*co*-(vinyl acetate)-*co*-(vinyl alcohol)
m45	(vinyl chloride)-*co*-(vinyl propionate)	(vinyl chloride)-*co*-(vinylidene chloride)-*co*-(acrylic acid ester)
m48	(vinyl chloride)-*co*-(vinyl acetate)-*co*-(vinyl alcohol)	(vinyl chloride)-*co*-(vinyl acetate)-*co*-(maleic acid)
m49	(vinyl chloride)-*co*-(vinyl acetate)-*co*-(vinyl alcohol)	(vinyl chloride)-*co*-(vinylidene chloride)-*co*-(acrylic acid ester)
m50	(vinyl chloride)-*co*-(vinyl acetate)-*co*-(maleic acid)	(vinyl chloride)-*co*-(vinylidene chloride)-*co*-(acrylic acid ester)
m51	(vinyl acetate)-*co*-(*N*-vinylpyrrolidone)	(vinyl chloride)-*co*-(vinyl acetate)-*co*-(vinyl alcohol)
m52	(vinyl acetate)-*co*-(*N*-vinylpyrrolidone)	(vinyl chloride)-*co*-(vinyl acetate)-*co*-(maleic acid)
m53	(vinyl acetate)-*co*-(*N*-vinylpyrrolidone)	(vinyl chloride)-*co*-(vinylidene chloride)-*co*-(acrylic acid ester)
m61	styrene-*co*-(maleic acid)	styrene-*co*-(maleic acid ester)
m63	styrene-*co*-(maleic acid)	(vinyl chloride)-*co*-(vinyl acetate)-*co*-(maleic acid)
m69	(ethyl acrylate)-*co*-(methyl methacrylate)	(butyl acrylate)-*co*-(methyl methacrylate)
m70	(ethyl acrylate)-*co*-(acrylic acid)	(ethyl acrylate)-*co*-(methacrylic acid)
m71	(ethyl acrylate)-*co*-(methacrylic acid)	(methyl methacrylate)-*co*-(methacrylic acid)

Table IV

Mixtures of Three Polymers That May Be Compatible or
Conditionally Compatible at Room Temperature

System no.	Polymer I poly ( )	Polymer II poly ( )	Polymer III poly ( )
n1	ethylene	vinyl chloride	ethylene-*co*-(*N*-methyl-*N*-vinyl acetamide)
n2	chloroprene	vinyl chloride	acrylonitrile
n3	chloroprene	vinyl chloride	vinylidene chloride
n5	chloroprene	vinyl acetate	vinylidene chloride
n6	chloroprene	vinyl acetate	acrylonitrile
n7	vinyl chloride	ε-caprolactone	ethylene-*co*-(*N*-methyl-*N*-vinyl acetamide)

IV. PREDICTION SCHEMES FOR POLYMER–POLYMER COMPATIBILITY

A. Review of the Literature

Scott [58] compared his theoretical predictions for polymer A–polymer B–solvent systems with earlier data obtained by Dobry and Boyer-Kawenoki [78] and found a general, but not detailed, agreement with his predictions. His prediction that $(\chi_{AB})_{cr}$ would be very small for two polymers having appreciable degrees of polymerization was borne out by the experimental fact that only a small percentage, about 10%, of all polymer pairs investigated were compatible. In most cases, Scott's prediction that a polymer pair that showed phase separation in one solvent would also show phase separation in all other solvents was also verified. His prediction that the critical concentrations of the polymers for phase separation would not depend on the solvent was, however, shown to be false. In one case, cellulose acetate plus nitrocellulose, the polymers underwent phase separation at a fairly low concentration of solids in one solvent, acetone, but not at any concentration in another, acetic acid. This is a contradiction of Scott's theoretical predictions, but it should not be surprising. This is a system in which hydrogen bonding is possible, and the Flory–Huggins theory of polymer solutions does *not* allow for specific interactions between molecules, such as hydrogen bonding. *It is best to expect odd instances of compatibility between polymers, outside the scope of the Flory–Huggins theory, whenever hydrogen bonding is possible.* The importance of the molecular weight of the polymers predicted by Scott had also been found by Dobry.

For example, cellulose acetate (I) and poly(vinyl acetate) (II), System **b1**, were much less compatible when I had $\overline{M}_w = 5.6 \times 10^4$ and II had $\overline{M}_w = 9.7 \times 10^4$ than when I had $\overline{M}_w = 1.6 \times 10^4$ and II had $\overline{M}_w = 3.9 \times 10^4$; the mixture of lower molecular weight polymers went into solution in acetone at about 3.8% solids or less, while the mixture of higher molecular weight polymers could go into solution only if it comprised 1.4% or less of the mixture.

Later work also indicated qualitative agreement with Scott's predictions; this includes Berek, Lath, and Ďurďovič's [55] investigation of polystyrene and polypropylene in toluene, and Bristow's [110] study of rubber and poly(methyl methacrylate) in benzene or *n*-butyl acetate, Berek *et al.* [55] and Bristow [110] give reasonably complete phase diagrams for the three-component mixtures; these are necessary for detailed comparison with the theoretical treatment. Phase diagrams also appear in the literature for the following polymer–polymer–solvent mixtures: polystyrene–polybutadiene–toluene or benzene or CCl_4 [155], or tetralin [158]; polystyrene–poly(*p*-chlorostyrene)–benzene [184]; polystyrene–nitrocellulose–methyl ethyl ketone [78]; poly(vinyl acetal)–polystyrene–methyl ethyl ketone or $CHCl_3$ [78], poly(vinyl acetate)–poly(vinyl acetal)–acetone or methyl ethyl ketone [78], polystyrene–poly(vinyl acetal)–$CHCl_3$ [78]; nitrocellulose–poly(vinyl acetal)–methyl ethyl ketone [78]; rubber–polystyrene–benzene [78], linear polyethylene–atactic or isotactic polypropylene–diphenyl ether [52]; poly(1,1,2-trichloro-1,3-butadiene)–chlorinated poly(*cis*-1,4-isoprene)–CCl_4 [230]; 1/1 mixtures of polystyrene–poly(dimethylsiloxane) in ethyl acetate and of polyisobutene–poly(dimethylsiloxane) in phenetole [202]. Phase diagrams have also appeared for mixtures of low and even high molecular weight polymers in the absence of solvent–polyisobutene–poly(dimethyl-siloxane) [227], polyisoprene–polystyrene [52, 107], polystyrene–poly(1-butene) [173], polystyrene–poly(vinyl methyl ether) [41, 54, 74], poly-butene–poly[vinyltoluene-*co*-(α-methylstyrene)] [241], poly(methyl meth-acrylate)–poly(styrene-*co*-acrylonitrile) [220], and polycaprolactone–poly-(styrene-*co*-acrylonitrile) [74]. Many phase diagrams are in qualitative agreement with Scott's predictions, but others show an LCST or binodal coexistence curves (see Figs. 1 and 2).

One phase diagram for a three-component polymer system appears in the literature: polystyrene–poly(methyl methacrylate)–poly[styrene-*co*-(methyl methacrylate)] [251].

One worker [265] used equation of state theory to calculate spinodals with LCSTs for the system poly(vinyl chloride)–poly(ε-caprolactone) from gas–liquid chromatography data using the polymer blends as supports.

Scott's other predictions [68], which involve the lack of compatibility of

random copolymers of the same two monomers having different compositions, have been verified by the work in most of the references cited for
Systems of Type **o**. If we consider as an example Molau's [260] work on
mixtures of various styrene–acrylonitrile copolymers, we find that these
mixtures do not quite conform with the assumption in Scott's simplified
equation, Eq. (13) in this chapter, nevertheless it is instructive to compare
its predictions with Molau's data. Equation (13) can be used to predict the
maximum spread of δ values allowable in a copolymer sample before phase
separation occurs. Molau's samples all had $\overline{M}_w \simeq 10^5$, and using $T = 300°$K
and $\rho \simeq 1$, we get $\Delta\delta = 0.18$. Using reasonable values for the solubility
parameters of the homopolymers, $\delta \simeq 9.0$ for polystyrene and $\delta \simeq 12.7$ for
polyacrylonitrile, and using Eq. (14) for calculating the solubility parameters
of Molau's copolymer samples, we find that $\Delta\delta \simeq 3.7\Delta\phi_A$, where $\Delta\phi_A$ is the
difference in the volume fraction of acrylonitrile in the two copolymer
samples mixed. Using the calculated maximum $\Delta\delta = 0.18$ for Molau's
polymers, the critical $\Delta\phi_A = 0.049$, that is, a difference of 4.9% in acrylonitrile
volume fraction between two of these copolymers, with $\overline{M}_w \simeq 10^5$, is sufficient
for phase separation to occur. Molau found experimentally that a difference
of 3.5–4.5 wt% acrylonitrile was sufficient to cause phase separation. If we
use Eqs. (3a), (12), and (14), that is, the critical conditions for a two-
component mixture of copolymers, to calculate the critical value of $\Delta\phi_A$
for the same polymers, we get $\Delta\phi_A = 0.030$. Agreement with Molau's
experimental data is good by either method of calculation, especially since
Molau's samples must have been quite polydisperse.

It may be noted here that this remarkable agreement between theory and
experiment was obtained in a system in which no solvent was present. In
systems containing solvent as well as polymers, agreement between experiment and theory becomes much more qualitative. There are two ways to
look at this problem. The first way was discussed in Section II.C, in connection with equation of state theories of polymer solutions. It was stated
there that Flory–Huggins theory works quite well when the components of
the mixture have similar thermal expansion coefficients. Since the thermal
expansion coefficients of most polymers are much smaller than those of most
solvents, Flory–Huggins theory should work better for polymer mixtures in
the absence of solvents.

The second way of looking at the solvent problem does not involve
consideration of the equation of state theories of polymer solutions. In
simple terms, we may say that the problem lies in Eq. (12) and the train
of thought that associates the interaction parameter χ_{AB} only with energies
of interaction between molecules. When the Flory–Huggins theory is
applied to mixtures of molecules whose size is extremely different, that is,

polymer–solvent mixtures, it becomes necessary, as Huggins [63] recognized, to include an entropy contribution in the definition of χ_{AB}, so that Eq. (12) is no longer correct. However, this entropy contribution does not exist for high polymer mixtures, and Eq. (12) becomes quite accurate. We may therefore say that *predictions made using Flory–Huggins theory for mixtures of high polymers will be much more precise than predictions made for mixtures of polymers with solvents.* When very low molecular weight polymers are mixed, problems may be encountered. For example, Allen *et al.* [227] found that χ_{AB} for poly(dimethylsiloxane) samples (molecular weights: 850, 1350, 17,000) mixed with polyisobutenes (molecular weights: 250, 440) contained an entropy contribution. The particular low molecular weight polyisobutenes they used must, however, be considered as solvents rather than polymers in the comparison of theory with experiment.

Many workers have found, on analyzing the two or three phases into which a polymer A–polymer B–solvent separated, that each phase contained both polymers as well as the solvent. Therefore, this type of phase separation cannot be used to separate two polymers completely from each other unless a number of stages is used. This behavior is also predicted by Scott's [58] theoretical treatment. The phases into which a mixture of two polymers separates in the absence of solvent are also not pure unless χ_{AB} is very large indeed. Equation (5) can be used to calculate the approximate volume fractions of A and B to be expected in the coexisting phases. Equations (6a) and (6b) must be used for exact calculations.

The presence of polymer A in the polymer B phase, and vice versa, is demonstrated by experimental observations of the glass transition temperatures of the coexisting phases. Often, the mixture does not exhibit the exact glass transition temperatures of the two homopolymers, but transition temperatures that are somewhat displaced, each in the direction of the second polymer. This is expected if the coexisting phases are each mixtures, and not the pure components. Boyer and Spencer [176] for example, noticed that the T_g of the polystyrene-rich phase in polystyrene–polyisobutylene mixtures was 4°C or more below the T_g of the pure polystyrene. Similarly, Bares and Pegoraro [266] explained a decrease in the T_g of poly(vinyl chloride) microphases in a copolymer consisting of poly(vinyl chloride) grafts on an ethylene–propylene backbone by assuming that the poly(vinyl chloride) microphases contained some of the ethylene–propylene backbone.

Schneier [267] has made some predictions of polymer–polymer compatibility using heats of mixing and solubility parameters in a way that is rather different from the manner in which heat of mixing is presented here. His method is not discussed further because the lack of entropy considerations preclude any consideration of the very important molecular weight dependence of compatibility.

B. A Simple Scheme Based on Flory–Huggins Theory

1. Solubility Parameters

In principle, Eqs. (1)–(14) could be used to calculate the compatibility of any two polymers of interest if two pieces of information are known for each polymer: solubility parameter and molecular weight. For many homo-polymers and some copolymers, solubility parameters have been listed in tables, some of which may be found in Brandrup and Immergut [268], Shvarts [269], Shvarts *et al.* [270], Gardon [271], and Burrell [272]. These tables have to be used with some caution. The experimental methods for determining solubility parameters of polymers involve studies of the polymers in solution or swollen by solvent, that is, the thermodynamics of the situation involves the entropy contribution to the polymer–solvent interaction parameter that, as mentioned above, complicates the theoretical treatment. It often turns out that experimental solubility parameters for the same polymer vary a great deal, depending both on the method used and on the experimenter. Two methods are generally used: (a) swelling of a cross-linked polymer, where it is assumed that the solubility parameter of the polymer is equal to that of the solvent which swells it most highly; and (b) intrinsic viscosity of a soluble polymer sample, where it is assumed that the solubility parameter of the polymer is equal to that of the solvent in which its intrinsic viscosity is greatest. For the same polymer, these methods often give different results depending on the nature of the solvents, which may be non-polar, polar, or hydrogen bonding.

Because of these problems, the solubility parameter concept has often been criticized, and, more often, been extended to include separate non-polar, polar, and, sometimes, hydrogen-bonding contributions. Although these ideas have been useful, they would complicate the simple scheme presented here, and, therefore, only the simple solubility parameter is used in this chapter. For those who wish to explore the solubility parameter concept more thoroughly, the following references may be helpful: a discussion of the strengths and weaknesses of the solubility parameter concept with some discussion of the hydrogen-bonding problem [273]; several discussions of polar and nonpolar contributions to solubility parameter [177, 274, 275]; several discussions of polar, nonpolar, and hydrogen-bonding contributions to solubility parameters [276, 277].

While searching for values of polymer solubility parameters to use in polymer–polymer compatibility calculations, I have found that the most accurate predictions are made when calculated rather than experimental values of δ are used for the predictions. These calculated values of solubility parameters are ultimately based on experimentally determined solubility parameters of pure solvents via some version of Small's tables [278]. Small found that the solubility parameters of solvents depended, in general, on the structure of the solvent molecules, their density, and their molecular weight. This also works for polymers if

$$\delta = \rho \sum F_i/M \tag{17}$$

where ρ is the density of the polymer at the temperature of interest, M the molecular weight of the repeat group in the polymer, and $\sum F_i$ is the sum of all the molar attraction constants of all the chemical groups in the polymer repeat unit. Table V is a revised form of Small's tables.

Recently, Hoy [279] reexamined a large amount of data on solvents and their solubility parameters and devised a new version of Small's molar attraction constants to use with Eq. (17). These revised molar attraction constants are given in Table VI. The solubility parameters of polystyrene, $\delta = 9.0$, and polyacrylonitrile, $\delta = 12.7$, which were used in my calculations

Table V

Group Molar Attraction Constants:
Revised Small's Tables[a]

Group	Molar attraction [(cal/cm^3)$^{1/2}$/mole]	Group	Molar attraction [(cal/cm^3)$^{1/2}$/mole]
Single bonded		H (variable)	80–100
—CH$_3$	214	O, ethers	70
—CH$_2$—	133	CO, ketones	275
—CH<	28	COO, esters	310
>C<	−93	CN	410
Double bonded		Cl (mean)	260
CH$_2$=	190	Cl, single	270
—CH=	111	Cl, twinned as in >CCl$_2$	260
>C=	19	Cl, triple as in —CCl$_3$	250
Triple bonded		Br, single	340
CH≡C—	285	I, single	425
—C≡C—	222	CF$_2$ } *n*-fluorocarbons only	150
Aromatic		CF$_3$	274
Phenyl	735	S, sulfides	225
Phenylene (o, m, p)	685	SH, thiols	315
Naphthyl	1146	ONO$_2$, nitrates	∼440
Other		NO$_2$, (aliphatic nitro	
Ring, 5-membered	105–115	compounds)	∼440
Ring, 6-membered	95–105	PO$_4$ (organic phosphates)	∼500
Conjugation	20– 30	Si (in silicones)	−38

[a] From Burrell [278a]. Reprinted by permission of the publisher, the Federation of Societies for Paint Technology. Reprinted by Brandrup and Immergut [268].

Table VI

Group Molar Attraction Constants According to Hoy[a]

Group	Molar attraction [(cal/cm^3)$^{1/2}$/mole]	Group or feature	Molar attraction [(cal/cm^3)$^{1/2}$/mole]
—CH$_3$	147.3	—S—	209.42
—CH$_2$—	131.5	Cl$_2$	342.67
>CH—	85.99	Cl, primary	205.06
—C—	32.03	Cl, secondary	208.27
		Cl, aromatic	161.0
CH$_2$=	126.54	Br	257.8
—CH=	121.53	Br aromatic	205.60
>C=	84.51	F	41.33
—CH=, aromatic	117.12		
—C=, aromatic	98.12	Structure feature	
—O— (ether, acetal)	114.98	Conjugation	23.26
—O— (epoxide)	176.20	Cis	−7.13
—COO—	326.58	Trans	−13.50
>C=O	262.96	4-Membered ring	77.76
—CHO	292.64	5-Membered ring	20.90
(CO)$_2$O	567.29	6-Membered ring	−23.44
—OH→	225.84	Ortho substitution	9.69
—H, acidic dimer	−50.47	Meta substitution	6.6
OH aromatic	170.99	Para substitution	40.33
NH$_2$	226.56		22.56
—NH—	180.03		
—N—	61.08		62.5
C≡N	254.56		
NCO	358.66	Base value	135.1

[a] From Hoy [279]. Reprinted by permission of the publisher, the Federation of Societies for Paint Technology.

in Section IV.A, were calculated using the constants in Table V or VI and densities of 1.05 and 1.18 for polystyrene and polyacrylonitrile, respectively, at 25°C. Incidentally, exactly the same solubility parameters are calculated by using either Table V or VI. Note that the structure features must be added in to give $\sum F_i$ *in addition to* the group contribution when using Table VI.

Hoy found that the only compounds that did not seem to adhere to the molar attraction constant scheme were acids, alcohols, and other compounds capable of association. However, these compounds could be made to fit into the scheme if the manner in which they associated was taken into account. Carboxylic acid, for example, may exist as a monomer or as dimer, example of acetic acid, whose solubility parameter as an unassociated molecule is 13.01, but, when considered as a dimer, the solubility parameter is 9.19. This allows a compound like acetic acid to have two different solubility characteristics; sometimes it behaves like a compound with $\delta = 13.01$, and sometimes like a compound with $\delta = 9.19$. Other types of compounds are capable of intramolecular hydrogen bonding and may or may not exist in that form, thus also having more than one solubility parameter. Hoy suggests that some polymers could have the same characteristics. This could be one explanation for the greater than usual compatibility with other polymers of some polymers which are capable of hydrogen bonding. Calculations made for such polymers should be undertaken carefully.

2. Steps Necessary to Predict Compatibility

a. *Calculate the solubility parameter of each polymer of interest.* This can be done for many homopolymers using Eq. (17) and Table V or VI if the density of the homopolymer is known at the temperature of interest. Both polymers, of course, need to have their solubility parameters calculated at the same temperature! If one or both of the polymers of interest are random copolymers, then the calculation according to Eq. (17) should be done for all the homopolymers corresponding to all the monomer units to be found in the copolymers. Solubility parameters of the copolymers can then be calculated according to Eq. (14). Weight fractions can be used in Eq. (14) in a pinch. (There is a scheme in the literature for calculating the solubility parameter of copolymers according to the molecular weights of the repeat units [280]. This scheme is probably not compatible with the ideas in this chapter.)

It should be noted that Eq. (17), because it contains the density of the polymer, will often lead to a prediction of different solubility parameters for polymers of different tacticity. This is probably a correct prediction, especially in light of some experimental evidence in the case of poly(methyl methacrylate) and of poly(α-methylstyrene) of different tacticities [281].

b. *Calculate the interaction parameter between the polymers of interest.* This is done by using Eq. (12), where δ_A and δ_B are the solubility parameters of the two polymers calculated in Step a, T is the temperature in degrees Kelvin, R is the gas constant equal to 1.987 cal/(deg-mole), and V_r is the reference volume in cm^3/mole, conveniently taken to be 100 cm^3/mole, as discussed earlier in the chapter. If the value is used, and the temperature of interest is 25°C, then Eq. (12) becomes

$$\chi_{AB} \simeq (\delta_A - \delta_B)^2/6 \qquad (18)$$

which is very convenient to use.

c. *If the word "compatible" is used to imply miscibility at all percentage compositions, use Eq. (3a) to calculate the critical* $(\chi_{AB})_{cr}$ *for the polymer of interest.* Equation (3a) involves the degree of polymerization of each polymer in terms of the reference volume V_r. There are two problems connected with the use of these degrees of polymerization. First, these degrees of polymerization are based on a polymer repeat unit whose molar volume equals the reference volume V_r, let us say 100 cm³/mole. Not all polymers have repeat units with this molar volume, however. It is, however, possible to calculate the degree of polymerization, say x_A needed for Eq. (3a), from the actual degree of polymerization x if the actual molar volume of the polymer repeat unit within the polymer $\overline{V}$, is known. Then

$$x_A = (\overline{V}/V_r)x \tag{19}$$

· It is often not worthwhile to use Eq. (19), and it is a good approximation to let $x_A = M_A/100$, where M_A is the molecular weight of the polymer. *It is necessary to know the approximate molecular weight of the polymers of interest in order to predict whether they will be compatible or not.* Table VII shows how $(\chi_{AB})_{cr}$ varies with degree of polymerization for two polymers having the same degree of polymerization, $x_A = x_B$. Also shown in Table VII is $|\delta_A - \delta_B|_{cr}$, the maximum difference between the solubility parameters of the two polymers that will allow compatibility at each value of degree of polymerization. In many cases, $x_A = x_B = 100$ refers to two polymers of molecular weight 10,000, while $x_A = x_B = 1000$ refers to two polymers of molecular weight 100,000. Values of $(\chi_{AB})_{cr}$ or $|\delta_A - \delta_B|_{cr}$ become very small when molecular weights are high.

d. *Compare* $(\chi_{AB})_{cr}$ *from Step c with* χ_{AB} *from Step b. If* $\chi_{AB} > (\chi_{AB})_{cr}$, the two polymers should be incompatible at some percentage compositions.

Table VII

$(\chi_{AB})_{cr}$ When $x_A = x_B$

$x_A = x_B$	$(\chi_{AB})_{cr}$	$|\delta_A - \delta_B|_{cr}$
50	0.040	0.49
100	0.020	0.35
200	0.010	0.25
300	0.0067	0.20
500	0.0040	0.15
1000	0.0020	0.11
2000	0.0010	0.077
3000	0.00067	0.063
5000	0.00040	0.049

The greater the difference between the two values of χ_{AB}, the smaller the range of compositions in which the polymers will be compatible.

e. *If compatibility is desired only at a particular composition, use Eq. (5) to calculate* $(\chi_{AB})_{sp}$. The compositions in Eq. (5) are written in terms of volume fractions, but weight fractions can usually be used. If $\chi_{AB} \geqslant (\chi_{AB})_{sp}$, phase separation will probably occur in the mixture at the composition of interest.

f. *Remember that these calculations are only approximate.* They can serve as a guide, but as usual with calculations, they are no substitute for experiment.

C. A Critique of the Simple Scheme and Comparison with Data

At this point, let us restate a number of comments on the Flory–Huggins theory of polymer solutions, and reconsider the precautions necessary for proper use of the "simple scheme." Finally, let us make some comments about the data in this chapter.

First, it is to be expected that the scheme will work best for high polymers in the absence of solvent, especially if the polymers have similar thermal expansion coefficients.

Second, the scheme will work best for polymers of similar polarity and similar hydrogen-bonding characteristics. The scheme will be farthest from supplying correct predictions when one polymer is a proton donor and the other is a proton acceptor; in such cases, complexes between the polymers are likely.

Third, even though it is necessary to consider the hydrogen-bonding characteristics of the polymers separately from their polarity, it is unnecessary, in almost all cases, to consider specific interactions between polymers based on polarity. This is taken care of in the solubility parameter concept.

These points can probably be illustrated by anyone dealing with polymer mixtures. However, there are a number of systems that do not seem to fit this scheme, as we shall see below. Furthermore, the use of this scheme is not as easy as it may seem. If one of the polymers to be mixed is glassy and the other is rubbery at the temperature of interest, should one use the actual densities of the polymers in Eq. (17) or should one try to figure out the state of the proposed mixture, say rubbery, and use the density that the glassy polymer would have if it were rubbery at that temperature? (This is done by extrapolation of the density of the polymer from temperatures at which it is in the rubbery state.) This is a difficult question for which I have no answer. One can only say that experiments will be necessary for the final determination of compatibility for many years to come, perhaps forever.

Now one might ask how many of the polymer pairs in Tables I and II would have been predicted to be compatible by the scheme shown in this chapter. Most of the systems involving copolymers could have been predicted because, depending on their composition, copolymers have solubility parameters that vary between those of their corresponding homopolymers. The solubility parameter of styrene–acrylonitrile copolymers, for example, varies between that of polystyrene, about 9.0, and that of polyacrylonitrile, about 12.7. It is expected that a styrene–acrylonitrile copolymer of some intermediate composition will be compatible with any polymer that has a solubility parameter in this range. In practice, a range of copolymer compositions may have to be tested in order to find one that is compatible.

Many of the homopolymer–homopolymer mixtures in Tables I and II would not have been predicted. Nitrocellulose, for example, is a hydrogen-bonding material, and I would hesitate to make any predictions concerning this polymer. Systems **g12**, poly(vinyl acetate)–poly(vinyl nitrate), **h30**, poly-styrene–poly(vinyl methyl ether), and **h31**, polystyrene–poly(2,6-dimethyl-1,4-phenylene ether), would be predicted to be incompatible; their disagreement with Flory–Huggins theory probably has to do with differences in the equation of state parameters of the polymer pairs—eventually, these parameters will all be known and better predictions should be possible.

Certain homopolymer–homopolymer pairs in these tables can be predicted perfectly by this scheme. For example, System **i18**, poly(isopropyl acrylate)–poly(isopropyl methacrylate), System **d29**, poly(vinyl chloride)–poly(ε-caprolactone), and System **h9**, polystyrene–poly(α-methylstyrene) are predicted to be compatible or, in the case of System **h9**, the actual conditions of compatibility can be predicted with surprising accuracy [181].

The comments made above lead to several conclusions:

1. The prediction scheme works only for some polymer systems, but the polarity of the polymers has little to do with the accuracy of the scheme.

2. Experimentation will probably continue to indicate additional compatible polymer pairs that cannot yet be predicted.

3. Someone will have to determine equation of state parameters for many of the polymers of interest; these include density, coefficient of thermal expansion, and coefficient of isothermal compressibility, preferably over a large temperature range.

Table VIII shows the calculated solubility parameters for some of the polymers just mentioned above, along with the values of density used in Eq. (17); it also shows whether Table V or VI was used for the molar attraction constants.

Table VIII

Some Calculated Solubility Parameters

Polymer poly ( )	Density used in calculation	δ
vinyl methyl ether	1.05	8.1
α-methylstyrene	1.066	8.93
styrene	1.05	9.05
isopropyl methacrylate	1.037	9.02
ethyl acrylate	1.06	9.16
methyl methacrylate	1.188	9.24
ε-caprolactone	1.10	9.41
vinyl chloride	1.37	9.44
vinyl acetate	1.19	9.48
2,6-dimethyl-1,4-phenylene ether	1.06	9.5 (Hoy)[a]
		10.21 (Small)[b]
vinyl nitrate	1.56	10.5
acrylonitrile	1.18	12.7

[a] Using group molar attraction constants from Table VI.
[b] Using group molar attraction constants from Table V.

APPENDIX: REVIEW OF DATA ON POLYMER MIXTURES IN THE LITERATURE

In Categories a–k, a polymer pair appears under the first category that fits; for example, the cellulose acetate–poly(vinyl acetate) system is in Category b, not g. However, most mixtures of one copolymer with another copolymer will appear in Category m, except for cellulose derivatives and acrylics. High-conversion copolymers, which have a very broad composition distribution, and mixtures of copolymers that contain the same monomers but in different proportions, are listed in Category o. Polymer mixtures are numbered within the categories, so that each polymer mixture is listed under a *system number*, which includes the category and the number within the category.

The statement *compatible, conditionally compatible, almost compatible, incompatible,* or *ambiguous* possibly with the words "may be" or "must be" appears next to each polymer mixture. This judgment is often obvious from the experimental data given, but on occasion it represents my judgment based on the data. The words *almost compatible* or *may be compatible* sometimes imply a conditional compatibility; if two polymers are almost compatible, then low molecular weight samples may be compatible, whereas higher molecular weight samples are incompatible. Perhaps the same mixture appeared compatible when films are cast from one solvent, whereas films of the composition appeared incompatible when cast from another solvent. Mixtures containing a small proportion of either polymer may be compatible, while mixtures containing more or less equal proportions of the two polymers appear incompatible. There is a further discussion of these points elsewhere in this chapter. The words *must be conditionally compatible* appear when at least one of the polymers in the mixture is a copolymer and the same monomer is present as a component of both of the polymers.

Some of the homopolymers cited are really copolymers, such as cellulose acetate and other cellulose derivatives, and various chlorinated rubbers. These polymers are listed in their appropriate sections rather than under "Copolymers and Other Copolymers," category. In most cases, information on composition of these copolymers was not given so that it is possible that different results could be obtained for different compositions. The word *ambiguous* sometimes appears in cases involving such copolymers, especially where different workers have obtained different results using copolymers having different compositions. This possibility is noted wherever appropriate. The word *ambiguous* also is used when experimental data obtained by different workers are contradictory.

Molecular weight or equivalent information is given wherever possible for the polymer mixtures. Compositions of copolymers, especially commercial copolymers, have been ascertained whenever possible. On occasion, as stated above (Section I.B), I was not able to ascertain the identity of a trademarked polymer, especially from Grosse and Friese [16]; these trademarked polymers have not been included in this chapter.

A. Cellulose Derivatives and Other Cellulose Derivatives

a1. *Cellulose acetate* (I)–*nitrocellulose* (II), *ambiguous:* No phase separation in acetone solution when I had ac. gr. 56%, $[\eta] = 2.00$ in acetone, and II had 11.9% N (slightly > dinitrate), $[\eta] = 1.90$ in acetone [76, 77]. Phase separation in acetone solution above 5.5% solids when I had $\overline{M}_n = 5.6 \times 10^4$, $[\eta] = 1.70$ in acetone, and II had $\overline{M}_n = 9.2 \times 10^4$, $[\eta] = 2.60$ in acetone; these samples were completely miscible in acetic acid in all proportions [78]. Phase separation in cyclohexanone for 1/1 mixture, and films were not clear when II was AS 5-6 sec, RS 15-20 sec, Hercules, Inc., and I was cellulose diacetate, 394-60, Eastman Chemical Co. [79]. Phase separation in acetone when I was 52.3% acetyl $\overline{M}_v = 8.1 \times 10^4$, and II had 11.9% N, $\overline{M}_v = 8.3 \times 10^4$ for I/II: 3/7 to 8.5/1.5 [80]. Nonhomogeneous solution and turbid film when I was Daiseru L-AC and II was Asahi Kasei Co. RS-1/2 [37]. Stated incompatible for I/II: 4/1 to 1/4 in solutions and films when II was Nitrocell E 620 and I was either Cellit T, cellulose triacetate, 60–61% acetic acid or Cellit L, cellulose diacetate, 53.5% acetic acid except for Cellit L when I/II: 4/1 [16].

a2. *Cellulose acetate* (I)–*ethylcellulose* (II), *incompatible:* Various samples showed phase separation in various solvents and in films [16, 37, 78, 79].

a3. *Cellulose acetate* (I)–*benzylcellulose* (II), *incompatible:* Several samples showed phase separation in solution and in films [16, 78].

a4. *Cellulose acetate* (I)–*cellulose acetate* (II), *conditionally compatible* (these are copolymers): Phase separation in acetone when I had ac. gr. 56%, $[\eta] = 2.00$ in acetone, and II had ac. gr. 48%, $[\eta] = 1.25$ in acetone [76, 77]. Stated compatible in ethylene chlorohydrin solution and in films for I/II: 4/1 to 1/4 when I was Cellit T, cellulose triacetate, 60–61% acid and II was Cellit L, cellulose diacetate, 53.5% acetic acid, but stated incompatible in films from $CHCl_3$–acetone: 1/1 [16].

a5. *Cellulose acetate* (I)–*cellulose acetate–propionate* (II), *incompatible:* Phase separation for 1/1 mixture in cyclohexane and film was not clear when I was cellulose diacetate 394-60, Eastman Chemical Co., and II was 319 E-40001-H6, Tennessee Eastman [79].

a6. *Cellulose acetate* (I)–*cellulose acetate–butyrate* (II), *incompatible:* Several samples showed phase separation in solution and in films [37, 79].

a7, *Nitrocellulose* (I)–*ethylcellulose* (II), *ambiguous:* Phase separation in acetone above 3.7% solids but miscible in acetic acid above 20% solids at 16–18°C when I had $\overline{M}_n = 9.2 \times 10^4$, $[\eta] = 2.60$ in acetone, and II had $\overline{M}_n = 3.5 \times 10^4$, $[\eta] = 1.10$ in acetone [78]. No phase separation for 1/1 mixture in cyclohexanone, and films were clear when I was AS 5-6 sec, RS 15-20 sec, Hercules, Inc., and II was 48.4% ethoxyl, Hercules, Inc [79]. Homogeneous solution and transparent film when I was Asahi Kasei Co. RS-1/2 and II was Dow Chemical Ethocell 50 cps [37]. Stated compatible in solution and in films for I/II: 4/1 to 1/4 when II was Nitrocell E 620 [16].

a8. *Nitrocellulose* (I)–*Benzyl cellulose* (II); *incompatible:* Several samples showed phase separation in solution and in films [16, 78].

a9. *Nitrocellulose* (I)–*cellulose acetate–propionate* (II), *may be compatible:* No phase separation in cyclohexanone for 1/1 mixture, and film was clear when I was AS 5-6 sec, RS 15-20 sec, Hercules, Inc., and II was 319 E-40001-H6, Tennessee Eastman [79].

a10. *Nitrocellulose* (I)–*cellulose acetate–butyrate* (II), *incompatible:* No phase separation in cyclohexanone for 1/1 mixture, but film was not clear when I was AS 5-6 sec, RS 15-20 sec, Hercules, Inc., and II was 17% butyryl, Tennessee Eastman [79].

a11. *Ethylcellulose* (I)–*Benzylcellulose* (II), *ambiguous:* Phase separation in $CHCl_3$ above 20% solids, in ethyl acetate above 4% solids at 16–18°C when I had $\overline{M}_n = 3.5 \times 10^4$, $[\eta] = 1.10$ in acetone [78]. Stated compatible for I/II: 4/1 to 1/4 in solution and in films [16].

a12. *Ethylcellulose* (I)–*cellulose acetate–propionate* (II), *incompatible:* No phase separation in cyclohexanone for 1/1 mixture, but film was not clear when I was 48.4% ethoxyl, Hercules, Inc., and II was 319E-40001-H6, Tennessee Eastman [79].

a13. *Ethylcellulose* (I)–*cellulose acetate–butyrate* (II), *incompatible:* Several mixtures had turbid films though solutions could be homogeneous [79] or nonhomogeneous [37].

a14. *Cellulose acetate–propionate* (I)–*cellulose acetate–butyrate* (II), *incompatible:* No phase separation in cyclohexanone for 1/1 mixture, but film was not clear when I was 319E-40001-H6, Tennessee Eastman, and II was 17% butyryl, Tennessee Eastman [79].

B. Cellulose Derivatives and Other Polymers

b1. *Cellulose acetate* (I)–*poly(vinyl acetate)* (II), *incompatible:* Phase separation found in solution and/or films for many different mixtures [29, 37, 76–81].

b2. *Cellulose acetate* (I)–*polystyrene* (II), *incompatible:* Phase separation found in solutions and/or films for many different mixtures [16, 37, 78, 79, 83].

b3. *Cellulose acetate* (I)–*poly(methyl methacrylate)* (II), *incompatible:* Phase separation found in solution and/or films for many different mixtures [16, 37, 76–79, 83].

b4. *Cellulose acetate* (I)–*poly(acrylic esters)* (II), *incompatible:* Stated to be incompatible if II was poly(methyl acrylate), poly(ethyl acrylate), poly(n-butyl acrylate), poly(ethyl methacrylate), or poly(n-butyl methacrylate) [83] (without proof).

b5. *Cellulose acetate* (I)–*polyacrylonitrile* (II), *incompatible:* Phase separation in dimethylformamide (DMF) at various polymer ratios above 3.6% solids [84]. Phase separation in DMF, but considered compatible because of a minimum in plot of activation energy for decomposition versus composition [81]. *Note:* There is little or no theoretical basis for the use of activation energy for decomposition to determine compatibility.

b6. *Cellulose acetate* (I)–*poly(vinyl acetal)* (II), *incompatible:* Phase separation in acetone above 2.1% solids at 16–18°C when I had $\overline{M}_n = 5.6 \times 10^4$, $[\eta] = 1.70$ in actone, and II had $\overline{M}_n = 3.8 \times 10^4$, $[\eta] = 0.75$ in acetone [78].

b7. *Cellulose acetate* (I)–*poly(vinyl butyral)* (II), *incompatible:* Nonhomogeneous solution and turbid film when I was Daiseru L-AC and II was Shekisui Chemical Co. BM-2 [37].

b8. *Cellulose acetate* (I)–*polyisobutylene* (II), *incompatible:* Stated incompatible in solution and in films for I/II: 4/1 to 1/4 when I was Cellit T, cellulose triacetate, 60–61% acetic acid, or Cellit L, cellulose diacetate, 53.5% acetic acid, and II was Oppanol B 100 [16].

b9. *Cellulose acetate* (I)–*polyisoprene* (II), *incompatible:* Same as System **b8** except that II was natural rubber [16].

b10. *Cellulose acetate* (I)–*polychloroprene* (II), *incompatible:* Same as System **b8** except that II was Neoprene AC.[16]

b11. *Cellulose acetate* (I)–*poly(vinyl chloride)* (II), *incompatible:* Phase separation found in solution and in films for several mixtures [16, 37].

b12. *Cellulose acetate* (I)–*chlorinated poly(vinyl chloride)* (II), *incompatible:* Same as System **b8** except that II was Vinoflex PC [16].

b13. *Cellulose acetate* (I)–*poly(vinylidene chloride)* (II), *incompatible:* Nonhomogeneous solution and turbid film when I was Daiseru L-AC and II was Asahi Dow Co. EX 5701 [37].

b14. *Cellulose acetate* (I)–*chlorosulfonated polyethylene* (II), *incompatible:* Same as System **b13** except that II was Showa Neoprene Hyperon 30 [37].

b15. *Cellulose acetate* (I)–*polyester* (II), *incompatible:* Same as system **b13** except that II was Toyobo Co. Ester Resin 20 [37].

b16. *Cellulose acetate* (I)–*polyurethane* (II), *incompatible:* Phase separation in solution and in films for three mixtures including polyester and polyether urethanes [37, 79].

b17. *Cellulose acetate* (I)–*polycarbonate* (II), *incompatible:* Phase separation in cyclohexane for 1/1 mixture and film was not clear when I was cellulose diacetate, 394-60, Eastman Chemical Co., and II was Lexan 125, General Electric Co., or Merlon M-50, Mobay Chemical Co. [79].

b18. *Cellulose acetate* (I)–*polyepichlorohydrin* (II), *incompatible:* Same as System **b17** except that II was Hydrin 100, B. F. Goodrich [79].

b19. *Cellulose acetate* (I)–*polysulfone* (II), *incompatible:* Same as System **b17** except that II was Union Carbide polysulfone [79].

b20. *Cellulose acetate* (I)–*poly(2,6-dimethyl-1,4-phenylene ether)* (II), *incompatible:* same as System **b17** except that II was General Electric Co. PPO [79].

b21. *Cellulose acetate* (I)–*poly[(vinyl chloride)-co-(vinyl acetate)]* (II), *incompatible:* Phase separation in solutions and in films for many mixtures [16, 37, 79].

b22. *Cellulose acetate* (I)–*poly[(vinyl chloride)-co-(vinyl propionate)]* (II), *incompatible:* Same as System **b13** except that II was Ryuron QS-430 [37].

b23. *Cellulose acetate* (I)–*poly[(vinyl chloride)-co-(vinyl acetate)-co-(vinyl alcohol)]*, *incompatible:* Same as System **b13** except that II was Shekisui Chemical Co. Esulex A [37].

b24. *Cellulose acetate* (I)–*poly*[(*vinyl chloride*)-*co*-(*vinylidene chloride*)-*co*-(*acrylic acid ester*)] (II), *incompatible:* Same as System **b13** except that II was Towa Gousei Co. Aron 321 [37].

b25. *Cellulose acetate* (I)–*poly*[(*N*-*vinylpyrrolidone*)-*co*-(*vinyl acetate*)] (II), *may be compatible:* Homogeneous solution and transparent film when I was Daiseru L-AC and II was General Aniline and Film S-630 [37].

b26. *Cellulose acetate* (I)–*poly*[*ethylene*-*co*-(*vinyl acetate*)] (II), *incompatible:* Same as System **b13** except that II was Nippon Goshei Co. Soalex R-CH or Soalex R-FH [37].

b27. *Cellulose acetate* (I)–*poly*(*butadiene*-*co*-*acrylonitrile*) (II), *ambiguous:* Phase separation in cyclohexanone for 1/1 mixture and film was not clear when I was cellulose diacetate, 394-60, Eastman Chemical Co., and II was B. F. Goodrich Hycar 1432 [79]. Nonhomogeneous solution and turbid film when I was Daiseru L-AC and II was Nippon Rubber Co. Hycar 1043 or 1432 [37]. Stated incompatible for I/II: 4/1 to 1/4 in solutions and films when I was Cellit T, cellulose triacetate, 60–61% acetic acid, and II was Buna NW, 28% acrylonitrile, but stated compatible for I/II: 4/1 to 1/4 when I was changed to Cellit L, cellulose diacetate, 53.5% acetic acid [16].

b28. *Cellulose acetate* (I)–*poly*(*styrene*-*co*-*acrylonitrile*) (II), *incompatible:* Same as System **b17** except that II was Dow Chemical Co. Tyril 767 [79].

b29. *Cellulose acetate* (I)–*poly*[*styrene*-*co*-(*methyl methacrylate*)] (II), *incompatible:* Same as System **b17** except that II was Dow Chemical Co. Zerlon 150 [79].

b30. *Cellulose acetate* (I)–*poly*[*styrene*-*co*-(*maleic acid*)] (II), *incompatible:* Same as System **b13** except that II was Daidou Kogyo Styrite CM-2 or CM-3 [37].

b31. *Cellulose acetate* (I)–*poly*[*styrene*-*co*-(*maleic acid ester*)] (II), *incompatible:* Same as System **b13** except that II was Daidou Kogyo Styrite HS-2 [37].

b32. *Cellulose acetate* (I)–*poly*[(*methyl vinyl ether*)-*co*-(*maleic anhydride*)] II, *incompatible:* Same as System **b17** except that II was B. F. Goodrich Gantrez AN-169 [79].

b33. *Cellulose acetate* (I)–*poly*[(*epichlorohydrin*)-*co*-(*ethylene oxide*)] (II), *incompatible:* Same as System **b17** except that II was B. F. Goodrich Hydrin 200 [79].

b34. *Cellulose acetate* (I)–*miscellaneous resins* (II), *incompatible:* Same as System **b13** except that II was Petroleum Resin, Nippon Petroleum Co., Nitsuseki Neopolymer 150, or Cumarone Resin, Mitsubishi TG, or Indene Resin, Fuji Iron Products Co. VM-1/2 [37].

b35. *Nitrocellulose* (I)–*polyisoprene* (II), *incompatible:* Phase separation in solutions and films for several mixtures [16, 30].

b36. *Nitrocellulose* (I)–*polybutadiene* (II), *incompatible:* Phase separation in solutions and films for several mixtures [85, 86].

b37. *Nitrocellulose* (I)–*polyisobutene* (II), *incompatible:* Stated incompatible for I/II: 4/1 to 1/4 in solutions and films when I was Nitrocell E 620 and II was Oppanol B 100 [16].

b38. *Nitrocellulose* (I)–*polychloroprene* (II), *incompatible:* Same as System **b37** except that II was Neoprene AC [16].

b39. *Nitrocellulose* (I)–*poly*(*vinyl chloride*) (II), *incompatible:* Phase separation in solution and films for several mixtures [16, 37].

b40. *Nitrocellulose* (I)–*chlorinated poly*(*vinyl chloride*) (II), *incompatible:* Phase separation in solutions and films for several mixtures [16, 36].

b41. *Nitrocellulose* (I)–*poly*(*vinylidene chloride*) (II), *incompatible:* Nonhomogeneous solution and turbid film when I was Asahi Kasei Co. RS-1/2, and II was Asahi Dow Co. EX 5701 [37].

b42. *Nitrocellulose* (I)–*poly*(*vinyl acetate*) (II), *compatible:* No phase separation observed in solutions and films for many mixtures [36, 37, 76–80]. Single T_g by DTA [36, 37]. Dynamic mechanical measurements showed one transition when mixtures contained 0–60% I, two transitions seen from 70 to 90% I, presumably not due to unmixing [87].

b43. *Nitrocellulose* (I)–*polystyrene* (II), *incompatible:* Phase separation in solvents and films for many mixtures [16, 37, 78, 79].

b44. *Nitrocellulose* (I)–*poly(methyl methacrylate)* (II), *compatible:* No phase separation observed in solutions and films for many mixtures [16, 37, 78, 79]. Stated to be compatible without proof [83].

b45. *Nitrocellulose* (I)–*poly(methyl acrylate)* (II), *compatible:* Mixture had only a single T_g [88]. Stated to be compatible without proof [83].

b46. *Nitrocellulose* (I)–*poly(acrylic esters)* (II), *may be compatible:* Listed as compatible if II was poly(ethyl acrylate), poly(n-butyl acrylate), poly(ethyl methacrylate), or poly(n-butyl methacrylate) (without proof) [83].

b47. *Nitrocellulose* (I)–*poly(vinyl acetal)* (II), *incompatible:* Phase separation in many solvents at 2.6% solids or less at 16–18°C when I had $\overline{M}_n = 9.2 \times 10^4$, $[\eta] = 2.60$ in acetone, and II had $\overline{M}_n = 3.8 \times 10^4$, $[\eta] = 0.75$ in acetone. The same samples were miscible in mesityl oxide to $> 5\%$ solids and in acetic acid to $> 20\%$ solids [78].

b48. *Nitrocellulose* (I)–*poly(vinyl butyral)* (II), *incompatible:* Same as System **b41** except that II was Shekisui Chemical Co. BM-2 [37].

b49. *Nitrocellulose* (I)–*poly(ε-caprolactone)* (II), *conditionally compatible:* Single T_g from dynamic mechanical measurements for I/II: 1/9 to 1/1; films of blends clear but crystallinity of I developed for blends with $> 50\%$ I; I contained 12% N [89].

b50. *Nitrocellulose* (I)–*chlorosulfonated polyethylene* (II), *incompatible:* Same as System **b41** except that II was Showa Neoprene Hyperon 30 [37].

b51. *Nitrocellulose* (I)–*polyurethane* (II), *may be conditionally compatible:* Homogeneous solution and transparent film when I was Asahi Kasei Co. RS-1/2, and II was Hodogaya Co. Pellet 22S [37]. No phase separation in cyclohexanone for 1/1 mixture, and film was clear when I was AS 5-6 sec, RS 15-20 sec, Hercules, Inc., and II was Estane 5707-Fl, B. F. Goodrich, a polyester urethane, but solutions showed phase separation and film was not clear when II was Estane 5740-X140, B. F. Goodrich, a polyether urethane [79].

b52. *Nitrocellulose* (I)–*polyester* (II), *may be compatible:* Homogeneous solution and transparent film when I was Asahi Kasei Co. RS-1/2 and II was Toyobo Co. Ester Resin 20 [37].

b53. *Nitrocellulose* (I)–*polyepichlorohydrin* (II), *incompatible:* Phase separation in cyclohexanone for 1/1 mixture, and film was not clear when I was AS 5-6 sec, RS 15-20 sec, Hercules, Inc., and II was B. F. Goodrich Hydrin 100 [79].

b54. *Nitrocellulose* (I)–*polycarbonate* (II), *incompatible:* Same as System **b53** except that II was General Electric Co. Lexan 125 or Mobay Chemical Co. Merlon M-50 [79].

b55. *Nitrocellulose* (I)–*poly(2,6-dimethyl-1,4-phenylene ether)*, *incompatible:* Same as System **b53** except that II was General Electric Co. PPO [79].

b56. *Nitrocellulose* (I)–*polysulfone* (II), *incompatible:* Same as System **b53** except that II was Union Carbide polysulfone [79].

b57. *Nitrocellulose* (I)–*poly(butadiene-co-styrene)* (II), *ambiguous:* Films considered weakened because of incompatibility when I had 12% N ($>$ dinitrate) and II was 30% styrene [90].

b58. *Nitrocellulose* (I)–*poly[butadiene-co-styrene-co-(methacrylic acid)]* (II), *incompatible:* Phase separation in 1–5% solutions in organic solvents except at 25/75 ratio of I/II when I had 12% N ($>$ dinitrate) and II was SKS-30-1 (30% styrene; 10–12% methacrylic acid) [30].

b59. *Nitrocellulose* (I)–*poly(butadiene-co-acrylonitrile)*, *conditionally compatible:* Phase separation in ethyl acetate/benzene/ethanol: 4/2/1 at 10% solids with 20–80% II in mixture when I had 12% N ($>$ dinitrate) and II had 18.4% acrylonitrile; no phase separation under above conditions if II had 28.6, 37.7, or 44.4% acrylonitrile [85]. Films cast from above mixtures at 50°C were transparent for all I/II ratios if II had 28.6 or 37.7% acrylonitrile; films had honeycomb structure if II had 18.4% acrylonitrile, and films were transparent with wavy surfaces if II had 44.4% acrylonitrile [86]. Viscosity data for 0.2, 0.7, and 2% solutions indicated compatibility at 20, 40, and 60°C when II was SKN-26 (26% acrylonitrile) [91]. Comments that tensile strength and elongation at break were inconclusive when I had 12% N and II had 11.7 to 36.9% acrylonitrile [90]. No phase separation in cyclohexanone for 1/1 mixture, but film was not clear when I was AS 5-6 sec, RS 15-20 sec, Hercules, Inc. and II was Hycar 1432, B. F. Goodrich [79]. Deformation–temperature studies were analyzed to determine incompatibility when II had 18.4% acrylonitrile but compatibility when II had 28.6%, 37.7%, or 44.4% acrylonitrile because these copolymers could "elasticize" I [92]. Homogeneous solution and transparent film when I was Asahi Kasei Co. RS-1/2, and II was Nippon Rubber Co. Hycar 1043 or 1432 [37]. Stated incompatible for I/II:4/1 to 1/4 in solution and films when I was Nitrocell E 620 and II was Buna NW, 28% acrylonitrile [16].

b60. *Nitrocellulose* (I)–*poly*[(*vinyl chloride*)-*co*-(*vinyl acetate*)] (II), *ambiguous:* Phase separation in cyclohexanone for 1/1 mixture, and film was not clear when I was AS 5-6 sec, RS 15-20 sec, Hercules, Inc., and II was B. F. Goodrich Geon 440 × 24 [79]. Nonhomogeneous solution and turbid film when I was Asahi Kasei Co. RS-1/2 and II was Shekisui Chemical Co. Esulex C, or Esulex CL, or Esulex CH-1, or Denki Chemical Co. Denka Vinyl 1000 AK [37]. Stated compatible for I/II: 4/1 to 1/4 in solutions and films when I was Nitrocell E620 and II was Vinalit MPS [16].

b61. *Nitrocellulose* (I)–*poly*[(*vinyl chloride*)-*co*-(*vinyl propionate*)] (II), *may be compatible:* Same as System **b52** except that II was Ryuron QS-430 [37].

b62. *Nitrocellulose* (I)–*poly*[(*vinyl chloride*)-*co*-(*vinyl acetate*)-*co*-(*vinyl alcohol*)] (II), *incompatible:* Same as System **b41** except that II was Shekisui Chemical Co. Esulex A [37].

b63. *Nitrocellulose* (I)–*poly*[(*vinyl chloride*)-*co*-(*vinyl acetate*)-*co*-(*maleic acid*)] (II), *incompatible:* Same as System **b41** except that II was Shekisui Chemical Co., Esulex M. [37].

b64. *Nitrocellulose* (I)–*poly*[(*vinyl chloride*)-*co*-(*vinylidene chloride*)-*co*-(*acrylic acid ester*)] (II), *incompatible:* Same as System **b41** except that II was Towa Gousei Co. Aron 321 [37].

b65. *Nitrocellulose* (I)–*poly*[*ethylene*-*co*-(*vinyl acetate*)] (II), *may be conditionally compatible* (probably depends on composition of II): Same as System **b52** except that II was Nippon Goshei Co. Soalex R-CH or Soalex R-FH; single T_g by DTA when II was Soalex R-FH [37].

b66. *Nitrocellulose* (I)–*poly*(*styrene*-*co*-*acrylonitrile*) (II), *may be conditionally compatible* (probably depends on composition of II): No phase separation in cyclohexanone for 1/1 mixture, and film was clear when I was AS 5-6 sec, RS 15-20 sec, Hercules, Inc. and II was Dow Chemical Co. Tyril 767 [79].

b67. *Nitrocellulose* (I)–*poly*[*styrene*-*co*-(*methyl methacrylate*)] (II), *may be conditionally compatible* (probably depends on composition of II): Same as System **b66** except that II was Dow Chemical Co. Zerlon 150 [79].

b68. *Nitrocellulose* (I)–*poly*[*styrene*-*co*-(*maleic acid*)] (II), *incompatible:* Same as System **b41** except that II was Daidou Kogyo Styrite CM-2 or CM-3 [37].

b69. *Nitrocellulose* (I)–*poly*[*styrene*-*co*-(*maleic acid ester*)] (II), *incompatible:* Same as System **b41** except that II was Daidou Kogyo Styrite HS-2 [37].

b70. *Nitrocellulose* (I)–*poly*[(*methyl vinyl ether*)-*co*-(*maleic anhydride*)] (II), *incompatible:* Same as System **b53** except that II was B. F. Goodrich Gantrez AN-169 [79].

b71. *Nitrocellulose* (I)–*poly*[*epichlorohydrin*-*co*-(*ethylene oxide*)] (II), *incompatible:* Same as System **b53** except that II was B. F. Goodrich Hydrin 200 [79].

b72. *Nitrocellulose* (I)–*indene resin* (II), *may be compatible:* Same as System **b52** except that II was Fuji Iron Products Co. VM-1/2 [37].

b73. *Nitrocellulose* (I)–*miscellaneous resins* (II), *incompatible:* Same as System **b41** except that II was Petroleum Resin, Nippon Petroleum Chemical Co., Nitsuseki Neopolymer 150, or Cumarone Resin, Mitsubishi Co. [37].

b74. *Cellulose acetate*–*propionate* (I)–*polyisoprene* (II), *incompatible:* Stated to be incompatible if I was Hercose AP, 15% acetyl, 30% propionyl, and II was rubber [93].

b75. *Cellulose acetate*–*propionate* (I)–*chlorinated rubber* (II), *incompatible:* Stated to be incompatible if I was Hercose AP, 15% acetyl, 30% propionyl [93].

b76. *Cellulose acetate–propionate* (I)–*poly(methyl methacrylate)* (II), *incompatible:* Stated to be incompatible [83]. Phase separation in cyclohexanone for 1/1 mixture, and film was not clear when I was 319E-40001-H6, Tennessee Eastman, and II was du Pont Lucite 147 or 148 [79].

b77. *Cellulose acetate–propionate* (I)–*polyacrylic esters* (II), *incompatible:* Stated to be partially compatible if II was poly(methyl acrylate), poly(ethyl acrylate), or poly(*n*-butyl acrylate). Stated to be incompatible if II was poly(ethyl methacrylate) or poly(*n*-butyl methacrylate) [83].

b78. *Cellulose acetate–propionate* (I)–*poly(vinyl acetate)* (II), *incompatible:* Phase separation in cyclohexanone for 1/1 mixture, and film was not clear when I was 319E-40001-H6, Tennessee Eastman, and II was Vinac B100, Air Reduction and Chemical Co. [79].

b79. *Cellulose acetate–propionate* (I)–*polystyrene* (II), *incompatible:* Same as System **b78** except that II was Dow Chemical Co. Styron 690 [79].

b80. *Cellulose acetate–propionate* (I)–*polycarbonate* (II), *incompatible:* Same as System **b78** except that II was General Electric Co. Lexan 125 or Mobay Chemical Co. Merlon M-50 [79].

b81. *Cellulose acetate–propionate* (I)–*polyepichlorohydrin* (II), *incompatible:* Same as System **b78** except that II was B. F. Goodrich Hydrin 100 [79].

b82. *Cellulose acetate–propionate* (I)–*poly(2,6-dimethyl-1,4-phenylene ether)* (II), *incompatible:* Same as System **b78** except that II was General Electric Co. PPO [79].

b83. *Cellulose acetate–propionate* (I)–*polysulfone* (II), *incompatible:* Same as System **b78** except that II was Union Carbide polysulfone [79].

b84 *Cellulose acetate–propionate* (I)–*polyurethane* (II), *incompatible:* Same as System **b78** except that II was B. F. Goodrich polyester urethane Estane 5707-Fl or polyether urethane Estane 5740-X140 [79].

b85. *Cellulose acetate–propionate* (I)–*poly(butadiene-co-acrylonitrile)* (II), *incompatible:* Same as System **b78** except that II was B. F. Goodrich Hycar 1432 [79].

b86. *Cellulose acetate–propionate* (I)–*poly[(vinyl chloride)-co-(vinyl acetate)]* (II), *incompatible:* Same as System **b78** except that II was B. F. Goodrich Geon 440-X24 [79].

b87. *Cellulose acetate–propionate* (I)–*poly[styrene-co-(methyl methacrylate)]* (II), *incompatible:* Same as System **b78** except that II was Dow Chemical Co. Zerlon 150 [79].

b88. *Cellulose acetate–propionate* (I)–*poly(styrene-co-acrylonitrile)* (II), *incompatible:* Same as System **b78** except that II was Dow Chemical Co. Tyril 767 [79].

b89. *Cellulose acetate–propionate* (I)–*poly[(methyl vinyl ether)-co-(maleic anhydride)]* (II), *incompatible:* Same as System **b78** except that II was B. F. Goodrich Gantrez AN-169 [79].

b90. *Cellulose acetate–propionate* (I)–*poly[epichlorohydrin-co-(ethylene oxide)]* (II), *incompatible:* No phase separation in cyclohexanone for 1/1 mixture, but film was not clear when I was 319E-40001-H6, Tennessee Eastman, and II was Hydrin 200, B. F. Goodrich [79].

b91. *Cellulose acetate–butyrate* (I)–*poly(vinyl chloride)* (II), *incompatible:* Nonhomogeneous solution and turbid film when I was Eastman EAB-381-2, and II was Nippon Carbide Co. P1050 [37].

b92. *Cellulose acetate–butyrate* (I)–*poly(vinylidene chloride)* (II), *incompatible:* Same as System **b91** except that II was Asahi Dow Co. EX5701 [37].

b93. *Cellulose acetate–butyrate* (I)–*polystyrene* (II), *incompatible:* Phase separation in solutions and films for several mixtures [37, 79].

b94. *Cellulose acetate–butyrate* (I)–*poly(vinyl acetate)* (II), *may be conditionally compatible* (may depend on composition of I): Stated to be miscible if I was Hercose C or Cellit [93]. Phase separation in cyclohexanone for 1/1 mixture when I was 17% butyryl, Tennessee Eastman, and II was Vinac B-100, Air Reduction and Chemical Co. [79].

b95. *Cellulose acetate–butyrate* (I)–*poly(methyl methacrylate)* (II), *incompatible:* Phase separation in solutions and films for several mixtures [37, 79]. Listed as incompatible [83].

b96.   *Cellulose acetate–butyrate* (I)–*poly*(*acrylic esters*)(II), *incompatible:* Stated to be partially compatible if II was poly(methyl acrylate), poly(ethyl acrylate), or poly(*n*-butyl acrylate). Stated to be incompatible if II was poly(ethyl methacrylate) or poly(*n*-butyl methacrylate) [83].

b97.   *Cellulose acetate–butyrate* (I)–*poly*(*vinyl butyral*)(II), *incompatible:* Same as System **b91** except that II was Shekisui Chemical Co. BM-2 [37].

b98.   *Cellulose acetate–butyrate* (I)–*polycarbonate* (II), *incompatible:* Phase separation in cyclohexanone for 1/1 mixture when I was 17% butyril, Tennessee Eastman, and II was General Electric Co. Lexan 125 or Mobay Chemical Co. Merlon M-50 [79].

b99.   *Cellulose acetate–butyrate* (I)–*polyepichlorohydrin* (II), *incompatible:* Same as System **b98** except that II was B. F. Goodrich Hydrin 100 [79].

b100.   *Cellulose acetate–butyrate* (I)–*poly*(*2,6-dimethyl-1,4-phenylene ether*) (II), *incompatible:* Same as System **b98** except that II was General Electric Co. PPO [79].

b101.   *Cellulose acetate–butyrate* (I)–*polysulfone* (II), *incompatible:* Same as System **b98** except that II was Union Carbide polysulfone [79].

b102.   *Cellulose acetate–butyrate* (I)–*chlorosulfonated polyethylene* (II), *incompatible:* Same as System **b91** except that II was Showa Neoprene Hyperon 30 [37].

b103.   *Cellulose acetate–butyrate* (I)–*polyurethane* (II), *incompatible:* Phase separation in solutions and films for mixtures including polyester and polyether urethanes [37, 79].

b104.   *Cellulose acetate–butyrate* (I)–*polyester* (II), *incompatible:* Same as System **b91** except that II was Toyobo Co. Ester Resin 20 [37].

b105.   *Cellulose acetate–butyrate* (I)–*poly*(*butadiene-co-styrene*) (II), *incompatible:* Same as System **b91** except that II was Nippon Rubber Co. Hycar 2057 [37].

b106.   *Cellulose acetate–butyrate* (I)–*poly*(*butadiene-co-acrylonitrile*) (II), *conditionally compatible:* Deformation–temperature curves gave two breaks or more when I had 17% butyrate and II was SKN-40 (40% acrylonitrile) and mixture contained 40–80% II. Only one break in curves when mixture contained 1–10% II or 90–99% II [94]. Phase separation in cyclohexanone for 1/1 mixture, and film not clear when I was 17% butyryl, Tennessee Eastman, and II was Hycar 1432, B. F. Goodrich [79]. Stated to form a homogeneous polyblend when II had 60% butadiene and mixture contained 20–90 wt % II [95]. Nonhomogeneous solution and turbid film when I was Eastman EAB 381-2, and II was Nippon Rubber Co. Hycar 1043 or 1432 [37].

b107.   *Cellulose acetate–butyrate* (I)–*poly*[*ethylene-co-(vinyl acetate)*] (II), *may be conditionally compatible:* Nonhomogeneous solution and turbid film when I was Eastman EAB-381-2, and II was Nippon Goshei Co. Soalex R-CH, but homogeneous solution and transparent film when II was Nippon Goshei Co. Soalex R-FH [37].

b108.   *Cellulose acetate–butyrate* (I)–*poly*[(*vinyl chloride*)-*co-(vinyl acetate)*] (II), *incompatible:* Phase separation in solutions and fibers for many mixtures [37, 79].

b109.   *Cellulose acetate–butyrate* (I)–*poly*[(*vinyl chloride*)-*co-(vinyl propionate)*] (II), *incompatible:* Same as System **b91** except that II was Ryuron QS-430 [37].

b110.   *Celluloseacetate–butyrate* (I)–*poly*[(*vinyl chloride*)-*co-(vinyl acetate)-co-(vinyl alcohol)*] (II), *incompatible:* Same as System **b91** except that II was Shekisui Chemical Co. Esulex A [37].

b111.   *Cellulose acetate–butyrate* (I)–*poly*[(*vinyl chloride*)-*co-(vinyl acetate)-co-(maleic acid)*] (II), *incompatible:* Same as System **b91** except that II was Shekisui Chemical Co. Esulex M [37].

b112.   *Cellulose acetate–butyrate* (I)–*poly*[(*vinyl chloride*)-*co-(vinylidene chloride)-co-(acrylic acid ester)*] (II), *incompatible:* Same as System **b91** except that II was Towa Gousei Co. Aron 321 [37].

b113.   *Cellulose acetate–butyrate* (I)–*poly*[*styrene-co-(methyl methacrylate)*] (II), *incompatible:* Same as System **b98** except that II was Dow Chemical Co. Zerlon 150 [79].

b114.   *Cellulose acetate–butyrate* (I)–*poly*(*styrene-co-acrylonitrile*)(II), *incompatible:* Same as System **b98** except that II was Dow Chemical Co. Tyril 767 [79].

b115. *Cellulose acetate–butyrate* (I)–*poly*[*styrene-co-(maleic acid)*] (II), *incompatible:* Same as System **b91** except that II was Daidou Kogyo Styrite CM-2 or CM-3 [37].

b116. *Cellulose acetate–butyrate* (I)–*poly*[*styrene-co-(maleic acid ester)*] (II), *incompatible:* Same as System **b91** except that II was Daidou Kogyo Styrite HS-2 [37].

b117. *Cellulose acetate–butyrate* (I)–*poly*[*(methyl vinyl ether)-co-(maleic anhydride)*] (II), *incompatible:* Same as System **b98** except that II was B. F. Goodrich Gantrez An-169 [79].

b118. *Cellulose acetate–butyrate* (I)–*poly*[*epichlorohydrin-co-(ethylene oxide)*] (II), *incompatible:* Same as System **b98** except that II was B. F. Goodrich Hydrin 200 [79].

b119. *Cellulose acetate–butyrate* (I)–*miscellaneous resins* (II), *incompatible:* Same as System **b91** except that II was Petroleum Resin, Nippon Petroleum Chemical Co., Nitsuseki Neopolymer 150, or Cumarone Resin, Mitsubishi TG, or Indene Resin, Fuji Iron Products Co. VM-1/2 [37].

b120. *Methylcellulose* (I)–*poly(vinyl alcohol)* (II), *incompatible:* Phase separation in water above 3.2% solids at 16–18°C if I had $\bar{M}_n = 1.6 \times 10^5$, $[\eta] = 3.80$ in water, and II was a fraction with $\bar{M}_n = 6 \times 10^4$, $[\eta] = 1.10$ in water [78].

b121. *Ethylcellulose* (I)–*polyisobutene* (II), *incompatible:* Stated incompatible for I/II: 4/1 to 1/4 in solutions and films when II was Oppanol B 100 [16].

b122. *Ethyl cellulose* (I)–*polyisoprene* (II), *incompatible:* Phase separation in solutions and films for several mixtures [16, 78].

b123. *Ethylcellulose* (I)–*poly(vinyl chloride)* (II), *incompatible:* Phase separation in solutions and films for several mixtures [16, 37].

b124. *Ethylcellulose* (I)–*chlorinated poly(vinyl chloride)* (II), *incompatible:* Same as System

b125. *Ethylcellulose* (I)–*poly(vinylidene chloride)* (II), *incompatible:* Nonhomogeneous solution and turbid film when I was Dow Chemical Ethocell 50 cps, and II was Asahi Dow Co. EX5701 [37].

b126. *Ethylcellulose* (I)–*polychloroprene* (II), *incompatible:* Same as System **b121** except that II was Neoprene AC [16].

b127. *Ethylcellulose* (I)–*poly(vinyl acetate)* (II), *incompatible:* Phase separation in solutions and films for several mixtures [37, 78] and in film though not in solution for another [79].

b128. *Ethylcellulose* (I)–*polystyrene* (II), *incompatible:* Phase separation in solutions and films for many mixtures [16, 37, 48, 78, 79]. UCST at 31.7°C for one mixture but this mixture looked homogeneous at 22°C in a velocity gradient of 200 sec^{-1} [46].

b129. *Ethylcellulose* (I)–*poly(methyl methacrylate)* (II), *incompatible:* Phase separation in solutions and films for many mixtures [16, 37, 78, 79].

b130. *Ethylcellulose* (I)–*poly(acrylic esters)* (II), *incompatible:* Stated to be incompatible if II was poly(methyl acrylate), poly(ethyl acrylate), poly(n-butyl acrylate), poly(ethyl methacrylate), or poly(n-butyl methacrylate) [83].

b131. *Ethylcellulose* (I)–*poly(vinyl acetal)* (II), *incompatible:* Phase separation in CHCl$_3$ above 4.0% solids at 16–18°C when I had $\bar{M}_n = 3.5 \times 10^4$, $[\eta] = 1.10$ in acetone, and II had $\bar{M}_n = 3.8 \times 10^4$, $[\eta] = 0.75$ in acetone [78].

b132. *Ethylcellulose* (I)–*poly(vinyl butyral)* (II), *incompatible:* Same as System **b125** except that II was Shekisui Chemical Co. BM-2 [37].

b133. *Ethylcellulose* (I)–*poly(vinylpyrrolidone)* (II), *incompatible:* Same as System **b125** except that II was General Aniline and Film Corp. K-30 [37].

b134. *Ethylcellulose* (I)–*chlorosulfonated polyethylene* (II), *incompatible:* Same as System **b125** except that II was Showa Neoprene, Hyperon 30 [37].

b135. *Ethylcellulose* (I)–*polycarbonate* (II), *incompatible:* Phase separation in cyclohexanone for 1/1 mixture, and film was not clear when I was 48.4% ethoxyl, Hercules, Inc., and II was Lexan 125, General Electric Co., or Merlon M-50, Mobay Chemical Co. [79].

b136. *Ethylcellulose* (I)–*polyepichlorohydrin* (II), *incompatible:* Same as System **b135** except that II was B. F. Goodrich Hydrin 100 [79].

b137. *Ethylcellulose* (I)–*poly*(*2,6-dimethyl-1,4-phenylene ether*) (II), *incompatible:* Same as System **b135** except that II was General Electric Co. PPO [79].

b138. *Ethylcellulose* (I)–*polysulfone* (II), *incompatible:* Same as System **b135** except that II was Union Carbide polysulfone [79].

b139. *Ethylcellulose* (I)–*polyurethane* (II), *incompatible:* Phase separation in solutions and films for several mixtures including polyester and polyether urethanes [37,79].

b140. *Ethylcellulose* (I)–*polyester* (II), *incompatible:* Same as System **b125** except that II was Toyobo Co. Ester Resin 20 [37].

b141. *Ethylcellulose* (I)–*poly*(*butadiene-co-acrylonitrile*)(II), *may be conditionally compatible:* Same as System **b135** except that II was B. F. Goodrich Hycar 1432 [79]. Same as System **b125** except that II was Nippon Rubber Co. Hycar 1043 or 1432 [37]. Stated compatible for I/II: 4/1 to 1/4 in solutions and films when II was Buna NW, 28% acrylonitrile [16].

b142. *Ethylcellulose* (I)–*poly*[*ethylene-co-(vinyl acetate)*] (II), *incompatible:* Same as System **b125** except that II was Nippon Goshei Co. Soalex R-CH or R-FH [37].

b143. *Ethylcellulose* (I)–*poly*[(*vinyl chloride*)*-co-(vinyl acetate)*] (II), *incompatible:* Phase separation in solutions and films for a number of mixtures [16, 37, 79].

b144. *Ethylcellulose* (I)–*poly*[(*vinyl chloride*)*-co-(vinyl propionate)*] (II), *incompatible:* Same as System **b125** except that II was Ryuron QS-430 [37].

b145. *Ethylcellulose* (I)–*poly*[(*vinyl chloride*)*-co-(vinyl acetate)-co-(vinyl alcohol)*] (II), *incompatible:* Same as System **b125** except that II was Shekisui Chemical Co. Esulex A. [37].

b146. *Ethylcellulose* (I)–*poly*(*vinyl chloride*)*-co-(vinyl acetate)-co-(maleic acid)*] (II), *incompatible:* Same as System **b125** except that II was Shekisui Chemical Co. Esulex M [37].

b147. *Ethylcellulose* (I)–*poly*[(*vinyl chloride*)*-co-(vinylidene chloride)-co-(acrylic acid ester)*] (II), *incompatible:* Same as System **b125** except that II was Towa Gousei Co. Aron 321 [37].

b148. *Ethylcellulose* (I)–*poly*[(*vinyl acetate*)*-co-(N-vinylpyrrolidone)*] (II), *incompatible:* Same as System **b125** except that II was General Aniline and Film S-630 [37].

b149. *Ethylcellulose* (I)–*poly*[*styrene-co-(methyl methacrylate)*] (II), *incompatible:* Same as System **b135** except that II was Dow Chemical Co. Zerlon 150 [79].

b150. *Ethylcellulose* (I)–*poly*(*styrene-co-acrylonitrile*)(II), *incompatible:* Same as System **b135** except that II was Dow Chemical Co. Tyril 767 [79].

b151. *Ethylcellulose* (I)–*poly*[*styrene-co-(maleic acid)*] (II), *may be conditionally compatible:* Homogeneous solution and transparent film when I was Dow Chemical Ethocell 50 cps and II was Daidou Kogyo Styrite CM-2 or CM-3 [37].

b152. *Ethylcellulose* (I)–*poly*[*styrene-co-(maleic acid ester)*] (II), *may be conditionally compatible:* Same as System **b151** except that II was Daidou Kogyo Styrite HS-2 [37].

b153. *Ethylcellulose* (I)–*poly*[(*methyl vinyl ether*)*-co-(maleic anhydride)*] (II), *incompatible:* Same as System **b135** except that II was B. F. Goodrich Gantrez AN-169 [79].

b154. *Ethylcellulose* (I)–*poly*[*epichlorohydrin-co-(ethylene oxide)*] (II), *incompatible:* Same as System **b135** except that II was B. F. Goodrich Hydrin 200 [79].

b155. *Ethylcellulose* (I)–*indene resin* (II), *may be compatible:* Same as System **b151** except that II was Fuji Iron Products Co. VM-1/2 [37].

b156. *Ethylcellulose* (I)–*miscellaneous resins* (II), *incompatible:* Same as System **b125** except that II was Petroleum Resin, Nippon Petroleum Chemical Co. Nitsuseki Neopolymer 150 or Cumarone Resin, Mitsubishi TG [37].

b157. *Benzylcellulose* (I)–*polyisobutene* (II), *incompatible:* Stated incompatible for I/II: 4/1 to 1/4 in solutions and films when II was Oppanol B100 [16].

b158. *Benzylcellulose* (I)–*polyisoprene* (II), *incompatible:* Same as System **b157** except that II was natural rubber [16].

b159. *Benzylcellulose* (I)–*polychloroprene*(II), *incompatible:* Same as System **b157** except that II was Neoprene AC [16].

b160. *Benzylcellulose* (I)–*poly(vinyl chloride)* (II), *incompatible:* Same as System **b157** except that II was PVC-G [16].

b161. *Benzylcellulose* (I)–*chlorinated poly(vinyl chloride)* (II), *incompatible:* Same as System **b157** except that II was Vinoflex PC [16].

b162. *Benzylcellulose* (I)–*poly(vinyl acetate)* (II), *incompatible:* Phase separation in $CHCl_3$ above 2.5% solids at 16–18°C when II was Rhodopas HH, $\overline{M}_n = 1.12 \times 10^5$, $[\eta] = 0.85$ in acetone [78].

b163. *Benzylcellulose* (I)–*polystyrene* (II), *ambiguous:* Miscible in all proportions in $CHCl_3$ at 16–18°C when II had $\overline{M}_n = 2.25 \times 10^5$, $[\eta] = 2.15$ in $CHCl_3$ [78]. Phase separation in cyclo-hexanone when I had $[\eta] = 1.40$ in $CHCl_3$ and II had $[\eta] = 2.90$ in benzene; phase separation in $CHCl_3$, same I, but II different [76, 77]. Stated incompatible for I/II: 4/1 to 1/4 in solutions and films when II was BW [16].

b164. *Benzylcellulose* (I)–*poly(methyl methacrylate)* (II), *ambiguous:* Miscible in dioxane above 10% solids at 16–18°C when II had $\overline{M}_n > 2 \times 10^6$, $[\eta] = 3.65$ in acetone [78]. Stated to be partially compatible [83]. Stated incompatible for I/II: 4/1 to 1/4 in solutions and films when II was Piacryl G. [16].

b165. *Benzylcellulose* (I)–*poly(ethyl methacrylate)* (II), *may be compatible:* Listed as compatible [83].

b166. *Benzylcellulose* (I)–*poly(acrylic esters)* (II), *incompatible:* List as partially compatible when II was poly(ethyl acrylate), and incompatible when II was poly(methyl acrylate), poly(*n*-butyl acrylate), and poly(*n*-butyl methacrylate) [83].

b167. *Benzylcellulose* (I)–*poly(vinyl acetal)* (II), *almost compatible:* Phase separation in $CHCl_3$ above 10.5% solids at 16–18°C when II had $\overline{M}_n = 3.8 \times 10^4$, $[\eta] = 0.75$ in acetone [78].

b168. *Benzylcellulose* (I)–*poly(butadiene-co-acrylonitrile)* (II), *incompatible:* Same as System **b157** except that II was Buna NW, 28% acrylonitrile [16].

b169. *Benzylcellulose* (I)–*poly[(vinyl chloride)-co-(vinyl acetate)]* (II), *incompatible:* Same as System **b157** except that II was Vinalit MPS [16].

C. Polyisoprene and Other Polymers

c1. *Polyisoprene* (I)–*polyisoprene* (II), *may be conditionally compatible* (if structure in polymer is different): Phase contrast and electron microscopy showed zones when I was natural rubber and II was synthetic [96].

c2. *Polyisoprene* (I)–*polyethylene* (II), *incompatible:* Phase contrast and electron microscopy showed zones when I was natural rubber [96]. Damping maximum of I appeared in all mixes [97] when I was natural rubber, pale crepe. Phase separation in solution when I was "rubber" [30].

c3. *Polyisoprene* (I)–*polyisobutene* (II), *may be compatible:* Stated compatible for I/II: 4/1 to 1/4 in solutions and films when I was natural rubber and II was Oppanol B 100 [16].

c4. *Polyisoprene* (I)–*polybutadiene* (II), *conditionally compatible:* A single T_g during linear expansion when I was NK and II was SKB [98]. No phase separation in benzene when I was NK, $[\eta] = 3.40$ in benzene, and II was SKB, $[\eta] = 1.70$ in benzene [76, 77]. No phase separation in gasoline at 5% solids for 1/1 ratio of I/II when I was NK and II was SKB [99]. Heterogeneous by phase contrast and electron microscopy for 3/1 and 1/3 mixtures when I and II were both cis-1,4 [100]. Single T_g by dilatometry or rolling ball loss spectro-meter for 3/1 to 1/3 mixtures when I was natural rubber and II was SKB, sodium poly-butadiene [101]. Phase contrast and electron microscopy showed zones when I was natural rubber and II was one of five samples, 92–94% cis, MW = 2.8 to 12×10^5, or Buna 85, or 98% cis, or trans [96]. Fine structure by phase contrast microscopy for 1/1 mixture and two transitions by dynamic mechanical measurements plus melting of II when I was Goodyear Natsyn, high cis, and II was Goodyear Budene, cis [102]. Two mechanical loss maxima for I/II: 1/9 to 1/1 when I was SKI-3 and II was SKD [103]. Stated homogeneous polyblends 0–100% II when I was natural rubber [95]. Two T_g's by radiothermoluminescence when I was SKI or natural rubber and II was SKD, 3% 1,2, stereoregular, but single T_g when II was SKB-30, 40–66% 1,2 [50]. Two dielectric and mechanical loss peaks for I/II: 1/4 to 4/1 when I was natural rubber, RSS # 1, and II was cis-1,4 [104]. Single T_g by radiothermoluminescence for 1/1 mixture but T_g's of I and II were very close together when I was SKI or natural rubber, smoked sheet, and II was sodium butadiene, SKB-60 or SKBM-50, but two T_g's when II was SKD [105]. Incompatible by phase contrast and electron microscopy and two T_g's by DTA for I/II: 1/3 to 3/1 when I was cis-1,4 and II was high cis-1,4 [106].

    c5. *Polyisoprene* (I)–*polychloroprene* (II), *ambiguous*: Incompatible by phase contrast and electron microscopy for I/II: 1/3 to 3/1 when I was cis-1,4 [106]. Two T_g's by dilatometry and rolling ball loss spectrometer when I was natural rubber and II was Neoprene GNA [101]. Phase contrast and electron microscopy showed zones when I was natural rubber and II was Neoprene W [96]. Stated compatible for I/II: 4/1 to 1/4 in solutions and films when I was natural rubber and II was Neoprene AC [16].

    c6. *Polyisoprene* (I)–*poly(vinyl chloride)* (II), *incompatible*: Stated incompatible for I/II: 4/1 to 1/4 in solutions and films when I was natural rubber and II was PVC-G [16].

    c7. *Polyisoprene* (I)–*chlorinate poly(vinyl chloride)* (II), *incompatible*: Same as System **c6** except that II was Vinoflex PC [16].

    c8. *Polyisoprene* (I)–*poly(vinyl acetate)* (II), *incompatible*: Phase separation in benzene above 2.8% solids at 16–18°C when I was "rubber" and II was Rhodopas HH, $\overline{M}_n = 1.12 \times 10^5$, $[\eta] = 0.85$ in acetone [78].

    c9. *Polyisoprene* (I)–*polystyrene* (II), *incompatible*: Various mixtures had two damping maxima [97], and phase separation in solution [16, 78] and films [16]. Coexistence curves for oligomers of MW = 2000–3000 were bimodal with UCST 135–175°C [52, 67], while for oligomers of MW = 1000–3000, they were *not* bimodal and had UCST from −30°C to +56°C [107]. Limits of solubility of II in I found by many methods and extrapolated, using I of mol wt 10^6 and II of many molecular weights; II would be completely soluble in I for MW of II < 500 [108].

    c10. *Polyisoprene* (I)–*poly(α-methylstyrene)* (II), *incompatible*: Limits of solubility of II in I fit on same curves as limits of solubility mentioned under System **c9** [108].

    c11. *Polyisoprene* (I)–*poly(methyl acrylate)* (II), *ambiguous*: A 1/1 mix with added silica was transparent when I was natural rubber [109].

    c12. *Polyisoprene* (I)–*poly(methyl methacrylate)* (II), *incompatible*: Phase separation in solutions and films found for a number of samples [16, 78, 110]. Two damping maxima [97, 16, 78, 110] observed and ~1% maximum mutual solubility when I was Kariflex type, $M = 9.0 \times 10^5$, and II was suspension, $M = 9 \times 10^4$ [27].

    c13. *Polyisoprene* (I)–*poly(vinyl acetal)* (II), *incompatible*: Phase separation in benzene +5% absolute alcohol at 16–18°C above 2.0% solids when I was "rubber" and II had $\overline{M}_n = 3.8 \times 10^4$, $[\eta] = 0.75$ in acetone [78].

    c14. *Polyisoprene* (I)–*poly(butadiene-co-styrene)* (II), *may be conditionally compatible*: Phase separation in benzene when I was NK, $[\eta] = 3.40$ in benzene and II was SKS-30 (30% styrene, $[\eta] = 1.50$ in benzene [76, 77]. Incompatible at ratios of I/II from 9/1 to 3/7 when I was natural rubber, MW = 3.1×10^5, and II was BUNA S-3 (~80% butadiene), MW = 1.86×10^5 [111]. Two damping maxima when I was natural rubber, pale crepe, and II was Duranit No. 10 (10% butadiene) or Duranit No. 30 (30% butadiene) [97]. Phase separation in ligroin at 50% solids if I/II ratio was between 2/8 and 8/2 but no phase separation if ratio was 1/9 or 9/1 and I was natural rubber and II was SKS-30A (30% styrene) [112]. Phase separation in gasoline at 5% solids for 1/1 mixture when I was NK and II was SKS-30 [99]. Microscopic investigation showed transparent films at room temperature, but two peaks by mechanical loss measurements below 0°C when I was natural rubber and II was SBR [113]. Incompatible by phase contrast and electron microscopy for I/II: 1/3 to 3/1 when I was natural rubber, smoked sheet No. 1, and II was SBR 1500 [106]. Heterogeneous by phase contrast and electron microscopy for I/II: 3/1 and 1/3 when I was cis-1,4 and II was SBR 1500 [100]. Single T_g by dilatometry and rolling ball loss spectrometer when I was natural rubber and II was Phillips Petroleum Co. Solprene 1204, solution SBR [101]. Phase contrast and electron microscopy showed zones when I was natural rubber and II was Krylene [96]. Stated compatible for I/II: 4/1 to 1/4 in solutions and films when I was natural rubber and II was Buna S3 (~80% butadiene) [16].

c15. *Polyisoprene* (I)–*poly*(*butadiene-co-acrylonitrile*) (II), *incompatible:* Two T_g's for many mixtures by various methods [50, 98, 99, 101, 105]. Phase separation in solution [16] and bulk [16, 106].

c16. *Polyisoprene* (I)–*poly*(*ethylene-co-propylene-co-*[*probably*] *diene*) (II), *incompatible:* Incompatible by phase contrast and electron microscopy when I was natural rubber, smoked sheet No. 1, and II was EPT [106].

c17. *Polyisoprene* (I)–*poly*[(*vinyl chloride*)-*co*-(*vinyl acetate*)] (II), *incompatible:* Same as System **c6** except that II was Vinalit MPS [16].

D. Poly(vinyl chloride) (PVC) and Other Polymers

d1. *PVC* (I)–*chlorinated PVC* (II), *ambiguous:* Stated to be incompatible when II was highly chlorinated [114]. Dynamic loss peak of I seen in I/II: 85/15 when I was Montecatini Sicron 548 FM and II was 68% by wt Cl [115]. Stated compatible for I/II: 4/1 to 1/4 in solutions and films but cloudy films observed for mixtures with >80% I when I was PVC-G and II was Vinoflex PC [16].

d2. *PVC* (I)–*polychloroprene* (II), *may be compatible:* Stated compatible for I/II: 4/1 to 1/4 in solutions and films when I was PVC-G and II was Neoprene AC [16].

d3. *PVC* (I)–*chlorinated poly*(*isoprene*) (II), *incompatible:* Stated incompatible for I/II: 1/9 to 9/1 by observation of films when II was chlorinated rubber [16].

d4. *PVC* (I)–*poly*(*vinylidene chloride*) (II), *incompatible:* Nonhomogeneous solution and turbid film when I was Nippon Carbide Co. P1050 and II was Asahi Dow Co. EX 5701 [37].

d5. *PVC* (I)–*polyethylene* (II), *incompatible* (probably): Cheesy mixtures when I contained 1.5 wt % ethylene and 2% stabilizer [116].

d6. *PVC* (I)–*chlorinated polyethylene* (II), *incompatible:* Same as System **d3** except that II was as above, with 30% Cl [16].

d7. *PVC* (I)–*chlorosulfonated polyethylene* (II), *incompatible:* Same as System **d3** except that II was as above, 29.5% Cl, 1.6% S [16].

d8. *PVC* (I)–*polybutadiene* (II), *incompatible:* Two T_g's by various methods for a number of mixtures [117–119]. Domains by electron microscopy [119].

d9. *PVC* (I)–*poly*(*vinyl acetate*) (II), *incompatible:* Phase separation for many mixtures in solution [37, 120–122]. Heterogeneous under polarizing microscope [123].

d10. *PVC* (I)–*polystyrene* (II), *incompatible:* Phase separation in solutions [16, 37, 121, 124] and films [16, 37]. Two T_g's observed by several methods [118, 125].

d11. *PVC* (I)–*poly*(*methyl acrylate*) (II), *incompatible:* Phase separation in THF at 10% solids and 1/1 weight ratio of I/II [121].

d12. *PVC* (I)–*poly*(*ethyl acrylate*) (II), *incompatible:* Phase separation in solutions [16, 121] and films [16] for several mixtures.

d13. *PVC* (I)–*Poly*(*n-butyl acrylate*) (II), *incompatible:* Two damping maxima for 1/1 mixture [126]. Phase separation in THF at 10% solids and 1/1 weight ratio of I/II [121].

d14. *PVC* (I)–*poly*(*2-ethylhexyl acrylate*) (II), *incompatible:* Stated to be incompatible when mixtures were prepared from the melt or from solution containing 50–90% I [4].

d15. *PVC* (I)–*poly*(*methyl methacrylate*) (II), *ambiguous:* Phase separation in THF at 10% solids and 1/1 weight ratio of I/II [121]. Stated incompatible for I/II: 4/1 to 1/4 in solutions and films when I was PVC-G and II was Piacryl G [16]. Nonhomogeneous solution and turbid film when I was Nippon Carbide Co. P1050 and II was from Mitsubishi Rayon Co. [37]. Phase separation in THF for I/II: 3/7 to 7/3 when I had $\overline{M}_v = 1.07 \times 10^5$ and II had $\overline{M}_v = 9.87 \times 10^4$ [80]. Electron microscopy showed two phases for 5–98% II but films seemed clear up to 20% II, and DTA showed two T_g's for mechanically mixed films for 33–80% II, T_g of I for 20–33% II and intermediate T_g for 50–80% II for coprecipitated fibers [127]. Two T_g's by DSC for 30–80 wt % II and by dynamic mechanical measurements for 25–75 wt % II and mixtures turbid when I was Pichiney & St. Gobain Lucoyl R8010, $\overline{M}_v = 5.5 \times 10^4$, and II was isotactic, $\overline{M}_v = 6.30 \times 10^5$, triads i/h/s: 92/3/5; single T_g by DSC and dynamic mechanical measurements for 10–60% II but two T_g's by DSC for 70–95% II when II was syndiotactic, $\overline{M}_v = 3.7 \times 10^5$, triads i/h/s: 2/4/94 or atactic, $\overline{M}_v = 8.0 \times 10^4$, triads i/h/s: 5/32/63 [128].

d16. *PVC* (I)–*poly*(*ethyl methacrylate*) (II), *may be compatible:* Compatible at 30% solids [121].

d17. *PVC* (I)–*poly*(*n-propyl methacrylate*) (II), *may be compatible:* Compatible at 30% solids [121].

d18. *PVC* (I)–*poly*(*isopropyl methacrylate*) (II), *incompatible:* Phase separation in THF at 10% solids and 1/1 weight ratio of I/II [121].

d19. *PVC* (I)–*poly*(*n-butyl methacrylate*) (II), *may be compatible:* Compatible at 30% solids [121].

d20. *PVC* (I)–*poly*(*isobutyl methacrylate*) (II), *may be compatible:* Compatible at 30% solids [121].

d21. *PVC* (I)–*Poly*(*2-ethylhexyl methacrylate*) (II), *incompatible:* Same as System **d18** [121].

d22. *PVC* (I)–*poly*(*n-octyl methacrylate*) (II), *incompatible:* Same as System **d18** [121].

d23. *PVC* (I)–*poly*(*n-dodecyl methacrylate*) (II), *incompatible:* Same as System **d18** [121].

d24. *PVC* (I)–*poly*(*n-octadecyl methacrylate*) (II), *incompatible:* Same as System **d18** [121].

d25. *PVC* (I)–*poly*(*vinyl butyral*) (II), *incompatible:* Same as System **d4** except that II was Shekisui Chemical Co. BM-2 [37].

d26. *PVC* (I)–*poly*(*vinyl isobutyl ether*) (II), *incompatible:* Same as System **d18** [121].

d27. *PVC* (I)–*poly*(*methyl vinyl ketone*) (II), *incompatible:* Same as System **d18** [121].

d28. *PVC* (I)–*poly*(*vinylpyrrolidone*) (II), *incompatible:* Same as System **d4** except that II was General Aniline and Film K-30 [37].

d29. *PVC* (I)–*poly*(*ε-caprolactone*) (II), *conditionally compatible:* Blends clear and had a single T_g up to 50% II [129]; they became crystalline at higher contents of II when I had $\eta_{sp} = 1.0$ at 0.2 g/dl in cyclohexanone at 30°C and II had $\overline{M}_v = 4.1 \times 10^4$. Stated compatible [89].

d30. *PVC* (I)–*polyurethane* (II), *may be compatible:* Homogeneous solution and transparent film when I was Nippon Carbide Co. P1050 and II was Nippon Rubber Co. 5740X1 or Hodogaya Co. Pellet 22S [37].

d31. *PVC* (I)–*polyester* (II), *incompatible:* Same as System **d4** except that II was Toyobo Co. Ester Resin 20 [37].

d32. *PVC* (I)–*poly*[(*vinyl chloride*)-co-(*vinyl acetate*)] (II), *must be conditionally compatible:* Two layers after 55 days for I/II: 2/8 to 5/5 at 10% solids in cyclohexanone when I had 59.5% Cl, $\overline{M}_w = 4.0 \times 10^4$, and II was Covicet, 80% vinyl chloride, 44.1% Cl, $\overline{M}_w = 2.9 \times 10^4$

[122]; the same I and II at I/II: 2/8 showed very small particles in polarizing microscope [123]. Homogeneous solutions and transparent films when I was Nippon Carbide Co. P1050 and II was Shekisui Chemical Co. Esulex C, CL, or CH-1, or Denki Chemical Co. Denka Vinyl 1000 AK [37]. Stated incompatible for I/II: 9/1 to 4/1 and compatible 4/1 to 1/9 for solutions and films when I was PVC-G and II was Vinalit MPS [16].

d33. *PVC* (I)–*poly*[(*vinyl chloride*)-*co*-(*vinyl propionate*)] (II), *must be conditionally compatible or compatible:* Same as System **d30** except that II was Ryuron QS-430 [37].

d34. *PVC* (I)–*poly*[(*vinyl chloride*)-*co*-(*vinyl acetate*)-*co*-(*vinyl alcohol*)] (II), *must be conditionally compatible or compatible:* Same as System **d30** except that II was Shekisui Chemical Co. Esulex A [37].

d35. *PVC* (I)–*poly*[(*vinyl chloride*)-*co*-(*vinylidene chloride*)-*co*-(*acrylic acid ester*)] (II), *must be conditionally compatible or compatible:* Same as System **d30** except that II was Towa Gousei Co. Aron 321 [37].

d36. *PVC* (I)–*poly*(*butadiene-cco-styrene*) (II), *incompatible:* Two transitions using torsional pendulum when II was Buna S3 (75% butadiene); stated that the same results were obtained when II was Buna S4 (like Buna S3 in composition) or Buna SS (~60% butadiene) [117]. Two phase from dynamic mechanical measurements and dilatometry when I had $\overline{M}_v = 6.6 \times 10^4$ and II was Japanese Synthetic Rubber Co. SBR, 24.5% styrene [118]. Stated conditionally compatible for I/II: 4/1 to 1/4 in solutions and films when I was PVC-G and II was Buna S3 (~80% butadiene) [16].

d37. *PVC* (I)–*poly*[*butadiene-co*-(*α-methylstyrene*)] (II), *incompatible:* Two dielectric absorption maxima for I/II: 1/4 to 4/1 when I had [η] = 0.91 in cyclohexane at 20°C and II was 22.5% α-methylstyrene, [η] = 0.56 in cyclohexanone at 20°C [130]. Two dielectric loss peaks for I/II: 9/1 to 4/6 but authors say compatible because of maxima in curves relating tensile strength and elongation at break to composition [131].

d38. *PVC* (I)–*poly*(*butadiene-co-acrylonitrile*) (II), *conditionally compatible:* A single T_g observed when II contained 36.9% acrylonitrile [132]. Torsional modulus gave one damping peak, somewhat broad, on evaporated film when II contained 8.9% N (33.7% acrylonitrile) [133]. One loss maximum in torsional modulus and dielectric measurements for ratios of I/II from 20/80 to 80/20 when II was Buna NW [134]. One transition on torsional pendulum when II was Perbunan (~28% acrylonitrile); stated that Perbunan Extra (40% acrylonitrile) and Perbunan W also gave only one transition when used as II [117]. Geon Polyblend 500 × 479, which had 76% I, formed a transparent film and had a single damping maximum [135]. Stated to be miscible at 140–150°C when II was SKN-26 (26% acrylonitrile) [91]. One damping region found [125]. Single T_g for I/II: 5/95 and 95/5 and two T_g's for 1/1 by dilatometry when I was L-5 and II was SKN-40 [136]. Stated homogeneous polyblend for 20–90 wt% I [95]. Single T_g by DSC and torsional pendulum and same appearance as I in phase contrast microscope for I/II: 7/3 by weight when I was Monsanto Chemical Co. Opalon 630 and II was Firestone FR-N-500, 77% butadiene by weight, or FR-N-504, 55% butadiene by weight, and for I/II: 9/1 to 1/1 by weight when II was FR-N-510, 69% butadiene by weight [137]. Single dynamic mechanical relaxation time for I/II: 4/1 to 1/4 when II was SKN-40, but two relaxation times when II was SKN-26 or SKN-18 [138]. Two T_g's from length expansion when I was Pevikan, Kema Nord AB, Sweden, and II was Hycar 1024, 21.7% acrylonitrile, but single T_g when II was Hycar 1043, 29.6% acrylonitrile, or Hycar 1041, 41.6% acrylonitrile [139]. Microphotograph of 1/1 mixture interpreted as one phase by authors and *Chemical Abstracts* reference states single dynamic mechanical relaxation time [140]. Single dynamic mechanical T_g and no structure in phase contrast microscopy for I/II: 2/3 to 13/7 when II had 33% acrylonitrile [102]. Stated compatible for I/II: 4/1 to 1/4 in solutions and films when I was PVC-G and II was Buna NW, 28% acrylonitrile [16]. Homogeneous solution and transparent film when I was Nippon Carbide Co. P1050 and II was Nippon

Rubber Co. Hycar 1043 or 1432 [37]. Electron microscopy showed heterogeneity after staining for I/II: 100/15 when I was the Japanese Geon Co. Geon 103, EP-8, $P = 800$, and II was the Japanese Geon Co., Hycar 1043, 30% acrylonitrile, or had 8% or 15% acrylonitrile; two dynamic loss peaks and domains in electron micrographs for I/II: 100/15 and 100/25 when II was B. F. Goodrich, Hycar 1014, 20% acrylonitrile; microheterogeneity in electron micrographs and dynamic loss peak of I shifts as % II increases but II peak seems present throughout although authors say it disappears for I/II: 100/10 to 100/50 when II was the Japanese Geon Co., Hycar 1041, 40% acrylonitrile [119].

d39. *PVC* (I)–*poly[butadiene-co-(vinylidene chloride)]* (II), *ambiguous:* Considered compatible because specific volume decreased on mixing when II was DVKhB-70 (70% vinylidene chloride) [30]. Considered compatible because of strength and extensibility of films when II was divinyl–vinylidene chloride rubber [141]. *Note:* Specific volume decrease probably has nothing to do with compatibility [31].

d40. *PVC* (I)–*poly[ethylene-co-(vinyl acetate)]* (II), *conditionally compatible:* 1/1 blends of I/II were clear when II contained 60–75% by weight vinyl acetate; loss peaks of both polymers observed when II contained 40% vinyl acetate; a single intermediate loss peak observed when II contained 65% vinyl acetate and the I/II ratio was 25/75 or 85/15 [142]. Single dielectric loss peak for I/II 1/4 to 4/1 when II had 45% vinyl acetate [143]. Wide-line NMR interpreted as phase separation for 3–14 wt % II when I was suspension polymerized and II was Levapren 450 M, Bayer AG, 45 wt % vinyl acetate [144]. One low temperature dynamic mechanical relaxation for I/II: 1/9 to 1/4 and two relaxations for 1/1 to 92.5/7.5 when I was Pevikan, Kema Nord AB, $\overline{M}_w = 7.4 \times 10^4$, and II was Wacker Chemie AG, VAE 661, 65 wt % vinyl acetate, $\overline{M}_n = 6.4 \times 10^4$ [145]. Two phases by electron microscopy at room temperature but says that homogeneity occurs when processed at 190°C when II was Levapren 450 P, Farbenfabriken Bayer ~45% vinyl acetate, mol wt ~1.0–1.2 × 10^5 [146]. Very turbid above 5% II when I was suspension polymerized and II had 45 wt % vinyl acetate; two dynamic mechanical loss peaks for I/II: 17/3 to 1/1 when II had 65 wt % vinyl acetate [147]. Two dynamic loss peaks for I/II: 1/9 to 3/1 when I was Sicron 548 FM and II was Levapren 450 P, 45 wt % vinyl acetate, $\overline{M}_n = 4.73 \times 10^4$ [148]. Homogeneous solution, transparent film, and single T_g by DTA when I was Nippon Carbide Co. P1050 and II was Nippon Goshei Co. Soalex R-CH, but nonhomogeneous solution and turbid film when II was Nippon Goshei Co. Soalex R-F-H [37].

d41. *PVC* (I)–*poly(chlorinated [ethylene-co-(vinyl acetate)])* (II), *ambiguous:* Two dynamic loss peaks for I/II: 1/4 to 4/1 when I was Sicron 548 FM and II was chlorinated Levapren 450 P, 45 wt % vinyl acetate, $\overline{M}_n = 4.73 \times 10^4$, 32% Cl; two dynamic loss peaks for I/II: 1/4, very broad single peak for 3/7, and single loss peak very close to that of I up to I/II: 4/1, but authors say compatible when II was as above but 38% Cl [148].

d42. *PVC* (I)–*poly[ethylene-co-(vinyl acetate)-co-(sulfur dioxide)]* (E–VA–SO₂) (II), *conditionally compatible:* Single T_g by DSC or torsion pendulum and clear by phase contrast microscopy for 60–90% II when I was Geon 103 with stabilizer 831, $\eta_{inh} = 0.98$ at 30°C, 0.5% in THF, and II had E/VA/SO₂: 72.7/18.5/8.8 mole %, $\eta_{inh} = 0.4$–0.5 at 30°C, 0.5% in THF; two T_g's by DSC and torsion pendulum and two phases by phase contrast microscopy for 30% II when II had E/VA/SO₂: 89.0/7.8/3.2 mole % [149].

d43. *PVC* (I)–*poly[ethylene-co-(vinyl alcohol)]* (II), *incompatible:* Blends hazy and opaque when I contained 1.5 wt % ethylene and 2% stabilizer and II had 22 wt % vinyl alcohol [116].

d44. *PVC* (I)–*poly(ethylene-co-acrylonitrile)* (II), *may be conditionally compatible:* Blend clear and transparent when I contained 1.5 wt % ethylene and 2% stabilizer and II had 13 wt % acrylonitrile [116].

d45. *PVC* (I)–*poly[ethylene-co-(N,N-dimethylacrylamide)]* (II), *may be conditionally compatible:* Transparent for I/II: 3/7 and 1/1 when I was as in System **d44** and II had 73 wt % ethylene [116].

d46. *PVC* (I)–*poly*[*ethylene-co-(N-methyl-N-vinyl acetamide)*] (II), *may be conditionally compatible:* Melt clear for I/II: 9/1 when I was as in System **d44** and II had 90.9% ethylene; blends clear and transparent for I/II: 1/1 when II had 81.6% ethylene [116].

d47. *PVC* (I)–*poly*[*ethylene-co-(n-butyl urethane)*] (II), *may be conditionally compatible:* Homogeneous, clear, transparent blends for 1/1 mixtures when I and II were as in System **d43** but II had been reacted with *n*-butyl isocyanate [116].

d48. *PVC* (I)–*poly*[*ethylene-co-(4-vinylpyridine)*] (II), *may be conditionally compatible:* Solution in THF at 60°C of 1/1 mixture gave a clear, one phase solution when I was as in System **d44** and II had 11 wt % 4-vinylpyridine [116].

d49. *PVC* (I)–*poly*[*ethylene-co-(N,N-dimethylaminoethyl methacrylate)*] (II), *may be conditionally compatible:* Blend clear for 1/1 mixture when I was as in System **d44** and II had 89 wt % ethylene [116].

d50. *PVC* (I)–*poly*[*(vinyl acetate)-co-(N-vinylpyrrolidone)*] (II), *may be conditionally compatible:* Same as System **d30** except that II was General Aniline and Film S-630 [37].

d51. *PVC* (I)–*poly*[*styrene-co-(maleic acid)*] (II), *incompatible:* Same as System **d4** except that II was Daidou Kogyo Styrite CM-2 or CM-3 [37].

d52. *PVC* (I)–*poly*[*(methyl methacrylate)-co-(ethyl methacrylate)*] (MMA–EMA) (II): *may be conditionally compatible:* Phase separation at 10% solids when II had mole ratio MMA/EMA of 73/27 or greater; miscible when II had mole ratio MMA/EMA of 55/45 or less [121].

d53. *PVC* (I)–*poly*[*acrylonitrile-co-(2-ethylhexyl acrylate)*] (II), *incompatible:* Somewhat hazy films and domains seen in electron micrographs for I/II: 3/2 when I was Goodyear Pliovic K656 or K906 or Dow 144 and II had 40% acrylonitrile [150].

d54. *PVC* (I)–*poly*[*α-methylstyrene-co-methacrylonitrile-co-(ethyl acrylate)*] (αMS-MAN-EA) (II), *compatible at composition of II specified:* Films and plaques transparent, single T_g for 1/1 mixture, and deflection T versus composition curve for I/II: 4/1 to 1/4 like T_g versus composition plot for αMS/MAN/EA: 58/40/2 wt % in II [151].

d55. *PVC* (I)–*cumarone resin* (II), *incompatible:* Same as System **d4** except that II was Mitsubishi TG [37].

d56. *PVC* (I)–*indene resin* (II), *may be compatible:* Same as System **d30** except that II was Fuji Iron Products Co. VM-1/2 [37].

E.   Polyethylene (PE) and Other Polymers

e1. *PE* (I)–*PE* (II), *ambiguous* (depends on molecular weight and molecular weight distribution): Two or three phases in diphenyl ether at various temperatures, in agreement with theoretical predictions when I had $\bar{M}_w = 5.40 \times 10^5$ and II had $\bar{M}_w = 1.2 \times 10^4$ or 2.5×10^4 [152].

e2. *PE* (I)–*polypropylene* (II), *conditionally compatible* (compatible if II is atactic): Incompatible region inside tenary phase diagram with diphenyl ether when I was linear and II was atactic but I and II were miscible, but very incompatible when II was isotactic [52, 66]. Complete miscibility with diphenyl ether above 157.5°C but complete miscibility of I and II alone at all temperatures investigated when I was Marlex 6050, $\bar{M}_n = 7.9 \times 10^3$, $\bar{M}_w = 7.6 \times 10^4$, and II was atactic, $[\eta] = 0.85$ at 135°C in decalin; incompatible in bulk or in diphenyl ether to above 164.2°C when II was isotactic, Carlona GM 21, $\bar{M}_n = 5.2 \times 10^4$, $\bar{M}_w = 6.4 \times 10^5$ [57].

e3. *PE* (I)–*polyisobutene* (II), *incompatible:* x-Ray diffraction showed both amorphous halos for 5–20% II when I was high density P-4007 ETA or P-4015 ETA and II was PIB-118 [153]. Two phases for I/II: 1/9 to 9/1 at 100°C when I was Mirathen 1311 and II was Oppanol B 100 [16].

e4. *PE* (I)–*polybutadiene* (II), *may be almost compatible:* Phase separation in xylene at 20%

solids for 1/1 weight ratio of I/II at 90°C [124]. Two transitions by radiothermoluminescence when II was SKD, 2.5% 1,2 [154].

e5. *PE* (I)–*polystyrene* (II), *incompatible:* Phase separation in xylene at 10% solids and 1/1 weight ratio of I/II at 90°C when II had $[\eta] = 1.06$ in toluene [124].

e6. *PE* (I)–*poly(ethyl acrylate)* (II), *ambiguous, probably incompatible:* Could not be milled unless silica was added; then transparent if I/II ratio was 4/1, 1/2, 1/3, or 1/5 [109].

e7. *PE* (I)–*poly(n-butyl acrylate)* (II); *ambiguous, probably incompatible:* Milky film on milling for 1/1 mixture; transparent if 82 silica was added [109].

e8. *PE* (I)–*poly[ethylene-co-(vinyl acetate)]* (II), *ambiguous, probably incompatible:* Milky on milling of 1/1 mixture; if silica was added, films were clear for I/II ratios of 3/1, 3.2/1. 3.5/1 when II contained 80 mole % ethylene [109].

e9. *PE* (I)–*poly[(ethyl acrylate)-co-acrylonitrile]*, *ambiguous; probably incompatible;* Could not be mixed by milling unless silica was added; then transparent if I/II ratio was 3/1, 1/1, 1/3 and II contained 80 mole % ethyl acrylate [109].

e10. *PE* (I)–*poly[butadiene-co-styrene-co-(methacrylic acid)]* (II), *incompatible:* Phase separation in organic solvents at 1–5% solids except when I/II ratio was 3/1 (no phase separation in 60 days) when I was high pressure PE and II was SKS-30-1 (30% styrene, 1.25 or 8% methacrylic acid) [30].

F.   Polybutadiene (PBD) and Other Polymers

f1. *PBD* (I)–*PBD* (II), *probably compatible:* Single T_g by radiothermoluminescence when I was SKD, stereoregular, 3% 1,2, and II was SKBM-50; single T_g for I/II: 1/9 to 9/1 by weight when II was SKB-30; single T_g except when cross-linked for I/II: 7/3 to 3/7 by weight when II was SKB-60 (all samples of II were 40–66% 1,2) [50]. Single transition by radiothermoluminescence, but closer to that of II, for I/II: 3/7 to 7/3 when I was SKD, 2.5% 1,2, and II was SKB, 60% 1,2 [154]. Single T_g by radiothermoluminescence, but T_g's of I and II were close together, when I was sodium butadiene SKBM-50 and II was sodium butadiene, SKB-60, I/II: 1/1; and when II was SKD, I/II: 1/1 and 4/1 [105].

f2. *PBD* (I)–*polystyrene* (II), *incompatible:* Stated incompatible [117], phase separation of many mixtures in many solvents [27, 51, 76, 77, 117, 155–158], and two T_g's found for several mixtures [118, 159, 160].

f3. *PBD* (I)–*poly(methyl methacrylate)* (II), *incompatible:* Two phases by dilatometry and dynamic mechanical properties when I was Firestone Co. No. 2004, gel content 85%, and II was from Mitsubishi Rayon Co. [118].

f4. *PBD* (I)–*poly(isoprene-co-isobutene)* (II), *incompatible:* Very turbid film for 1/1 mixture when I was BWH Buna CB10, *cis*-BR, and II was Polysar Butyl 301 [161].

f5. *PBD* (I)–*poly(butadiene-co-styrene)* (BD–S) (II), *must be conditionally compatible:* Phase separation in benzene or gasoline when I was SKB, $[\eta] = 1.70$ in benzene, and II was SKS-30 (30% styrene), $[\eta] = 1.50$ in benzene [76, 77]. Phase separation in gasoline at 5% solids for 1/1 ratio of I/II when I was SKB and II was SKS-30 [99]. Dielectric and mechanical loss showed single peak but mechanical loss peak had shoulder for I/II: 1/4 to 4/1 when I was cis-1,4 and II was JSR 1500 [104]. Single loss peak by dynamic and dielectric measurements for I/II: 1/4 to 4/1 when II had <25% S and I was Diene NF-35, low cis content; I/II: 2/3 homogeneous by phase contrast microscopy when II had <30% S; two loss peaks for I/II: 1/4 to 4/1 when II had >41.4% S; phase separation by phase contrast microscopy for I/II: 2/3 when II had >57% S [162]. Compatible by phase contrast and electron microscopy for I/II: 1/3 to 3/1 when I was high *cis*-1,4-BR, or emulsion *cis*-1,4-EBR, and II was SBR 1500 [106]. Single damping maximum near that of one of the components when I was *cis*-BR, 97% cis, and II was Texas–

U.S. Chemical Co. Synpol 1500, 23% S; in addition, two T_g's by DTA when I was EBR and II was Synpol 8000, 40% S [163]. No heterogeneity seen by phase contrast and electron microscopy when I was cis-1,4 and II was SBR 1500 [100]. Two dynamic mechanical transitions for unvulcanized and single transition for vulcanized I/II: 4/1 to 1/4 when I was Japan Synthetic Rubber Co. BR01 or EBR and II was SBR 1500 [45]. Single T_g by dilatometry or rolling ball loss spectrometer when I was Firestone Tire and Rubber Co. Diene 55NF and II was SBR 1712 [101]. Transparent film for 1/1 mixture; crystalline torsional transition seen up to 50% II; single T_g for amorphous samples and portions of samples for I/II: 1/4 to 4/1 [161]. Two dynamic mechanical transitions for 1/1 mixture except after much mixing; solution blends had two transitions which became one after milling and two again after dissolving and solvent removal; I/II: 1/9 to 1/1 had single T_g and 1/1 to 9/1 also had T_m, 1/1 mixture had fine structure by phase contrast microscopy which disappeared on heating at 80°C [102].

f6. *PBD* (I)–*poly(butadiene-co-acrylonitrile)* (II), *incompatible* (but should be conditionally compatible, e.g., at small acrylonitrile content): Phase separation in benzene at 5% solids for a 1/1 mixture of I/II when I was SKB and II was SKN-18 [99]. Two T_g's by a number of methods for a number of mixtures with II of varying acrylonitrile content [50, 101, 105]. Transparent film but refractive index of I and II differed only in third decimal when I was BWH Buna CB10, cis-BR, and II was Bayer Perbunan 3805 [161].

f7. *PBD* (I)–*poly(styrene-co-acrylonitrile)* (II), *incompatible:* Two T_g's by several methods for a number of mixtures in which II varied from 25 to 50% acrylonitrile [118, 133].

f8. *PBD* (I)–*poly(ethylene-co-propylene)* (II), *incompatible:* Two T_g's by radiothermolumines- cence for I/II: 97/3 to 3/97 by weight when I was SKD, stereoregular, 3% 1,2, and II was SKEP [50].

f9. *PBD* (I)–*poly(ethylene-co-propylene-co-diene)* (II), *conditionally compatible:* Stated com- patible at room temperature but immiscible when heated when I was butyl rubber and II was EPDM [164]. Miscible in bulk but there was an incompatible region inside the ternary phase diagram with benzene when I was butyl rubber and II was EPDM [52]. Two T_g's from torsional modulus for vulcanized I/II: 1/4 to 4/1, 1/1 film very turbid, and I crystalline up to 50% II, when I was BWH Buna CB 10, cis-BR, and II was EPDM, DSM Kelton 512 [161].

f10. *PBD* (I)–*poly(ethylene-co-propylene-co-?)* (II), *incompatible:* Incompatible by phase contrast and electron microscopy when I was high cis-1,4 and II was EPT [106].

G. Poly(vinyl acetate) (PVA) and Other Polymers

g1. *PVA* (I)–*cross-linked PVA* (II), *compatible:* Swelling of II in the presence of I plus ethyl acetate yielded a negative interaction parameter for I with II when I had $[\eta] = 0.089$ or 0.139 in ethyl acetate or degree of polymerization (cryoscopic) = 48 or 94 [165].

g2. *PVA* (I)–*polystyrene* (II), *incompatible:* Phase separation for many mixtures in many solvents [37, 78, 79, 86, 120, 124, 166, 167] and two T_g's by several methods for another mixture [118].

g3. *PVA* (I)–*poly(p-methyl styrene)* (II), *incompatible:* Phase separation at 20% solids in benzene at 25°C for 1/1 weight ratio of I/II when I had MW = 5×10^4 or 1.5×10^5 and II had $\eta_{sp} = 0.17$ at 0.1% in benzene [124].

g4. *PVA* (I)–*poly(p-chlorostyrene)* (II), *incompatible:* Same as System **g3** except that II had $\eta_{sp} = 0.06$ at 0.1% in toluene [124].

g5. *PVA* (I)–*poly(methyl acrylate)* (II), *ambiguous:* Evaporated films of 1/1 mixtures were transparent or very clear but torsional modulus gave T_g of II and sometimes T_g of I [34]. Stated to be compatible without proof [83]. No phase separation at any concentration in acetone, dioxane, or benzene for a 1/1 mixture at 25°C when I had MW = 5×10^4 or 1.5×10^5 and

II had $\eta_{sp} = 0.15$ at 0.1% in acetone [124]. Stated to be homogeneous polyblend for 1/1 by weight [95]. Clear by front illumination and clear film but nonhomogeneous by side illumination for 6 gm of 1/1 mixture in 10 ml ethyl acetate and nonhomogeneous in toluene when I had $\overline{M}_w = 2.75 \times 10^5$ and II had $\overline{M}_w = 3.65 \times 10^5$ [168].

g6. *PVA* (I)–*poly*(*ethyl acrylate*) (II), *may be compatible*: Stated to be compatible without proof [83].

g7. *PVA* (I)–*poly*(*butyl acrylate*) (II), *ambiguous*: Phase separation in acetone when I had $[\eta] = 1.77$ in acetone and II had $[\eta] = 1.00$ in acetone [76, 77]. Stated to be compatible without proof [83].

g8. *PVA* (I)–*poly*(*methyl methacrylate*) (II), *ambiguous*: Phase separation in acetone when I had $[\eta] = 1.92$ in acetone and II had $[\eta] = 1.30$ in acetone [76, 77]. Stated to be compatible without proof [83]. Films containing cross-linked II showed selective swelling with alcohol vapors [169]. Films containing 2–3% $CHCl_3$ were clear but had two damping maxima and two T_g's on refractive index–temperature curves when I had degree of polymerization 850 and II had degree of polymerization 3000, both industrial materials; films containing 22% $CHCl_3$ and a 3/7 ratio of I/II had a single damping maximum [170]. Phase separation in acetone above 4.5% solids and in ethyl acetate above 8.5% solids at 16–18°C when I was Rhodopas H, $\overline{M}_n = 5.6 \times 10^4$, $[\eta] = 0.60$ in acetone, and II had $\overline{M}_n > 2 \times 10^6$, $[\eta] = 3.65$ in acetone; the same I and II were miscible in dioxane and in acetic acid to above 10% solids [78]. Phase separation in acetone at 20% solids at 25°C for 1/1 mixture when I had MW $= 5 \times 10^4$ or 1.5×10^5, $[\eta] = 1.11$ in acetone and II had $[\eta] = 0.28$ in acetone [124]. No phase separation in cyclohexanone for 1/1 mixture, and film was clear when I was Vinac B-100, Air Reduction and Chemical Co., and II was Lucite 147 or 148, du Pont [79]. Transparent films from freeze-dried samples and dilatometric T_g versus percent composition gave smooth curve but heat treatment at 130°C caused two T_g's, that is, phase separation, when I had $\overline{M}_v = 1.4 \times 10^5$ and II had $\overline{M}_v = 3.3 \times 10^5$ [42]. Solutions and films homogeneous and transparent from chloroform, toluene, and chlorobenzene, but phase separated from ethyl acetate, DMF, and benzyl acetate [36]. Homogeneous solution and transparent film but two T_g's by DTA when I was Nippon Goshei Co. NZ-5 and II was from Mitsubishi Rayon Co. [37].

g9. *PVA* (I)–*poly*(*ethyl methacrylate*) (II), *incompatible*: Stated to be incompatible without proof [83].

g10. *PVA*(I)–*poly*(*n-butyl methacrylate*)(II), *incompatible*: Stated to be incompatible without proof [83].

g11. *PVA* (I)–*polyacrylonitrile* (II), *incompatible*: Phase separation of 1/1 mixture after 50 days in DMF or dimethyl sulfoxide but considered compatible by authors because of minimum in activation energy for decomposition [81] (see Note for System **b5**).

g12. *PVA* (I)–*poly*(*vinyl nitrate*) (II), *compatible*: Single T_g by dilatometry but T_g's of I and II are less than 1°C apart; however, mixtures had higher T_g's than homopolymers for I/II: 9/1 to 1/9 by weight when I had DP $= 1800$ and II was nitrated poly(vinyl alcohol) with DP $= 1700$, 94.8% nitrated [171]. More structure visible in homopolymers than in blends by phase contrast microscopy; IR spectra and x-ray scattering curves were intermediate between those of I and II, and refractive index and mechanical loss T_g's had a maximum at an intermediate composition [172].

g13. *PVA* (I)–*poly*(*vinylpyrrolidone*) (II), *incompatible*: Nonhomogeneous solution and turbid film when I was Nippon Goshei Co. NZ-5, and II was General Aniline and Film K-30 [37].

g14. *PVA* (II)–*poly*(*methyl vinyl ketone*) (II), *incompatible*: Same as System **g3** except that II had $\eta_{sp} \sim 1.6$ at 0.1% in ethyl acetate [124].

g15. *PVA* (I)–*poly*(*vinyl acetal*) (II), *incompatible*: Phase separation for several mixtures in a number of solvents [78].

g16. *PVA* (I)–*poly(vinyl butyral)* (II), *incompatible:* Same as System **g13** except that II was Shekisui Chemical Co. BM-2 [37].

g17. *PVA* (I)–*poly(vinylidene chloride)* (II), *incompatible:* Same as System **g13** except that II was Asahi Dow Co. EX5701 [37].

g18. *PVA* (I)–*poly(ε-caprolactone)* (II), *incompatible:* Two T_g's from shear modulus at I/II: 9/1 and 1/1 [89].

g19. *PVA* (I)–*polycarbonate* (II), *incompatible:* Phase separation in cyclohexanone for 1/1 mixture, and film was not clear, when I was Vinac B-100, Air Reduction and Chemical Co., and II was General Electric Co. Lexan 125 or Mobay Chemical Co. Merlon M-50 [79].

g20. *PVA* (I)–*polyepichlorohydrin* (II), *incompatible:* Same as System **g19** except that II was B. F. Goodrich Hydrin 100 [79].

g21. *PVA* (I)–*poly(2,6-dimethyl-1,4-phenylene ether)* (II); *incompatible:* Same as System **g19** except that II was General Electric Co. PPO [79].

g22. *PVA* (I)–*polysulfone* (II), *incompatible:* Same as System **g19** except that II was Union Carbide polysulfone [79].

g23. *PVA* (I)–*polyurethane* (II), *incompatible:* Phase separation in solutions and films for several mixtures including polyester and polyether urethanes [37, 79].

g24. *PVA* (I)–*polyester* (II), *incompatible:* Same as System **g13** except that II was Toyobo Co. Ester Resin 20 [37].

g25. *PVA* (I)–*chlorosulfonated polyethylene* (II), *incompatible:* Same as System **g13** except that II was Showa Neoprene Hyperon 30 [37].

g26. *PVA* (I)–*poly[ethylene-co-(vinyl acetate)]* (II), *must be conditionally compatible:* Nonhomogeneous solution and turbid film when I was Nippon Goshei Co. NZ-5 and II was Nippon Goshei Co. Soalex R-CH, but homogeneous solution and transparent film when II was Soalex R-FH [37].

g27. *PVA* (I)–*poly[(vinyl acetate)-co-(vinyl chloride)]* (II), *ambiguous, but must be conditionally compatible:* Two damping maxima for 50/50 and 40/60 mixtures of I/II and one damping maximum for 20/80 and 60/40 mixtures when I was Vinalit SP 60 and II was Vinalit MPS [134]. Phase separation in cyclohexanone for 1/1 mixture when I was Vinac B-100, Air Reduction and Chemical Co., and II was Geon 440 × 24, B. F. Goodrich [79]. Stated to be homogeneous polyblend for 40–50 by weight I when II had 90% vinyl chloride [95]. Two or three phases after 55 days for 10% solution in cyclohexanone of I/II: 2/3 to 4/1 when I had $\overline{M}_w = 4.85 \times 10^5$, 85% acetate, and II was Covicet, $\overline{M}_w = 2.9 \times 10^4$, 44.1% Cl, 80% vinyl chloride [122]; small particles under polarizing microscope for I/II: 3/2 [123]. Nonhomogeneous solutions and turbid films when I was Nippon Goshei Co. NZ-5, and II was Shekisui Chemical Co. Esulex C or CL or CH-1 or Denki Chemical Co. Denka Vinyl 1000 AK [37].

g28. *PVA* (I)–*poly[(vinyl acetate)-co-(vinyl chloride)-co-(vinyl alcohol)]* (II), *incompatible:* Same as System **g13** except that II was Shekisui Chemical Co. Esulex A [37].

g29. *PVA* (I)–*poly[(vinyl acetate)-co-(vinyl chloride)-co-(maleic acid)]* (II), *incompatible:* Same as System **g13** except that II was Shekisui Chemical Co. Esulex M [37].

g30. *PVA* (I)–*poly[(vinyl acetate)-co-(N-vinylpyrrolidone)]* (II), *incompatible:* Same as System **g13** except that II was General Aniline & Film S-630 [37].

g31. *PVA* (I)–*poly(butadiene-co-styrene)* (II), *incompatible:* Same as System **g13** except that II was Nippon Rubber Co. Hycar 2057 [37].

g32. *PVA* (I)–*poly(butadiene-co-acrylonitrile)* (II), *incompatible:* Phase separation in solutions and films for several mixtures [37, 79].

g33. *PVA* (I)–*poly[(vinyl chloride)-co-(vinyl propionate)]* (II), *incompatible:* Same as System **g13** except that II was Ryuron QS-430 [37].

g34. *PVA* (I)–*poly[(vinyl chloride)-co-(vinylidene chloride)-co-(acrylic acid ester)]* (II), *incompatible:* Same as System **g13** except that II was Towa Gousel Co. Aron 321 [37].

g35. *PVA* (I)–*poly*[*styrene-co-(methyl methacrylate)*] (II), *incompatible:* Same as System **g19** except that II was Dow Chemical Co. Zerlon 150 [79].

g36. *PVA* (I)–*poly*(*styrene-co-acrylonitrile*) (II), *incompatible:* Same as System **g19** except that II was Dow Chemical Co. Tyril 767 [79].

g37. *PVA* (I)–*poly*[*styrene-co-(maleic acid)*] (II), *incompatible:* Homogeneous solutions and transparent films, but two T_g's by DTA when I was Nippon Goshei Co. NZ-5, and II was Daidou Kogyo Styrite CM-2 or CM-3 [37].

g38. *PVA* (I)–*poly*[*styrene-co-(maleic acid ester)*] (II), *may be conditionally compatible:* Homogeneous solution and transparent film when I was Nippon Goshei Co. NZ-5, and II was Daidou Kogyo Styrite HS-2 [37].

g39. *PVA* (I)–*poly*[*methyl vinyl ether*)-*co-(maleic anhydride*] (II); *may be compatible or conditionally compatible:* No phase separation in cyclohexanone for 1/1 mixture, and film was clear when I was Vinac B-100, Air Reduction and Chemical Co., and II was Gantrez AN-169, B. F. Goodrich [79].

g40. *PVA* (I)–*poly*[*epichlorophydrin-co-(ethylene oxide)*] (II), *incompatible:* Same as System **g19** except that II was B. F. Goodrich Hydrin 200 [79].

g41. *PVA* (I)–*miscellaneous resins* (II), *incompatible:* Same as System **g13** except that II was Petroleum Resin, Nippon Petroleum Chemical Co. Nitsuseki Neopolymer 150, or Cumarone Resin, Mitsubishi TG, or Indene Resin, Fuji Iron Products Co. VM-1/2 [37].

H. Polystyrene (PS) and Other Polymers

h1. *PS* (I)–*polypropylene* (II), *incompatible:* Phase separation in toluene from 20 to 60°C for various mixtures of fractions where I had $\bar{M}_v = 6 \times 10^4$ to 7×10^5 and II had $\bar{M}_v = 2 \times 10^4$ to 10^5 [55].

h2. *PS* (I)–*poly*(1-*butene*) (II), *incompatible:* Phase diagrams in CCl_4 showed phase separation up to 60–70% solvent when I was USS chemicals PS-200, $\bar{M}_n = 8.5 \times 10^4$, $\bar{M}_w/\bar{M}_n = 2$, and II was Witco Chemicals PB-001, $\bar{M}_n = 9.7 \times 10^4$, $\bar{M}_w/\bar{M}_n = 6$, or PB-003, $\bar{M}_w = 3.0 \times 10^4$, $\bar{M}_w/\bar{M}_n = 6$, or PB-004; phase separation up to 78% solvent when I was USS Chemicals PS-209, $\bar{M}_n = 9.6 \times 10^4$, $\bar{M}_w/\bar{M}_n = 2$, and II was Witco Chemicals PB-004 [173].

h3. *PS* (I)–*polyisobutene* (II), *incompatible:* Light scattering of mixtures of fractions gave a very small interaction parameter indicating at least partial compatibility [174]. Phase separation of a number of mixtures in different solvents [16, 156, 175]. Specific heat–temperature curves showed T_g of I affected slightly by the presence of II up to 20% II and then a larger effect at 30% II when II had MW $= 1.2 \times 10^4$ and I had been polymerized to give a molecular mixture of I and II [176].

h4. *PS* (I)–*polychloroprene* (II), *incompatible:* Stated incompatible for I/II: 4/1 to 1/4 in solutions and films when I was BW and II was Neoprene AC [16].

h5. *PS* (I)–*chlorinated poly*(*vinyl chloride*) (II), *incompatible:* Same as System **h4** except that II was Vinoflex PC [16].

h6. *PS* (I)–*poly*(*vinylidene chloride*) (II), *incompatible:* Nonhomogeneous solution and turbid film when I was Asahi Dow Co. Styron 666 and II was Asahi Dow Co. EX 5701 [37].

h7. *PS* (I)–*poly*(2-*vinylpyridine*) (II), *incompatible:* Precipitated from CH_2Cl_2 when I was Union Carbide SGN-3000, reduced viscosity 0.85 dl/gm at 0.2 gm/dl in CH_2Cl_2 and II had reduced viscosity 1.095 dl/gm at 0.2 gm/dl in CH_2Cl_2 [177].

h8. *PS* (I)–*poly*(*vinyltoluene*) (II), *incompatible:* When polymerization occurred in a solution which was 10% in monomer of II, the mixture was opaque during the whole polymerization when II was 60% meta, 40% para [178].

h9. *PS* (I)–*poly*(α-*methylstyrene*) (II), *conditionally compatible:* Mixtures were cloudy and had two damping maxima although only a single T_g was observed in block copolymers; unfractionated block copolymer was clear but block copolymer fractions appeared cloudy [179]. Two T_g's on thermomechanical analyzer and no turbidity for 30% solution in benzene [180]. Single T_g by DTA for 1/1 mixture when I had $\bar{M}_n = 5.1 \times 10^4$ and II had $\bar{M}_w = 1.85 \times 10^5$, $\bar{M}_n = 1.47 \times 10^5$, and for I/II: 1/4 and 2/3 when I had $\bar{M}_n = 1.6 \times 10^5$, but two T_g's for 1/1 mixture when I had $\bar{M}_n = 1.6 \times 10^5$ [181]. Two dynamic mechanical T_g's for I/II: 1/3 when I was Union Carbide SMD-3500, $\bar{M}_n = 1.10 \times 10^5$, $\bar{M}_w = 2.70 \times 10^5$, and II had $\bar{M}_w = 3.10 \times 10^5$, and when I had $\bar{M}_w = 3.6 \times 10^4$ and II had $\bar{M}_n = 1.60 \times 10^5$, $\bar{M}_w = 1.65 \times 10^5$ [182]. DSC indicated compatibility and small angle neutron scattering showed $(\bar{M}_w)_{app} = \bar{M}_w$ of I, that is, no clustering of I to show incompatibility, when I had $\bar{M}_w = 5.4 \times 10^4$ and II was deuterated [183].

h10. *PS* (I)–*poly*(o-*methylstyrene*) (II), *ambiguous:* No phase separation in $CHCl_3$ for 1/1 by weight mixtures at 25°C at any concentration when I had $[\eta] = 1.06$ in toluene, and II had $\eta_{sp} = 0.15$ at 0.1% in benzene [124]. Phase separation in $CHCl_3$ for 1/1 by weight mixtures above 0.25 gm solids/cm³ $CHCl_3$ when I had $[\eta] = 0.9$ in benzene, and II had $[\eta] = 1.1$ in $CHCl_3$ [184].

h11. *PS* (I)–*poly*(m-*methylstyrene*) (II), *incompatible:* Phase separation for several mixtures in $CHCl_3$ [124, 184].

h12. *PS* (I)–*poly*(p-*methylstyrene*) (II), *incompatible:* Phase separation for several mixtures in $CHCl_3$ [124, 184].

h13. *PS* (I)–*poly*(p-*chlorostyrene*) (II), *incompatible:* Phase separation for several mixtures in benzene [124, 184].

h14. *PS* (I)–*poly*(*chlorostyrene*) (II), *incompatible:* Same as System **h7** except that II was mixed isomers, reduced viscosity 0.639 dl/gm at 0.2 gm/dl in CH_2Cl_2 [177].

h15. *PS* (I)–*poly*(*dichlorostyrene*) (II), *incompatible:* Same as System **h7** except that II was mixed isomers, reduced viscosity 0.647 dl/gm at 0.2 gm/dl in CH_2Cl_2 [177].

h16. *PS* (I)–*poly*(p-*methoxystyrene*) (II), *incompatible:* Phase separation in benzene above 12.4% solids for 1/1 by volume mixture when I had $[\eta] = 0.9$ in benzene, and II had $[\eta] = 0.2$ in benzene [184].

h17. *PS* (I)–*poly*(p-*tert-butylstyrene*) (II), *incompatible:* When polymerization occurred in a solution which was 10% I in monomer of II, the mixture was opaque during the whole polymerization [178].

h18. *PS* (I)–*poly*(*methyl acrylate*) (II), *incompatible:* Phase separation for several mixtures in several solvents [124, 185]. Evaporated films were whitish and had two T_g's [78].

h19. *PS* (I)–*poly*(*ethyl acrylate*) (II), *incompatible:* Demixing occurred early in a polymerization of monomer of I in II [178]; phase separation in solution [185]; interpenetrating network was opalescent and had two dynamic mechanical T_g's when II contained 1% butadiene [186].

h20. *PS* (I)–*poly*(*butyl acrylate*) (II), *incompatible:* Phase separation in $CHCl_3$ above 5 gm solids/dl at 25°C when I had $\bar{M}_v = 5.0 \times 10^5$ and II had $\bar{M}_v = 1.5 \times 10^6$ [185].

h21. *PS* (I)–*poly*(*cyclododecyl acrylate*) (II), *incompatible:* Whitish, opaque films and two T_g's by DSC for 29% and 50% II, when I had $\bar{M}_n = 9.0 \times 10^4$ and II had $\bar{M}_n = 1.84 \times 10^5$, $\bar{M}_w/\bar{M}_n = 1.94$ [187].

h22. *PS* (I)–*poly*(*methyl methacrylate*) (II), *incompatible:* Interaction parameter between I and II small but indicated incompatibility [188]. Phase separation early in polymerization of monomer of I in II [178]. Phase separation in many solutions and films [16, 37, 78, 79, 124, 166, 170, 184, 185, 189]. Two T_g's by various methods [118, 170]. Various studies of limiting solubility of I in II or vice versa showed low solubility [27, 108, 190–192].

h23. *PS* (I)–*poly(butyl methacrylate)* (II), *incompatible:* Same as System **h20**, but no information given for II [185].

h24. *PS* (I)–*poly(benzyl methacrylate)* (II), *incompatible:* Same as System **h20**, except that II had $[\eta] = 0.629$ in benzene at 30°C [185].

h25. *PS* (I)–*poly(isobornyl methacrylate* (II), *incompatible:* Same as System **h20**, but no information given for II [185].

h26. *PS* (I)–*poly(vinylpyrrolidone)* (II), *incompatible:* Same as System **h6** except that II was General Aniline and Film K-30 [37].

h27. *PS* (I)–*poly(vinyl acetal)* (II), *incompatible:* Phase separation in $CHCl_3$ above 3.2% solids at 16–18°C when I had $\bar{M}_n = 2.25 \times 10^5$, $[\eta] = 2.15$ in $CHCl_3$, and II had $\bar{M}_n = 3.8 \times 10^4$, $[\eta] = 0.75$ in acetone [78].

h28. *PS* (I)–*poly(vinyl butyral)* (II), *incompatible:* Same as System **h6** except that II was Shekisui Chemical Co. BM-2 [37].

h29. *PS* (I)–*poly(ethylene oxide)* (II), *incompatible:* Stated incompatible when I had $\bar{M}_w = 1.5 \times 10^5$ and II was Polyglycol E 4000, Dow Chemical Co., low molecular weight, and II was Union Carbide WSR-35, high molecular weight [193].

h30. *PS* (I)–*poly(vinyl methyl ether)* (II), *conditionally compatible:* Clear films with a single T_g, intermediate between those of I and II, when cast from toluene, but cloudy films with T_g's characteristic of the homopolymers when cast from trichloroethylene or $CHCl_3$, when I was bulk polymer, $\bar{M}_n = 1.04 \times 10^5$, or Dow-Corning PS-690, $\bar{M}_n = 1.5 \times 10^5$, and II was a sample from GAF, precipitated twice and dried, $\bar{M}_n = 5.24 \times 10^5$, close to atactic; films were cast at 25°C onto Al or Hg and vacuum dried at 110°C; T_g's from differential scanning calorimetry [38]. LCST about 82°C by phase contrast microscopy when I had $M = 2.0 \times 10^5$, $\bar{M}_w/\bar{M}_n = 1.06$, and II had $\bar{M}_v = 5.5 \times 10^5$ [54]. LCST about 120°C by film turbidity when I was Union Carbide SMD-3500, $\bar{M}_n = 7.84 \times 10^4$, $\bar{M}_w = 2.37 \times 10^5$, and II was GAF Gantrez MO93, $\bar{M}_n = 7.7 \times 10^3$, $\bar{M}_w = 1.33 \times 10^4$ [74]. Film became opaque at room temperature after several months for 0.7 weight fraction II; film became opaque after 3 weeks, transparent after heating to 47°C for 40 hr, and opaque at 150°C with pulsed NMR indicating phase separation for 0.75 weight fraction II; NMR data were as follows: no phase separation according to T_1 for I/II: 3/1, two T_1's above 140°C and two T_2's above 25°C for 1/1 mixture, and two T_1's above 140°C for I/II: 1/3, when I was Sinclair-Koppers Co. Dylene 8, $[\eta] = 0.94$ at 25°C in benzene and II was from Cellomer Associates, Inc., $[\eta] = 0.51$ at 25°C in benzene; T_2 indicated micro-heterogeneous system, that is, clustered, above 25°C [40]. Films visually clear for I/II: 4/1 to 1/4 when cast from toluene but had LCST about 125°C; when phase separated mixtures were quenched in liquid N_2, two T_g's were seen by DSC where a single T_g was seen ordinarily [194]. Films cast from toluene showed LCST by turbidity measurements ranging from 98°C when I had $M = 2 \times 10^5$ and II was from Cellomer Associates, $\bar{M}_w = 5.15 \times 10^4$ to 210°C when I had $M = 10^4$ and LCST also decreased when I had $M > 2 \times 10^5$; films cast at room temperature from trichloroethylene when I had $M = 2 \times 10^4$ for 0.25–0.97 weight fraction I were hetero-geneous but could be annealed to give same data as toluene cast films, so that UCST was postulated for this mixture [41].

h31. *PS* (I)–*poly(2,6-dimethyl-1,4-phenylene ether)* (II), *compatible* (slightly ambiguous): Mixtures had a single T_g and heat distortion temperature, varying smoothly between those of the two polymers when I was Styron 666 or Lustrex HT 88-I rubber modified high impact polystyrene [195]. Stated to be compatible [196]. Single T_g by dielectric or mechanical properties or DSC when I was Monsanto atactic, additive free, $\bar{M}_n = 1.73 \times 10^5$, and II was General Electric additive free, $\bar{M}_n = 5.8 \times 10^4$ [197]. Films optically clear, single transition by DSC, and two overlapping dynamic mechanical loss peaks where the lower peak corresponded to the DSC peak for I/II: 1/3 to 3/1 [43]. Diffusivity, permeability, and solubility of Ne, Ar, and Kr in 1/3 to 3/1 blends gave results inconclusive as to number of phases present when I was

Styron 662 U, $\overline{M}_w = 2.5 \times 10^5$, $\overline{M}_n = 1.0 \times 10^5$, and II was GE Grade 531-801 PPO, $\overline{M}_w = 4.0 \times 10^4$, $\overline{M}_n = 2.0 \times 10^4$ [198]. Single T_g by DSC and dynamic mechanical measurements when I had $M = 9.72 \times 10^4$, $\overline{M}_w/\overline{M}_n \leqslant 1.06$, and II was PPO, $\overline{M}_w = 3.72 \times 10^4$, $\overline{M}_n = 1.85 \times 10^4$ [35]. Single transition using DSC and thermooptical analysis for I/II: 1/9 to 9/1 except 1/1 and 3/7, when I and II were as above for Shultz and Beach [35] [49]. Viscoelastic response showed single T_g for 0.049 to 0.604 weight fraction II when I had $\overline{M}_w = 9.72 \times 10^4$, $\overline{M}_w/\overline{M}_n = 1.04$ or MW $= 4.11 \times 10^5$, and II was PPO, $\overline{M}_v = 6.9 \times 10^4$, $\overline{M}_w/\overline{M}_n = 2.1$ [199]. Stated to be homogeneous polyblend at all compositions [95]. Single T_g by DSC for I/II: 1/3 to 3/1 when I was "monodisperse," $\overline{M}_w = 2 \times 10^4$, and II was PPO [200]. Phase separation in cyclohexanone for 1/1 mixture, and film not clear when I was Styron 690, Dow Chemical Co., and II was PPO, General Electric Co. [79].

h32. *PS* (I)–*poly(2-methyl-6-phenyl-1,4-phenylene ether)* (II); *incompatible:* Milky films and two T_g's by DSC and thermo-optical analysis for I/II: 4/1 to 1/4 when I had $M = 9.72 \times 10^4$, $\overline{M}_w/\overline{M}_n = 1.06$, and II had $[\eta] = 0.88$ dl/g at 30°C in CHCl$_3$ [201].

h33. *PS* (I)–*other substituted phenylene ether polymers* (II), *compatible:* Claim that mixtures had a single T_g and heat distortion temperature when I was Styron 666 and II was poly-(2,6-diethyl-1,4-phenylene ether), poly(2-methyl-6-ethyl-1,4-phenylene ether), poly(2-methyl-6-propyl-1,4-phenylene ether), poly(2,6-dipropyl-1,4-phenylene ether), or poly(2-ethyl-6-propyl-1,4-phenylene ether) [195].

h34. *PS* (I)–*poly(oxycarbonyloxy-1,4-phenyleneisopropylidene-1,4-phenylene)* (II), *incompatible:* Same as System **h7** except that II was General Electric Lexan 101, reduced viscosity 0.855 dl/gm at 0.2 gm/dl in CH$_2$Cl$_2$ [177].

h35. *PS* (I)–*polycarbonate* (II), *incompatible:* Phase separation in cyclohexanone for 1/1 mixture and film was not clear when I was Styron 690, Dow Chemical Co., and II was General Electric Lexan 125, or Mobay Chemical Co. Merlon M-50 [79].

h36. *PS* (I)–*poly[oxycarbonyloxy(2,6-dimethyl-1,4-phenylene)isopropylidene(3,5-dimethyl-1,4-phenylene)]* (II); *compatible:* Stated "quite compatible" at room temperature and single dynamic mechanical T_g after processing at 50°C but two T_g's after processing at 300°C when I was as in System **h7** and II had reduced viscosity 0.62 dl/gm at 0.2 gm/dl in CH$_2$Cl$_2$ [177].

h37. *PS* (I)–*poly[oxycarbonyloxy(2,2,4-tetramethyl-1,3-cyclobutylene)]* (II), *incompatible:* Same as System **h7** except that II had reduced viscosity 1.17 dl/gm at 0.2 gm/dl in CH$_2$Cl$_2$ [177].

h38. *PS* (I)–*polyepichlorohydrin* (II), *incompatible:* Same as System **h35** except that II was B. F. Goodrich Hydrin 100 [79].

h39. *PS* (I)–*polysulfone* (II), *incompatible:* Same as System **h35** except that II was Union Carbide polysulfone [79].

h40. *PS* (I)–*chlorosulfonated polyethylene* (II), *incompatible:* Same as System **h6** except that II was Showa Neoprene Hyperon 30 [37].

h41. *PS* (I)–*polyurethane* (II), *incompatible:* Phase separation in solutions and films for a number of samples including polyester and polyether urethanes [37, 79].

h42. *PS* (I)–*polyester* (II), *incompatible:* Same as System **h6** except that II was Toyobo Co. Ester Resin 20 [37].

h43. *PS* (I)–*poly(dimethylsiloxane)* (II), *incompatible:* Phase separation for 1/1 mixture in ethyl acetate when I had $\overline{M}_n = 6.7 \times 10^5$ and II had $\overline{M}_v = 3.9 \times 10^5$ or 6.3×10^5; stated that there is an UCST $\sim 35°C$ when I and II were approximately hexamers [202].

h44. *PS* (I)–*poly(styrene-co-butadiene)* (II), *must be conditionally compatible:* Stated to be incompatible when II was Buna S3 or Buna S4 (77.5 to 80% butadiene) or Buna SS (57.5–60% butadiene) [117]. Two damping maxima obtained when II contained 23% styrene and the mixture contained 5,10, or 15% II, and when II contained 30% styrene and the mixture contained 10% II [160]. Two damping maxima and films were cloudy [135]. Two damping maxima when II contained 60% styrene and the mixture contained 10–60% II [203]. Incompatible by phase contrast microscopy when latexes were blended and dried and I had $\overline{M}_w =$ 3.68×10^5, and II was 90% conversion at 50°C (divinylbenzene added) and had 22, 30, or 60% styrene content [204]. Two phases by dynamic mechanical measurements and dilatometry when I had $\overline{M}_v = 8.9 \times 10^4$ and II was Japan Synthetic Rubber Co. SBR, 24.5% styrene [118]. Stated to be homogeneous polyblend when wt % I $\geqslant 40$ and II had 35.7% styrene [95]. Non-homogeneous solution and turbid film when I was Asahi Dow Co. Styron 666, and II was Nippon Rubber Co. Hycar 2057 [37]. Stated incompatible for I/II: 4/1 and 1/1 but compatible for 1/4 when I was BW and II was Buna S3 ($\sim$80% butadiene) [16].

h45. *PS* (I)–*poly[styrene-co-(ethyl acrylate)]* (II), *must be conditionally compatible*: Phase separation in $CHCl_3$ at higher percent solids for I with $\overline{M}_v = 5.0 \times 10^5$ and II containing 50% styrene. $\overline{M}_v \simeq 10^6$, than when II was poly(ethyl acrylate), $\overline{M}_v = 1.2 \times 10^6$; no phase separation at 3 gm solids/dl when I/II was 1/1 or 1/3 [185].

h46. *PS* (I)–*poly(styrene-co-acrylonitrile)* (II), *must be conditionally compatible*: When polymerization occurred in a solution which was 10% I in a 75/25 styrene–acrylonitrile monomer mixture, the mixture was opaque during the whole polymerization [178]. Phase separation in cyclohexanone for 1/1 mixture, and film was not clear when I was Styron 690, Dow Chemical Co., and II was Tyril 767, Dow Chemical Co. [79].

h47. *PS* (I)–*poly[styrene-co-(methyl methacrylate)]* (II), *must be conditionally compatible*: Same as System **h35** except that II was Dow Chemical Co. Zerlon 150 [79].

h48. *PS* (I)–*poly[styrene-co-(maleic acid)]* (II), *must be conditionally compatible*: Same as System **h6** except that II was Daidou Kogyo Styrite CM-2 or CM-3 [37].

h49. *PS* (I)–*poly[styrene-co-(maleic acid ester)]* (II), *must be conditionally compatible*: Same as System **h6** except that II was Daidou Kogyo Styrite HS-2 [37].

h50. *PS* (I)–*poly[(vinyl chloride)-co-(vinyl acetate)]* (II), *incompatible*: Phase separation in solutions and films for a number of samples [16, 37, 79]. Two phases by dynamic mechanical properties and dilatometry [118].

h51. *PS* (I) *poly[(vinyl chloride)-co-(vinyl propionate)]* (II), *incompatible*: Same as System **h6** except that II was Ryuron QS-430 [37].

h52. *PS* (I)–*poly[(vinyl chloride)-co-(vinyl acetate)-co-(vinyl alcohol)]* (II), *incompatible*: Same as System **h6** except that II was Shekisui Chemical Co. Esulex A [37].

h53. *PS* (I)–*poly[(vinyl chloride)-co-(vinyl acetate)-co-(maleic acid)]* (II), *incompatible*: Same as System **h6** except that II was Shekisui Chemical Co. Esulex M [37].

h54. *PS* (I)–*poly[(vinyl chloride)-co-(vinylidene chloride)-co-(acrylic acid ester)]* (II), *incompatible*: Same as System **h6** except that II was Towa Gousei Co. Aron 321 [37].

h55. *PS* (I)–*poly(butadiene-co-acrylonitrile)* (II), *incompatible*: Stated incompatible [117]; two T_g's by various methods [118, 125]; phase separation in solutions and films for a number of mixtures [16, 37, 79].

h56. *PS* (I)–*poly[(vinyl acetate)-co-(N-vinylpyrrolidone)]* (II), *incompatible*: Same as System **h6** except that II was General Aniline & Film S-630 [37].

h57. *PS* (I)–*poly[ethylene-co-(vinyl acetate)]* (II), *incompatible*: Same as System **h6** except that II was Nippon Goshei Co. Soalex R-CH or R-FH [37].

h58. *PS* (I)–*poly[epichlorohydrin-co-(ethylene oxide)]* (II), *incompatible*: Same as System **h35** except that II was B. F. Goodrich Hydrin 200 [79].

h59. *PS* (I)–*poly[(methyl vinyl ether)-co-(maleic anhydride)]* (II), *incompatible*: Same as System **h35** except that II was B. F. Goodrich Gantrez AN-169 [79].

h60. *PS* (I)–*miscellaneous resins* (II), *may be compatible*: Homogeneous solutions and transparent films when I was Asahi Dow Co. Styron 666 and II was Petroleum Resin, Nippon Petroleum Chemical Co. Nitsuseki Neopolymer 150, or Cumarone Resin, Mitsubishi TG, or Indene Resin, Fuji Iron Products Co. VM-1/2 [37].

I. Acrylic Polymers and Other Acrylic Polymers

i1. *Poly(methyl acrylate)* (I)–*poly(ethyl acrylate)* (II), *incompatible:* Phase separation for many mixtures in many solvents [124, 168, 185].

i2. *Poly(methyl acrylate)* (I)–*poly(butyl acrylate)* (II), *incompatible:* Phase separation in solution when II was just called "butyl" [185] or was the *n*-butyl ester [168].

i3. *Poly(methyl acrylate)* (I)–*poly(methyl methacrylate)* (II), *incompatible:* Phase separation in many solvents for many mixtures [76, 77, 124, 168, 185]. Films mostly hazy and torsional modules showed two transitions or that of I alone [34].

i4. *Poly(methyl acrylate)* (I)–*poly(ethyl methacrylate)* (II), *incompatible:* Phase separation in benzene for 1/1 mixture when I had $\bar{M}_w = 5.0 \times 10^5$ or 3.65×10^5 and II had $\bar{M}_w = 1.25 \times 10^5$ or 1.9×10^5 [168].

i5. *Poly(methyl acrylate)* (I) –*poly(butyl methacrylate)* (II), *incompatible:* Phase separation in solution for mixtures containing unspecified butyl ester [76, 77], or the *n*-butyl or isobutyl ester [168].

i6. *Poly(methyl acrylate)* (I)–*poly(hexyl methacrylate)* (II), *incompatible:* Same as System **i4** except that II had $\bar{M}_w = 2.1 \times 10^5$ or 2.4×10^5 [168].

i7. *Poly(ethyl acrylate)* (I)–*poly(isopropyl acrylate)* (II), *incompatible:* Phase separation in $CHCl_3$ above 5 gm solids at 25°C when I had $\bar{M}_w = 1.2 \times 10^6$ and II had $\bar{M}_v = 1.3 \times 10^6$ [185].

i8. *Poly(ethyl acrylate)* (I)–*poly(butyl acrylate* (II), *incompatible:* Phase separation in solution for mixtures containing unspecified butyl ester [185] or the *n*-butyl ester [168].

i9. *Poly(ethyl acrylate)* (I)–*poly(methyl methacrylate)* (II), *incompatible:* Phase separation for many mixtures in many solvents [16, 168, 185, 205] and in films [16, 34]; two damping maxima [34, 205]. Interpenetrating networks were formed by swelling I with monomer of II and photopolymerizing; clear, transparent materials with one broad transition in the modulus–temperature curve were obtained; clear, homogeneous materials were obtained on mixing and heating I with $\bar{M}_n = 6200$ and II with $\bar{M}_n = 6400$; softening temperature of mixture was in line with results obtained on interpenetrating network [39]. (*Note:* The modulus–temperature curve of the interpenetrating network could be resolved to give two transitions; the miscibility of low molecular weight polymers is to be expected for many polymer–polymer systems.) Other interpenetrating networks were optically clear, with one damping maximum interpreted as composition distribution for I/II: 1/3 to 3/1 [186].

i10. *Poly(ethyl acrylate)* (I)–*poly(ethyl methacrylate)* (II), *incompatible:* Phase separation in several solvents for several mixtures [168, 185].

i11. *Poly(ethyl acrylate)* (I)–*poly(n-butyl methacrylate)* (II), *incompatible:* Phase separation in solution for several mixtures [168, 185]. Films transparent but showed T_g of I and sometimes of II [34].

i12. *Poly(ethyl acrylate)* (I)–*poly(isobutyl methacrylate)* (II), *incompatible:* Phase separation in benzene when I had $\bar{M}_w = 3.35 \times 10^5$ or 4.6×10^5 and II had $\bar{M}_w = 1.27 \times 10^5$ or 1.9×10^5 [168].

i13. *Poly(ethyl acrylate)* (I)–*poly(hexyl methacrylate)* (II), *incompatible:* Same as System **i12** except that II had $\bar{M}_w = 2.1 \times 10^5$ or 2.4×10^5 [168].

i14. *Poly(ethyl acrylate)* (I)–*poly(isobornyl methacrylate)* (II), *incompatible:* Phase separation in $CHCl_3$ above 5 gm/dl at 25°C when I had $\bar{M}_v = 1.2 \times 10^6$ [185].

i15. *Poly(ethyl acrylate)* (I)–*poly[(ethyl acrylate)-co-(methyl methacrylate)]* (II), *must be conditionally compatible:* Phase separation in $CHCl_3$ for 1/1 ratio of I/II at 6.7 gm/dl but no phase separation at 4.0 gm/dl at 25°C when I had $\bar{M}_v = 1.2 \times 10^6$ and II contained 50% ethyl acrylate and had $\bar{M}_v \sim 2 \times 10^6$; phase separation began at higher concentration than when II was polymethyl methacrylate with $\bar{M}_v = 8.48 \times 10^5$ [185].

i16. *Poly(isopropyl acrylate)* (I)–*poly(butyl acrylate)* (II), *incompatible:* Phase separation in CHCl$_3$ above 5 gm/dl at 25°C when I had $\bar{M}_v = 1.3 \times 10^6$ and II had $\bar{M}_v = 1.5 \times 10^6$ [185].

i17. *Poly(isopropyl acrylate)* (I)–*poly(tert-butyl acrylate)* (II), *incompatible:* Same as System **i16** but II was not specified [185].

i18. *Poly(isopropyl acrylate)* (I)–*poly(isopropyl methacrylate)* (II), *compatible:* A single T_g for all proportions of I/II by dilatometry; miscible in all proportions in CHCl$_3$ and evaporated films were clear unless the CHCl$_3$ was evaporated rapidly [206].

i19. *Poly(butyl acrylate)* (I)–*poly(sec-butyl acrylate)* (II), *incompatible:* Phase separation in CHCl$_3$ above 5 gm/dl at 25°C when I had $\bar{M}_v = 1.5 \times 10^6$ [185].

i20. *Poly(butyl acrylate)* (I)–*poly(tert-butyl acrylate)* (II), *incompatible:* Same as System **i19** [185].

i21. *Poly(butyl acrylate)* (I)–*poly(methyl methacrylate)* (II), *incompatible:* Phase separation in several solvents by several mixtures, in one of which I was stated to be the *n*-butyl ester [168, 185].

i22. *Poly(n-butyl acrylate)* (I)–*poly(ethyl methacrylate)* (II), *incompatible:* Phase separation in benzene when I had $\bar{M}_w = 3.65 \times 10^5$ or 5.3×10^5, and II had $\bar{M}_w = 1.25 \times 10^5$ or 1.9×10^5 [168].

i23. *Poly(butyl acrylate)* (I)–*poly(butyl methacrylate)* (II), *incompatible:* Phase separation in several solvents for several mixtures, some of which were specified as mixtures of *n*-butyl esters [76, 77, 168].

i24. *Poly(butyl acrylate)* (I)–*poly(isobutyl methacrylate)* (II), *incompatible:* Phase separation in several solvents for several mixtures, in one of which I was specified as the *n*-butyl ester [168, 185].

i25. *Poly(n-butyl acrylate)* (I)–*poly(hexyl methacrylate)* (II), *incompatible:* Same as System **i22** except that II had $\bar{M}_w = 2.1 \times 10^5$ or 2.4×10^5 [168].

i26. *Poly(butyl acrylate)* (I)–*poly[(butyl acrylate)-co-(methyl methacrylate)]* (II), *must be conditionally compatible):* Phase separation in CHCl$_3$ above 5 gm/dl for 1/1 and 1/3 weight ratios of I/II at 25°C when I had $\bar{M}_v = 1.5 \times 10^6$ and II contained 50% butyl acrylate and had $\bar{M}_v \sim 3 \times 10^6$; phase separation at higher concentration than when II was polymethyl methacrylate, $\bar{M}_v = 8.48 \times 10^5$ [185].

i27. *Poly(sec-butyl acrylate)* (I)–*poly(tert-butyl acrylate)* II, *incompatible:* Phase separation in CHCl$_3$ above 5 gm/dl at 25°C [185].

i28. *Poly(acrylic acid)* (I)–*poly(sodium acrylate)* (II), *may be compatible:* No phase separation in water at 25°C when I had $\bar{M}_v = 1.6 \times 10^5$ and II was the sodium salt of I [185].

i29. *Poly(acrylic acid)* (I)–*poly(methacrylic acid)* (II), *almost compatible:* Phase separation in water at 25°C for 20 gm solids/dl but no phase separation at 13 gm solids/dl when I had $\bar{M}_v = 1.6 \times 10^5$ and II had $\bar{M}_v = 3.3 \times 10^4$ or 2.82×10^4; no phase separation for 1/1 weight ratio in DMF up to 40 gm solids/dl or in dimethyl sulfoxide at 20 gm solids/dl; no phase separation in water for the sodium salts of I and II; phase separation in water at 20 gm solids/dl for I with the sodium salt of II or for II with the sodium salt of I but no phase separation at 10 gm solids/dl [185].

i30. *Poly(acrylic acid)* (I)–*poly[(acrylic acid)-co-methacrylic acid)]* (II), *incompatible (must be conditionally compatible):* Phase separation in water at 20 gm solids/dl at 25°C when I had $\bar{M}_v = 1.6 \times 10^5$ and II contained 10% acrylic acid [185].

i31. *Isotactic poly(methyl methacrylate)* (I)–*syndiotactic poly(methyl methacrylate)* (II), *compatible* (complex formation): 1/1 and 1/2 Complexes of I/II were formed [207]. Intrinsic viscosity minimum at 1/2 ratio of I/II indicated complex formation [208]. No phase separation in CHCl$_3$ or in bulk at any proportions; films had a single T_g by dilatometry when I had $\bar{M}_n = 2.97 \times 10^4$ and II had $\bar{M}_n = 9.52 \times 10^4$ [206]. A 1/1 blend from solution gave the T_g's of the two components by dilatometry when I had $\bar{M}_v = 1.07 \times 10^6$ and II was "conventional"

PMMA (probably 75% syndiotactic), $\overline{M}_v = 1.96 \times 10^6$ [209].

i32. *Poly(methyl methacrylate)* (I)–*poly(ethyl methacrylate)* (II), *incompatible:* Phase separation in several solvents for several mixtures [168, 185].

i33. *Poly(methyl methacrylate)* (I)–*poly(propyl methacrylate)* (II), *incompatible:* Stated incompatible for I/II: 1/9 to 9/1 in solutions and films when I was Piacryl G [16].

i34. *Poly(methyl methacrylate)* (I)–*poly(butyl methacrylate)* (II), *incompatible:* Phase separation in several solvents for several mixtures [16, 76, 77, 168]; two phases in microphotograph and two dynamic mechanical maxima [140]; II specified as *n*-butyl ester in some cases [140, 168].

i35. *Poly(methyl methacrylate)* (I)–*poly(isobutyl methacrylate)* (II), *incompatible:* Phase separation in several solvents for several mixtures [168, 185].

i36. *Poly(methyl methacrylate)* (I)–*poly(hexyl methacrylate)* (II), *incompatible:* Phase separation in a number of solvents [16, 168] and in films [16] for several mixtures.

i37. *Poly(methyl methacrylate)* (I)–*poly(octyl methacrylate)* (II), *incompatible:* Same as System **i33** [16].

i38. *Poly(methyl methacrylate)* (I)–*poly(benzyl methacrylate)* (II), *incompatible:* Phase separation in CHCl$_3$ above 5 gm solids/dl at 25°C when I had $\overline{M}_v = 8.48 \times 10^5$ and II had $[\eta] = 0.629$ in benzene at 30°C [185].

i39. *Poly(methyl methacrylate)* (I)–*poly(isobornyl methacrylate)* (II), *incompatible:* Same as System **i38** except that II was unspecified [185].

i40. *Poly(methyl methacrylate)* (I)–*polymethacrylonitrile* (II), *incompatible:* Phase separation in acetone at 15% solids for 1/1 weight ratio at 25°C when I had $[\eta] = 0.28$ in acetone and II had $\eta_{sp} = 0.35$ or 0.034 at 0.1% in acetone [124].

i.41. *Poly(methyl methacrylate)* (I)–*poly[(ethyl acrylate)-co-methyl methacrylate]* (II), *must be conditionally compatible:* Phase separation in CHCl$_3$ above 5 gm solids/dl at 25°C for 1/1 and 1/3 weight ratio of I/II when I had $\overline{M}_v = 8.48 \times 10^5$ and II contained 50% MMA, $\overline{M}_v \sim 2 \times 10^6$; phase separation occurred at higher concentrations than when II was poly(ethyl acrylate), $\overline{M}_v = 1.2 \times 10^6$ [185].

i42. *Poly(methyl methacrylate)* (I)–*poly[(butyl acrylate)-co-(methyl methacrylate)]* (II), *must be conditionally compatible:* Phase separation in CHCl$_3$ above 5 gm solids/dl for 1/1 and 1/3 weight ratio of I/II at 25°C when I had $\overline{M}_v = 8.48 \times 10^5$ and II contained 50% MMA, $\overline{M}_v \sim 3 \times 10^6$; phase separation occurred at higher concentrations than when II was poly-(butyl acrylate), $\overline{M}_v = 1.5 \times 10^6$ [185].

i43. *Poly(ethyl methacrylate)* (I)–*poly(n-butyl methacrylate)* (II), *incompatible:* Phase separation in benzene above 15–20 gm/100 ml when I had $\overline{M}_w = 1.25 \times 10^5$ or 1.9×10^5, and II had $\overline{M}_w = 1.36 \times 10^5$ or 2.25×10^5 [168].

i44. *Poly(ethyl methacrylate)* (I)–*poly(isobutyl methacrylate)* (II), *incompatible:* Same as System **i43** except that II had $\overline{M}_w = 1.27 \times 10^5$ or 1.9×10^5 [168].

i45. *Poly(ethyl methacrylate)* (I)–*poly(hexyl methacrylate)* (II), *incompatible:* Phase separation in benzene above 7–12 gm/100 ml when I was as in System **i43** and II had $\overline{M}_w = 2.1 \times 10^5$ or 2.4×10^5 [168].

i46. *Poly(propyl methacrylate)* (I)–*poly(butyl methacrylate)* (II), *may be conditionally compatible:* Stated compatible for I/II: 1/9 to 9/1 in solutions and films except in acetone [16].

i47. *Poly(propyl methacrylate)* (I)–*poly(hexyl methacrylate)* (II), *may be conditionally compatible:* Same as System **i46** [16].

i48. *Poly(propyl methacrylate)* (I)–*poly(octyl methacrylate)* (II), *incompatible:* Stated incompatible for I/II: 1/9 to 9/1 in solutions and films [16].

i49. *Poly(n-butyl methacrylate)* (I)–*poly(isobutyl methacrylate)* (II), *almost compatible:* Phase separation in benzene above 80–85 gm/100 ml when I had $\overline{M}_w = 1.36 \times 10^5$ or 2.25×10^5 and II had $\overline{M}_w = 1.27 \times 10^5$ or 1.9×10^5 [168].

i50. *Poly(butyl methacrylate)* (I)–*poly(hexyl methacrylate)* (II), *may be conditionally compatible (depending on isomers, possibly):* Stated compatible for I/II: 1/9 to 9/1 in solutions and films [16]. Phase separation in benzene above 18–27 gm/100 ml when I was the *n*-butyl ester with $\overline{M}_w = 1.36 \times 10^5$ or 2.25×10^5, or the isobutyl ester with $\overline{M}_w = 1.27 \times 10^5$ or 1.9×10^5, and II had $\overline{M}_w = 2.1 \times 10^5$ or 2.4×10^5 [168].

i51. *Poly(butyl methacrylate)* (I)–*poly(hexyl methacrylate)* (II), *incompatible:* Same as System **i48** [16].

i52. *Poly(hexyl methacrylate)* (I)–*poly(octyl methacrylate)* (II), *may be compatible:* Same as System **i46** [16].

i53. *Poly(benzyl methacrylate)* (I)–*poly(isobornyl methacrylate)* (II), *incompatible:* Phase separation in CHCl₃ above 5 gm solids/dl at 25°C when I had $[\eta] = 0.629$ in benzene at 30°C [185].

i54. *Poly(methacrylic acid)* (I)–*poly(sodium methacrylate)* (II), *may be compatible:* No phase separation in water at 25°C when I had $\overline{M}_v = 3.3 \times 10^4$ and II was its sodium salt [185].

i55. *Poly(methacrylic acid)* (I)–*poly[(methacrylic acid)-co-acrylic acid)]* (II), *may be compatible* (must be at least conditionally compatible): No phase separation in water at 20 gm solids/dl at 25°C when I had $\overline{M}_v = 2.82 \times 10^4$ and II contained 90% methylacrylic acid [185].

J. Acrylic Polymers and Other Polymers

j1. *Poly(methyl acrylate)* (I)–*polypropylene* (II), *ambiguous:* Formed an opaque film in 1/1 ratio but transparent when silica was added [109].

j2. *Poly(methyl acrylate)* (I)–*poly(tert-butyl vinyl ether)* (II), *incompatible:* Phase separation in CHCl₃ above 5 gm solids/dl at 25°C when I had $\overline{M}_v = 6.0 \times 10^5$ and II had $[\eta] = 0.551$ in benzene at 30°C [185].

j3. *Poly(methyl acrylate)* (I)–*chlorinated polyisoprene* (II), *incompatible:* Stated incompatible when II was chlorinated rubber [83].

j4. *Poly(ethyl acrylate)* (I)–*chlorinated polyisoprene* (II), *ambiguous:* Stated compatible when II was chlorinated rubber [83]. Stated incompatible for I/II: 1/9 to 9/1 in solutions and films when I was Acrylit L and II was chlorinated rubber [16].

j5. *Poly(ethyl acrylate)* (I)–*chlorinated poly(vinyl chloride)* (II), *incompatible:* Solutions and films homogeneous and transparent from ethyl acetate but phase separated from CHl₃ when I was Schkopau L370 and II was Ekalit PC, 61.6% Cl [36]. Stated incompatible for I/II: 1/4 to 9/1 but compatible for 1/9 in solutions and films when I as Acrylit L [16].

j6. *Poly(ethyl acrylate)* (I)–*chlorinated polyethylene* (II), *incompatible:* Stated incompatible for I/II: 1/9 to 9/1 in solutions and films when I was Acrylit L and II had 30% Cl [16].

j7. *Poly(ethyl acrylate)* (I)–*chlorosulfonated polyethylene* (II), *incompatible:* Same as System **j6** except that II had 29.5% Cl, 1.6% S [16].

j8. *Poly(ethyl acrylate)* (I)–*poly[(ethyl acrylate)-co-styrene]* (II), *must be conditionally compatible:* Phase separation in CHCl₃ above 5 gm solids/dl at 25°C for 1/1 and 1/3 ratios of I/II, but no phase separation at 3 gm solids/dl when I had $\overline{M}_v = 1.2 \times 10^6$ and II contained 50% ethyl acrylate, $\overline{M}_v \sim 10^6$; phase separation at higher concentration than when II was polystyrene, $\overline{M}_v = 5.0 \times 10^5$ [185].

j9. *Poly(ethyl acrylate)* (I)–*poly[(vinyl chloride)-co-(vinyl acetate)]* (II), *almost compatible (probably):* Stated compatible for I/II: 1/9, 4/1, and 9/1 in solutions and films, incompatible for 1/1, cloudy film for 4/1, when I was Acrylit L and II was Vinalit MPS [16].

j10. *Poly(ethyl acrylate)* (I)–*poly[styrene-co-(methyl methacrylate)]* (II), *incompatible:* Three mechanical damping maxima and domain structure by electron microscopy for inter-penetrating networks when I had 1% butadiene comonomer and II had styrene–methyl meth-acrylate: 2/1 to 1/3 by weight [186].

j11. *Poly(n-butyl acrylate)* (I)–*chlorinated polyisoprene* (II), *may be compatible:* Stated to be compatible when II was chlorinated rubber [83].

j12. *Poly(acrylic acid)* (I)–*poly(vinyl alcohol)* (II), *incompatible:* Phase separation when I had $\bar{M}_v = 1.6 \times 10^5$ [185].

j13. *Poly(acrylic acid)* (I)–*poly(ethylene oxide)* (II), *conditionally compatible* (complex formation): A 1/1 complex precipitated below pH 3.8 and above pH 12 but remained in solution at intermediate pH, when I was either commercial or prepared in the laboratory and II was Carbowax 6000, MW = 5500, or other products up to Polyox, MW = 10^6 [210] Precipitated in water but formed a clear film with only one transition when the mixture contained $\geqslant 50\%$ I; when the mixture contained 25% I, crystals of II coexisted with the amorphous phase [211].

j14. *Poly(acrylic acid)* (I)–*poly(vinylpyrrolidone)* (II), *ambiguous* (complex formation possible): Precipitated in water solution when a 2% solution of I was mixed with a 1, 1.5, or 2% solution of II [212].

j15. *Sodium polyacrylate* (I)–*poly(4-vinyl-N-n-butylpyridonium bromide)* (II), *ambiguous* (complex formation): Precipitation observed by light scattering for a solution 10^{-8} M in poly-electrolyte [213].

j16. *Poly(methyl methacrylate)* (I)–*polyisobutene* (II), *incompatible:* Stated incompatible for I/II: 4/1 to 1/4 in solutions and films when I was Piacryl G and II was Oppanol B-100 [16].

j17. *Poly(methyl methacrylate)* (I)–*polychloroprene* (II), *incompatible:* Same as System **j16** except that II was Neoprene AC [16].

j18. *Poly(methyl methacrylate)* (I)–*chlorinated polyethylene* (II), *incompatible:* Stated in-compatible for I/II: 1/9 to 9/1 in solutions and films when I was Piacryl G and II had 30% Cl [16].

j19. *Poly(methyl methacrylate)* (I)–*chlorosulfonated polyethylene* (II), *incompatible:* Non-homogeneous solution and turbid film when I was from Mitsubishi Rayon Co. and II was Showa Neoprene Hyperon 30 [37].

j20. *Poly(methyl methacrylate)* (I)–*chlorinated polyisoprene* (II), *incompatible:* Same as System **j18** except that II was chlorinated rubber [16]. Stated partially compatible when II was chlorinated rubber [83].

j21. *Poly(methyl methacrylate)* (I)–*chlorinated poly(vinyl chloride)* (II), *may be compatible or conditionally compatible:* Solutions and films homogeneous and transparent from ethyl acetate, acetone, DMF, bromobenzene, benzyl acetate, and tetralin but phase separated from CHCl₃, ethyl chloroacetate, and toluene; two T_g's for I/II: 2/1 film from ethyl acetate [36]. Stated compatible for I/II: 1/9 to 9/1 in solutions and films when I was Piacryl G and II was Vinoflex PC [16].

j22. *Poly(methyl methacrylate)* (I)–*poly(vinylidene chloride)* (II), *incompatible:* Same as System **j19** except that II was Asahi Dow Co. EX 5701 [37].

j23. *Poly(methyl methacrylate)* (I)–*poly(vinylidene fluoride)* (II), *conditionally compatible:* Transparent, with a single T_g when blended in the melt, but II crystallized when II/I by weight exceeded 1/1 [214]. Single T_g by DTA and dynamic mechanical properties for wt % II $\leqslant 50\%$; crystallinity present for wt % II $\geqslant 60\%$ with single T_g in amorphous regions when I was Rohm & Haas Plexiglas V(811)–100, and II was Pennwalt Kynar 301 [215]. Films clear with a single T_g by DTA and dilatometry for 15–70% II, and >35% II had a crystalline phase, when I was American Cyanamide Acrylite H-12 and II was Pennwalt Kynar 401 [216].

j24. *Poly(methyl methacrylate)* (I)–*poly(2-vinylpyridine)* (II), *incompatible:* Precipitated from CH₂Cl₂ when I was du Pont Lucite 140, reduced viscosity 0.415 dl/gm at 0.2 gm/dl in CH₂Cl₂ [177].

j25. *Poly(methyl methacrylate)* (I)–*poly(α-methylstyrene)* (II), *may be compatible:* Film transparent for 1/1 mixture when I had $\overline{M}_w = 1.6 \times 10^6$, $\overline{M}_w/\overline{M}_n \sim 2$, and II had $\overline{M}_w = 5.4 \times 10^5$, $\overline{M}_w/\overline{M}_n < 1.06$ [217].

j26. *Poly(methyl methacrylate)* (I)–*polychlorostyrene* (II), *incompatible:* Same as System **j24** except that II was mixed isomers, reduced viscosity 0.639 dl/gm at 0.2 gm/dl in CH_2Cl_2 [177],

j27. *Poly(methyl methacrylate)* (I)–*polydichlorostyrene* (II), *incompatible:* Same as System **j24** except that II was mixed isomers, reduced viscosity 0.647 dl/gm at 0.2 gm/dl in CH_2Cl_2 [177].

j28. *Poly(methyl methacrylate)* (I)–*poly(vinyl pyrrolidone)* (II), *incompatible:* Same as System **j19** except that II was General Aniline and Film K-30 [37].

j29. *Poly(methyl methacrylate)* (I)–*poly(vinyl acetal)* (II), *incompatible:* Phase separation in acetone above 2.2% solids at 16–18°C when I had $\overline{M}_n > 2 \times 10^6$, $[\eta] = 3.65$ in acetone, and II had $\overline{M}_n = 3.8 \times 10^4$, $[\eta] = 0.75$ in acetone [78].

j30. *Poly(methyl methacrylate)* (I)–*poly(vinyl butyral)* (II), *incompatible:* Same as System **j19** except that II was Shekisui Chemical Co. BM-2 [37].

j31. *Poly(methyl methacrylate)* (I)–*polycarbonate* (II), *incompatible:* Phase separation in cyclohexanone for 1/1 mixture, and film was not clear when I was Lucite 147 or 148, du Pont, and II was General Electric Co. Lexan 125 or Mobay Chemical Co. Merlon M-50 [79].

j32. *Poly(methyl methacrylate)* (I)–*poly(oxycarbonyloxy-1,4-phenyleneisopropylidene-1,4-phenylene)* (II), *incompatible:* Same as System **j24** except that II was General Electric Lexan 101, reduced viscosity 0.855 dl/gm at 0.2 gm/dl in CH_2Cl_2 [177].

j33. *Poly(methyl methacrylate)* (I)–*poly[oxycarbonyloxy(2,2,4,4-tetramethyl-1,3-cyclobutylene)]* (II), *incompatible:* Same as System **j24** except that II had reduced viscosity 1.17 dl/gm at 0.2 gm/dl in CH_2Cl_2 [177].

j34. *Poly(methyl methacrylate)* (I)–*polyepichlorohydrin* (II), *incompatible:* Same as System **j31** except that II was B. F. Goodrich Hydrin 100 [79].

j35. *Poly(methyl methacrylate)* (I)–*poly(2,6-dimethyl-1,4-phenylene ether)* (II), *incompatible:* Same as System **j31** except that II was General Electric Co. PPO [79].

j36. *Poly(methyl methacrylate* (I)–*polysulfone* (II), *incompatible:* Same as System **j31** except that II was Union Carbide polysulfone [79].

j37. *Poly(methyl methacrylate)* (I)–*polyurethane* (II), *incompatible:* Phase separation in several solutions and films for several mixtures involving polyester and polyether urethanes [37, 79].

j38. *Poly(methyl methacrylate)* (I)–*polyester* (II), *incompatible:* Same as System **j19** except that II was Toyobo Co. Ester Resin 20 [37].

j39. *Poly(methyl methacrylate)* (I)–*poly[(methyl methacrylate)-co-styrene]* (II), *should be conditionally compatible:* No phase separation in cyclohexanone for 1/1 mixture, and film was clear when I was Lucite 147 or 148, du Pont, and II was Zerlon150, Dow Chemical Co. [79].

j40. *Poly(methyl methacrylate)* (I)–*poly[ethylene-co-(vinyl acetate)]* (II), *incompatible:* Same as System **j19** except that II was Nippon Goshei Co. Soalex R-CH or R-FH [37].

j41. *Poly(methyl methacrylate)* (I)–*poly(butadiene-co-styrene)* (II), *incompatible:* Same as System **j19** except that II was Nippon Rubber Co. Hycar 2057 [37]. Two dynamic mechanical and dilatometric T_g's when I was from Mitsubishi Rayon Co. and II was Japanese Synthetic Rubber Co. SBR, 24.5% styrene [118].

j42. *Poly(methyl methacrylate)* (I)–*poly(butadiene-co-acrylonitrile)* (II), *incompatible:* Phase separation for many mixtures in many solvents and films [16, 37, 79].

j43. *Poly(methyl methacrylate)* (I)–*poly[(vinyl chloride)-co-(vinyl acetate)]* (II), *conditionally compatible:* Same as System **j31** except that II was B. F. Goodrich Geon 440X24 [79]. Solutions and films homogeneous and transparent from ethyl acetate but phase separated from $CHCl_3$ when I was Piacryl G and II was Schkopau MPS-SP [36]. Nonhomogeneous solution and turbid film when I was from Mitsubishi Rayon Co. and II was Shekisui Chemical Co. Esulex C or CL; homogeneous solution and transparent film when II was Esulex CH–1 or

Denki Chemical Co. Denka Vinyl 1000 AK [37]. Stated compatible for I/II: 1/9 to 9/1 in solutions and films when I was Piacryl G and II was Vinalit MPS [16].

j44. *Poly(methyl methacrylate)* (I)–*poly[(vinyl chloride)-co-(vinyl propionate)]* (II), *incompatible:* Same as System **j19** except that II was Ryuron QS-430 [37].

j45. *Poly(methyl methacrylate)* (I)–*poly[(vinyl chloride)-co-(vinyl acetate)-co-(vinyl alcohol)]* (II); *may be conditionally compatible:* Homogeneous solution and transparent film when I was from Mitsubishi Rayon Co. and II was Shekisui Chemical Co. Esulex A [37].

j46. *Poly(methyl methacrylate)* (I)–*poly[(vinyl chloride)-co-(vinyl acetate)-co-(maleic acid)]* (II), *incompatible:* Same as System **j19** except that II was Shekisui Chemical Co. Esulex M [37].

j47. *Poly(methyl methacrylate)* (I)–*poly[(vinyl chloride)-co-(vinylidene chloride)-co-(acrylic acid ester)]* (II), *incompatible:* Same as System **j19** except that II was Towa Gousei Co. Aron 321 [37].

j48. *Poly(methyl methacrylate)* (I)–*poly[(vinyl acetate)-co-(N-vinylpyrrolidone)]* (II), *may be conditionally compatible:* Same as System **j45** except that II was General Aniline & Film S-630 [37].

j49. *Poly(methyl methacrylate)* (I)–*poly(styrene-co-acrylonitrile)* (II), *conditionally compatible:* Phase separation in cyclohexanone for 1/1 mixture and film was not clear when I was Lucite 147 or 148, du Pont, and II was Dow Chemical Co. Tyril 767 [79]. Homogeneous clear films with no evidence for phase separation in a year for I/II: 1/4 to 4/1 when I was du Pont Elvacite 2009 and II was Union Carbide RMD 4500 [218]. Two T_g's, opaque samples, and phases of order 1 μm in electron micrographs of 1/1 mixtures with 6.5 and 8% by weight acrylonitrile in II when I was Degalan P8(LP51/03), $\bar{M}_v = 1.1$–1.2×10^5, 0.5–1.1% monomer, and II were low conversion, $\bar{M}_v = 1.5$–2.5×10^5, $<0.2\%$ monomer; single T_g, transparent films, and no phases in electron micrographs for 9–27% by weight acrylonitrile in II; single T_g for $>28\%$ acrylonitrile in II [219]. LCST about 150°C when I was du Pont Lucite 140, $M_n = 4.56 \times 10^4$, $M_w = 9.2 \times 10^4$, and II was Union Carbide RMD-4511, 28% acrylonitrile, $\bar{M}_n = 8.86 \times 10^4$, $\bar{M}_w = 2.23 \times 10^5$ [220]. Demixing for copolymers with $<9\%$ and $>29\%$ acrylonitrile and compatibility at intermediate compositions [221]. Soluble in CH_2Cl_2 when II contained 12.7% and 18.0% acrylonitrile but precipitated when II had 6.8% or 20.6–42% acrylonitrile when I was as in System **j24** [177].

j50. *Poly(methyl methacrylate)* (I)–*poly[styrene-co-(maleic acid)]* (II), *incompatible:* Homogeneous solution and transparent film but two T_g's by DTA when I was from Mitsubishi Rayon Co. and II was Daidou Kogyo Styrite CM-2 but nonhomogeneous solution and turbid film when I was Styrite CM-3 [37].

j51. *Poly(methyl methacrylate)* (I)–*poly[styrene-co-(maleic acid ester)]* (II), *may be compatible:* Same as System **j45** except that II was Daidou Kogyo Styrite HS-2 [37].

j52. *Poly(methyl methacrylate)* (I)–*poly[(methyl vinyl ether)-co-(maleic anhydride)]* (II), *may be conditionally compatible:* No phase separation in cyclohexane for 1/1 mixture, and film was clear when I was Lucite 147 or 148 du Pont, and II was Gantrez AN-169, B. F. Goodrich [79].

j53. *Poly(methyl methacrylate)* (I)–*poly[epichlorohydrin-co-(ethylene oxide)]* (II), *incompatible:* No phase separation in cyclohexanone for 1/1 mixture, but film was not clear when I was Lucite 147 or 148, du Pont, and II was Hydrin 200, B. F. Goodrich [79].

j54. *Poly(methyl methacrylate)* (I)–*indene resin* (II), *may be compatible:* Same as System **j45** except that II was Fuji Iron Products Co. VM-1/2 [37].

j55. *Poly(methyl methacrylate)* (I)–*miscellaneous resins* (II), *incompatible:* Same as System **j19** except that II was Petroleum Resin, Nippon Petroleum Co. Nitsuseki Neopolymer 150, or Cumarone Resin, Mitsubishi TG [37].

j56. *Poly(ethyl methacrylate)* (I)–*chlorinated polyisoprene* (II), *may be compatible:* Stated compatible when II was chlorinated rubber [83].

j57. *Poly(ethyl methacrylate)* (I)–*poly(vinylidene fluoride)* (II), *conditionally compatible:*

Transparent, with a single T_g when blended in the melt, but II crystallized when II/I by weight exceeded $1/2$ [214]. Blends clear if II $\leqslant 30\%$ and translucent if II $\geqslant 40\%$ because of crystallinity of II; single T_g by DSC and dynamic mechanical properties for 10–80% II plus mp for II $\geqslant 40\%$, when I had $\overline{M}_v = 7.60 \times 10^5$ and II was Pennwalt Kynar 301 [222]. Amorphous up tp 50% II; clear films and single T_g by DTA and dilatometry for 10–40% II when II was Pennwalt Kynar 401 [216].

j58. *Poly(n-butyl methacrylate)* (I)–*chlorinated polyisoprene* (II), *may be compatible:* Stated compatible when II was chlorinated rubber [83].

j59. *Poly(methacrylic acid)*(I)–*poly(vinyl alcohol)* (II), *almost compatible* (complex formation): Gels formed when I/II weight ratio was 9/1, 7/3, 5/5 at room temperature; no gels formed near 0°C; gels formed at lower ratios of I/II at 80°C but disappeared at room temperature; gels dissolved in alkaline solution and in acetone or dioxane [223]. Gels precipitated in water at 7% solids at room temperature when I had MW = 87,000 and II had MW = 84,000; completely miscible below -2°C [224]. Gels formed in solution; gels had fibrillar structure, and spectra indicated H bonds when I had MW = 87,000 and II had MW = 83,000 [225]. Phase separation when I had $\overline{M}_v = 3.3 \times 10^4$ [185].

j60. *Poly(methacrylic acid)* (I)–*poly(ethylene oxide)* (II), *ambiguous* (complex formation): Associates looked like I/II ratio was 1/3 when II ranged from Carbowax 6000, MW = 5500, to Polyox, MW = 10^6 [210].

K. Miscellaneous Homopolymers

k1. *Polypropene* (I)–*poly(1-hexene)* (II), *incompatible:* Two dynamic mechanical loss peaks in amorphous region when II had $\overline{M}_n = 1.6 \times 10^5$; samples partially crystalline for 20–50% II [226].

k2. *Polyisobutene*(I)–*polychloroprene*(II), *incompatible:* Stated incompatible for I/II; 4/1 to 1/4 in solutions and films when I was Oppanol B100 and II was Neoprene AC [16].

k3. *Polyisobutene* (I)–*chlorinated poly(vinyl chloride)* (II), *incompatible:* Same as System k2 except that II was Vinoflex PC [16].

k4. *Polyisobutene* (I)–*poly(dimethylsiloxane)* (II), *incompatible:* Phase separation in solution [156, 202] and also in the absence of solvent even when I and II were both oligomers [227].

k5. *Polyisobutene* (I)–*acetylene resin* (II), *compatible or ambiguous:* 1/1 by volume solution in xylene at 50% solids was dark but transparent; transparent in bulk [228].

k6. *Polychloroprene* (I)–*chlorinated poly(vinyl chloride)* (II), *incompatible:* Same as System k2 except that I was Neoprene AC and II was Vinoflex PC [16].

k7. *Polychloroprene* (I)–*poly(1,1,2-trichloro-1,3-butadiene)* (II), *ambiguous:* Authors state single dielectric maximum for I/II: 4/1 to 1/4 but looks like two to me; they say compatible for some compositions [229].

k8. *Chlorinated polyisoprene* (I)–*poly(1,1,2-trichloro-1,3-butadiene)* (II), *incompatible:* Phase separation in CCl_4 when I was 96% 1,4-*cis*, and II was emulsion polymer [230]. Dielectric loss peaks for both components for I/II: 1/4 to 4/1 when I had MW = 8.0×10^4 and II had MW = 2.5×10^5; authors state compatible for $>40\%$ I [229].

k9. *Chlorinated polyisoprene* (I)–*chlorinated poly(vinyl chloride)* (II), *incompatible or ambiguous:* Stated compatible for I/II: 1/9 to 1/4 but incompatible for 1/1 to 9/1 in solutions and films [16].

k10. *Chlorinated polyisoprene* (I)–*chlorinated polyethylene* (II), *almost compatible:* Stated compatible for I/II: 1/9 to 9/1 except incompatible for 1/1 in solutions and films when II had 30% Cl [16].

k11. *Chlorinated polyisoprene* (I)–*chlorosulfonated polyethylene* (II), *incompatible:* Stated incompatible for I/II: 1/9 to 9/1 in solutions and films when II had 29.5% Cl, 1.6% S [16].

k12. *Chlorinated poly(vinyl chloride)* (I)–*chlorinated polyethylene* (II), *almost compatible:* Stated compatible for I/II: 1/9 to 9/1 except cloudy film at 1/1 in solutions and films [16].

k13. *Chlorinated poly(vinyl chloride)* (I)–*chlorosulfonated polyethylene* (II), *incompatible:* Stated compatible for I/II: 1/9 but incompatible for 1/4 to 9/1 in solutions and films when II had 29.5% Cl, 1.6% S [16].

k14. *Poly(vinylidene chloride)* (I)–*poly(vinyl butyral)* (II), *incompatible:* Nonhomogeneous solution and turbid film when I was Asahi Dow Co. EX 5701 and II was Shekisui Chemical Co. BM-2 [37].

k15. *Poly(vinylidene chloride)* (I)–*chlorosulfonated polyethylene* (II), *incompatible:* Same as System **k14** except that II was Showa Neoprene Hyperon 30 [37].

k16. *Poly(vinylidene chloride)* (I)–*polyurethane* (II), *may be compatible:* Homogeneous solution and transparent film when I was Asahi Dow Co. EX5701 and II was Nippon Rubber Co. 5740X1 or Hodogaya Co. Pellet 22S [37].

k17. *Poly(vinylidene chloride)* (I)–*polyester* (II), *incompatible:* Same as System **k14** except that II was Toyobo Co. Ester Resin 20 [37].

k18. *Poly(vinylidene chloride)* (I)–*indene resin* (II), *may be compatible:* Same as System **k16** except that II was Fuji Iron Products Co. VM-1/2 [37].

k19. *Poly(vinylidene chloride)* (I)–*miscellaneous resins* (II), *incompatible:* Same as System **k14** except that II was Petroleum Resin, Nippon Petroleum Chemical Co. Nitsuseki Neopolymer 150, or Cumarone Resin, Mitsubishi TG [37].

k20. *Chlorosulfonated polyethylene* (I)–*polyurethane* (II), *incompatible:* Nonhomogeneous solution and turbid film when I was Showa Neoprene Hyperon 30, and II was Nippon Rubber Co. 5740X1 or Hodogaya Co. Pellet 22S [37].

k21. *Chlorosulfonated polyethylene* (I)–*polyester* (II), *incompatible:* Same as System **k20** except that II was Toyobo Co. Ester Resin 20 [37].

k22. *Chlorosulfonated polyethylene* (I)–*petroleum resin* (II), *incompatible:* Same as System **k20** except that II was Nippon Petroleum Chemical Co. Nitsuseki Neopolymer 150 [37].

k23. *Chlorosulfonated polyethylene* (I)–*poly(vinyl butyral)* (II), *incompatible:* Same as System **k20** except that II was Shekisui Chemical Co. BM-2 [37].

k24. *Poly(vinyl fluoride)* (I)–*poly(vinylidene fluoride)* (II), *incompatible:* Two or three T_g's found by dynamic mechanical measurements for 25, 50, or 75% II [231].

k25. *Poly(2-vinylpyridine)* (I)–*poly(chlorostyrene)* (II), *incompatible:* Precipitated from CH_2Cl_2 when I had reduced viscosity 1.095 dl/gm at 0.2 gm/dl in CH_2Cl_2 and II was mixed isomers, reduced viscosity 0.639 dl/gm at 0.2 gm/dl in CH_2Cl_2 [177].

k26. *Poly(2-vinylpyridine)* (I)–*poly(dichlorostyrene)* (II), *incompatible:* Same as System **k25** except that II was mixed isomers, reduced viscosity 0.647 dl/gm at 0.2 gm/dl in CH_2Cl_2 [177].

k27. *Poly(2-vinylpyridine)* (I)–*poly(oxycarbonyloxy-1,4-phenyleneisopropylidene-1,4-phenylene)* (II), *incompatible:* Same as System **k25** except that II was General Electric Lexan 101 reduced viscosity 0.855 dl/gm at 0.2 gm/dl in CH_2Cl_2 [177].

k28. *Poly(2-vinylpyridine)* (I)–*poly[oxycarbonyloxy(2,2,4,4-tetramethyl-1,3-cyclobutylene)]* (II), *incompatible:* Same as System **k25** except that II had reduced viscosity 1.17 dl/gm at 0.2 gm/dl in CH_2Cl_2 [177].

k29. *Poly(vinylpyrrolidone)* (I)–*poly(vinyl butyral)* (II), *may be compatible:* Homogeneous solution and transparent film when I was General Aniline & Film K-30 and II was Shekisui Chemical Co. BM-2 [37].

k30. *Poly(vinyl butyral)* (I)–*polyurethane* (II), *incompatible:* Nonhomogeneous solution and turbid film when I was Shekisui Chemical Co. BM-2 and II was Nippon Rubber Co. 5740X1 or Hodogaya Co. Pellet 22S [37].

k31. *Poly(vinyl butyral)* (I)–*polyester* (II), *incompatible:* Same as System **k30** except that II was Toyobo Co. Ester Resin 20 [37].

k32. *Poly(vinyl butyral)* (I)–*miscellaneous resins* (II), *incompatible:* Same as System **k30** except that II was Petroleum Resin, Nippon Petroleum Chemical Co. Nitsuseki Neopolymer 150, or Cumarone Resin, Mitsubishi TG, or Indene Resin, Fuji Iron Products Co. VM-1/2 [37].

k33. *Poly(vinylbiphenyl)*(I)–*poly(ethylene oxide)*(II), *incompatible:* A1/1 ratio blend gave two transitions when I had $\bar{M}_w = 6.8 \times 10^4$; II was crystalline [232].

k34. *Poly(vinylnaphthalene)* (I)–*poly(ethylene oxide)* (II), *conditionally compatible:* Blends containing 25, 40, or 50% I gave two transitions when I had $\bar{M}_w \sim 9.5 \times 10^5$ or $\bar{M}_w = 4 \times 10^5$ [232]. Amorphous region contained 46% II and II crystallized for 50% II and 75% II when II was Dow Chemical Co. Polyglycol E4000 or Union Carbide WSR-35 and I had MW = 5.1×10^5 or 7.2×10^5; 25% II blends were amorphous; postulated I/II: 1/3 (repeat units) complex [193, 233].

k35. *Poly(α-methylstyrene)* (I)–*poly(2,6-dimethyl-1,4-phenylene ether)* (II), *compatible:* Claimed a single T_g and heat distortion temperature varying smoothly between those of I and II for various compositions [195]. Singly dynamic mechanical T_g when I was General Electric PPO and II had $\bar{M}_w = 3.1 \times 10^5$ [182].

k36. *Poly(o-methylstyrene)* (I)–*poly(m-methylstyrene)* (II), *incompatible:* Phase separation in CHCl₃ at 20% solids for 1/1 weight ratio at 25°C when I had $\eta_{sp} = 0.15$ at 0.1% in benzene, and II and $\eta_{sp} = 0.27$ at 0.1% in benzene [124].

k37. *Poly(o-methylstyrene)* (I)–*poly(p-methylstyrene)* (II), *incompatible:* Same as System **k36** except that II had $\eta_{sp} = 0.17$ in benzene [124].

k38. *Poly(m-methylstyrene)* (I)–*poly(p-methylstyrene)* (II), *almost compatible:* No phase separation in CHCl₃ at any concentration for 1/1 weight ratio at 25°C when I had $\eta_{sp} = 0.27$ at 0.1% in benzene, and II had $\eta_{sp} = 0.17$ at 0.1% in benzene [124]. Phase separation above 0.40 gm solids/cm³ CHCl₃ for 1/1 weight ratio when I had $[\eta] = 1.2$ in CHCl₃ and II had $[\eta] = 1.9$ in CHCl₃ [184].

k39. *Poly(p-methylstyrene)* (I)–*poly(p-chlorostyrene)* (II) *incompatible:* Phase separation in benzene for several mixtures [124, 184].

k40. *Poly(p-chlorostyrene)*(I)–*poly(p-methoxystyrene)*(II), *incompatible:* Phase separation in benzene at 2.8 volume % solids for 1/1 volume ratio when I had $[\eta] = 0.4$ in benzene, and II had $[\eta] = 0.2$ in benzene [184].

k41. *Poly(p-chlorostyrene)* (I)–*poly(2,6-dimethyl-1,4-phenylene ether)* (II), *incompatible:* Blends hazy [35] and two T_g's by a variety of measurements [35, 234].

k42. *Polychlorostyrene* (I)–*polydichlorostyrene* (II), *incompatible:* Precipitated from CH_2Cl_2 when I was mixed isomers, reduced viscosity 0.639 dl/gm at 0.2 gm/dl in CH_2Cl_2, and II was mixed isomers, reduced viscosity 0.647 dl/gm at 0.2 gm/dl in CH_2Cl_2 [177].

k43. *Polychlorostyrene* (I)–*poly(oxycarbonyloxy-1,4-phenyleneisopropylidene-1,4-phenylene)* (II), *incompatible:* Same as System **k42** except that II was General Electric Lexan 101 polycarbonate, reduced viscosity 0.855 dl/gm at 0.2 gm/dl in CH_2Cl_2 [177].

k44. *Polychlorostyrene* (I)–*poly[oxycarbonyloxy(2,2,4,4-tetramethyl-1,3-cyclobutylene)]* (II), *incompatible:* Same as System **k42** except that II had reduced viscosity 1.17 dl/gm at 0.2 gm/dl in CH_2Cl_2 [177].

k45. *Polydichlorostyrene* (I)–*poly(oxycarbonyloxy-1,4-phenyleneisopropylidene-1,4-phenylene)*(II), *incompatible:* Same as System **k43** except that I was the same as II in System **k42** [177].

k46. *Polydichlorostyrene* (I)–*poly[oxycarbonyloxy(2,2,4,4-tetramethyl-1,3-cyclobutylene)]* (II), *incompatible:* Same as System **k44** except that I was the same as II in System **k42** [177].

k47. *Polyacrylonitrile* (I)–*partially hydrolyzed polyacrylonitrile* (II), *ambiguous* (depends on hydrolysis products): Miscible in solution and appeared homogeneous in fibers although I and II crystallized separately when II contained 21.9% N as amide and imide groups; not very miscible in solution when II contained acid groups [235].

k48. *Polyepichlorohydrin* (I)–*polycarbonate* (II), *incompatible:* Phase separation in cyclohexanone for 1/1 mixture, and film was not clear when I was Hydrin 100, B. F. Goodrich, and II was Lexan 125, General Electric Company, or Merlon M-50, Mobay Chemical Co. [79].

k49. *Polyepichlorohydrin* (I)–*poly*(*2,6-dimethyl-1,4-phenylene ether*) (II), *incompatible:* Same as System **k48** except that II was General Electric Co. PPO [79].

k50. *Polyepichlorohydrin* (I)–*polysulfone* (II), *incompatible:* Same as System **k48** except that II was from Union Carbide [79].

k51. *Polyepichlorohydrin* (I)–*polyurethane* (II), *incompatible:* Same as System **k48** except that II was B. F. Goodrich polyester urethane, Estane 5707-F1 or polyether urethane, Estane 5740-X140 [79].

k52. *Polyepichlorohydrin* (I)–*poly*(*propylene glycol*) (II), *may be compatible:* Stated to be completely miscible when I was Union Carbide PPG 2025 and II was Shell Chemical Co. 1500, or when I was Dow Chemical Co. PPG 4000 and II was Shell Chemical Co. 2000 [236].

k53. *Poly*(*ethylene glycol*) (I)–*poly*(*propylene glycol*) (II), *may be compatible:* Stated to be completely miscible when I was Union Carbide PPG 425 and II was Union Carbide PEG 300, $\bar{X}_n \approx 7$ [236].

k54. *Poly*(*ethylene oxide*) (I)–*poly*(*propylene oxide*) (II), *conditionally compatible:* Stated immiscible for I/II: 1/1 and 9/1 by volume, when I had MW = 4000, 91 units, commercial sample, and II had MW = 2000; stated immiscible for I/II: 4/1 but miscible for 9/1 when II had MW = 1000; stated immiscible for I/II: 1/1 and 3/2 but miscible for 7/3 to 9/1 when II had MW = 800; stated miscible for I/II: 1/1 and 9/1 by volume when II had MW = 400 or 230 [237].

k55. *Poly*(*ethylene oxide*) (I)–*poly*(*4-vinylbiphenyl*) (II), *incompatible:* Stated incompatible when I was Dow Chemical Co. Polyglycol E 4000, low molecular weight, or Union Carbide WSR-35, high molecular weight, and II and $\bar{M}_w = 4.5 \times 10^5$ [193].

k56. *Poly*(*2,6-dimethyl-1,4-phenylene ether*) (I)–*poly*(*2-methyl-6-phenyl-1,4-phenylene ether*) (II), *compatible:* Single T_g by thermooptical analysis or DSC for I/II: 1/4 to 4/1 when I had $\bar{M}_n = 1.85 \times 10^4$, $\bar{M}_w = 3.72 \times 10^4$, and II had $[\eta] = 0.88$ dl/gm in CHCl₃ at 30°C [201].

k57. *Poly*(*2,6-dimethyl-1,4-phenylene ether*) (I)–*poly*(*2-methyl-6-benzyl-1,4-phenylene ether*) (II), *compatible:* Films optically clear and single T_g by thermooptical analysis or DSC for I/II: 1/9 to 9/1 when I was PR 5255, $\bar{M}_w = 3.72 \times 10^4$, $\bar{M}_n = 1.85 \times 10^4$, and II had $[\eta] = 0.88$ or 1.15 in CHCl₃ at 25°C [238].

k58. *Poly*(*2,6-dimethyl-1,4-phenylene ether*) (I)–*polycarbonate* (II), *incompatible:* Phase separation in cyclohexanone for 1/1 mixture, and film was not clear when I was General Electric Co. PPO, and II was General Electric Co. Lexan 125 or Mobay Chemical Co. Merlon M-50 [79].

k59. *Poly*(*2,6-dimethyl-1,4-phenylene ether*) (I)–*polysulfone* (II), *incompatible:* Same as System **k58** except that II was from Union Carbide [79].

k60. *Poly*(*2,6-dimethyl-1,4-phenylene ether*) (I)–*polyurethane* (II), *incompatible:* Same as System **k58** except that II was B. F. Goodrich polyester urethane, Estane 5707-F1, or polyether urethane, Estane 5740-X140 [79].

k61. *Poly*(*isobutyl vinyl ether*) (I)–*poly*(*tert-butyl vinyl ether*) (II), *incompatible:* Phase separation in CHCl₃ above 5 gm solids/dl at 25° C when I had $[\eta] = 0.569$ in benzene at 30°C and II had $[\eta] = 0.551$ in benzene at 30°C [185].

k62. *Poly*(*ε-caprolactone*)(I)–*polyepichlorohydrin*(II), *may be conditionally compatible:* Single T_g from shear modulus; stated compatible at 20–30% I but not at "high levels" of I [89].

k63. *Poly*(*ε-caprolactone*) (I)–*phenoxy* (II), *conditionally compatible:* Quenched samples gave single T_g by shear modulus for I/II: 1/9 to 9/1, when II was from Union Carbide, by condensation of bisphenol A and epichlorohydrin; crystallization if > 50% I [89].

k64. *Poly*(*ε-caprolactone*) (I)–*chlorinated polyether* (II), *incompatible:* Two T_g's from shear modulus for I/II: 1/9 and 1/1 when II was Hercules Penton [89].

k65. *Poly(ε-caprolactone)* (I)–*poly* (*ε-methyl-ε-caprolactone*) (II), *incompatible:* Two mechanical loss peaks for I/II: 1/9 to 9/1 when I was Union Carbide PCL-700, reduced viscosity 0.65 dl/gm at 0.2 gm/dl in benzene at 30°C, and II had reduced viscosity 0.92 dl/gm at 0.2 gm/dl at 30°C in benzene [239].

k66. *Poly(ε-caprolactone)* (I)–*poly(β,δ-mixed methyl-ε-caprolactone)* (II), *incompatible:* Two mechanical loss peaks when I was as in System **k65** and II had reduced viscosity 0.96 dl/gm at 0.2 gm/dl in benzene at 30°C [239].

k67. *Poly(ε-caprolactone)*(I)–*poly(γ-methyl-ε-caprolactone)*(II), *incompatible:* Same as System **k66** except that II had reduced viscosity 0.72 dl/gm at 0.2 gm/dl in benzene at 30°C [239].

k68. *Poly(ε-caprolactone)* (I)–*poly(α,ε-dimethyl-ε-caprolactone)* (II), *incompatible:* Same as System **k66** except that II had reduced viscosity 0.45 dl/gm at 0.2 gm/dl in benzene at 30°C [239].

k69. *Poly(oxycarbonyloxy-1,4-phenyleneisopropylidene-1,4-phenylene)* (I)–*poly(oxyethylene)* *carbonate* (II), *incompatible:* Milky, rough films from mixtures in CHCl₃, possibly crystalline, when I was poly(bisphenol-A–carbonate) and II was prepared by phosgenation of 4000 Carbowax in pyridine [240].

k70. *Poly(oxycarbonyloxy-1,4-phenyleneisopropylidene-1,4-phenylene)* (I)–*poly(oxyethylene* *glycol)* (II), *incompatible:* Milky rough films from mixtures in CHCl₃, 7% I plus 14–43% II, when I was poly(bisphenol A–carbonate) and II was 4000 Carbowax [240].

k71. *Polycarbonate* (I)–*polysulfone* (II), *incompatible:* Phase separation in cyclohexanone for 1/1 mixture, and film was not clear when I was Lexan 125, General Electric Co., or Mobay Chemical Co. Merlon-50, and II was Union Carbide polysulfone [79].

k72. *Polycarbonate* (I)–*polyurethane* (II), *incompatible:* Same as System **k71** except that II was B. F. Goodrich polyester urethane, Estane 5707-F1, or polyether urethane, Estane 5740-X140 [79].

k73. *Poly(oxycarbonyloxy-1,4-phenylenisopropylidene-1,4-phenylene)* (I)–*poly[oxycarbonyloxy(2,2,4,4-tetramethyl-1,3-cyclobutylene)]* (II), *incompatible:* Precipitated from CH₂Cl₂ when I was General Electric Lexan 101, reduced viscosity 0.855 dl/gm at 0.2 gm/dl in CH₂Cl₂, and II had reduced viscosity 1.17 dl/gm at 0.2 gm/dl in CH₂Cl₂ [177].

k74. *Polysulfone* (I)–*polyurethane* (II), *incompatible:* Same as System **k72** except that I was Union Carbide polysulfone [79].

k75. *Polyurethane* (I)–*polyurethane* (II), *may be conditionally compatible:* Same as System **k72** except that I and II were both listed as II in that System [79]. Homogeneous solution and transparent film when I was Nippon Rubber Co. 5740X1, and II was Hodogaya Co. Pellet 22S [37].

k76. *Polyurethane* (I)–*polyester* (II), *incompatible:* Nonhomogeneous solution and turbid film when I was Nippon Rubber Co. 5740X1, or Hodogaya Co. Pellet 22S, and II was Toyobo Co. Ester Resin 20 [37].

k77. *Polyurethane* (I)–*miscellaneous resins* (II), *incompatible:* Same as System **k76** except that II was Petroleum Resin, Nippon Petroleum Chemical Co. Nitsuseki Neopolymer 150, or Cumarone Resin, Mitsubishi TG, or Indene Resin, Fuji Iron Products Co. VM-1/2 [37].

k78. *Polyester* (I)–*miscellaneous resins* (II), *incompatible:* Same as System **k77** except that I was Toyobo Co. Ester Resin 20 [37].

k79. *Miscellaneous resins* (I and II), *may be compatible:* Homogeneous solutions and transparent films when I and II were any of II from System **k77** [37].

k80. *Coal tar lacquer* (I)–*phenol–formaldehyde resin* (II), *incompatible:* Opaque and dirty solution, 1/1 volume ratio, 50% solids, when II was No. 101 [228].

k81. *Perchlorovinyl resin* (I)–*rosin* (II), *ambiguous:* No phase separation at 15% solids in 3/1: xylene/butyl acetate for 1/1 volume ratio when I was SPS-V [228].

k82. *Sodium poly(styrene sulfonate)* (I)–*poly(pyridonium bromide)* (II), *ambiguous* (complex formation): Precipitation at very low concentration observed by light scattering [213].

L.   Miscellaneous Homopolymers and Miscellaneous Copolymers

I1.   *Polybutene* (I)–*poly*(α-*methylstyrene-co-vinyltoluene*) (II), *incompatible:* UCST around room temperature when I was Indopol L10, MW = 320, and II was Piccotex 100; UCST ranging up to >200°C when I was Indopol H300, MW = 1290; phase diagrams had two maxima [241].

I2.   *Polyisobutene* (I)–*poly*[*ethylene-co-*(*vinyl acetate*)] (II), *ambiguous:* 1/1 mixture was opaque when milled, but transparent when 82 silica was added [109].

I3.   *Polyisobutene* (I)–*poly*(*butadiene-co-styrene*) (II), *may be conditionally compatible:* Transition of II seen in mixtures containing 50–90% II when I was Vistanex MD 333, MW = 3×10^5, and II was SBR-1000 [242]. Stated compatible for I/II:4/1 to 1/4 in solutions and films when I was Oppanol B100 and II was Buna S3, ~80% butadiene [16].

I4.   *Polyisobutene* (I)–*poly*(*butadiene-co-acrylonitrile*) (II), *incompatible:* Stated incompatible for I/II: 4/1 to 1/4 in solutions and films when I was Oppanol B100 and II was Buna NW, 28% acrylonitrile [16]

I5.   *Polyisobutene* (I)–*poly*[(*vinyl chloride*)-*co-*(*vinyl acetate*)] (II), *incompatible:* Same as System I4 except that II was Vinalit MPS [16].

I6.   *Chlorinated polyisoprene* (I)–*poly*[*ethylene-co-*(*vinyl acetate*)] (II), *conditionally compatible:* 1/1 Mixtures at 10% solids at 25°C were compatible in octane, toluene, benzenes, and CCl_4 and formed clear films from these solvents when I had $\overline{M}_n = 5.2 \times 10^4$ and II had $\overline{M}_n = 1.28 \times 10^4$; the same mixtures were incompatible in trichloroethylene, chloroform, tetrachloroethane, and dichloroethylene and formed opalescent films from these solvents [44]. Films from trichloroethylene, CH_2Cl_2, and $CHCl_3$ were opalescent, films from xylene and tetrachloroethylene were clear, I/II: 1/3 to 3/1 became opalescent with LCST ~ 120°C, and DSC showed single T_g when I was Hercules Powder Co., 63.2 wt % Cl, $\overline{M}_n = 5.2 \times 10^4$, and II was du Pont Elvax-210, 27–29 wt % vinyl acetate, $\overline{M}_n = 1.28 \times 10^4$; films from $CHCl_3$, CH_2Cl_2 were hazy, films from xylene, toluene, and tetrachloroethylene were clear, films became opalescent with LCST ~ 188°C, DSC showed single T_g but crystallinity of II for 60 wt % II when II was du Pont Elvax-40, 39–42 wt % vinyl acetate [243].

I7.   *Chlorinated polyisoprene* (I)–*poly*[(*vinyl chloride*)-*co-*(*vinyl acetate*)] (II), *incompatible:* Stated incompatible for I/II: 1/9 to 9/1 in solutions and films when II was Vinalit MPS [16].

I8.   *Polychloroprene* (I)–*poly*(*butadiene-co-styrene*) (II), *may be conditionally compatible:* Phase contrast and electron microscopy showed zones when I was Neoprene W and II was Krylene [96]. Stated compatible for I/II: 4/1 to 1/4 in solutions and films when I was Neoprene AC and II was Buna S3, ~80% butadiene [16].

I9.   *Polychloroprene* (I)–*poly*(*butadiene-co-acrylonitrile*) (II), *may be conditionally compatible:* Two T_g's on linear expansion when II was SKN-40 (40% acrylonitrile) [98]. Two T_g's for I/II ratios from 1/4 to 4/1 when I was Neoprene and II was Hycar OR 15 (40% acrylonitrile) [244]. Stated compatible for I/II: 4/1 to 1/4 in solutions and films when I was Neoprene AC and II was Buna NW, 28% acrylonitrile [16].

I10.   *Polychloroprene* (I)–*poly*[(*vinyl chloride*)-*co-*(*vinyl acetate*)] (II), *incompatible:* Stated incompatible for I/II: 4/1 to 1/4 in solutions and films when I was Neoprene AC and II was Vinalit MPS [16].

I11.   *Chlorinated polyethylene* (I)–*poly*[(*vinyl chloride*)-*co-*(*vinyl acetate*)] (II), *incompatible:* Stated compatible for I/II: 1/9 to 9/1 but incompatible in between in solutions and films when I had 30% Cl, and II was Vinalit MPS [16].

I12.   *Chlorosulfonated polyethylene* (I)–*poly*[*ethylene-co-*(*vinyl acetate*)] (II), *may be conditionally compatible:* Homogeneous solution and transparent film when I was Showa Neoprene Hyperon 30 and II was Nippon Goshei Co. Soalex R-CH, but nonhomogeneous solution and turbid film when II was Soalex R-FH [37].

I13.   *Chlorosulfonated polyethylene* (I)–*poly*(*butadiene-co-styrene*) (II), *incompatible:*

Nonhomogeneous solution and turbid film when I was Showa Neoprene Hyperon 30 and II was Nippon Rubber Co. Hycar 2057 [37].

114. *Chlorosulfonated polyethylene* (I)–*poly*[(*vinyl chloride*)-*co*-(*vinyl acetate*)] (II), *may be conditionally compatible:* Nonhomogeneous solution and turbid film when I was Showa Neoprene Hyperon 30 and II was Shekisui Chemical Co. Esulex C or CL, but homogeneous solution and transparent film when II was Esulex CH-1 or Denka Vinyl 1000 AK, Denki Chemical Co. [37]. Stated compatible for I/II: 9/1 and incompatible for 4/1 to 1/9 in solutions and films when I had 29.5% Cl, 1.6% S, and II was Vinalit MPS [16].

115. *Chlorosulfonated polyethylene* (I)–*poly*[(*vinyl chloride*)-*co*-(*vinyl propionate*)] (II), *may be conditionally compatible:* Homogeneous solution and transparent film when I was Showa Neoprene Hyperon 30 and II was Ryuron QS-430 [37].

116. *Chlorosulfonated polyethylene* (I)–*poly*[(*vinyl chloride*)-*co*-(*vinyl acetate*)-*co*-(*vinyl alcohol*)] (II), *incompatible:* Nonhomogeneous solution and turbid film when I was Showa Neoprene Hyperon 30 and II was Shekisui Chemical Co. Esulex A [37].

117. *Chlorosulfonated polyethylene* (I)–*poly*[(*vinyl chloride*)-*co*-(*vinyl acetate*)-*co*-(*maleic acid*)] (II), *incompatible:* Same as System **116** except that II was Shekisui Chemical Co. Esulex M. [37].

118. *Chlorosulfonated polyethylene* (I)–*poly*[(*vinyl chloride*)-*co*-(*vinylidene chloride*)-*co*-(*acrylic acid ester*)] (II), *may be conditionally compatible:* Same as System **115** except that II was Towa Gousei Co. Aron 321 [37].

119. *Chlorosulfonated polyethylene* (I)–*poly*[(*vinyl acetate*)-*co*-(*N*-*vinylpyrrolidone*)] (II), *incompatible:* Same as System **116** except that II was General Aniline & Film S-630 [37].

120. *Chlorosulfonated polyethylene* (I)–*poly*[*styrene*-*co*-(*maleic acid*)] (II), *incompatible:* Same as System **116** except that II was Daidou Kogyo Styrite CM-2 or CM-3 [37].

121. *Chlorosulfonated polyethylene* (I)–*poly*[*styrene*-*co*-(*maleic acid ester*)] (II), *incompatible:* Same as System **116** except that II was Daidou Kogyo Styrite HS-2 [37].

122. *Chlorinated poly*(*vinyl chloride*) (I)–*poly*(*butadiene*-*co*-*styrene*) (II), *may be conditionally compatible:* Stated compatible for I/II: 4/1 to 1/4 in solutions and films when I was Vinoflex PC and II was Buna S3, ~80% butadiene [16].

123. *Chlorinated poly*(*vinyl chloride*) (I)–*poly*(*butadiene*-*co*-*acrylonitrile*) (II), *may be conditionally compatible:* Same as System **122** except that II was Buna NW, 28% acrylonitrile [16].

124. *Chlorinated poly*(*vinyl chloride*) (I)–*poly*[(*vinyl chloride*)-*co*-(*vinyl acetate*)] (II), *ambiguous:* Stated compatible for I/II: 4/1 to 1/4 in solutions and films when I was Vinoflex PC and II was Vinalit MPS but cloudy film for 1/1 and 4/1 [16].

125. *Poly*(*vinylidene chloride*) (I)–*poly*[*ethylene*-*co*-(*vinyl acetate*)] (II), *may be conditionally compatible:* Homogeneous solution and transparent film when I was Asahi Dow Co. EX5701 and II was Nippon Goshei Co. Soalex R-CH [37].

126. *Poly*(*vinylidene chloride*) (I)–*poly*(*butadiene*-*co*-*acrylonitrile*) (II), *incompatible:* Nonhomogeneous solutions and turbid films when I was Asahi Dow Co. EX5701 and II was Nippon Rubber Co. Hycar 1043 or 1432 [37].

127. *Poly*(*vinylidene chloride*) (I)–*poly*[(*vinyl chloride*)-*co*-*vinyl acetate*)] (II), *incompatible:* Same as System **126** except that II was Shekisui Chemical Co. Esulex C, or CL, or CH-1 [37].

128. *Poly*(*vinylidene chloride*) (I)–*poly*[(*vinyl chloride*)-*co*-(*vinyl proprionate*)] (II), *incompatible:* Same as System **126** except that II was Ryuron QS-430 [37].

129. *Poly*(*vinylidene chloride*) (I)–*poly*[(*vinyl chloride*)-*co*-(*vinyl acetate*)-*co*-(*vinyl alcohol*)] (II), *incompatible:* Same as System **126** except that II was Shekisui Chemical Co. Esulex A [37].

130. *Poly*(*vinylidene chloride*) (I)–*poly*[(*vinyl chloride*)-*co*-(*vinylidene chloride*)-*co*-(*acrylic acid ester*)] (II), *incompatible but must be conditionally compatible:* Same as System **126** except that II was Towa Gousei Co. Aron 321 [37].

131. *Poly*(*vinylidene chloride*) (I)–*poly*[*styrene*-*co*-(*maleic acid*)] (II), *incompatible:* Same as System **126** except that II was Daidou Kogyo Styrite CM-2 or CM-3 [37].

132. *Poly(vinylidene chloride)* (I)–*poly[styrene-co-(maleic acid ester)]* (II), *incompatible:* Same as System **126** except that II was Daidou Kogyo Styrite HS-2 [37].

133. *Poly(vinyl fluoride)* (I)–*poly[tetrafluoroethylene-co-(vinylidene fluoride)]* (II), *incompatible:* Two T_g's from dynamic mechanical measurements [231].

134. *Poly(vinylidene fluoride)* (I)–*Poly[tetrafluoroethylene-co-(vinylidene fluoride)]* (II), *incompatible but must be conditionally compatible:* Two or three T_g's by dynamic mechanical measurements for 25, 50, or 75% II [231].

135. *Poly(vinyl butyral)* (I)–*poly[ethylene-co-(vinyl acetate)]* (II), *incompatible:* Nonhomogeneous solution and turbid film when I was Shekisui Chemical Co. BM-2 and II was Nippon Goshei Co. Soalex R-CH or R-FH [37].

136. *Poly(vinyl butyral)* (I)–*poly(butadiene-co-styrene)* (II), *incompatible:* Same as System **135** except that II was Nippon Rubber Co. Hycar 2057 [37].

137. *Poly(vinyl butyral)* (I)–*poly(butadiene-co-acrylonitrile)* (II), *incompatible:* Same as System **135** except that II was Nippon Rubber Co. Hycar 1043 or 1432 [37].

138. *Poly(vinyl butyral)* (I)–*poly[(vinyl chloride)-co-(vinyl acetate)]* (II), *incompatible:* Same as System **135** except that II was Shekisui Chemical Co. Esulex C or CL or CH-1 or Denki Chemical Co. Denka Vinyl 1000 AK [37].

139. *Poly(vinyl butyral)* (I)–*poly[(vinyl chloride)-co-(vinyl proprionate)]* (II), *incompatible:* Same as System **135** except that II was Ryuron QS-430 [37].

140. *Poly(vinyl butyral)* (I)–*poly[(vinyl chloride)-co-(vinyl acetate)-co-(vinyl alcohol)]* (II), *incompatible:* Same as System **135** except that II was Shekisui Chemical Co. Esulex A [37].

141. *Poly(vinyl butyral)* (I)–*poly[(vinyl chloride)-co-(vinyl acetate)-co-(maleic acid)]* (II), *incompatible:* Same as System **135** except that II was Shekisui Chemical Co. Esulex M [37].

142. *Poly(vinyl butyral)* (I)–*poly[(vinyl chloride)-co-(vinylidene chloride)-co-(acrylic acid ester)]* (II), *incompatible:* Same as System **135** except that II was Towa Gousei Co. Aron 321 [37].

143. *Poly(vinyl butyral)* (I)–*poly[(vinyl acetate)-co-(N-vinylpyrrolidone)]* (II), *may be compatible:* Homogeneous solution and transparent film when I was Shekisui Chemical Co. BM-2 and II was General Aniline & Film S-630 [37].

144. *Poly(vinyl butyral)* (I)–*poly[styrene-co-(maleic acid)]* (II), *compatible (probably conditionally):* Same as System **143** except that II was Daidou Kogyo Styrite CM-2 or CM-3 and single T_g by DTA when II was Styrite CM-3 [37].

145. *Poly(vinyl butyral)* (I)–*poly[styrene-co-(maleic acid ester)]* (II), *may be conditionally compatible:* Same as System **143** except that II was Daidou Kogyo Styrite HS-2 [37].

146. *Poly(vinylpyrrolidone)* (I)–*poly[ethylene-co-(vinyl acetate)]* (II), *incompatible:* Nonhomogeneous solution and turbid film when I was General Aniline & Film K-30 and II was Nippon Goshei Co. Soalex R-CH or R-FH [37].

147. *Poly(vinylpyrrolidone)* (I)–*poly[(vinyl chloride)-co-(vinyl acetate)]* (II), *incompatible:* Same as System **146** except that II was Shekisui Chemical Co. Esulex C [37].

148. *Poly(vinylpyrrolidone)* (I)–*poly[(vinyl chloride)-co-(vinyl acetate)-co-(vinyl alcohol)]* (II), *incompatible:* Same as System **146** except that II was Shekisui Chemical Co. Esulex A [37].

149. *Poly(vinylpyrrolidone)* (I)–*poly[(vinyl chloride)-co-(vinyl acetate)-co-(maleic acid)]* (II), *incompatible:* Same as System **146** except that II was Shekisui Chemical Co. Esulex M [37].

150. *Poly(vinylpyrrolidone)* (I)–*poly[(vinyl chloride)-co-(vinylidene chloride)-co-(acrylic acid ester)]* (II), *may be conditionally compatible:* Homogeneous solution and transparent film when I was General Aniline & Film K-30 and II was Towa Gousei Co. Aron 321 [37].

151. *Poly(vinylpyrrolidone)* (I)–*poly[(N-vinylpyrrolidone)-co-(vinyl acetate)]* (II), *incompatible (but must be conditionally compatible):* Same as System **146** except that II was General Aniline & Film S-630 [37].

152. *Poly(vinylpyrrolidone)* (I)–*poly[styrene-co-(maleic acid ester)]* (II), *incompatible:* Same as System **146** except that II was Daidou Kogyo Styrite HS-2 [37].

153. *Polyepichlorohydrin* (I)–*poly*[(*vinyl chloride*)-*co*-(*vinyl acetate*)] (II), *incompatible:* Phase separation in cyclohexanone for 1/1 mixture and film was not clear when I was Hydrin 100, B. F. Goodrich, and II was Geon 440 × 24, B. F. Goodrich [79].

154. *Polyepichlorohydrin* (I)–*poly*[*styrene*-*co*-(*methyl methacrylate*)] (II), *may be conditionally compatible:* No phase separation in cyclohexanone for 1/1 mixture, and film was clear when I was Hydrin 100, B. F. Goodrich, and II was Zerlon 150, Dow Chemical Co. [79].

155. *Polyepichlorohydrin* (I)–*poly*(*styrene*-*co*-*acrylonitrile*) (II), *may be conditionally compatible:* Same as System **154** except that II was Dow Chemical Co. Tyril 767 [79].

156. *Polyepichlorohydrin* (I)–*poly*(*methyl vinyl ether*)-*co*-(*maleic anhydride*) (II), *incompatible:* Same as System **153** except that II was B. F. Goodrich Gantrez AN-169 [79].

157. *Poly*(ε-*caprolactone*) (I)–*poly*(*styrene*-*co*-*acrylonitrile*) (II), *conditionally compatible:* LCST about 85°C by film turbidity when I was Union Carbide PCL-700, $\bar{M}_n = 2.24 \times 10^3$, $\bar{M}_w = 3.5 \times 10^4$, and II was Union Carbide RMD-4511, 28% acrylonitrile, $\bar{M}_n = 8.86 \times 10^4$ [74]. Two T_g's from shear modulus for I/II: 1/9 and 1/1 [89].

158. *Poly*(2,6-*dimethyl*-1,4-*phenylene ether*) (I)–*poly*(*butadiene*-*co*-*acrylonitrile*) (II), *incompatible:* Phase separation in cyclohexanone for 1/1 mixture, and film was not clear when I was General Electric Co. PPO and II was B. F. Goodrich Hycar 1432 [79].

159. *Poly*(2,6-*dimthyl*-1,4-*phenylene ether*) (I)–*poly*[(*vinyl chloride*)-*co*-(*vinyl acetate*)] (II), *incompatible:* Same as System **158** except that II was B. F. Goodrich Geon 440 × 24 [79].

160. *Poly*(2,6-*dimethyl*-1,4-*phenylene ether*) (I)–*poly*[*styrene*-*co*-(*methyl methacrylate*)] (II), *incompatible:* Same as System **158** except that II was Dow Chemical Co. Zerlon 150 [79].

161. *Poly*(2,6-*dimethyl*-1,4-*phenylene ether*) (I)–*poly*[*styrene*-*co*-(*p*-*chlorostyrene*)] (S–CS) (II), *conditionally compatible:* Single T_g by various methods for I/II: 1/7 to 7/1 when I had $\bar{M}_w = 3.72 \times 10^4$, $\bar{M}_n = 1.85 \times 10^4$ and II was six samples with mole fraction S from 0.454 to 0.347, $\bar{M}_w = 2.9 - 4.0 \times 10^5$, $\bar{M}_w/\bar{M}_n = 1.8$–2.2; two T_g's by thermooptical analysis except after annealing for II with mole fraction S = 0.334 ± 0.012; two T_g's by various methods when II had mole fraction S from 0.320 to 0.279, concluded compatibility when II had mole fraction S ⩾ 0.347; mole fraction S = 0.334 borderline; incompatibility for mole fraction S ⩽ 0.320 [35]. Two T_g's by DSC and dielectric measurement when I was General Electric PPO, $\bar{M}_n = 5.8 \times 10^4$, and II had 32 mole % S, $\bar{M}_n = 1.1 \times 10^5$, $\bar{M}_w/\bar{M}_n = 2.1$; single T_g when II had 53 mole %S, $\bar{M}_n = 9.2 \times 10^4$, $\bar{M}_w/\bar{M}_n = 2.5$ [234].

162. *Poly*(2,6-*dimethyl*-1,4-*phenylene ether*) (I)–*poly*(*styrene*-*co*-*acrylonitrile*) (II), *incompatible* (*but should be conditionally compatible*): Same as System **158** except that II was Dow Chemical Co. Tyril 767 [79]. Two phase blend when I was General Electric Co. PPO and II was Dow Chemical Co. Tyril 767, 27% acrylonitrile [195].

163. *Poly*(2,6-*dimethyl*-1,4-*phenylene ether*) (I)–*poly*[*epichlorohydrin*-*co*-(*ethylene oxide*)] (II), *incompatible:* Same as System **158** except that II was B. F. Goodrich Hydrin 200 [79].

164. *Poly*(2,6-*dimethyl*-1,4-*phenylene ether*)(I)–*poly*[(*methyl vinyl ether*)-*co*-(*maleic anhydride*)] (II), *incompatible:* Same as System **158** except that II was B. F. Goodrich Gantrez AN-169 [79].

165. *Polycarbonate* (I)–*poly*(*butadiene*-*co*-*acrylonitrile*) (II), *incompatible:* Phase separation in cyclohexanone for 1/1 mixture and film was not clear when I was Lexan 125, General Electric Co., or Mobay Chemical Co. Merlon M-50 and II was B. F. Goodrich Hycar 1432 [79].

166. *Polycarbonate* (I)–*poly*[(*vinyl chloride*)-*co*-(*vinyl acetate*)] (II), *incompatible:* Same as System **165** except that II was B. F. Goodrich Geon 440X24 [79].

167. *Polycarbonate* (I)–*poly*[*styrene*-*co*-(*methyl methacrylate*)] (II), *incompatible:* Same as System **165** except that II was Dow Chemical Co. Zerlon 150 [79].

168. *Polycarbonate* (I)–*poly*(*styrene*-*co*-*acrylonitrile*) (II), *incompatible:* Same as System **165** except that II was Dow Chemical Co. Tyril 767 [79].

169. *Polycarbonate* (I)–*poly*[*epichlorohydrin*-*co*-(*ethylene oxide*)] (II), *incompatible:* Same as System **165** except that II was B. F. Goodrich Hydrin 200 [79].

170. *Polycarbonate* (I)–*poly[(methyl vinyl ether)-co-(maleic anhydride)]* (II), *incompatible:* Same as System **165** except that II was B. F. Goodrich Gantrez AN-169 [79].

171. *Polysulfone* (I)–*poly(butadiene-co-acrylonitrile)* (II), *incompatible:* Phase separation in cyclohexanone for 1/1 mixture, and film was not clear when I was from Union Carbide and II was B. F. Goodrich Hycar 1432 [79].

172. *Polysulfone* (I)–*poly[(vinyl chloride)-co-(vinyl acetate)]* (II), *incompatible:* Same as System **171** except that II was B. F. Goodrich Geon 440X24 [79].

173. *Polysulfone* (I)–*poly[styrene-co-(methyl methacrylate)]* (II), *incompatible:* Same as System **171** except that II was Dow Chemical Co. Zerlon 150 [79].

174. *Polysulfone* (I)–*poly(styrene-co-acrylonitrile)* (II), *incompatible:* Same as System **171** except that II was Dow Chemical Co. Tyril 767 [79].

175. *Polysulfone* (I)–*poly[epichlorohydrin-co-(ethylene oxide)]* (II), *incompatible:* Same as System **171** except that II was B. F. Goodrich Hydrin 200 [79].

176. *Polysulfone* (I)–*poly[(methyl vinyl ether)-co-(maleic anhydride)]* (II), *incompatible:* Same as System **171** except that II was B. F. Goodrich Gantrez AN-169 [79].

177. *Poly[oxy(2,5-dimethy-1,4-phenylene)sulfonyl(2,5-dimethyl-1,4-phenylene)oxy-1,4-phenylene-isopropylidene-1,4-phenylene]* (I)–*poly(styrene-co-acrylonitrile)* (II), *incompatible:* Precipitated in CH_2Cl_2 when I had reduced viscosity 0.32 dl/gm at 0.2 gm/dl in CH_2Cl_2 and II was Union Carbide RMD-4511, 28% acrylonitrile, reduced viscosity 1.15 dl/gm at 0.2 gm/dl in CH_2Cl_2, or one of eight samples ranging from 6.8 to 42% acrylonitrile, reduced viscosity 0.604 to 1.694 dl/gm at 0.2 gm/dl in CH_2Cl_2; single dynamic mechanical T_g when II had 12.7% acrylonitrile; two T_g's when II was RMD-4511 [177].

178. *Poly(oxy-1,4-phenylsulfonyl-1,4-phenyleneoxy-1,4-phenyleneisopropylidene-1,4-phenylene)* (I)–*poly(styrene-co-acrylonitrile)* (II), *incompatible:* Precipitated from CH_2Cl_2 when I was Union Carbide P-1700 polysulfone, reduced viscosity 0.409 dl/gm at 0.2 gm/dl in CH_2Cl_2 and II were the same as in System **177** [177].

179. *Poly[oxy(tetrachloro-1,4-phenylene)oxy-1,4-phenyleneisopropylidene-1,4-phenylene]* (I)–*poly(styrene-co-acrylonitrile)* (II), *incompatible:* Same as System **178** except that I had reduced viscosity 0.45 dl/gm at 0.2 gm/dl in CH_2Cl_2 [177].

180. *Poly[oxy-1,4-phenylenesulfonyl-1,4-phenyleneoxy(2,6-diisopropyl-1,4-phenylene)isopropylidene(3,5-diisopropyl-1,4-phenylene)]* (I)–*poly(styrene-co-acrylonitrile)* (II), *conditionally compatible:* Compatible by modulus–temperature measurements if acrylonitrile content in II was 13–16% [245].

181. *Poly[oxy-1,4-phenylenesulfonyl-1,4-phenyleneoxy(2,6-dimethyl-1,4-phenylene)(1-methyl-4-isopropyl-1,2-cyclohexylene)(3,5-dimethyl-1,4-phenylene)]* (I)–*poly(styrene-co-acrylonitrile)* (II), *incompatible:* Same as System **178** except that I contained other isomers, reduced viscosity 0.9 dl/gm at 0.2 gm/dl in CH_2Cl_2 [177].

182. *Polyurethane* (I)–*poly[ethylene-co-(vinyl acetate)]* (II), *incompatible:* Nonhomogeneous solutions and turbid films when I was Nippon Rubber Co. 5740X1 or Hodogaya Co. Pellet 22S and II was Nippon Goshei Co. Soalex R-CH or R-FH [37].

183. *Polyurethane* (I)–*poly(butadiene-co-acrylonitrile)* (II), *incompatible:* Phase separation in several solvents for several mixtures including polyester and polyether urethanes [37, 79].

184. *Polyurethane* (I)–*poly[(vinyl chloride)-co-(vinyl acetate)]* (II), *may be conditionally compatible:* Homogeneous solution and transparent film when I was Nippon Rubber Co. 5740X1 (IA) or Hodogaya Co. Pellet 22S (IB), and II was Shekisui Chemical Co. Esulex C (IIA), or IB with Shekisui Chemical Co. Esulex CL (IIB); nonhomogeneous solution and turbid film with IA plus IIB or with IA or IB plus Shekisui Chemical Co. Esulex CH-1 [37]. No phase separation in cyclohexanone for 1/1 mixture, and film was clear when II was Geon 440 × 24, B. F. Goodrich and I was polyester–urethane, Estane 5707-F1 or polyether–urethane, Estane 5740-X140, B. F. Goodrich [79].

185.   *Polyurethane* (I)–*poly*[(*vinyl chloride*)-*co*-(*vinyl propionate*)] (II), *may be conditionally compatible:* Nonhomogeneous solution and turbid film when I was Nippon Rubber Co. 5740X1 and II was Ryuron QS-430; homogeneous solution and transparent film when I was Hodogoya Co. Pellet 22S [37].

186.   *Polyurethane* (I)–*poly*[(*vinyl chloride*)-*co*-(*vinyl acetate*)-*co*-(*vinyl alcohol*)] (II), *may be conditionally compatible:* Homogeneous solution and transparent film when I was as in **182** and II was Shekisui Chemical Co. Esulex A [37].

187.   *Polyurethane* (I)–*poly*[(*vinyl chloride*)-*co*-(*vinylidene chloride*)-*co*-(*acrylic acid ester*)] (II), *may be conditionally compatible:* Same as System **186** except that I was Hodogoya Co. Pellet 22S and II was Towa Gousei Co. Aron 321 [37].

188.   *Polyurethane* (I)–*poly*[(*vinyl acetate*)-*co*-(*N-vinylpyrrolidone*)] (II), *may be conditionally compatible:* Same as System **186** except that II was General Aniline & Film S-630 [37].

189.   *Polyurethane* (I)–*poly*[*styrene*-*co*-(*methyl methacrylate*)] (II), *incompatible:* Phase separation in cyclohexanone for 1/1 mixture, and film was not clear when II was Zerlon 150, Dow Chemical Co., and I was polyester-urethane, Estane 5707-F1 or polyether–urethane, Estane 5740-X140, B. F. Goodrich [79].

190.   *Polyurethane* (I)–*poly*(*styrene*-*co*-*acrylonitrile*) (II), *incompatible:* Same as System **189** except that II was Dow Chemical Co. Tyril 767 [79].

191.   *Polyurethane* (I)–*poly*[*styrene*-*co*-(*maleic acid*)] (II), *incompatible:* Same as System **182** except that II was Daidou Kogyo Styrite CM-2 or CM-3 [37].

192.   *Polyurethane* (I)–*poly*[*styrene*-*co*-(*maleic acid ester*)] (II), *incompatible:* Same as System **182** except that II was Daidou Kogyo Styrite HS-2 [37].

193.   *Polyurethane* (I)–*poly*[*epichlorohydrin*-*co*-(*ethylene oxide*)] (II), *incompatible:* Same as System **189** except that II was B. F. Goodrich Hydrin 200 [79].

194.   *Polyurethane* (I)–*poly*[(*methyl vinyl ether*)-*co*-(*maleic anhydride*)] (II), *incompatible:* Same as System **189** except that II was B. F. Goodrich Gantrez AN-169 [79].

195.   *Polyester* (I)–*poly*[*ethylene*-*co*-(*vinyl acetate*)] (II), *incompatible:* Nonhomogeneous solution and turbid film when I was Toyobo Co. Ester Resin 20 and II was Nippon Goshei Co. Soalex R-CH or R-FH [37].

196.   *Polyester* (I)–*poly*(*butadiene*-*co*-*acrylonitrile*) (II), *incompatible:* Same as System **195** except that II was Nippon Rubbers Co. Hycar 1043 or 1432 [37].

197.   *Polyester* (I)–*poly*[(*vinyl chloride*)-*co*-(*vinyl acetate*)] (II), *incompatible:* Same as System **195** except that II was Shekisui Chemical Co. Esulex C or CL or CH-1 [37].

198.   *Polyester*(I)–*poly*[(*vinyl chloride*)-*co*-(*vinyl propionate*)](II),*incompatible:*Same as System **195** except that II was Ryuron QS-430 [37].

199.   *Polyester* (I)–*poly*[(*vinyl chloride*)-*co*-(*vinyl acetate*)-*co*-(*vinyl alcohol*)] (II), *may be conditionally compatible:* Homogeneous solution and transparent film when I was Toyobo Co. Ester Resin 20 and II was Shekisui Chemical Co. Esulex A [37].

1100.   *Polyester* (I)–*poly*[(*vinyl chloride*)-*co*-(*vinylidene chloride*)-*co*-(*acrylic acid ester*)] (II), *may be conditionally compatible:* Same as System **199** except that II was Towa Gousei Co. Aron 321 [37].

1101.   *Polyester* (I)–*poly*[(*vinyl acetate*)-*co*-(*N-vinylpyrrolidone*)] (II), *may be conditionally compatible:* Same as System **199** except that II was General Aniline & Film S-630 [37].

1102.   *Polyester* (I)–*poly*[*styrene*-*co*-(*maleic acid*)] (II), *incompatible:* Same as System **195** except that II was Daidou Kogyo Styrite CM-2 or CM-3 [37].

1103.   *Polyester* (I)–*poly*[*styrene*-*co*-(*maleic acid ester*)] (II), *incompatible:* Same as System **195** except that II was Daidou Kogyo Styrite HS-2 [37].

1104.   *Perchlorovinyl resin* (I)–*poly*[(*vinyl chloride*)-*co*-(*vinyl acetate*)] (II), *may be compatible:* No phase separation in 3/1: xylene/butyl acetate at 15% solids for 1/1 volume ratio when I was SPS-V and II was A-15 [228].

1105. *Perchlorovinyl resin* (I)–*poly(butadiene-co-acrylonitrile)* (II), *may be compatible:* Same as System **1104** except that experiment was done at 10% solids and II was SKN-26 (26% acrylonitrile [228].

1106. *Perchlorovinyl resin* (I)–*carboxylated poly(butadiene-co-acrylonitrile* (II), *may be compatible:* Same as System **1105** except that II was SKN-26-1 (26% acrylonitrile, 5% methacrylic acid) [228].

1107. *Colophony rosin* (I)–*poly(butadiene-co-acrylonitrile)* (II), *may be compatible:* Same as System **1105** except for difference in I [228].

1108. *Rosin* (I)–*carboxylated poly(butadiene-co-acrylonitrile)* (II), *may be compatible:* Same as System **1106** except for difference in I [228].

1109. *Indene resin* (I)–*poly[ethylene-co-(vinyl acetate)]* (II), *may be compatible:* Homogeneous solution and transparent film when I was Fuji Iron Products Co. VM-1/2 and II was Nippon Goshei Co. Soalex R-CH and R-FH [37].

1110. *Miscellaneous resins* (I)–*poly[ethylene-co-(vinyl acetate)]* (II), *incompatible:* Nonhomogeneous solution and turbid film when I was Petroleum Resin, Nippon Petroleum Chemical Co. Nitsuseki Neopolymer 150, or Cumarone Resin, Mitsubishi TG, and II was as in System **1109** [37].

1111. *Indene resin* (I)–*poly(butadiene-co-acrylonitrile)* (II), *may be compatible:* Same as System **1109** except that II was Nippon Rubber Co. Hycar 1043 or 1432 [37].

1112. *Miscellaneous resins* (I)–*poly(butadiene-co-acrylonitrile)* (II), *incompatible:* Same as System **1110** except that II was as in System **1111** [37].

1113. *Phenolic resin* (I)–*poly(butadiene-co-acrylonitrile)* (II), *may be compatible:* Single dynamic mechanical transition, but much like vulcanized rubber, when I was Durez 12687, Durez Plastics and Chemicals, and II was Polyser 350, 35% acrylonitrile [246].

1114. *Indene resin* (I)–*poly[(vinyl chloride)-co-(vinyl acetate)]* (II), *may be compatible:* Homogeneous solution and transparent film when I was Fuji Iron Products Co. VM-1/2 and II was Shekisui Chemical Co. Esulex CL or CH-1 [37].

1115. *Miscellaneous resins* (I)–*poly[(vinyl chloride)-co-(vinyl acetate)]* (II), *incompatible:* Same as System **1110** except that II was Shekisui Chemical Co. Esulex C or CL or CH-1; Petroleum Resin was not tested with Esulex CL [37].

1116. *Miscellaneous resins* (I)–*poly[(vinyl chloride)-co-(vinyl propionate)]* (II), *incompatible:* Nonhomogeneous solutions and turbid films when I was as in Systems **1109** and **1110** and II was Ryuron QS-430 [37].

1117. *Indene resin* (I)–*poly[(vinyl chloride)-co-(vinyl acetate)-co-(vinyl alcohol)]* (II), *may be compatible:* Same as System **1109** except that II was Shekisui Chemical Co. Esulex A [37].

1118. *Miscellaneous resins* (I)–*poly[(vinyl chloride)-co-(vinyl acetate)-co-(vinyl alcohol)]* (II), *incompatible:* Same as System **1110** except that II was Shekisui Chemical Co. Esulex A [37].

1119. *Miscellaneous resins* (I)–*poly[(vinyl chloride)-co-(vinylidene chloride)-co-(acrylic acid ester)]* (II), *may be compatible:* Same as System **1109** except that I was as in Systems **1109** and **1110** and II was Towa Gousei Co. Aron 321 [37].

1120. *Miscellaneous resins* (I)–*poly[(vinyl acetate)-co-(N-vinylpyrrolidone)]* (II), *incompatible:* Same as System **1110** except that I was as in Systems **1109** and **1110** and II was General Aniline & Film S-630 [37].

1121. *Miscellaneous resins* (I)–*poly[styrene-co-(maleic acid)]* (II), *incompatible:* Same as System **1110** except that I was as in Systems **1109** and **1110** and II was Daidou Kogyo Styrite CM-2 or CM-3 [37].

1122. *Miscellaneous resins* (I)–*poly[styrene-co-(maleic acid ester)]* (II), *incompatible:* Same as System **1110** except that I was as in Systems **1109** and **1110** and II was Daidou Kogyo Styrite HS-2 [37].

M. Copolymers and Other Copolymers

m1. *Poly*[*ethylene-co-(maleic acid)*] (I)–*poly*[*ethylene-co-(N-n-octadecylmaleamic acid)*] (II), *incompatible:* Surface tension measurements indicated migration of II to the surface, that is, phase separation, for I/II: 500/1 to 1/1, when I was Monsanto EMA-21, alternating copolymer, $\overline{M}_w = 1.5$–2.0×10^4, and II was alternating copolymer [247].

m2. *Poly*[*ethylene-co-(vinyl acetate)*] (I)–*poly*(*butadiene-co-styrene*) (II), *incompatible:* Non-homogeneous solution and turbid film when I was Nippon Goshei Co. Soalex R-CH or R-FH and II was Nippon Rubber Co. Hycar 2057 [37].

m3. *Poly*[*ethylene-co-(vinyl acetate)*] (I)–*poly*(*butadiene-co-acrylonitrile*) (II), *incompatible:* Same as System **m2** except that II was Nippon Rubber Co. Hycar 1043 or 1432 [37].

m4. *Poly*[*ethylene-co-(vinyl acetate)*] (I)–*poly*[(*vinyl chloride*)-*co-(vinyl acetate)*] (II), *may be conditionally compatible:* Homogeneous solution and transparent film when I was Nippon Goshei Co. Soalex R-CH (IA) or R-FH (IB), and II was Shekisui Chemical Co. Esulex C or CL; same when I was IA and II was Shekisui Chemical Co. Esulex CH-1 or Denki Chemical Co. Denka Vinyl 1000 AK, but nonhomogeneous solution and turbid film when these were mixed with IB [37].

m5. *Poly*[*ethylene-co-(vinyl acetate)*] (I)–*poly*[(*vinyl chloride*)-*co-(vinyl propionate*)] (II), *incompatible:* Same as System **m2** except that I was Nippon Goshei Co. Soalex R-CH and II was Ryuron QS-430 [37].

m6. *Poly*[*ethylene-co-(vinyl acetate)*] (I)–*poly*[(*vinyl chloride*)-*co-(vinyl acetate*)-*co-(vinyl alcohol*)] (II), *may be conditionally compatible:* Homogeneous solution and transparent film when I was Nippon Goshei Co. Soalex R-CH and R-FH and II was Shekisui Chemical Co. Esulex A [37].

m7. *Poly*[*ethylene-co-(vinyl acetate)*] (I)–*poly*[(*vinyl chloride*)-*co-(vinyl acetate*)-*co-(maleic acid*)] (II), *may be conditionally compatible:* Same as System **m6** except that II was Shekisui Chemical Co. Esulex M [37].

m8. *Poly*[*ethylene-co-(vinyl acetate)*] (I)–*poly*[(*vinyl chloride*)-*co-vinylidene chloride*)-*co-(acrilic acid ester*)] (II), *may be conditionally compatible:* Same as System **m6** except that II was Towa Gousei Co. Aron 321 [37].

m9. *Poly*[*ethylene-co-(vinyl acetate)*] (I)–*poly*[(*vinyl acetate*)-*co-(N-vinylpyrrolidone*)] (II), *incompatible:* Same as System **m2** except that II was General Aniline & Film S-630 [37].

m10. *Poly*[*ethylene-co-(vinyl acetate)*] (I)–*poly*[*styrene-co-(maleic acid*)] (II), *may be conditionally compatible:* Nonhomogeneous solution and turbid film when I was Nippon Goshei Co. Soalex R-CH and II was Daidou Kogyo Styrite CM-2 (IIA) or Styrite CM-3 (IIB); homogeneous solution and transparent film when I was Soalex R-FH; single T_g by DTA when I was Soalex R-FH and II was IIB [37].

m11. *Poly*[(*ethylene-co-(vinyl acetate*))](I)–*poly*[*styrene-co-(maleic acid ester*)] (II), *may be conditionally compatible:* Nonhomogeneous solution and turbid film when I was Nippon Goshei Co. Soalex R-CH and II was Daidou Kogyo Styrite HS-2, but homogeneous solution and transparent film when I was Soalex R-FH [37].

m12. *Poly*(*isobutene-co-isoprene*) (I)–*poly*(*butadiene-co-styrene*) (II), *incompatible:* Transition of I seen for 50–90% I in mixture when II was SBR-1000 and I was Butyl 218 (1.5–2.0% unsaturated) [242].

m13. *Poly*(*isobutene-co-isoprene*) (I)–*poly*(*butadiene-co-acrylonitrile*) (II), *incompatible:* Transition of I and II seen when II was Hycar 1042 NBR (Goodrich), medium acrylonitrile content, and I was Butyl 218 (1.5–2.0% unsaturated) [242].

m14. *Poly*(*butadiene-co-styrene*) (I)–*poly*(*butadiene-co-acrylonitrile*) (II), *ambiguous* (*must be conditionally compatible):* Two dynamic mechanical and dielectric loss maxima for 1/1 mixture when I was SKS-30 and II was SKN-26 [248]. Incompatible by phase contrast and electron microscopy when I was SBR 1500 and II was high or medium nitrile [106]. Stated compatible for I/II: 4/1 to 1/4 in solutions and films when I was Buna S3, ~80% butadiene and II was Buna NW, 28% acrylonitrile [16]. Phase separation in benzene at 5% solids for 1/1 mixture when I was SKN-18 (18% acrylonitrile) and II was SKS-30 (30% styrene) [99].

m15. *Poly(butadiene-co-styrene)* (I)–*poly[(vinyl chloride)-co-(vinyl acetate)]* (II), *may be conditionally compatible:* Nonhomogeneous solution and turbid film when I was Nippon Rubber Co. Hycar 2057 and II was Shekisui Chemical Co. Esulex C or CL or CH-1 [37]. Stated compatible for I/II: 4/1 to 1/4 in solutions and films when I was Buna S3, ~80% butadiene, and II was Vinalit MPS [16].

m16. *Poly(butadiene-co-styrene)* (I)–*poly[(vinyl acetate)-co-(N-vinylpyrrolidone)]* (II), *incompatible:* Nonhomogeneous solution and turbid film when I was Nippon Rubber Co. Hycar 2057 and II was General Aniline and Film S-630 [37].

m17. *Poly(butadiene-co-styrene)* (I)–*poly(styrene-co-acrylonitrile)* (II), *incompatible* (but must be conditionally compatible): Two phases by dilatometry or dynamic mechanical properties when I was Japanese Synthetic Rubber Co. SBR, 24.5% styrene, and II had 25% or 50% acrylonitrile [118].

m18. *Poly(butadiene-co-styrene)* (I)–*poly[styrene-co-(maleic acid)]* (II), *incompatible* (but must be conditionally compatible): Same as System **m16** except that II was Daidou Kogyo Styrite CM-2 or CM-3 [37].

m19. *Poly(butadiene-co-styrene)* (I)–*poly[styrene-co-(maleic acid ester)]* (II), *incompatible* (but must be conditionally compatible): Same as System **m16** except that II was Daidou Kogyo Styrite HS-2 [37].

m20. *Poly(butadiene-co-styrene)* (I)–*polysulfide polymer* (II), *ambiguous:* Variations of T_g in mixtures that might indicate compatibility when I was SBR-1000 and II was thiokol FA (ethylene dichloride, di-2-chloroethyl formal, and $Na_2S_{1.8}$) [242].

m21. *Poly(butadiene-co-acrylonitrile)*(I)–*poly[(vinyl chloride)-co-(vinyl acetate)]* (II), *probably conditionally compatible:* No phase separation in solutions and films for a number of mixtures of commercial polymers [37, 79]. Stated conditionally compatible for I/II: 4/1 to 1/4 when I was Buna NW, 28% acrylonitrile, and II was Vinalit MPS [16].

m22. *Poly(butadiene-co-acrylonitrile)* (I)–*poly[(vinyl chloride)-co-(vinyl acetate)-co-(vinyl alcohol)]* (II), *may be conditionally compatible:* Homogeneous solution and transparent film when I was Nippon Rubber Co. Hycar 1043 or 1432 and II was Shekisui Chemical Co. Esulex A. [37].

m23. *Poly(butadiene-co-acrylonitrile)* (I)–*poly[(vinyl chloride)-co-(vinyl acetate)-co-(maleic acid)]* (II), *may be conditionally compatible:* Same as System **m22** except that II was Shekisui Chemical Co. Esulex M [37].

m24. *Poly(butadiene-co-acrylonitrile)* (I)–*poly[(vinyl chloride)-co-(vinyl propionate)]* (II), *may be conditionally compatible:* Same as System **m22** except that II was Ryuron QS-430 [37].

m25. *Poly(butadiene-co-acrylonitrile)* (I)–*poly[(vinyl chloride)-co-(vinylidene chloride)-co-(acrylic acid ester)]* (II), *may be conditionally compatible:* Same as System **m22** except that II was Towa Gousei Co. Aron 321 [37].

m26. *Poly(butadiene-co-acrylonitrile)* (I)–*poly[(vinyl acetate)-co-(N-vinylpyrrolidone)]* (II), *incompatible:* Nonhomogeneous solution and turbid film when I was Nippon Rubber Co. Hycar 1043 or 1432 and II was General Aniline and Film S-630 [37].

m27. *Poly(butadiene-co-acrylonitrile)* (I)–*poly[styrene-co-(methyl methacrylate)]* (II), *incompatible:* Phase separation in cyclohexanone for 1/1 mixture, and film was not clear when II was Zerlon 150, Dow Chemical Co., and I was Hycar 1432, B. F. Goodrich [79].

m28. *Poly(butadiene-co-acrylonitrile)* (I)–*poly(styrene-co-acrylonitrile)* (II), *must be conditionally compatible:* Same as System **m27** except that II was Dow Chemical Co. Tyril 767 [79]. Stated to give homogeneous blend for 70–100 wt % II when I had 65% butadiene and II had 80% styrene [95].

m29. *Poly(butadiene-co-acrylonitrile)* (I)–*poly[styrene-co-(maleic acid)]* (II), *incompatible:*

Same as System **m26** except that II was Daidou Kogyo Styrite CM-2 or CM-3 [37].

m30. *Poly(butadiene-co-acrylonitrile)* (I)–*poly[styrene-co-(maleic acid ester)]* (II), *incompatible:* Same as System **m26** except that II was Daidou Kogyo Styrite HS-2 [37].

m31. *Poly(butadiene-co-acrylonitrile)* (I)–*carboxylated poly(butadiene-co-acrylonitrile)* (II), *must be conditionally compatible:* No phase separation in 3/1 xylene/butyl acetate at 10% solids for 1/1 volume ratio when I was SKN-26 (26% acrylonitrile) and II was SKN-26-1 (26% acrylonitrile, 5% methacrylic acid) [228].

m32. *Poly[(vinyl chloride)-co-(vinyl acetate)]* (I)–*poly[(vinyl chloride)-co-(vinyl propionate)]* (II), *must be conditionally compatible:* Homogeneous solution and transparent film when I was Shekisui Chemical Co. Esulex C or CL or CH-1 and II was Ryuron QS-430 [37].

m33. *Poly[(vinyl chloride)-co-(vinyl acetate)]* (I)–*poly[(vinyl chloride)-co-(vinyl acetate)-co-(vinyl alcohol)]* (II), *must be conditionally compatible:* Homogeneous solution and transparent film when I was Shekisui Chemical Co. Esulex C or CL or CH-1 or Denki Chemical Co. Denka Vinyl 1000 AK, and II was Shekisui Chemical Co. Esulex A [37].

m34. *Poly[(vinyl chloride)-co-(vinyl acetate)]* (I)–*poly[(vinyl chloride)-co-(vinyl acetate)-co-(maleic acid)]* (II), *must be conditionally compatible:* Same as System **m33** except that II was Shekisui Chemical Co. Esulex M [37].

m35. *Poly[(vinyl chloride)-co-(vinyl acetate)]* (I)–*poly[(vinyl chloride)-co-(vinylidene chloride)-co-(acrylic acid ester)]* (II), *must be conditionally compatible:* Same as System **m33** except that II was Towa Gousei Co. Aron 321 [37].

m36. *Poly[(vinyl chloride)-co-(vinyl acetate)]* (I)–*poly[(vinyl acetate)-co-(N-vinylpyrrolidone)]* (II), *must be conditionally compatible:* Nonhomogeneous solution and turbid film when I was Shekisui Chemical Co. Esulex C or CL or Denki Chemical Co. Denka Vinyl 1000 AK, and II was General Aniline and Film S-630, but homogeneous solution and transparent film when I was Esulex CH-1 [37].

m37. *Poly[(vinyl chloride)-co-(vinyl acetate)]* (I)–*poly[styrene-co-(methyl methacrylate)]* (II), *incompatible:* Phase separation in cyclohexanone for 1/1 mixture, and film was not clear when I was Geon 440 × 24, B. F. Goodrich, and II was Zerlon 150, Dow Chemical Co. [79].

m38. *Poly[(vinyl chloride)-co-(vinyl acetate)]* (I)–*poly(styrene-co-acrylonitrile)* (II), *incompatible:* Same as System **m37** except that II was Dow Chemical Co. Tyril 767 [79].

m39. *Poly[(vinyl chloride)-co-(vinyl acetate)]* (I)–*poly[styrene-co-(maleic acid)]* (II), *incompatible:* Nonhomogeneous solution and turbid film when I was as in System **m33** and II was Daidou Kogyo Styrite CM-2 or CM-3 [37].

m40. *Poly[(vinyl chloride)-co-(vinyl acetate)]* (I)–*poly[(styrene-co-(maleic acid ester)]* (II), *incompatible:* Same as System **m39** except that II was Daidou Kogyo Styrite HS-2 [37].

m41. *Poly[(vinyl chloride)-co-(vinyl acetate)]* (I)–*poly[epichlorohydrin-co-(ethylene oxide)]* (II), *incompatible:* Same as System **m37** except that II was B. F. Goodrich Hydrin 200 [79].

m42. *Poly[(vinyl chloride)-co-(vinyl acetate)]* (I)–*poly[(methyl vinyl ether)-co-(maleic anhydride)]* (II), *incompatible:* Same as System **m37** except that II was B. F. Goodrich Gantrez AN-169 [79].

m43. *Poly[(vinyl chloride)-co-(vinyl acetate)]* (I)–*carboxylated poly(butadiene-co-acrylonitrile)* (II), *may be conditionally compatible:* No phase separation in 3/1 xylene/butyl acetate at 10% solids for 1/1 volume ratio when II was SKN-26-1 (26% acrylonitrile, 5% methacrylic acid) and I was A-15 [228].

m44. *Poly[(vinyl chloride)-co-(vinyl propionate)]* (I)–*poly[(vinyl chloride)-co-(vinyl acetate)-co-(vinyl alcohol)]* (II), *must be conditionally compatible:* Homogeneous solution and transparent film when I was Ryuron QS-430 and II was Shekisui Chemical Co. Esulex A [37].

m45. *Poly[(vinyl chloride)-co-(vinyl propionate)]* (I)–*poly[(vinyl chloride)-co-(vinylidene chloride)-co-(acrylic acid ester)]* (II), *must be conditionally compatible:* Same as System **m44** except that II was Towa Gousei Co. Aron 321 [37].

m46. *Poly[(vinyl chloride)-co-(vinyl propionate)]* (I)–*poly[styrene-co-(maleic acid)]* (II), *incompatible:* Nonhomogeneous solution and turbid film when I was Ryuron QS-430 and II was Daidou Kogyo Styrite CM-2 or CM-3 [37].

m47. *Poly[(vinyl chloride)-co-(vinyl propionate)]* (I)–*poly[(styrene-co-(maleic acid ester)]* (II), *incompatible:* Same as System **m46** except that II was Daidou Kogyo Styrite HS-2 [37].

m48. *Poly[(vinyl chloride)-co-(vinyl acetate)-co-(vinyl alcohol)]* (I)–*poly[(vinyl chloride)-co-(vinyl acetate)-co-(maleic acid)]* (II), *must be conditionally compatible:* Homogeneous solution and transparent film when I was Shekisui Chemical Co. Esulex A and II was Shekisui Chemical Co. Esulex M [37].

m49. *Poly[(vinyl chloride)-co-(vinyl acetate)-co-(vinyl alcohol)]* (I)–*poly[(vinyl chloride)-co-(vinylidene chloride)-co-(acrylic acid ester)]* (II), *must be conditionally compatible:* Same as System **m48** except that II was Towa Gousei Co. Aron 321 [37].

m50. *Poly[(vinyl chloride)-co-(vinyl acetate)-co-(maleic acid)]* (I)–*poly[(vinyl chloride)-co-(vinylidene chloride)-co-(acrylic acid ester)]* (II), *must be conditionally compatible:* Same as System **m49** except that I was Shekisui Chemical Co. Esulex M [37].

m51. *Poly[(vinyl acetate)-co-(N-vinylpyrrolidone)]* (I)–*poly[(vinyl acetate)-co-(vinyl chloride)-co-(vinyl alcohol)]* (II), *must be conditionally compatible:* Homogeneous solution and transparent film when I was General Aniline and Film S-630 and II was Shekisui Chemical Co. Esulex A [37].

m52. *Poly[(vinyl acetate)-co-(N-vinylpyrrolidone)]* (I)–*poly[(vinyl acetate)-co-(vinyl chloride)-co-(maleic acid)]* (II), *must be conditionally compatible:* Same as System **m51** except that II was Shekisui Chemical Co. Esulex M [37].

m53. *Poly[(vinyl acetate)-co-(N-vinylpyrrolidone)]* (I)–*poly[(vinyl chloride)-co-(vinylidene chloride)-co-(acrylic acid ester)]* (II), *must be conditionally compatible:* Same as System **m51** except that II was Towa Gousei Co. Aron 321 [37].

m54. *Poly[(vinyl acetate)-co-(N-vinylpyrrolidone)]* (I)–*poly[styrene-co-(maleic acid)]* (II), *incompatible:* Nonhomogeneous solution and turbid film when I was General Aniline and Film S-630 and II was Daidou Kogyo Styrite CM-2 or CM-3 [37].

m55. *Poly[(vinyl acetate)-co-(N-vinylpyrrolidone)]* (I)–*poly[styrene-co-(maleic acid ester)]* (II), *incompatible:* Same as System **m54** except that II was Daidou Kogyo Styrite HS-2 [37].

m56. *Poly[styrene-co-(methyl methacrylate)]* (I)–*poly(styrene-co-acrylonitrile)* (II), *incompatible (but must be conditionally compatible:* Phase separation in cyclohexanone for 1/1 mixture, and film was not clear when II was Tyril 767, Dow Chemical Co. and I was Zerlon 150, Dow Chemical Co. [79]. Stated not miscible for 1/1 mixtures when I had 87% methyl methacrylate and II had 65% or 75% styrene [219].

m57. *Poly[styrene-co-(methyl methacrylate)]* (I)–*poly[epichlorohydrin-co-(ethylene oxide)]* (II), *incompatible:* No phase separation in cyclohexanone for 1/1 mixture, but film was not clear when I was Zerlon 150, Dow Chemical Co, and II was Hydrin 200, B. F. Goodrich [79].

m58. *Poly[styrene-co-(methyl methacrylate)]* (I)–*poly[(methyl vinyl ether)-co-(maleic anhydride)]* (II), *incompatible:* Same as System **m57** except that II was B. F. Goodrich Gantrez AN-169 [79].

m59. *Poly(styrene-co-acrylonitrile)* (I)–*poly[epichlorohydrin-co-(ethylene oxide)]* (II), *incompatible:* Same as System **m57** except that I was Dow Chemical Co. Tyril 767 [79].

m60. *Poly(styrene-co-acrylonitrile)* (I)–*poly[(methyl vinyl ether)-co-(maleic anhydride)]* (II), *incompatible:* Same as System **m58** except that I was Dow Chemical Co. Tyril 767 [79].

m61. *Poly[styrene-co-(maleic acid)]* (I)–*poly[styrene-co-(maleic acid ester)]* (II), *must be conditionally compatible:* Homogeneous solution and transparent film when I was Daidou Kogyo Styrite CM-2 or CM-3 and II was Styrite HS-2 [37].

m62. *Poly[styrene-co-(maleic acid)]* (I)–*Poly[(vinyl chloride)-co-(vinyl acetate)-co-(vinyl alcohol)]* (II), *incompatible:* Nonhomogeneous solution and turbid film when I was Daidou

Kogyo Styrite CM-2 or CM-3 and II was Shekisui Chemical Co. Esulex A [37].

m63. *Poly[styrene-co-(maleic acid)]* (I)–*poly[(vinyl chloride)-co-(vinyl acetate)-co-(maleic acid)]* (II), *incompatible* (*but must be conditionally compatible*)*: Same as System **m62** except that II was Shekisui Chemical Co. Esulex M [37].

m64. *Poly[styrene-co-(maleic acid)]* (I)–*poly[(vinyl chloride)-co-(vinylidene chloride)-co-(acrylic acid ester)]* (II), *incompatible*: Same as System **m62** except that II was Towa Gousei Co. Aron 321 [37].

m65. *Poly[styrene-co-(maleic acid ester)]* (I)–*poly[(vinyl chloride)-co-(vinyl acetate)-co-(vinyl alcohol)]* (II), *incompatible*: Nonhomogeneous solution and turbid film when I was Daidou Kogyo Styrite HS-2 and II was Shekisui Chemical Co. Esulex A [37].

m66. *Poly[styrene-co-(maleic acid ester)]* (I)–*poly[(vinyl chloride)-co-(vinyl acetate)-co-(maleic acid)]* (II), *incompatible*: Same as System **m65** except that II was Shekisui Chemical Co. Esulex M [37].

m67. *Poly[styrene-co-(maleic acid ester)]* (I)–*poly[(vinyl chloride)-co-(vinylidene chloride)-co-(acrylic ester)]* (II), *incompatible*: Same as System **m65** except that II was Towa Gousei Co. Aron 321 [37].

m68. *Poly[styrene-co-(butyl acrylate)-co-(methacrylic acid)-co-(hydroxyethl methacrylate)-co-(butylated melamine)-co-formaldehyde)]* (I)–*poly(urethane-co-silicone-co-oxalkylene)* (II), *may be conditionally compatible*: One T_g by DSC for I/II: 1/3 to 3/1 for cross-linked interpenetrating network when I was Inmont Corp. Acrylic 342-CD-725 cross-linked with Inmont Corp. Melamine RU 522, and II was prepared by reaction of 4,4′-methylenebis(cyclohexyl isocyanate) with di-*n*-butylamine plus Union Carbide Silicone L-522, a poly(dimethylsiloxane)–poly(oxalkylene) copolymer [249].

m69. *Poly[(ethyl acrylate)-co-(methyl methacrylate)]* (I)–*poly[(butyl acrylate)-co-(methacrylate)]* (II), *incompatible* (*but must be conditionally compatible*): Phase separation in $CHCl_3$ above 5 gm solids/dl at 25°C when I and II each contained 50% methyl methacrylate and I had $\overline{M}_v \sim 2 \times 10^6$, and II had $\overline{M}_v \sim 3 \times 10^6$; phase separation at higher threshold concentration than when I was poly(ethyl acrylate), $\overline{M}_v = 1.2 \times 10^6$, and II was poly(butyl acrylate), $\overline{M}_v = 1.5 \times 10^6$ [185].

m70. *Poly[(ethyl acrylate)-co-(acrylic acid)]* (I)–*poly[(ethyl acrylate)-co-(methacrylic acid)]* (II), *incompatible* (*but must be conditionally compatible*): Phase separation in methanol at 10 gm solids/dl for 1/1 mixture at 25°C, but no phase separation in DMF or dimethyl sulfoxide at 20 gm solids/dl when I contained 50% acrylic acid, $\overline{M}_v \sim 8 \times 10^5$, and II contained 53% methacrylic acid, $\overline{M}_v \sim 9 \times 10^5$; phase separation when sodium or ammonium salts of I and II were used [185].

m71. *Poly[(ethyl acrylate)-co-(methacrylic acid)]* (I)–*poly[(methyl methacrylate)-co-(methacrylic acid)]* (II), *incompatible* (*but must be conditionally compatible*): Phase separation in methanol or ethanol at 10 gm/dl at 25°C but no phase separation in DMF or dimethyl sulfoxide at 20 gm/dl when I and II each contained 53% MAA, and I had $\overline{M}_v \sim 9 \times 10^5$, and II had $\overline{M}_v \sim 5 \times 10^5$; the sodium salts of I and II showed phase separation at 15 gm/dl [185]. Phase separation in acetone at 15 gm/dl at 25°C; phase separation in DMF at 20 gm/dl but not at 10 gm/dl; and phase separation in dimethyl sulfoxide at 20 gm/dl when I and II each contained 28% MAA and I had $\overline{M}_v \sim 10^6$ and II had $\overline{M}_v \sim 5 \times 10^5$ [185].

N. Mixtures of Three Polymers

n1. *Polyethylene* (I)–*poly(vinyl chloride)* (II)–*poly[ethylene-co-(N-methyl-N-vinyl acetamide)]* (III), *may be conditionally compatible*: Homogeneous blend for I/II/III/glass fibers: 2/10/1/3 when II contained 1.5 wt % ethylene and 2% stabilizer, and III was 81.6 wt % ethylene [116].

n2. *Polychloroprene* (I)–*poly(vinyl chloride)* (II)–*polyacrylonitrile* (III), *ambiguous, may be conditionally compatible:* Stated to be compatible when I was Nairit NT at some ratios of I/II/III but only mechanical properties given [250].

n3. *Polychloroprene* (I)–*poly(vinyl chloride)* (II)–*poly(vinylidene chloride)* (III), *ambiguous, may be conditionally compatible:* Stated to be compatible when I was Nairit NT at some ratios of I/II/III but only mechanical properties given [250].

n4. *Polychloroprene* (I)–*poly(vinyl chloride)* (II)–*poly(butadiene-co-acrylonitrile)* (III), *ambiguous:* Stated to be compatible because η_{rel} of mixtures in solution are higher than those expected by adding η_{rel} of the components [91].

n5. *Polychloroprene* (I)–*poly(vinylidene chloride)* (II)–*poly(vinyl acetate)* (III), *ambiguous, may be conditionally compatible:* Stated to be compatible when I was Nairit NT at some ratios of I/II/III but only mechanical properties given [250].

n6. *Polychloroprene* (I)–*poly(vinyl acetate)* (II)–*polyacrylonitrile* (III), *ambiguous, may be conditionally compatible:* Stated to be compatible when I was Naitrit NT at some ratios of I/II/III but only mechanical properties given [250].

n7. *Poly(vinyl chloride)* (I)–*poly(ε-caprolactone)* (II)–*poly[ethylene-co-(N-methyl-N-vinyl acetamide)]* (III), *may be conditionally compatible:* Melt clear and transparent for I/II/III: 3/1/6 when I contained 1.5 wt % ethylene and 2% stabilizer and III had 81.6 wt % ethylene [116].

n8. *Polystyrene* (I)–*poly(methyl methacrylate)* (II)–*poly[styrene-co-(methyl methacrylate)]* (III), *incompatible (except near 100% III):* Films cast from CHCl$_3$ or benzene all hazy even when heated to 110°C for 2 weeks or 140°C for 1 week when I had $\overline{M}_v = 4.7 \times 10^4$, II had $\overline{M}_v = 3.9 \times 10^4$, and III was 18% conversion, $\overline{M}_v = 1.36 \times 10^6$, 30% styrene; films were all hazy except for compositions near 100% III when III was same as above, but I had $\overline{M}_v = 4.50 \times 10^5$ and II had $\overline{M}_v = 8.7 \times 10^5$; films were all hazy except for compositions near 100% III or II when I had $\overline{M}_v = 4.7 \times 10^4$, II had $\overline{M}_v = 3.30 \times 10^5$ and III was 12.3% conversion, $\overline{M}_v = 2.5 \times 10^5$, 50% styrene [251].

O. Same Copolymer but Different Compositions

o1. *Poly[ethylene-co-(vinyl acetate)]* (I, II different compositions), *compatible, probably conditionally:* Homogeneous solution and transparent film when I was Nippon Goshei Co. Soalex R-CH and II was Soalex R-FH [37].

o2. *Chlorinated polyethylene* (I, II different compositions), *conditionally compatible:* A single loss peak when I contained 66 wt % Cl and II contained 62 wt % Cl or 68 wt % Cl; incompatible when I contained 62 wt % Cl and II contained 68 wt % Cl [252]. *Note:* The T_g's of copolymers with similar compositions may be so close together that only a single loss peak is observed even when the system is incompatible.

o3. *Poly(isoprene-co-acrylonitrile)* (high conversion), *conditionally compatible:* Two T_g's for high conversion, 19.6% acrylonitrile, single T_g for high conversion, 32.5% acrylonitrile [253].

o4. *Poly(butadiene-co-styrene)* (I, II, etc. different compositions), *conditionally compatible:* No phase separation in gasoline at 5% solids for 1/1 mixture when I was SKS-30 (30% styrene) and II was SKS-10 (10% styrene) [99]. Two-phase structures by dynamic mechanical measurements and photoelastic methods when random copolymers, as prepared by anionic polymerization, differed in composition by 20% or more [254]; five samples used, 9.7% to 49.3% styrene, 32–45% cis, 40–53% trans, 14–15% vinyl, $\overline{M}_w = 1.2 - 2.0 \times 10^5$, $\overline{M}_w/\overline{M}_n = 1.15 - 1.21$, antioxidant added, two transitions for mixture of I with 9.7% styrene and II with 49.3% styrene, single broad dispersion for mixture of III with 19.6% styrene and IV with 39.6% styrene or for mixture of I, II, III, IV and V with 29.5% styrene, a single narrow dispersion for mixture of III, IV, and V [255]. Damping maximum of II only for I/II: 1/3 to

3/1 when I was Texas-U.S. Chemical Co. Synpol 8107, 5% styrene, and II was U.S. Rubber Co. Kralac A-EP, 85% styrene [163]. Phase contrast and electron microscopy showed zones when I was SBR-40, 40% styrene, and II was SBR-10, 10% styrene [96]. Single, sharp, dynamic mechanical transition and no structure by phase contrast microscopy when I had 16% styrene and II had 23.5% styrene; two distinct dynamic mechanical transitions and structure by phase contrast microscopy when I or II were mixed with III, 50% styrene; single dynamic mechanical transition but structure by phase contrast microscopy for mixtures of I, II, or III with IV, 37.5% styrene, or I/III/IV, or II/III/IV or I/II/III/IV [102].

   o5. *Poly(butadiene-co-acrylonitrile)* (I, II different compositions, and high conversion samples), *conditionally compatible:* A single T_g for 1/3 to 3/1 ratio of I/II by linear expansion when I was SKN-18 (18% acrylonitrile) and II was SKN-40 (40% acrylonitrile) [98]. Two T_g's for emulsion polymerized, high conversion samples containing from 20 to 35% acrylonitrile; one T_g for samples containing > 35% acrylonitrile [256]. Shear modulus, DTA, and electron microscopy indicated a single T_g plus a melting range for *cis*-1,4-polybutadiene sequences for Hycar 1014, 20 wt % acrylonitrile, or for Hycar 1041, 40 wt % acrylonitrile [257]. Two T_g's in all commercial and high conversion laboratory samples when percent acrylonitrile < 35% [253]. Two T_g's for ⩽ 33% acrylonitrile, but single T_g for > 33% acrylonitrile for 18 commercial polymers; samples prepared to have narrow composition distribution had single T_g even when percent acrylonitrile ⩽ 33%; two T_g's when two samples with narrow composition distribution were mixed; for 20% acrylonitrile copolymer, a single T_g up to 50% conversion but two T_g's at ⩾ 65% conversion [258]. Single dynamic mechanical and dielectric loss maximum for I/II: 1/4 to 4/1 when I was SKN-18 and II was SKN-40 [248]. Single T_g for high conversion samples with 11% or 42–63% acrylonitrile, but two T_g's for high conversion samples with 20–35% acrylonitrile [259]. Homogeneous solution and transparent film when I and II were Nippon Rubber Co. Hycar 1043 and 1432 [37]. Single T_g by dilatometry or rolling ball loss spectrometer for 1/1 mixture when I was SKN-18, 18% acrylonitrile and II was SKN-40, 40% acrylonitrile [101].

   o6. *Poly[(vinyl chloride)-co-(vinyl acetate)]* (I, II different compositions), *may be conditionally compatible:* Homogeneous solution and transparent film when I and II were any of the following: Shekisui Chemical Co. Esulex C or CL or CH-1, or Denki Chemical Co. Denka Vinyl 1000 AK [37].

   o7. *Poly[(vinyl chloride)-co-(methyl acrylate)]* (high-conversion sample), *compatible:* A single damping maximum obtained for 100% conversion sample, 59% vinyl chloride [η] = 2.37, heterogeneous in composition by fractionation [135].

   o8. *Poly[(vinyl chloride)-co-(cyanoethoxyethyl acrylate)]* (high-conversion sample), *incompatible:* Two damping maxima for 100% conversion, 49% vinyl chloride [135].

   o9. *Poly[styrene-co-(methyl acrylate)]* (high-conversion sample); *compatible:* A single damping maxima obtained for 100% conversion samples, 55% styrene, [η] = 2.40, heterogeneous in composition by fractionation [135].

   o10. *Poly[styrene-co-(methyl methacrylate)]* (high-conversion sample), *probably conditionally compatible:* Cloudy film with a single, broadened damping maximum for a high conversion sample [170].

   o11. Poly(*styrene-co-acrylonitrile*) (I, II or I, II, III different compositions or high conversion), *conditionally compatible:* Phase separation in absence of solvent when I and II differed by 3.5–4.5% acrylonitrile; phase separation at 40% solids in 2-butanone when I and II differed by 8–9.5% acrylonitrile; phase separation at 20% solids in 2-butanone when I and II differed by 10–13% acrylonitrile and all $\overline{M}_n$ were from 8×10^4 to 1.10×10^5, all $\overline{M}_w$ were from 1.40×10^5 to 1.70×10^6, all $\overline{M}_w/\overline{M}_n$ were from 1.3 to 2.0, all mixtures were 1/1 weight ratio of I/II; addition of copolymer of intermediate composition to a mixture did *not* compatibilize the mixture [260]. When I, II, and III all have $\overline{M}_n = (8.0–11.0) \times 10^4$, $\overline{M}_w = (1.4–1.7) \times 10^5$, $\overline{M}_w/\overline{M}_n = 1.3–2.0$, phase separation at 25% solids in 2-butanone for 1/1/1 ratio of I/II/III, when the compositions of any two of the three copolymers differed by more than 13.4% acrylonitrile; phase separation for the same copolymers at 25% solids in 2-butanone for 4.2% I: 16.6% II: 4.2% III in solution when the compositions of I and II differed by more than 19.4% acrylonitrile, even when II had an intermediate composition [260]. Azeotropic copolymers had one T_g by DSC and torsional pendulum; 28% acrylonitrile copolymer polymerized at 50°C had one T_g at 57% conversion and two T_g's ≥ 79% conversion; two T_g's for 1/1 mixtures of single T_g copolymers where I had 24,5% acrylonitrile and II had 31% acrylonitrile [261].

o12. *Poly[styrene-co-(maleic acid)* (I, II different compositions), *may be conditionally compatible:* Homogeneous solution and transparent film when I was Daidou Kogyo Styrite CM-2 and II was Styrite CM-3 [37].

o13. *Poly[styrene-co-acrylonitrile-co-(α-methylstyrene)]* (I, II different compositions), *conditionally compatible:* Samples of the following ratios of styrene/acrylonitrile/α-methylstyrene were observed: No. 2 had 65/20/15, No. 3 had 60/20/15, No. 4 had 55/33/15, No. 15 had 76/24/0, No. 16 had 70/25/5, No. 17 had 5/30/65, and No. 18 had 0/31/69; transparent films were observed for mixtures of 2 + 3, 4 + 3, 15 + 3, 16 + 3, 17 + 3, 18 + 3, 15 + 16, 15 + 17, 15 + 18, 16 + 17, 16 + 18, 17 + 18, and 15 + 16 + 17 + 18, a hazy film was obtained for 2 + 4 [262].

o14. *Poly[(methyl acrylate)-co-(methyl methacrylate)]* (*MA–MMA*) (I, II different compositions), *conditionally compatible:* Film turbidity used to determine that minimum Δ_{MMA} between I and II for phase separation varied from 28 mole % when copolymer with smallest % MMA (I) had 0% MMA, to 48 mole % when I had 40% MMA at degree of polymerization 700; Δ_{MMA} varied from 20 to 28 mole % when I had 0–40% MMA to 35 mole % when I had 58% MMA at degree of polymerization 3000 [263].

o15. *Poly[(ethyl acrylate)-co-(methyl methacrylate)]* (EA–MMA); (I, II different compositions), *conditionally compatible:* Film turbidity used to determine that minimum Δ_{MMA} between I and II for phase separation varied from 20 to 27% when the copolymer with smallest percent MMA had 0–60% MMA [263].

o16. *Poly[(ethyl acrylate)-co-(methacrylic acid)]* (I, II different compositions), *incompatible* (but must be conditionally compatible): Phase separation in methanol at 10–20 gm/dl at 25°C when I contained 47% EA, $\overline{M}_v \sim 9 \times 10^5$, and II contained 72% EA, $\overline{M}_v \sim 10^6$; the sodium salts of I and II gave opaque solutions in water at 20 gm/dl [185].

o17. *Poly[(butyl acrylate)-co-(methyl methacrylate)]* (*BA–MMA*) (I, II different compositions and high-conversion polymer), *conditionally compatible:* When I and II were prepared each with 3 mole % scatter in composition and all mixtures were made to contain 50 mole % MMA, then clear films became turbid when Δ_{MMA} between I and II ≥13%; stated when more than one phase present when Δ_{MMA} ≥ 10%; torsional modulus showed two T_g's at $\Delta_{MMA} = 20\%$; "natural" copolymers showed phase separation at degree of polymerization 2000 but not at 600 [264]. When I and II were 8–10% conversion copolymers, cast films of blends became turbid when Δ_{MMA} varied from 17 mole % when the copolymer with the smaller percent MMA (I) had 65% MMA, to 25 mole % when I had 20% MMA, at degree of polymerization 700; Δ_{MMA} varied from 10 mole % when I had 82% MMA to 22 mole % when I had 8% MMA at degree of polymerization 3000; Δ_{MMA} varied from 8 mole % when I had 82% MMA to 18 mole % when I had 10% MMA at degree of polymerization 4000; two phase in electron micrographs when films were not clear; two T_g's by torsional modulus where films were very opaque; high-conversion polymers with 50% MMA clear when degree of polymerization < 1000 but opaque and inhomogeneous by torsional modulus when degree of polymerization 2000 [263].

o18. *Poly[(methyl methacrylate)-co-(butyl methacrylate)]* (*MMA–BMA*) (I, II different compositions), *conditionally compatible:* Film turbidity used to determine that minimum Δ_{MMA} between I and II for phase separation varied from 30% when the copolymer with the smaller percent MMA (I) had 0% MMA, to 20% when I had 70% MMA, for degree of polymerization 3000 or 5000 [263].

o19. Poly[(methyl methacrylate)-co-(methacrylic acid)] (MMA–MAA) (I, II different compositions), must be conditionally compatible: Phase separation in DMF at 10 and 15 gm/dl at 25°C but no phase separation at 20 gm/dl in dimethyl sulfoxide when I contained 47% MMA, $\overline{M}_v \sim 5 \times 10^5$, and II contained 72% MMA, $\overline{M}_v \sim 5 \times 10^5$, the sodium salts of I and II showed phase separation at 12 gm/dl [129].

ACKNOWLEDGMENTS

I would like to acknowledge help in deciphering papers in the Slavic languages from Ms. Maria Novak, and in the Japanese language from Mr. N. Toyota and Dr. Y. Obata. In some cases in which I have referenced a paper printed in two different languages, I am in possession only of the English version of the paper. I wish to take responsibility for any mistakes in deciphering those papers that I was unable to read without help. Persons who sent me papers of interest have been acknowledged in Section I of this Chapter. I would also like to acknowledge support from the National Institutes of Health under a Career Development Award and from the National Science Foundation under Research Grant No. DMR76-19488.

REFERENCES

1.   S. Krause, J. Macromol. Sci.-Rev. Macromol. Chem. **C7**, 251 (1972).
2.   P. J. Flory, "Principles of Polymer Chemistry." Cornell Univ. Press, Ithaca, New York, 1953.
3.   H. Tompa, "Polymer Solutions." Butterworth, London, 1956.
4.   L. Bohn, Kolloid Z. Z. Polym. **213**, 55 (1966); Rubber Chem. Technol. **41**, 495 (1968).
5.   E. M. Fettes and W. N. Maclay, J. Appl. Polym. Sci. Symp. **7**, 3 (1968).
6.   M. J. Voorn, Fortschr. Hochpolym. Forsch. **1**, 192 (1959).
7.   K. Friese, Plaste Kaut. **12**, 90 (1965).
8.   K. Friese, Plaste Kaut. **13**, 65 (1966).
9.   K. Thinius, Plaste Kaut. **15**, 164 (1968).
10. H. Gerrens, Chem. Ing. Tech. **39**, 1053 (1967).
11. K. Schmieder, in "Kunststoffe" (R. Nitsche and K. A. Wolf, eds.), Vol. 1, Structure and Physical Behavior of Plastics, p. 791. Springer Verlag, Berlin, 1962.
12. P. J. Corish and B. D. W. Powell, Rubber Chem. Technol. **47**, 481 (1974).
13. H. Tlusta and J. Zelinger, Chem. Listy **65**, 1143 (1971).
14. T. Pazonyi and M. Dimitrov, Magyar Kem. Lapja **7**, 335 (1963); Rubber Chem. Technol. **40**, 1119 (1967).
15. V. N. Kuleznev and L. S. Krokhina, Akad. Nauk SSSR Usp. Khim. **42**, 1278 (1973); Russ. Chem. Rev. **42**, 570 (1973).
16. G. Grosse and K. Friese, unpublished work, quoted in Friese [7].
17. S. S. Voyutsky, A. D. Zaionchkovsky, and R. A. Reznikova, Kolloid Zh. **18**, 515 (1956); Colloid J. USSR **18**, 511 (1956).
18. J. Zelinger and V. Heidingsfeld, Sb. Vys. Šk. Chem. Technol. Praze, Org. Technol. **9**, 63 (1966).
19. G. R. Williamson and B. Wright, J. Polym. Sci. Part A **3**, 3885 (1965).

20. D. Feldman and M. Rusu, *Eur. Polym. J.* **6**, 627 (1970).
21. C. Hugelin and A. Dondos, *Makromol. Chem.* **126**, 206 (1969).
22. V. E. Gul, E. A. Penskaya and V. N. Kuleznev, *Kolloid Zh.* **27**, 341 (1965); *Colloid J. USSR* **27**, 283 (1965).
23. H. Schnecko and R. Caspary, *Kaut. Gummi Kunst.* **25**, 309 (1972).
24. V. N. Kuleznev, I. V. Konyukh, G. V. Vinogradov, and I. P. Dimitrieva, *Kolloid Zh.* **27**, 540 (1965); *Colloid J. USSR* **27**, 459 (1965).
25. R. A. Reznikova, A. D. Zaionchkovsky, and S. S. Voyutsky, *Kolloid Zh.* **15**, 108 (1953); *Colloid J. USSR* **15**, 111 (1953).
26. R. A. Reznikova, S. S. Voyutsky, and A. D. Zaionchkovsky, *Kolloid Zh.* **16**, 204 (1954); *Colloid J. USSR* **16**, 207 (1954).
27. V. N. Kuleznev, V. D. Klykova, E. I. Chernin, and Y. V. Evreinov, *Kolloid Zh.* **37**, 267 (1975); *Colloid J. USSR* **37**, 237 (1975).
28. C. Vasile, N. Asandei, and I. A. Schneider, *Rev. Roum. Chim.* **11**, 1247 (1966).
29. C. Vasile, F. Sandru, I. A. Schneider, and N. Asandei, *Makromol. Chem.* **110**, 20 (1967).
30. V. I. Alekseenko, *Vysokomol. Soedin.* **2**, 1449 (1960); *Polym. Sci. USSR* **3**, 367 (1962).
31. V. A. Kuleznev and K. M. Igosheva, *Vysokomol. Soedin.* **4**, 1858 (1962).
32. N. V. Chirkova, V. G. Epshtein and N. D. Zakharov, *Kolloid Zh.* **32**, 912 (1970); *Colloid J. USSR* **32**, 767 (1970).
33. C. Vasile and I. A. Schneider, *Eur. Polym. J.* **9**, 1063 (1973).
34. L. J. Hughes and G. L. Brown, *J. Appl. Polym. Sci.* **5**, 580 (1961).
35. A. R. Shultz and B. M. Beach, *Macromolecules* **7**, 902 (1974).
36. F. Friese, *Plaste Kaut.* **15**, 646 (1968).
37. K. Kosai and T. Higashino, *Nippon Setchaku Kyokai Shi* **11**, 2 (1975).
38. M. Bank, J. Leffingwell, and C. Thies, *Polym. Prepr. Amer. Chem. Soc. Div. Polym. Chem.* **10**, 622 (1969); *Macromolecules* **4**, 43 (1971).
39. L. H. Sperling, D. W. Taylor, M. L. Kirkpatrick, H. F. George, and D. R. Bardman, *J. Appl. Polym. Sci.* **14**, 73 (1970).
40. T. K. Kwei, T. Nishi, and R. F. Roberts, *Macromolecules* **7**, 667 (1974).
41. T. Nishi and T. K. Kwei, *Polymer* **16**, 285 (1975).
42. S. Ichihara, A. Komatsu, and T. Hata, *Polym. J.* **2**, 640 (1971).
43. J. Stoelting, F. E. Karasz, and W. J. MacKnight, *Polym. Prepr. Amer. Chem. Soc. Div. Polym. Chem.* **10**, 628 (1969); *Polym. Eng. Sci.* **10**, 133 (1970).
44. A. Purcell and C. Thies, *Polym. Prepr. Amer. Chem. Soc. Div. Polym. Chem.* **9**, 115 (1968).
45. N. Yoshimura and K. Fujimoto, *Nippon Gomu Kyokaishi* **41**, 161 (1968); *Rubber Chem. Technol.* **42**, 1009 (1969).
46. A. Silberberg and W. Kuhn, *Nature (London)* **170**, 450 (1952).
47. A. Silberberg and W. Kuhn, *J. Polym. Sci.* **13**, 21 (1954).
48. F. Burkhardt, H. Majer, and W. Kuhn, *Helv. Chim. Acta* **43**, 1192 (1960).
49. A. R. Shultz and B. M. Gendron, *J. Appl. Polym. Sci* **16**, 461 (1972).
50. L. Y. Zlatkevich and V. G. Nikolskii, *Rubber Chem. Technol.* **46**, 1210 (1973).
51. D. G. Welygan and C. M. Burns, *J. Polym. Sci. Polym. Lett. Ed.* **11**, 339 (1973).
52. R. Koningsveld, L. A. Kleintjens, and H. M. Schoffeleers, *Pure Appl. Chem.* **39**, 1 (1974).
53. B. E. Sundquist and R. A. Oriani, *J. Chem. Phys.* **36**, 2604 (1962).
54. T. Nishi, T. T. Wang, and T. K. Kwei, *Macromolecules* **8**, 227 (1975).
55. D. Berek, D. Lath, and V. Ďurďovič, *J. Polym. Sci. Part C* **16**, 659 (1967).
56. S. Krause, unpublished work.
57. R. Koningsveld, H. A. G. Chermin and M. Gordon, *Proc. Roy. Soc.* **A319**, 331 (1970).
58. R. L. Scott, *J. Chem. Phys.* **17**, 279 (1949).

59. H. Tompa, *Trans. Faraday Soc.* **45**, 1142 (1949).
60. P. J. Flory, *J. Chem. Phys.* **9**, 660 (1941).
61. P. J. Flory, *J. Chem. Phys.* **10**, 51 (1942).
62. M. L. Huggins, *J. Chem. Phys.* **9**, 440 (1941).
63. M. L. Huggins, *Ann. N.Y. Acad. Sci.* **43**, 1 (1942).
64. L. Zeman and D. Patterson, *Macromolecules* **5**, 513 (1972).
65. C. C. Hsu and J. M. Prausnitz, *Macromolecules* **7**, 320 (1974).
66. R. Koningsveld, *Chem. Zvest.* **26**, 263 (1972).
67. R. Koningsveld and L. A. Kleintjens, *Macromol. Chem.* **8**, 197 (1973) (supplement to *Pure Appl. Chem.*).
68. R. L. Scott, *J. Polym. Sci.* **9**, 423 (1952).
69. J. H. Hildebrand and R. L. Scott, "The Solubility of Nonelectrolytes," 3rd ed. Van Nostrand–Reinhold, Princeton, New Jersey, 1950; reprinted, Dover, New York, 1964.
70. J. H. Hildebrand and R. L. Scott, "Regular Solutions." Prentice-Hall, Englewood Cliffs, New Jersey, 1962.
71. S. Krause, A. L. Smith, and M. G. Duden, *J. Chem. Phys.* **43**, 2144 (1965).
72. I. Prigogine, "The Molecular Theory of Solutions." Wiley (Interscience), New York, 1959.
73. P. J. Flory, B. E. Eichinger, and R. A. Orwoll, *Macromolecules* **1**, 287 (1968).
74. L. P. McMaster, *Macromolecules* **6**, 760 (1973).
75. J. Biroš, L. Zeman, and D. Patterson, *Macromolecules* **4**, 30 (1971).
76. G. L. Slonimskii, *J. Polym. Sci.* **30**, 625 (1958).
77. G. V. Struminskii and G. L. Slonimskii, *Zh. Fiz. Khim.* **30**, 1941 (1956).
78. A. Dobry and F. Boyer-Kawenoki, *J. Polym. Sci.* **2**, 90 (1947).
79. R. J. Peterson, R. D. Corneliussen, and L. T. Rozelle, *Polym. Prepr. Amer. Chem. Soc. Div. Polym. Chem.* **10**, 385 (1969).
80. A. A. Tager, T. I. Scholokhovich, and Y. S. Bessonov, *Eur. Polym. J.* **11**, 321 (1975).
81. C. Vasile, S. Ioan, N. Asandei, and I. A. Schneider, *Angew. Makromol. Chem.* **6**, 24 (1969).
82. S. M. Lipatov, *Kolloid Zh.* **22**, 639 (1960), (Lipatov [167] is the translation of Lipatov [82].)
83. J. F. H. Van Eijnsbergen, *Chim. Peintures* **4**, 253 (1941).
84. D. M. Cates and H. J. White Jr., *J. Polym. Sci* **20**, 155 (1956).
85. L. E. Kalinina, V. I. Alekseenko, and S. S. Voyutsky, *Kolloid Zh.* **18**, 691 (1956); *Colloid J. USSR* **18**, 689 (1956).
86. L. E. Kalinina, V. I. Alekseenko, and S. S. Voyutsky, *Kolloid Zh.* **18**, 180 (1956); *Colloid J. USSR* **18**, 171 (1956).
87. T. Kawai, *Kogyo Kagaku Zasshi* **59**, 779 (1956).
88. V. A. Kargin, *J. Polym. Sci. Part C* **4**, 1601 (1963).
89. G. L. Brode and J. V. Koleske, *J. Macromol. Sci., Chem.* **A6**, 1109 (1972).
90. V. I. Alekseenko, N. U. Mishustin, and S. S. Voyutsky, *Kolloid Zh.* **17**, 3 (1955); *Colloid J. USSR* **17**, 1 (1955).
91. V. I. Alekseenko and N. U. Mishustin, *Vysokomol. Soedin.* **1**, 1593 (1959); *Polym. Sci. USSR* **1**, 63 (1960).
92. L. E. Kalinina, V. I. Alekseenko and S. S. Voyutskii, *Kolloid Zh.* **19**, 51 (1957); *Colloid J. USSR* **19**, 57 (1957).
93. J. Wool, *Chim. Peintures* **4**, 134 (1941).
94. R. M. Asimova, P. V. Kozlov, V. A. Kargin, and S. M. Vtorygin, *Vysokomol. Soedin.* **4**, 554 (1962).

95. B. D. Gesner, *Encycl. Polym. Sci. Technol.* **10**, 694 (1969).
96. M. H. Walters and D. N. Keyte, *Rubber Chem. Technol.* **38**, 62 (1965).
97. R. Ecker, *Kaut. Gummi Kunstst.* **9**, 153 (1956); *Rubber Chem. Technol.* **30**, 200 (1957).
98. G. M. Bartenev and G. S. Kongarov, *Vysokomol. Soedin.* **2**, 1962 (1960); *Rubber Chem. Technol.* **36**, 668 (1963).
99. G. L. Slonimskii and N. F. Komskaia, *Zh. Fiz. Khim.* **30**, 1746 (1956).
100. P. A. Marsh, A. Voet, and L. D. Price, *Rubber Chem. Technol.* **40**, 359 (1967).
101. P. J. Corish, *Rubber Chem. Technol.* **40**, 324 (1967).
102. D. I. Livingston and R. L. Rongone, *Proc. Int. Rubber Technol. Conf., 5th, Brighton, (May 1967)*, p. 337 (Pub. 1968).
103. A. I. Marei and E. A. Sidorovich, *Medkhan. Polim.* **1**, (No. 5), 85 (1965); *Polym. Mechan.* **1** (No. 5), 55 (1965).
104. K. Fujimoto and N. Yoshimiya, *Nippon Gomu Kyokaishi* **38**, 284 (1965); *Rubber Chem. Technol.* **41**, 669 (1968).
105. N. Y. Buben, V. J. Goldanskii, L. Y. Zlatkevich, V. G. Nikolskii, and V. G. Raevski, *Vysokomol. Soedin.* **A9**, 2275 (1967); *Polym. Sci. USSR* **9**, 2575 (1967).
106. P. A. Marsh, A. Voet, and L. D. Price, *Rubber Chem. Technol.* **41**, 344 (1968).
107. D. McIntyre, N. Rounds, and E. Campos-Lopez, *Polym. Prepr. Amer. Chem. Soc. Div. Polym. Chem.* **10**, 531 (1969).
108. V. N. Kuleznev, L. S. Krokhina, V. G. Oganesov, and L. M. Zlatsen, *Kolloid Zh.* **33**, 98 (1971); *Colloid J. USSR* **33**, 82 (1971).
109. British Patent 857,507 (December 29, 1960), E. I. du Pont de Nemours & Co. C. (1962), 19035.
110. G. M. Bristow, *J. Appl. Polym. Sci.* **2**, 120 (1959).
111. I. Mladenov, P. Nikolinski, and S. Vassileva, *C.R. Acad. Bulgare Sci.* **16**, 837 (1963).
112. B. A. Dogadkin, V. N. Kuleznev, and F. Priakhina, *Kolloid. Zh.* **21**, 174 (1959); *Colloid J. USSR* **21**, 161 (1959).
113. J. W. A. Sleijpen, M. F. J. Pijpers, and H. C. Booij, unpublished results, quoted in Koningsveld *et al.* [52].
114. G. Ajroldi, G. Gatta, P. D. Gugelmetto, R. Rettore, and G. P. Talamini, *Polym. Prepr. Amer. Chem. Soc. Div. Polym. Chem.* **11**, 357 (1970).
115. G. Ajroldi, G. Gatta, P. D. Gugelmetto, R. Rettore, and G. P. Talamini, *Adv. Chem. Ser.* **99**, 119 (1971).
116. J. E. McGrath and M. Matzner, U.S. Patent 3, 798, 289 (March 19, 1974).
117. W. Breurs, W. Hild, H. Wolff, W. Burmeister, and H. Hoyer, *Plaste Kaut.* **1**, 170 (1954).
118. S. Manabe, R. Murakami, M. Takayanagi, and S. Uemura, *Int. J. Polym. Mater.* **1**, 47 (1971).
119. M. Matsuo, C. Nozaki, and Y. Jyo, *Polym. Eng. Sci.* **9**, 197 (1969).
120. O. Fuchs, *Angew. Makromol. Chem.* **1**, 29 (1967).
121. R. J. Kern, *J. Polym. Sci.* **33**, 524 (1958).
122. I. A. Schneider and C. Vasile, *Eur. Polym. J.* **6**, 695 (1970).
123. I. A. Schneider and C. Vasile, *Eur. Polym. J.* **6**, 687 (1970).
124. R. J. Kern and R. J. Slocombe, *J. Polym. Sci.* **15**, 183 (1955).
125. R. Buchdahl and L. E. Nielsen, *J. Polym. Sci.* **15**, 1 (1955).
126. K. A. Wolf, *J. Polym. Sci. Part C* **4**, 1626 (1963).
127. I. N. Razinskaya, L. I. Vidyakina, T. I. Radbil', and B. P. Shtarkman, *Vysokomol. Soedin.* **A14**, 968 (1972); *Polym. Sci. USSR* **14**, 1079 (1972).
128. J. W. Schurer, A. DeBoer, and G. Challa, *Polymer* **16**, 201 (1975).
129. J. V. Koleske and R. D. Lundberg, *J. Polym. Sci. Part A-2* **7**, 795 (1969).
130. D. Feldman and M. Rusu, *Bul. Inst. Politeh. Iasi* **18**, 105 (1972).

131. D. Feldman and M. Rusu, *J. Polym. Sci. Symp. No. 42*, p. 639 (1973).
132. R. A. Reznikova, A. D. Zaionchkovskii, and S. S. Voyutsky, *Zh. Tekh Fiz.* **25**, 1045 (1955).
133. M. Takayanagi, H. Harima, and Y. Iwata, *Mem. Faculty Eng. Kyushu Univ.* **23**, 1 (1963).
134. H. Wolff, *Plaste Kaut.* **4**, 244 (1957).
135. L. E. Nielsen, *J. Am. Chem. Soc.* **75**, 1435 (1953).
136. Y. G. Oganesov, V. S. Osipchuk, K. G. Mindiyarov, V. G. Rayevskii, and S. S. Voyutskii, *Vysokomol. Soedin.* **A11**, 896 (1969); *Polym. Sci. USSR* **11**, 1012 (1969).
137. G. A. Zakrzewski, *Polymer* **14**, 347 (1973).
138. A. B. Aivazov, K. G. Mindiyarov, Y. V. Zelenev, Y. G. Oganesov, and V. G. Raevskii, *Vysokomol. Soedin.* **B12**, 10 (1970).
139. Y. J. Shur and B. Rånby, *J. Appl. Polym. Sci.* **19**, 2143 (1975).
140. D. Dimitrova, A. Aivazov, and Y. Zelenev, *God. Vissh. Khim. Tekhnol. Inst., Burgas, Bulg.* **8**, 57 (1971); *Chem. Abstr.* **79**, 127188d (1973).
141. V. I. Alekseenko and I. U. Mishustin, *Kolloid. Zh.* **18**, 257 (1956); *Colloid J. USSR* **18**, 247 (1956).
142. C. F. Hammer, *Macromolecules* **4**, 69 (1971).
143. D. Feldman and M. Rusu, *Eur. Polym. J.* **10**, 41 (1974).
144. C. Elmquist and S. E. Svanson, *Eur. Polym. J.* **11**, 789 (1975).
145. Y. J. Shur and B. Rånby, *J. Appl. Polym. Sci.* **19**, 1337 (1975).
146. D. Hardt, *Brit. Polym. J.* **1**, 225 (1969).
147. B. G. Rånby, *J. Polym., Sci. Polym. Symp.* **51**, 89 (1975).
148. K. Marcincin, A. Ramonov, and V. Pollak, *J. Appl. Polym. Sci.* **16**, 2239 (1972).
149. J. J. Hickman and R. M. Ikeda, *J. Polym. Sci. Polym. Phys. Ed.* **11**, 1713 (1973).
150. R. G. Bauer and M. S. Guillod, *Adv. Chem. Ser.* **142**, 231 (1975).
151. J. F. Kenney, *J. Polym., Sci. Polym. Chem. Ed.* **14**, 123 (1976).
152. R. Koningsveld and A. J. Staverman, *Kolloid Z.Z. Polym.* **220**, 31 (1967).
153. M. A. Martynov, V. M. Yuzhin, A. I. Malushin, and G. F. Tkachenko, *Plast. Massy* **10**, 6 (1965); *Sov. Plast.* **10**, 12 (1965).
154. N. Y. Buben, V. J. Goldanskii, L. Y. Zlatkevich, V. G. Nikolskii, and V. G. Raevskii, *Dokl. Akad. Nauk. USSR* **162**, 370 (1965).
155. T. R. Paxton, *J. Appl. Polym. Sci.* **7**, 1499 (1963).
156. G. Allen, G. Gee, and J. P. Nicholson, *Polymer* **1**, 56 (1960).
157. R. L. Kruse, *Adv. Chem. Ser.* **142**, 141 (1975).
158. D. G. Welygan and C. M. Burns, *J. Appl. Polym. Sci.* **18**, 521 (1974).
159. R. J. Angelo, R. M. Ikeda, and M. L. Wallack, *Polymer* **6**, 141 (1965).
160. S. G. Turley, *J. Polym. Sci. Part C* **1**, 101 (1963).
161. R. Caspary, *Kautschuk Gummi Kunst.* **25**, 249 (1972).
162. K. Fujimoto and N. Yoshimura, *Nippon Gomu Kyokaishi* **39**, 919 (1966).
163. H. K. deDekker and D. J. Sabatine, *Rubber Age* **99**(4), 73 (1967).
164. L. A. Kleintjens, unpublished results, quoted in Koningsveld *et al.* [52].
165. I. Sakurada, A. Nakajima, and H. Aoki, *J. Polym. Sci.* **35**, 507 (1959).
166. O. Fuchs, *Makromol. Chem.* **90**, 293 (1966).
167. S. M. Lipatov, *Colloid J. USSR* **22**, 637 (1960). (Lipatov [82] is the original Russian of Lipatov [167].)
168. O. Fuchs, *Angew Makromol. Chem.* **6**, 79 (1969).
169. I. Mladenov, P. Nikolinski, V. Gul, and N. Petrov, *C. R. Acad. Sci. Bulgare* **14**, 615 (1961).
170. E. Jenckel and H. U. Herwig, *Kolloid Z.* **148**, 57 (1956).
171. S. Akiyama, *Bull. Chem. Soc. Jpn.* **45**, 1381 (1972).

172. S. Akiyama, N. Inaba, and R. Kaneko, *Chem. High Polym. Jpn.* **26**, 529 (1969).
173. B. H. Clampitt, *Polym. Eng. Sci.* **14**, 827 (1974).
174. A. J. Hyde and A. G. Tanner, *J. Colloid Interface Sci.* **28**, 179 (1968).
175. V. N. Kuleznev, L. S. Krokhina, and B. A. Dogadkin, *Kolloid Zh.* **31**, 853 (1969); *Colloid J. USSR* **31**, 685 (1969).
176. R. F. Boyer and R. S. Spencer, *J. Appl. Phys.* **15**, 398 (1944).
177. M. T. Shaw, *J. Appl. Polym. Sci.* **18**, 449 (1974).
178. G. E. Molau, *J. Polym. Sci. Part A* **3**, 1267 (1965).
179. M. Baer, *J. Polym. Sci. Part A* **2**, 417 (1964).
180. P. Black and D. J. Worsfold, *J. Appl. Polym. Sci.* **18**, 2307 (1974).
181. D. J. Dunn and S. Krause, *J. Polym. Sci. Polym. Lett. Ed.* **12**, 591 (1974).
182. L. M. Robeson, M. Matzner, L. J. Fetters, and J. E. McGrath, *in* "Recent Advances in Polymer Blends, Grafts, and Blocks" (L. H. Sperling, ed.), p. 281. Plenum, New York, 1974.
183. J. Schelten, W. Schamtz, D. G. H. Ballard, G. W. Longman, M. G. Rayner, and G. D. Wignall, *Prepr. Int. Microsymp. Crystall. Fusion Polym., Louvain-la-Neuve, June 8–11* p. I (1976).
184. R. J. Kern, *J. Polym. Sci.* **21**, 19 (1956).
185. L. J. Hughes and G. E. Britt, *J. Appl. Polym. Sci.* **5**, 337 (1961).
186. V. Huelck, D. A. Thomas, and L. H. Sperling, *Macromolecules* **5**, 340, 348 (1972).
187. H. Daimon, H. Okitsu, and J. Kumanotani, *Polym. J.* **7**, 460 (1975).
188. W. H. Stockmayer and H. E. Stanley, *J. Chem. Phys.* **18**, 153 (1950).
189. S. D. Hong and C. M. Burns, *J. Appl. Polym. Sci.* **15**, 1995 (1971).
190. R. Kuhn, H. J. Cantow, and W. Burchard, *Angew. Makromol. Chem.* **2**, 157 (1968).
191. P. Kuhn, H. J. Cantow, and W. Burchard, *Angew. Makromol. Chem.* **2**, 146 (1968).
192. H. K. Yuen and J. B. Kinsinger, *Macromolecules* **7**, 329 (1974).
193. E. Cuddihy, J. Moacanin, and A. Rembaum, *J. Appl. Polym. Sci.* **9**, 1385 (1965).
194. M. Bank, J. Leffingwell, and C. Thiess, *J. Polym. Sci. Part A* **2**, 1097 (1972).
195. E. P. Cizek, U.S. Patent No. 3,383,435 (May 14, 1968), assigned to General Electric Co.
196. F. E. Karasz, W. J. MacKnight, and J. Stoelting, *Polym. Prepr. Amer. Chem. Soc. Div. Polym. Chem.* **11**, 357 (1970).
197. W. J. MacKnight, J. Stoelting, and F. E. Karasz, *Adv. Chem. Ser.* **99**, 29 (1971).
198. R. L. Stallings, H. B. Hopfenberg, and V. Stannett, *J. Polym. Sci. Polym. Symp.* **41**, 23 (1973).
199. W. M. Prest Jr. and R. S. Porter, *J. Polym. Sci. Part A-2* **10**, 1639 (1972).
200. H. E. Bair, *Polym. Sci. Eng.* **10**, 247 (1970).
201. A. R. Shultz and B. M. Gendron, *Polym. Prepr. Amer. Chem. Soc. Div. Polym. Chem.* **14**, 571 (1973).
202. T. Okazawa, *Macromolecules* **8**, 371 (1975).
203. R. Buchdahl and L. E. Nielsen, *J. Appl. Phys.* **21**, 482 (1950).
204. E. H. Merz, G. C. Claver, and M. Baer, *J. Polym. Sci.* **22**, 325 (1956).
205. P. Bauer, J. Hennig, and G. Schreyer, *Angew. Makromol. Chem.* **11**, 145 (1970).
206. S. Krause and N. Roman, *J. Polym. Sci. Part A* **3**, 1631 (1965).
207. T. Miyamoto and H. Inagaki, *Polym. J.* **1**, 46 (1970).
208. A. M. Liquori, M. DeSantis Savino, and M. D'Alagni, *J. Polym. Sci. Part B* **4**, 943 (1966).
209. R. G. Bauer and N. C. Bletso, *Polym. Prepr. Amer. Chem. Soc. Div. Polym. Chem.* **10**, 632 (1969).
210. F. E. Bailey, R. D. Lundberg, and R. W. Callard, *J. Polym. Sci. Part A* **2**, 845 (1964).
211. K. L. Smith, A. E. Winslow, and D. E. Peterson, *Ind. Eng. Chem.* **51**, 1361 (1959).

212. J. Néel and B. Sebille, *C. R. Acad. Sci. Paris* **250**, 1052 (1960).
213. R. M. Fuoss and H. Sadek, *Science* **110**, 552 (1949).
214. J. S. Noland, N. N. Hsu, R. Saxon, and J. M. Schmitt, *Polym. Prepr. Amer. Chem. Soc. Div. Polym. Chem.* **11**, 355 (1970).
215. D. R. Paul and J. O. Altamirano, *Adv. Chem. Ser.* **142**, 371 (1975).
216. J. S. Noland, N. N. C. Hsu, R. Saxon, and J. M. Schmitt, *Adv. Chem. Ser.* **99**, 15 (1971).
217. R. Kuhn, H. J. Cantow, and S. B. Liang, *Angew. Makromol. Chem.* **18**, 93 (1971).
218. V. Ramaswamy and H. P. Weber, *Appl. Opt.* **12**, 1581 (1973).
219. D. J. Stein, R. H. Jung, K. H. Illers, and H. Hendus, *Angew. Makromol. Chem.* **36**, 89 (1974).
220. L. P. McMaster, *Polym. Prepr. Amer. Chem. Soc. Div. Polym. Chem.* **15**, 254 (1974); *Adv. Chem. Ser.* **142**, 43 (1975).
221. R. H. Jung and D. J. Stein, *Prepr. IUPAC Symp. Aberdeen* No. G10, p. 411 (1973).
222. R. L. Imken, D. R. Paul, and J. W. Barlow, *Polym. Eng. Sci.* **16**, 593 (1976).
223. I. S. Okhrimenko and E. B. D'Yakonova, *Vysokomol. Soedin* **6**, 1891 (1964); *Polym. Sci. USSR* **6**, 2095 (1964).
224. E. B. D'Yakonova, I. S. Okhrimenko, and I. F. Yefremov, *Vysokomol. Soedin.* **7**, 1016 (1965); *Polym. Sci. USSR* **7**, 1123 (1965).
225. G. I. Distler, E. B. D'Yakonova, I. F. Yefremov, Y. I. Kortukova, I. S. Okhrimenko, and P. S. Sotnikov, *Vysokomol. Soedin.* **8**, 1737 (1966); *Polym. Sci USSR* **8**, 1917 (1966).
226. A. Piloz, J. Y. Decroix, and J. F. May, *Angew. Makromol. Chem.* **44**, 77 (1975).
227. G. Allen, G. Gee, and J. P. Nicholson, *Polymer* **2**, 8 (1961).
228. E. S. Gurevich and A. M. Frost, *Lakokrasoch. Mater. Prim.* (3), 11 (1961).
229. V. S. Aslizadyan, Y. V. Zelenev, and Y. K. Kabalyan, *Plaste Kaut.* **22**, 717 (1975).
230. M. E. Ovsepyan, A. V. Gevorkyan, A. S. Safarov, and L. K. Simonyan, *Collect. Czech. Chem. Commun.* **38**, 1764 (1973).
231. M. P. Zverev, R. A. Bychkov, and A. A. Konkin, *Vysokomol. Soedin. Part B* **11**, 438 (1969).
232. A. Rembaum, J. Moacanin, and E. Cuddihy, *J. Polym. Sci. Part C* **4**, 529 (1963).
233. J. Moacanin, E. Cuddihy, and A. Rembaum, *Adv. Chem. Ser.* **48**, 159 (1965).
234. F. E. Karasz, W. J. MacKnight, and J. J. Tkacik, *Polym. Prepr. Amer. Chem. Soc. Div. Polym. Chem.* **15**, 415 (1974).
235. J. Cypryk, M. Laczowski, and S. Piechuchi, *Faserforsch. Textiltech.* **14**, 265 (1963).
236. A. K. Rastogi and L. E. St. Pierre, *J. Colloid Interface Sci.* **31**, 168 (1969).
237. C. Booth and C. J. Pickles, *J. Polym. Sci. Polym. Phys. Ed.* **11**, 595 (1973).
238. A. R. Shultz and B. M. Gendron, *J. Polym. Sci. Symp.* **43**, 89 (1973).
239. C. G. Seefried Jr. and J. V. Koleske, *J. Polym. Sci. Polym. Phys. Ed.* **13**, 851 (1975).
240. E. P. Goldberg, *J. Polym. Sci. Part C* **4**, 707 (1963).
241. P. O. Powers, *Polym. Prepr. Amer. Chem. Soc. Div. Polym. Chem.* **15**(2), 528 (1974).
242. T. H. Meltzer, W. J. Dermody, and A. V. Tobolsky, *J. Appl. Polym. Sci.* **8**, 765 (1964).
243. J. Leffingwell, C. Thies, and H. Gertzman, *Polym. Prepr. Amer. Chem. Soc. Div. Polym. Chem.* **14**, 596 (1973).
244. K. L. Floyd, *Brit. J. Appl. Phys.* **3**, 373 (1952).
245. L. M. Robeson, Private communication, March 19, 1971, quoted in Shaw [177].
246. S. Nohara, *Kobunshi Kagaku* **12**, 47 (1955).
247. A. Schwarcz, *J. Polym. Sci. Polym. Phys. Ed.* **12**, 1195 (1974).
248. Y. V. Zelenev and G. M. Bartenev, *Vysokomol. Soedin.* **6**, 1047 (1964); *Polym. Sci. USSR* **6**, 1152 (1964).
249. K. C. Frisch, D. Klempner, S. Migdal, H. L. Frisch, and H. Ghiradella, *Polym. Eng. Sci.* **14**, 76 (1974).

250. V. I. Alekseenko and I. U. Mishustin. *Zh. Fiz. Khim.* **33**, 757 (1959).
251. G. Riess, J. Kohler, C. Tournut, and A. Banderet, *Makromol. Chem.* **101**, 58 (1967).
252. H. J. Oswald and E. T. Kubu, *SPE Trans.* **3**, 168 (1963).
253. A. H. Jorgensen, L. A. Chandler, and E. A. Collins, *Rubber Chem. Technol.* **46**, 1087 (1973).
254. G. Kraus and K. W. Rollman, *Polym. Prepr. Amer. Chem. Soc. Div. Polym. Chem.* **11**, 377 (1970).
255. G. Kraus and K. W. Rollman, *Adv. Chem. Ser.* **99**, 189 (1971).
256. L. A. Chandler and E. A. Collins, *Polym. Prepr. Amer. Chem. Soc. Div. Polym. Chem.* **9**, 1416 (1968).
257. F. S. Cheng and J. L. Kardos, *Polym. Prepr. Amer. Chem. Soc. Div. Polym. Chem.* **10**, 615 (1969).
258. M. R. Ambler, *J. Polym. Sci. Polym. Chem. Ed.* **11**, 1505 (1973).
259. L. A. Chandler and E. A. Collins, *J. Appl. Polym. Sci.* **13**, 1585 (1969).
260. G. E. Molau, *J. Polym. Sci. Part B* **3**, 1007 (1965).
261. V. R. Landi, *Rubber Chem. Technol.* **45**, 222 (1972).
262. R. J. Slocombe, *J. Polym. Sci.* **26**, 9 (1957).
263. F. Kollinsky and G. Markert, *Makromol. Chem.* **121**, 117 (1969).
264. F. Kollinsky and G. Markert, *Adv. Chem. Ser.* **99**, 175 (1971).
265. O. Olabisi, *Macromolecules* **8**, 316 (1975).
266. J. Bareš and M. Pegoraro, *J. Polym. Sci. Part A-2* **9**, 1287 (1971).
267. B. Schneier, *J. Appl. Polym. Sci.* **17**, 3175 (1973).
268. J. Brandrup and E. H. Immergut (eds.), "Polymer Handbook," Snd ed. Wiley (Interscience), New York, 1975.
269. A. G. Shvarts, *Kolloid Zh.* **18**, 755 (1956); *Colloid J. USSR* **18**, 753 (1956).
270. A. G. Shvarts, E. K. Chefranova, and L. A. Iotkovskaya, *Kolloid Zh.* **32**, 603 (1970); *Colloid J. USSR* **32**, 506 (1970).
271. J. L. Gardon, *Encycl. Polym. Sci. Technol.* **3**, 833 (1965).
272. H. Burrell, *Encycl. Polym. Sci. Technol.* **12**, 618 (1970).
273. H. Burrell, *J. Paint Technol.* **40**, 197 (1968).
274. L. A. Utracki, *J. Appl. Polym. Sci.* **16**, 1167 (1972).
275. J. L. Gardon, *J. Paint Technol.* **38**, 43 (1966).
276. A. F. M. Barton, *Chem. Rev.* **75**, 731 (1975).
277. C. Hansen and A. Beerbower, *Kirk–Othmer Encycl. Chem. Technol.* 2nd Ed. 1971, Suppl. p. 889.
278. P. A. Small, *J. Appl. Chem.* **3**, 71 (1953).
278a. Burnell, H., *Off. Dig. Fed. Soc. Paint Technol.* **27**, 726 (1955).
279. K. L. Hoy, *J. Paint Technol.* **42**, 76 (1970).
280. B. Schneier, *J. Polym. Sci. Polym. Lett. Ed.* **10**, 245 (1972).
281. J. M. G. Cowie, *Polymer* **10**, 708 (1969).

Chapter 3

Statistical Thermodynamics of Polymer Blends

Isaac C. Sanchez[†]

Materials Research Laboratory
and
Department of Polymer Science and Engineering
University of Massachusetts
Amherst, Massachusetts

I.   Introduction.   115
II. Flory Theory   117
     A. Equation of State   117
     B. Mixtures   119
III. Lattice Fluid   121
     A. Equation of State   121
     B. Mixtures   125
     C. Phase Stability and the Spinodal   128
     D. Physical Significance of the Ratio v_1^*/v_2^*   131
IV. Predicting Polymer Blend Compatibility   132
     A. Compatibility Criteria   132
     B. Comparison of Theoretical Predictions   135
     References   139

I. INTRODUCTION

Since 1960 there have been two especially important developments in polymer solution thermodynamics: one experimental and one theoretical. Freeman and Rowlinson [1] observed experimentally that many hydrocarbon polymers dissolved in hydrocarbon solvents phase separated at high temperatures. These nonpolar polymer solutions exhibited what are known as lower critical solution temperatures (LCST), a critical point phenomena that is relatively rare among low molecular weight solutions. It was recognized that the common appearance of LCST behavior in polymer solutions

† *Present address*: National Bureau of Standards, Polymer Division 311.00, Washington, D.C.

must be related to the large size difference between polymer and solvent molecules. Soon after the discovery of the universality of LCST behavior in polymer solutions, Flory and co-workers [2–5] developed a new theory of solutions that considers the "equation of state" properties of the pure components. This new theory of solutions, hereafter referred to as the Flory theory, demonstrates that mixture thermodynamic properties depend on the thermodynamic properties of the pure components. In particular LCST behavior can be understood in terms of the dissimilarity of the equation of state properties of polymer and solvent. Patterson [6–9] has also shown that LCST behavior is related to the dissimilarity in polymer–solvent properties by using the general corresponding states theory of Prigogine and collaborators [10]. The older Flory–Huggins theory [11], which ignores the equation of state properties of the pure components, completely fails to describe LCST behavior.

In the Flory theory each pure component is characterized by three equation of state parameters: a characteristic temperature T^*, a characteristic pressure p^*, and a characteristic specific volume v_{sp}^*. Since these parameters can be determined from the pressure–volume–temperature (PVT) properties of the pure components, it is hoped that thermodynamic properties of polymer solutions, including polymer–polymer blends, can largely be determined from the equation of state parameters.

More recently, a new equation of state theory of pure fluids [12, 13] and their solutions [14] has been formulated by this author and Lacombe. This theory has been characterized as an Ising or lattice fluid theory (hereafter referred to as the lattice fluid theory). In general, it departs markedly from a corresponding states theory. It also does not require separation of internal and external degrees of freedom as does the Flory theory and Prigogine corresponding state theories. Nevertheless, the lattice fluid theory has much in common with the Flory theory. Both theories require three equation of state parameters for each pure component. For mixtures, both reduce to the Flory–Huggins theory at very low temperatures.

Several other theoretical studies [15–20] have been devoted to the determination of a pure polymer equation of state. These theories as well as the Flory theory use a partition function which requires a separation of internal and external degrees of freedom. External degrees of freedom attributable to a "mer" in a r-mer chain are less than those for a similar small molecule. Thus, a mer in a r-mer will only have $3c$ ($c < 1$) instead of three degrees of freedom. External degrees of freedom are assumed to depend only on intermolecular forces, whereas internal degrees of freedom are associated with intramolecular chemical bond forces. By analogy with the form of a cell model partition function for monomers, it is assumed that the configurational partition function of a system of N r-mers, each with $3cr$ degrees of freedom, is given by [10]

$$Z(T, V) = z_{int}(T)[z_{ext}(T, V)]^{3crN} \exp(-E_0/kT) \tag{1}$$

where E_0 is the configurational or mean potential energy when all mers are at cell centers. The partition function associated with internal degrees of freedom z_{int} is assumed to be density independent, and thus, only z_{ext} and the density-dependent E_0 contribute to the PVT equation of state.

The Flory theory is based on a partition function of the form given in Eq. (1) and is the simplest c parameter theory to have been developed. As already mentioned, the lattice fluid theory is not based on the ideas inherent to Eq. (1), but it also yields a mathematically simple equation of state. Because of their simplicity, the Flory and lattice fluid theories can be easily generalized to mixtures.

The Flory and lattice fluid theories represent promising alternatives to the solubility parameter approach for predicting polymer compatibility (compatibility means thermodynamic miscibility when used here). This chapter is devoted to developing, comparing, and applying these two theories.

II. FLORY THEORY

A. Equation of State

In the spirit of Eq (1), Flory and co-workers [2–4] suggested the following partition function for a system of N r-mers:

$$Z(T, V) = \text{const}(v^{1/3} - v^{*1/3})^{3crN} \exp(-E_0/kT) \tag{2a}$$

or in reduced variables

$$Z(T, V) = \text{const}\, v^{*\,crN}(\tilde{v}^{1/3} - 1)^{3crN} \exp(crN\tilde{\rho}/\tilde{T}) \tag{2b}$$

where V is total volume; $v \equiv V/rN$, volume per mer; v^* is close-packed or hard core mer volume; $\tilde{v} \equiv V/rNv^* \equiv v/v^*$, reduced volume; $\tilde{\rho} \equiv 1/\tilde{v}$, reduced density; and E_0 is the mean intermolecular energy. Flory assumed a van der Waals form for E_0, that is, an energy that is inversely proportional to volume:

$$-E_0 = (rN)\frac{(s\eta/2)}{v} = (rNs/2)\frac{(\eta/v^*)}{\tilde{v}} \tag{3a}$$

or

$$-E_0/kT = crN\tilde{\rho}/\tilde{T} \tag{3b}$$

where s is number of intermolecular contact sites per mer; $rNs/2$ total number of pair interactions; η/v^* energy characteristic of a mer–mer interaction (η has units of energy $\times$ volume); $\varepsilon^* = s\eta/2v^*$, the interaction energy per mer in the close-packed configuration ($\tilde{v} = 1$); $\tilde{T} = T/T^*$, reduced temperature; and $T^* = \varepsilon/ck$.

The equation of state is obtained in the usual way from the partition function:

$$p = kT\, \partial \ln Z(T, V)/\partial V \,|_T \tag{4}$$

which yields

$$\tilde{p}\tilde{v}/\tilde{T} = \tilde{v}^{1/3}/(\tilde{v}^{1/3} - 1) - 1/\tilde{v}\tilde{T} \tag{5a}$$

or

$$\tilde{\rho}^2 + \tilde{p} - \tilde{T}\tilde{\rho}(1 - \tilde{\rho}^{1/3})^{-1} = 0 \tag{5b}$$

where $\tilde{p} = p/p^*$, reduced pressure, and $p^* \equiv \varepsilon^*/v^* = ckT^*/v^*$.

The reduced density and volume can be more conveniently expressed in terms of the *mass* density ρ or the specific volume v_{sp} ($\rho = 1/v_{sp}$):

$$\tilde{\rho} \equiv \rho/\rho^* \equiv v_{sp}^*/v_{sp} \equiv 1/\tilde{v} \tag{6}$$

where v_{sp}^* is the close-packed volume per unit mass and

$$\rho^* \equiv 1/v_{sp}^* \tag{7}$$

Differentiation of the equation of state with respect to temperature at constant pressure yields at zero pressure:

$$v_{sp}^* = v_{sp}\left[\frac{1 + T\alpha}{1 + 4T\alpha/3}\right]^3 \tag{8}$$

where $\alpha = \partial \ln V/\partial T\,|_p$ is the thermal expansion coefficient.

Differentiation of the equation of state with respect to temperature at constant volume yields at zero pressure:

$$p^* = \tilde{v}^2 T\gamma = \tilde{v}^2 T\alpha/\beta \tag{9}$$

where $\gamma = \partial p/\partial T\,|_v$ is the thermal pressure coefficient and $\beta = -\partial \ln V/\partial p\,|_T$ is the isothermal compressibility.

From the equation of state itself there obtains at zero pressure

$$T^* = T\tilde{v}^{4/3}/(\tilde{v}^{1/3} - 1)$$
$$= T/\tilde{\rho}(1 - \tilde{\rho}^{1/3}) \tag{10}$$

Equations (8)–(10), combined with a knowledge of experimental values of $v_{sp}(T)$, $\alpha(T)$, and $\gamma(T)$ or $\beta(T)$, are sufficient for the determination of v_{sp}^*, p^*, and T^*. Notice that the four fundamental molecular parameters r, v^*, ε^*, and c cannot be determined from the equation of state parameters

v_{sp}^*, p^*, and T^*, although the products rv^*, $r\varepsilon^*$, and rc are easily obtained:

$$rv^* = Mv_{sp}^* \tag{11}$$

$$r\varepsilon^* = Mp^*v_{sp}^* \tag{12}$$

$$rc = Mp^*v_{sp}^*/kT^* \tag{13}$$

where M is the molecular weight of the r-mer.

For future reference, the Gibbs free energy per mole μ° is, to within an additive constant, given by

$$\mu^\circ \equiv (-kT\ln Z + pV)/N \tag{14a}$$

$$\mu^\circ = r\varepsilon^*[-\tilde\rho + \tilde p\tilde v - 3\tilde T\ln(\tilde v^{1/3} - 1)] \tag{14b}$$

B. Mixtures

For the sake of clarity, only the results for binary mixtures are presented. Application of the theory to a mixture requires the adoption of a set of combining rules:

1. The mer volumes v_1^* and v_2^* are chosen so that

$$v_1^* = v_2^* = v^*$$

2. The close-packed volume of the mixture V^* is equal to the sum of the pure component close-packed volumes:

$$V^* = r_1 N_1 v_1^* + r_2 N_2 v_2^* = rNv^*$$

$$= N(x_1 M_1 v_{sp;\,1}^* + x_2 M_2 v_{sp;\,2}^*)$$

where

$$N = N_1 + N_2$$

$$x_1 = 1 - x_2 = N_1/N \qquad \text{(mole fraction)}$$

$$r = x_1 r_1 + x_2 r_2$$

3. The total number of pair interactions in the mixture is equal to the sum of the pure component pair interactions:

$$\tfrac{1}{2}(s_1 r_1 N_1 + s_2 r_2 N_2) = srN/2$$

$$= N_{11} + N_{12} + N_{22}$$

where

$$s \equiv \phi_1 s_1 + \phi_2 s_2$$

$$\phi_1 = r_1 N_1/rN = 1 - \phi_2$$

$$N_{ij} = \text{number of } i, j \text{ pairs}$$

The partition function of the mixture is then given by

$$Z(T, V) = \text{const} \, Z_{\text{comb}}(v^{1/3} - v^{*1/3})^{3(c_1 r_1 N_1 + c_2 r_2 N_2)} \exp(-E_0/kT) \quad (15a)$$

and $k \ln Z_{\text{comb}}$ is the usual Flory–Huggins combinatorial entropy:

$$k \ln Z_{\text{comb}} = -k(N_1 \ln \phi_1 + N_2 \ln \phi_2). \quad (16)$$

Application of the third combining rule allows for the calculation of the intermolecular energy of the mixture:

$$-E_0 = (N_{11}\eta_{11} + N_{12}\eta_{12} + N_{22}\eta_{22})/v. \quad (17a)$$

To evaluate the N_{ij}, it is assumed that among the $srN/2$ pair interactions, the interactions are random; thus

$$N_{11} = \tfrac{1}{2}(s_1 r_1 N_1 \theta_1); \quad N_{22} = \tfrac{1}{2}(s_2 r_2 N_2 \theta_2);$$

$$N_{12} = s_1 r_1 N_1 \theta_2 = s_2 r_2 N_2 \theta_1 \quad (18)$$

$$\theta_1 = 1 - \theta_2 = s_1 r_1 N_1/srN \quad (19)$$

$$-E_0/rN = s(\theta_1^2 \eta_{11} + 2\theta_1 \theta_2 \eta_{12} + \theta_2^2 \eta_{22})/2v \quad (17b)$$

$$= s(\theta_1 \eta_{11} + \theta_2 \eta_{22} - \theta_1 \theta_2 \Delta\eta)/2v \quad (17c)$$

$$\Delta\eta \equiv \eta_{11} + \eta_{22} - 2\eta_{12} \quad (20)$$

Now let

$$X_{12} \equiv s_1 \Delta\eta/2v^{*2} \neq X_{21} \quad (21)$$

then

$$-E_0/rN = (\phi_1 p_1^* + \phi_2 p_2^* - \phi_1 \theta_2 X_{12})(v^*/\tilde{v}) \quad (17d)$$

$$-E_0/rN = p^* v^*/\tilde{v} = ckT^*/\tilde{v} \quad (17e)$$

where

$$c \equiv \phi_1 c_1 + \phi_2 c_2 \quad (22)$$

$$p^* \equiv \phi_1 p_1^* + \phi_2 p_2^* - \phi_1 \theta_2 X_{12} \quad (23)$$

$$T^* \equiv p^* v^*/ck = p^*/(\phi_1 p_1^*/T_1^* + \phi_2 p_2^*/T_2^*). \quad (24)$$

The partition function (15a) can now be expressed in terms of reduced variables:

$$Z(T, V) = \text{const} \, v^{*crN} Z_{\text{comb}}(\tilde{v}^{1/3} - 1)^{3crN} \exp(crN/\tilde{v}\tilde{T}). \quad (15b)$$

Since partition function (15b) for the mixture is formally identical to that for the pure component (2b), the mixture satisfies the same equation of state (5) with p^* and T^* defined by Eqs. (23) and (24).

The chemical potential is related to the Helmholtz free energy A by

$$\mu_1 \equiv \partial A/\partial N_1|_{T,V,N_2} = -kT(\partial \ln Z)/\partial N_1|_{T,V,N_2} \tag{25}$$

or

$$\mu_1 - \mu_1{}^\circ = kT[\ln\phi_1 + (1-r_1/r_2)\phi_2] + r_1 v_1^* \tilde{\rho} X_{12}\theta_2{}^2$$
$$+ r_1\varepsilon_1^*\{\tilde{\rho}_1 - \tilde{\rho} + \tilde{p}_1(\tilde{v} - \tilde{v}_1) + 3\tilde{T}_1\ln[(\tilde{v}_1^{1/3}-1)/(\tilde{v}^{1/3}-1)]\} \tag{26}$$

where Eq. (14b) has been used for $\mu_1{}^\circ$ and $r_1\varepsilon_1^*$ and $r_1 v_1^*$ are given by Eqs. (11) and (12). The $\tilde{p}_1\tilde{v}$ term above, which is very small at atmospheric pressure, arises from a term involving the following derivative:

$$\partial\tilde{v}/\partial N_1|_{T,V,N_2} = V\,\partial(rNv^*)^{-1}/\partial N_1 = -\phi_1\tilde{v}/N_1 \tag{27}$$

This derivative was incorrectly evaluated previously [21] because it was not taken at constant volume.

III. LATTICE FLUID

A. Equation of State

The Gibbs free energy G is related to the configurational partition function Z in the pressure ensemble by

$$G = -kT\ln Z(T,p) \tag{28}$$

$$Z(T,p) = \sum_V \sum_E \Omega(E,V,N)\exp[-(E+pV)/kT] \tag{29}$$

where $\Omega(E,V,N)$ is the number of configurations available to a system of N molecules whose configurational energy and volume are E and V, respectively. The summations extend over all values of E and V. In the ensemble of systems under consideration the temperature and pressure are fixed.

A lattice is used to approximate Ω. In the lattice formulation, the number of configurations available to a system of N r-mers and N_0 vacant sites is approximately given by [12]

$$\Omega \simeq (1/f_0)^{N_0}(\omega/f)^N \tag{30}$$

where ω is the number of configurations available to a r-mer in the close-packed state; $f_0 = N_0/(N_0+rN)$, fraction of empty sites; and $f = rN/(N_0+rN)$, fraction of occupied sites. The derivation of Eq. (30) assumes that the coordination number z of the lattice is large and that all configurations of the N r-mers and N_0 empty sites are energetically equivalent.

As in the Flory theory, v^* is the close-packed mer volume and rv^* is close-packed molecular volume. They are related to the molecular weight M, the close-packed mass density ρ^*, and the close-packed specific volume by

$$rv^* = M/\rho^* = Mv_{sp}^*$$ (31)

The reduced volume $\tilde{v}$ and density $\tilde{\rho}$ are defined as

$$\tilde{v} = v_{sp}/v_{sp}^*$$ (32)

$$\tilde{\rho} \equiv 1/\tilde{v} = \rho/\rho^*$$ (33)

where v_{sp} is specific volume and $\rho = 1/v_{sp}$ is the mass density.

The total volume V is given by

$$V = (N_0 + rN)v^*$$ (34)

or

$$\tilde{v} \equiv V/V^* \equiv v/v^*; \qquad V^* \equiv rNv^*$$ (35)

Using Eqs. (31) and (35), the fraction of occupied sites f can be related to the reduced density $\tilde{\rho}$:

$$\rho = \frac{\text{mass}}{\text{volume}} = \frac{NM}{(N_0 + rN)v^*} = \frac{rN\rho^*}{(N_0 + rN)} = f\rho^*$$ (36)

or

$$f = \rho/\rho^* = \tilde{\rho}$$ (37)

The total number of pair interactions is $(z/2)(N_0 + rN)$. If random mixing of mers and vacant sites is assumed, then the probability that two sites are occupied by mers is equal to f^2. Denoting the mer–mer interaction energy by ε, the mean intermolecular energy is

$$E = -(z\varepsilon/2)(N_0 + rN)f^2$$ (38a)

or

$$E/rN = -\tilde{\rho}\varepsilon^*$$ (38b)

where $\varepsilon^* \equiv z\varepsilon/2$ is the total interaction energy per mer.

Since E and Ω are functions of a single parameter, the number of empty sites in the lattice, the double sum over E and V required in the evaluation of the partition function can be replaced by a single sum over N_0:

$$Z(T, p) = \sum_{N_0 = 0}^{\infty} \Omega \exp[-(E + pV)/kT]$$ (39)

In statistical mechanics the standard procedure is to approximate the above sum by its maximum term; the maximum term is overwhelmingly larger than any other for a macroscopic system. Mathematically, this is equivalent to equating the free energy to the logarithm of the generic term in the partition function and then finding the minimum value of the free energy; thus,

$$G = E + pV - kT \ln \Omega \tag{40}$$

or in reduced variables

$$G/(Nr\varepsilon^*) \equiv \tilde{G} = -\tilde{\rho} + \tilde{p}\tilde{v} + \tilde{T}[(\tilde{v}-1)\ln(1-\tilde{\rho}) + (1/r)\ln(\tilde{\rho}/\omega)] \tag{41}$$

where $\tilde{T}$ and $\tilde{p}$ are the reduced temperature and pressure:

$$\tilde{T} \equiv T/T^*; \qquad T^* \equiv \varepsilon^*/k \tag{42}$$

$$\tilde{p} \equiv p/p^*; \qquad p^* \equiv \varepsilon^*/v^* \tag{43}$$

The minimum value of the free energy is found in the usual way:

$$\partial \tilde{G}/\partial \tilde{v}\big|_{\tilde{T},\tilde{p}} = 0 \tag{44}$$

which yields

$$\tilde{\rho}^2 + \tilde{p} + \tilde{T}[\ln(1-\tilde{\rho}) + (1-1/r)\tilde{\rho}] = 0 \tag{45}$$

Equation (45) is the equation of state for the lattice fluid. It must be kept in mind that in the pressure ensemble $\tilde{\rho}$ is the dependent variable and $\tilde{p}$ and $\tilde{T}$ are the independent variables. Therefore, Eq. (45) defines the value of $\tilde{\rho}$ at given $(\tilde{T}, \tilde{p})$ that minimizes the free energy. Equation (45) can also be obtained directly from the standard thermodynamic equation of state relation, $V = \partial G/\partial p)_T$.

Equations (41) and (45) contain the complete thermodynamic description of the model fluid; all other thermodynamic properties follow from the standard thermodynamic formulas. The thermal expansion coefficient α, the isothermal compressibility β, and the thermal pressure coefficient γ are given by

$$T\alpha = \frac{1 + \tilde{p}\tilde{v}^2}{\tilde{T}\tilde{v}[1/(\tilde{v}-1) + 1/r] - 2} \tag{46}$$

$$p\beta = \frac{\tilde{p}\tilde{v}^2}{\tilde{T}\tilde{v}[1/(\tilde{v}-1) + 1/r] - 2} \tag{47}$$

$$T\gamma = p^*(\tilde{\rho}^2 + \tilde{p}) \tag{48}$$

A fluid is completely characterized by three molecular parameters ε^*, v^*, and r, or equivalently, the three equation of state parameters T^*, p^*, and ρ^*. Unlike the Flory theory, the molecular parameters can be obtained from the equation of state parameters:

$$\varepsilon^* = kT^* \tag{49}$$

$$v^* = kT^*/p^* \tag{50}$$

$$r = Mp^*/kT^*\rho^* = M/\rho^* v^* \tag{51}$$

Since r remains explicit in the reduced equation of state, Eq. (45), a simple corresponding states principle is not, in general, satisfied. However, for a polymer liquid $r \to \infty$ and the equation of state reduces to a corresponding states equation:

$$\tilde{\rho}^2 + \tilde{p} + \tilde{T}[\ln(1-\tilde{\rho}) + \tilde{\rho}] = 0 \tag{52}$$

At atmospheric pressure (zero pressure), Eqs. (46) and (48) become ($r \to \infty$)

$$T\alpha = 1/[\tilde{T}(1-\tilde{\rho}) - 2] \tag{53}$$

and

$$p^* = \tilde{v}^2 T\gamma = \tilde{v}^2 T\alpha/\beta \tag{54}$$

Table I lists values of $\tilde{T}$ and $\tilde{v}$ and the corresponding $T\alpha$ product at atmospheric pressure. From an experimental value of $T\alpha$, $\tilde{T}$ and $\tilde{v}$ can be determined with sufficient accuracy by straight line interpolation of Table I.

Table I

Theoretical Values of the $T\alpha$ Product at
Atmospheric Pressure for Several Reduced
Temperatures[a]

$\tilde{T}$	$\tilde{v}$	$T\alpha$
0.35	1.025	0.0831
0.40	1.039	0.1163
0.45	1.056	0.1534
0.50	1.075	0.1944
0.55	1.098	0.2388
0.60	1.123	0.2872
0.65	1.151	0.3396
0.70	1.183	0.3964
0.75	1.218	0.4581
0.80	1.258	0.5249
0.85	1.301	0.5976
0.90	1.349	0.6772

[a] This table can be used to determine T^*, ρ^*, and v_{sp}^* for polymer liquids if experimental values of ρ, α, and β (or γ) are available (see text).

Thus, T^* and v_{sp}^* can be determined from the known temperature and specific volume ($T^* = T/\tilde{T}$ and $v_{sp}^* = v_{sp}/\tilde{v} = 1/\rho^*$). From an experimental

compressibility or thermal pressure coefficient, the characteristic pressure
p^* can be determined from Eq. (54). These procedures for determining the
equation of state parameters are only valid for high molecular weight liquids;
procedures for determining these parameters for low molecular weight
fluids are described by Sanchez and Lacombe [12], and here a table of
parameters for 60 different fluids, including many common polymer solvents,
can be found.

B. Mixtures

Again, for the sake of clarity, only the results for binary mixtures are
presented; more generalized results are given by Lacombe and Sanchez [14].
The combining rules are as follows:

1. The close-packed molecular volume of each component is conserved.
If an i molecule occupies r_i^0 sites in the pure state and has a close-packed
molecular volume of $r_i^0 v_i^*$, then it will occupy r_i sites in the mixture, where

$$r_i = r_i^0(v_i^*/v^*)$$

This rule guarantees additivity of the close-packed volumes:

$$V^* = r_1^0 N_1 v_1^* + r_2^0 N_2 v_2^*$$

$$= (r_1 N_1 + r_2 N_2) v^*$$

2. The total number of *pair interactions* in the close-packed mixture is
equal to the sum of the pair interactions of the components in their close-
packed pure states, that is,

$$(z/2)(r_1^0 N_1 + r_2^0 N_2) = (z/2)(r_1 N_1 + r_2 N_2)$$

$$= (z/2) rN$$

where

$$r \equiv x_1 r_1^0 + x_2 r_2^0$$

$$\equiv x_1 r_1 + x_2 r_2$$

$$N \equiv N_1 + N_2$$

The combining rules yield the following relationship for the *average* close-
packed mer volume:

$$v^* = \phi_1^0 v_1^* + \phi_2^0 v_2^* \tag{55}$$

where

$$\phi_1^0 = r_1^0 N_1/(r_1^0 N_1 + r_2^0 N_2) = r_1^0 N_1/rN = 1 - \phi_2^0 \tag{56}$$

The volume of the mixture is

$$V = (N_0 + r_1 N_1 + r_2 N_2) v^*$$

$$= (N_0 + rN) v^* \tag{57}$$

and the reduced volume is

$$\tilde{v} \equiv V/rNv^* \equiv (N_0 + rN)/rN = 1/\tilde{\rho} \tag{58}$$

The number of configurations available to a system of $N_1 r_1$-mers, and $N_2 r_2$-mers and N_0 empty sites is approximately given by (cf. Eq. 30):

$$\Omega \simeq (1/f_0)^{N_0} (\omega_1/f_1)^{N_1} (\omega_2/f_2)^{N_2} \tag{59}$$

where

$$f_0 = N_0/(N_0 + rN) = 1 - \tilde{\rho}; \qquad f_1 = \phi_1 \tilde{\rho}; \qquad f_2 = \phi_2 \tilde{\rho} \tag{60}$$

$$\phi_1 = r_1 N_1/rN = (v_1^*/v^*)\phi_1^\circ = 1 - \phi_2 \tag{61}$$

and ω_i is the number of configurations available to a r_i-mer in the close-packed pure state.

If random mixing of the mer and vacant sites is assumed, the intermolecular energy can be written as

$$E = -(z/2)(N_0 + rN)(f_1^2 \varepsilon_{11} + 2 f_1 f_2 \varepsilon_{12} + f_2^2 \varepsilon_{22}) \tag{62a}$$

or

$$E/rN = -\tilde{\rho}\varepsilon^* \tag{62b}$$

where

$$\varepsilon^* = \phi_1 \varepsilon_{11}^* + \phi_2 \varepsilon_{22}^* - \phi_1 \phi_2 kT\chi \tag{63}$$

$$\chi = (\varepsilon_{11}^* + \varepsilon_{22}^* - 2\varepsilon_{12}^*)/kT \tag{64}$$

and $\varepsilon_{ij}^* \equiv (z/2)\varepsilon_{ij}$ is the interaction energy of a mer belonging to component i when it is surrounded by z mers belonging to component j ($\varepsilon_{ij}^* = \varepsilon_{ji}^*$).

The partition function for the mixture is given by Eq. (39). The Gibbs free energy can be expressed in terms of reduced variables:

$$\tilde{G} \equiv G/rN\varepsilon^*$$

$$\tilde{G} = -\tilde{\rho} + \tilde{p}\tilde{v} + \tilde{T}[(\tilde{v} - 1)\ln(1 - \tilde{\rho}) + (1/r)\ln \tilde{\rho} \tag{65}$$

$$+ (\phi_1/r_1)\ln(\phi_1/\omega_1) + (\phi_2/r_2)\ln(\phi_2/\omega_2)]$$

$$\tilde{T} = T/T^*; \qquad T^* = \varepsilon^*/k \tag{66}$$

$$\tilde{p} = p/p^*; \qquad p^* = \varepsilon^*/v^* \tag{67}$$

Minimization of the free energy with respect to N_0, or equivalently the reduced volume $\tilde{v}$, yields the same equation of state as before, Eq. (45), with T^* and p^* defined as above.

Each component satisfies the following relation (cf. Eq. 31):

$$\rho_i^* v_i^* = M_i/r_i^0 \tag{68}$$

With this equation $\phi_1{}^0$ and ϕ_1 can be related to the equation of state parameters:

$$\phi_1{}^0 \equiv r_1{}^0 N_1/rN = (m_1/\rho_1{}^* v_1{}^*)/(m_1/\rho_1{}^* v_1{}^* + m_2/\rho_2{}^* v_2{}^*) \tag{69}$$

$$\phi_1 \equiv r_1 N_1/rN = (m_1/\rho_1{}^*)/(m_1/\rho_1{}^* + m_2/\rho_2{}^*) \tag{70}$$

where m_1 and m_2 are mass fractions:

$$m_1 = x_1 M_1/(x_1 M_1 + x_2 M_2) = 1 - m_2 \tag{71}$$

By definition the close-packed mass density ρ^* of the mixture is equal to the total mass divided by the total close-packed volume; thus,

$$1/\rho^* = m_1/\rho_1{}^* + m_2/\rho_2{}^* = v_{sp}^* \tag{72}$$

The chemical potential μ_1 is given by

$$\mu_1 = \partial G/\partial N_1|_{T,p,N_2}$$

$$\mu_1 = kT\{\ln\phi_1 + (1 - r_1/r_2)\phi_2 + r_1{}^0\tilde{\rho}[\chi + (1 - v_1{}^*/v_2{}^*)\lambda_{12}]\phi_2{}^2\} \tag{73}$$

$$+ r_1{}^0 kT_1{}^*\{-\tilde{\rho} + \tilde{p}_1\tilde{v} + \tilde{T}_1[(\tilde{v} - 1)\ln(1 - \tilde{\rho}) + (1/r_1{}^0)\ln(\tilde{\rho}/\omega_1)]\}$$

where

$$\lambda_{12} \equiv \partial(\varepsilon^*/kT)/\partial\phi_1$$

$$= 1/\tilde{T}_1 - 1/\tilde{T}_2 + (\phi_1 - \phi_2)\chi = -\lambda_{21} \tag{74}$$

$$\tilde{T}_i = T/T_i^*; \qquad \tilde{p}_i = p/p_i^* \tag{75}$$

The expression for μ_2 is easily obtained by interchanging the indices 1 and 2. The chemical potentials have the following properties:

1. They reduce correctly to their appropriate molar pure state values; for example,

$$\lim_{\phi_1 \to 1} \mu_1 \equiv \mu_1{}^0 = r_1{}^0 \varepsilon_{11}^* \tilde{G}(\phi_1 = 1) = r_1{}^0 \varepsilon_{11}^* \tilde{G}_1 \tag{76}$$

2. At low temperatures or high pressures the reduced densities approach their maximum value of unity. In this limit with $v_1{}^* = v_2{}^*$, the Flory–Huggins chemical potentials are recovered:

$$\lim_{\tilde{\rho}_1, \tilde{\rho} \to 1} (\mu_1 - \mu_1{}^0) = kT[\ln\phi_1 + (1 - r_1/r_2)\phi_2 + r_1\chi\phi_2{}^2]. \tag{77}$$

3. There is only one parameter, ε_{12}^* or χ, that characterizes a binary mixture. All other parameters are known from the pure components. Knowledge of ω_1 (or ω_2) is not required since it is absent in the difference $\mu_1 - \mu_1^{\,0}$. It is convenient to express interaction energies in terms of a dimensionless parameter ζ, which measures the deviation of the interaction energy from the geometric mean:

$$\zeta \equiv \varepsilon_{12}^* / (\varepsilon_{11}^* \varepsilon_{22}^*)^{1/2} \tag{78}$$

and thus

$$T\chi = T_1^* + T_2^* - 2\zeta (T_1^* T_2^*)^{1/2} \tag{79}$$

For mixtures of nonpolar fluids one expects $\chi > 0$, which has been established for many binary mixtures containing at least one nonpolar component [14].

C. Phase Stability and the Spinodal

The stability of a homogeneous phase in a binary mixture requires that the chemical potentials possess the following positive property:

$$\partial \mu_1 / \partial x_1 > 0; \qquad \partial \mu_2 / \partial x_2 > 0 \tag{80}$$

It can be shown that these two conditions are satisfied if [14]

$$\tilde{\rho} \{ 2 [(\phi_1 + v\phi_2)\chi + (1 - v)\lambda_{12}] + \tilde{T}\psi^2 p^* \beta (\phi_1 + v\phi_2) \}$$
$$< 1/r_1^{\,0}\phi_1 + v/r_2^{\,0}\phi_2 \tag{81a}$$

where λ_{12} is defined by Eq. (74),

$$\psi \equiv \tilde{\rho}\lambda_{12} - \frac{v}{(\phi_1 + v\phi_2)^2} \left(\frac{1}{r_1^{\,0}} - \frac{1}{r_2^{\,0}} \right) + \frac{\tilde{p}\tilde{v}}{\tilde{T}} \frac{(v - 1)}{\phi_1 + v\phi_2} \tag{82}$$

$$v \equiv v_1^* / v_2^* \tag{83}$$

and β is the isothermal compressibility of the mixture. The definition of ψ given above differs by a factor $v/(\phi_1 + v\phi_2)^2$ than that given by Lacombe and Sanchez [14]; the inequality (58a) of Lacombe and Sanchez [14] can be reduced to the inequality (81a) above by using the following identities:

$$\phi_1/\phi_1^{\,0} \equiv \phi_1 + v\phi_2; \qquad \phi_2/\phi_2^{\,0} \equiv (\phi_1 + v\phi_2)/v \tag{84}$$

If the above inequality (81a) is not satisfied, a binary liquid mixture will phase separate into two liquid phases. The boundary separating the one-phase and the two-phase regions is called the spinodal and is defined when the above inequality becomes an equality.

The general character of the liquid–liquid phase diagram can be deduced by studying the spinodal. To facilitate the analysis, it is assumed that $v = 1$. The condition for phase stability then simplifies to

$$\tilde{\rho}(2\chi + \tilde{T}\psi^2 p^*\beta) < 1/r_1\phi_1 + 1/r_2\phi_2 \tag{81b}$$

The above spinodal reduces to the Flory–Huggins result when $\tilde{\rho} = 1$ ($\beta = 0$ when $\tilde{\rho} = 1$).

The stability of a solution with respect to a liquid–vapor transition requires $\beta > 0$; the condition $1/\beta = 0$ defines the spinodal for the liquid–vapor transition. Thus β is relatively large and positive near a liquid–vapor transition and becomes infinite at a liquid–vapor spinodal temperature T_s.

All quantities in (81b) are inherently positive except for χ, which can be negative. All of the temperature-dependent quantities are on the left-hand side of the inequality, and thus, the general features of the temperature–composition phase diagram can be ascertained by characterizing the temperature dependence of these quantities. The term χ is inversely proportional to temperature and $\tilde{T}\psi^2 p^*\beta$ is positive or zero in the temperature interval $(0, T_s)$. The term ψ can have, at most, one zero in $(0, T_s)$. The temperature T_0 at which the zero occurs when $v = 1$ and $r_1 \neq r_2$ is

$$T_0 = \tilde{\rho}[r_1 r_2/(r_2 - r_1)] T\lambda_{12} \tag{85}$$

If T_0 is not $(0, T_s)$, then $\tilde{T}\psi^2 p^*\beta$ increases monotonically with temperature.

By studying the spinodal inequality, four general phase diagrams in the temperature–composition plane are obtained (Types I–IV). These four diagrams as well as a fifth type (V) are shown in Fig. 1. Type V is a special case of I, II, or IV that is of special importance. The four basic phase diagrams are characterized below:

I. $\chi > 0$, T_0 not in $(0, T_s)$;
II. $\chi > 0$, T_0 in $(0, T_s)$;
III. $\chi \leqslant 0$, T_0 not in $(0, T_s)$;
IV. $\chi \leqslant 0$, T_0 in $(0, T_s)$.

Similar phase diagrams can also be obtained from Flory's theory [21]. Some general comments concerning these phase diagrams follow:

1. Notice that lower critical solution temperatures (LCST) are predicted in all cases. The term involving the isothermal compressibility (β) in Eq. (81b) diverges to plus infinity at the liquid–vapor spinodal temperature T_s. It is this term that also dominates LCST behavior.

2. LCST behavior is characteristic of polymer solutions but not of low molecular weight (MW) solutions. The right-hand side (RHS) of the spinodal inequality is very sensitive to MW, whereas the left-hand side (LHS) is not. Since the RHS decreases sharply (at a given composition) as the MW of

Isaac C. Sanchez

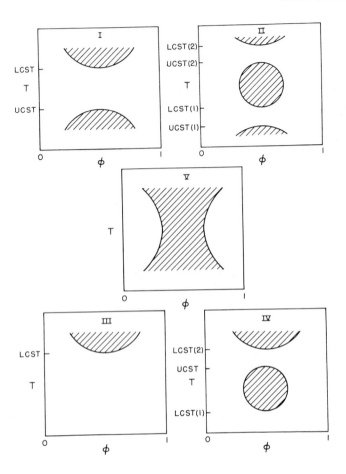

Fig. 1 Schematic liquid–liquid temperature–composition diagram constructed from the spinodal inequality (81b). Shade areas represent T–ϕ regimes of solution instability where phase separation will occur.

either component increases, the possibility of LCST behavior in polymer solutions is more probable.

3. Both the LCST and UCST should be very sensitive to MW. The UCST should *increase* and the LCST should *decrease* as the MW of one or both of the components increases. When a LCST and an UCST merge, an "hourglass" type of phase diagram is obtained as illustrated by Type V.

4. LCST behavior is predicted even if $\varepsilon_{11}^* = \varepsilon_{22}^*$, $v_1^* = v_2^*$, and $\zeta = 1$. The only requirement is a size difference, $r_1 \neq r_2$. Thus, polymer–oligomer mixtures are capable of exhibiting a LCST.

5. An LCST should, in general, be more sensitive to pressure than a UCST because the pressure sensitive compressibility β dominates LCST behavior; $\tilde{\rho}\beta$ decreases with increasing pressure and approaches zero as $p \to \infty$. Thus, the general prediction is that an *LCST should initially increase with increasing pressure*, but the sign of the pressure coefficient may change with increasing pressure. Analysis of the pressure dependence of the UCST is more difficult; it may initially rise or fall with increasing pressure, depending on the relative magnitudes of the two terms on the LHS of Eq. (81b). The pressure coefficient of the UCST can also change sign (once).

D. Physical Significance of the Ratio v_1^*/v_2^*

Mer volumes can be related to molecular surface to volume ratios by approximating the close-packed molecular volume of a r-mer rv^*, by a parallelepiped of length $r(v^*)^{1/3}$ and width $(v^*)^{1/3}$. The molecular surface to volume ratio is then given by (neglecting end surface areas)

$$s = 4r(v^*)^{2/3}/rv^* = 4/(v^*)^{1/3} \tag{86}$$

and

$$s_2/s_1 = (v_1^*/v_2^*)^{1/3} \equiv v^{1/3} \tag{87}$$

In Flory's theory, molecular surface area "corrections" are introduced explicitly via the ratio s_2/s_1. Molecular surface-to-volume ratios can be calculated using the tabulations of group surface areas and volumes given by Bondi [22] or, as Flory and co-workers have often done, by approximating the shape of a small solvent molecule (component 1) by a sphere and a polymer molecule (component 2) by a cylinder [4, 23, 24]. The Bondi method usually yields a value of s_1/s_2 that is lower than the sphere/cylinder approximation [23, 24]. For example, $s_1/s_2 = 1.2$ for methyl ethyl ketone/polystyrene by the Bondi approximation and 2.1 by the sphere/cylinder approximation [23]. The quantity $(v_2^*/v_1^*)^{1/3}$ for this pair is equal to 1.21, where $v_2^* = 17.1$ cm^3/mole (polystyrene [13]) and $v_1^* = 9.54$ cm^3/mole (methyl ethyl ketone [12]). For the pair benzene/poly(dimethyl siloxane) [24] the Bondi value of s_1/s_2 is 1.14 and the sphere/cylinder value is 1.67; $(v_2^*/v_1^*)^{1/3} = 1.10$ for this pair, where $v_2^* = 13.1$ cm^3/mole [poly(dimethyl siloxane) [13]] and $v_1^* = 9.80$ cm^3/mole (benzene [12]). These two examples illustrate that the s_1/s_2 ratio as calculated from Eq. (87) agrees favorably with the Bondi estimates of this ratio and support the assertion that the molecular surface to volume ratio is inversely proportional to $(v^*)^{1/3}$.

In the next section it will be seen that within the framework of the lattice fluid theory, the ratio $v = v_1^*/v_2^*$ plays an important role in determining polymer compatibility.

IV. PREDICTING POLYMER BLEND COMPATIBILITY

A. Compatibility Criteria

1. Flory Theory

McMaster [21] has carried out an extensive numerical analysis of the Flory theory for polymer blends. His findings are summarized below:

a. Negative or very small positive values of the interaction parameter X_{12} favor compatibility. Small positive values yield phase diagrams similar to I in Fig. 1, whereas larger positive values yield hourglass-type diagrams like Type V. Negative values of X_{12} yield Type III phase behavior.

b. Low molecular weight by one or both polymers favors compatibility.

c. Similar thermal expansion coefficients favor compatibility. This is tantamount to requiring that the characteristic temperatures be similar. When $X_{12} = 0$ and both polymers have molecular weights of about 2×10^5, they must have α values within 4% of one another (or the T^* temperatures must be within $200°$K of one another) to achieve significant compatibility.

d. If $\alpha_1 < \alpha_2$, or equivalently, if $T_1^* > T_2^*$, then it is desirable to have the thermal pressure coefficients such that $\gamma_1 > \gamma_2$, or equivalently, $p_1^* > p_2^*$.

In summary, since X_{12} is expected to be positive for nonpolar polymers, the conditions that favor polymer compatibility are low molecular weights and similar characteristic temperatures. The combination $T_1^* > T_2^*$, $p_1^* > p_2^*$ is favorable and the combination $T_1^* > T_2^*$, $p_1^* < p_2^*$ is very unfavorable. The effect of the ratio s_1/s_2 on polymer compatibility was not systematically investigated.

2. Lattice Fluid Theory

The conditions for polymer compatibility can be determined by examining the spinodal inequality Eq. (81a). The right-hand side (RHS) is positive and has a maximum value of

$$(\text{RHS})_{\text{max}} = [(vr_1^0)^{1/2} + (r_2^0)^{1/2}]^2/(r_1^0 r_2^0) \tag{88}$$

at a critical concentration ϕ_1^c of

$$\phi_1^c = (r_2^0)^{1/2}/[(vr_1^0)^{1/2} + (r_2^0)^{1/2}] \tag{89}$$

For a polymer blend, both r_1^0 and r_2^0 are very large and $(\text{RHS})_{\text{max}}$ approaches zero as r_1^0 and r_2^0 approach infinity. Since the term $\tilde{T}\psi^2 p^*\beta$ on the LHS is always positive, polymer compatibility for high molecular weight polymers can only be achieved if the other term on the LHS is negative. If this term is negative, then this is a necessary and sufficient condition for polymer compatibility over the entire composition range at low temperatures. At high temperatures the positive equation of state term $\tilde{T}\psi^2 p^*\beta$ will overwhelm the negative term, and a lower critical solution temperature-type behavior (III) will result.

The conditions for which the term

$$f = (\phi_1 + v\phi_2)\chi + (1-v)\lambda_{12} \tag{90}$$

is negative over the entire composition range can be easily determined. It is a linear function of the composition and will attain its maximum value at either $\phi_1 = 0$ or 1. Previous studies on low molecular weight solutions have yielded positive χ values for solutions involving nonpolar or weakly polar components [14]. In terms of the dimensionless interaction parameter ζ defined by Eq. (78), positive χ values usually mean that $\zeta < 1$. The exact condition on ζ is

$$\zeta < (1+\tau)/(2\tau^{1/2}); \qquad \tau \equiv T_1^*/T_2^* > 1 \tag{91}$$

Thus, attention will be restricted to conditions for which f is negative and χ is positive; when χ is positive, ζ satisfies the inequality (91). There are only two cases:

Case I: $\tau > 1$; $\quad v > 1$

Inspection of $\partial f/\partial \phi_1$ shows that it is negative under these conditions, and f has its maximum value at $\phi_1 = 0$. Evaluating f at $\phi_1 = 0$ yields the following inequality for v:

$$v > v_{min} = 2(1 - \zeta\tau^{1/2})/(3 + \tau - 4\zeta\tau^{1/2}) \tag{92}$$

Inspection of Eq. (92) shows further that finite positive values of v greater than unity will only obtain if

$$\zeta > \zeta_{min} = (3 + \tau)/(4\tau^{1/2}) \tag{93}$$

Thus, the chances of polymer compatibility are optimized when v and ζ satisfy the above inequalities, Eqs. (92) and (93). The variation of v_{min} with τ and ζ is shown in Table II.

Case II: $\tau > 1$; $\quad v < 1$

In this case f_{max} occurs at $\phi_1 = 1$ and no values of $v < 1$ exist for which f is negative for all ϕ_1. Thus, compatibility over the entire composition range is very unlikely under these conditions.

Table II

Values of $v_{min}{}^a$ as a Function of τ and the Interaction Parameter ζ

τ	ζ: 0.95	0.96	0.97	0.98	0.99	1.00
1.05	∞	∞	∞	∞	3.71	1.01
1.10	∞	∞	∞	4.91	1.44	1.03
1.15	∞	∞	7.42	1.90	1.28	1.04
1.20	∞	15.86	2.49	1.56	1.22	1.05
1.25	∞	3.39	1.92	1.44	1.20	1.06
1.30	5.09	2.41	1.71	1.39	1.20	1.08
1.35	3.18	2.07	1.61	1.36	1.20	1.09
1.40	2.58	1.89	1.55	1.34	1.20	1.10
1.45	2.29	1.79	1.51	1.33	1.21	1.11
1.50	2.12	1.73	1.49	1.33	1.21	1.13
1.55	2.02	1.69	1.48	1.33	1.22	1.14
1.60	1.95	1.67	1.47	1.34	1.23	1.15
1.65	1.91	1.65	1.47	1.34	1.24	1.17
1.70	1.87	1.64	1.48	1.35	1.26	1.18
1.75	1.85	1.64	1.48	1.36	1.27	1.19

^{a} v_{min} is calculated from Eq. (92). For example, if $\tau = 1.25$ and $\zeta = 0.98$, v must exceed 1.44 if a polymer pair is to achieve compatibility over the entire composition range (see text).

In summary, I favors compatibility whereas II does not. The explanation for this result is embodied in the heat of mixing ΔH_m:

$$\Delta H_m = rNkT[\tilde{\rho}\phi_1\phi_2\chi + (\tilde{\rho}_1\phi_1{}^0 - \tilde{\rho}\phi_1)/\tilde{T}_1 + (\tilde{\rho}_2\phi_2{}^0 - \tilde{\rho}\phi_2)/\tilde{T}_2] + p\Delta V_m$$

$$(94)$$

and in the high-density limit $(T \to 0, p \to \infty)$ reduces to

$$\Delta H_m(\tilde{\rho} = \tilde{\rho}_1 = \tilde{\rho}_2 = 1) = rNkT\phi_1\phi_2[\chi - (v-1)(\tau-1)/\tilde{T}_2(\phi_1 + v\phi_2)] \quad (95)$$

The latter differs from classical theory by the presence of the term containing $v - 1$. This term is a direct consequence of the combining rules and, as pointed out earlier, v appears to be a measure of the dissimilarity in molecular surface area properties. Thus, Eq. (95) shows that even if $\chi = 0$, positive or negative values of ΔH_m are possible. A negative contribution occurs when $v > 1$ and $\tau > 1$, that is, Case I conditions favor a negative heat of mixing at low temperatures. A Type III phase diagram is expected when $\Delta H_m < 0$ at low temperatures.

Surface area effects on ΔH_m can be interpreted as follows: if $v = 1$, forming two 1—2 bonds requires breaking a 2—2 bond and a 1—1 bond (classical view); if $v > 1$, forming two 1—2 bonds requires breaking a 2—2 bond as before, but only a fraction of a 1—1 bond. If 1—1 bonds are stronger than 2—2 bonds ($\tau > 1$), the latter can be an energetically favorable process and may yield a negative ΔH_m even if $\chi > 0$.

B. Comparison of Theoretical Predictions

Predicting polymer compatibility by either the Flory or lattice fluid theories requires accurate values of the equation of state parameters. Therefore, it is very important that these parameters for a given polymer pair be evaluated from experimental values of ρ, α, and β (or γ) determined at the same temperature and pressure. Procedures for determining the equation of state parameters were discussed in Sections II and III.

Table III lists the equation of state parameters for ten different polymers determined at the indicated temperature and at atmospheric pressure. In addition, solubility parameters for each polymer have been calculated under the same conditions by using the relation

$$\delta^2 = T\alpha/\beta \qquad (96)$$

At zero pressure the internal pressure $(\partial E/\partial V)_T$ is equal to $T\alpha/\beta$ and for a van der Waals fluid the cohesive energy density δ^2 is equal to the internal pressure. Also, since $\delta^2 \sim V^2$ for a van der Waals fluid, it is easy to show that

$$\partial \ln \delta/\partial T|_p = -\alpha \qquad (97)$$

The temperature dependence of solubility parameters can be estimated from Eq. (97). A good rule of thumb is that δ decreases 0.1 unit, in units of $(\text{cal/cm}^3)^{1/2}$, for a 25°C temperature increase.

Solubility Parameter Method

According to the solubility parameter method of predicting compatibility, the difference in solubility parameters $\Delta\delta$ must be very small. Inspection of Table III shows that there are four polymer pairs for which $\Delta\delta < 0.2$. These pairs are PE–PIB, PVA–PEO, PVA–PnBMA, and PEO–PMMA. For the latter two pairs, temperature corrections were employed by using Eq. (97) or the general rule of thumb mentioned above. Of these four pairs, PEO–PMMA has the smallest $\Delta\delta$ (less than 0.1). Thus, of the 45 possible polymer pairs in Table III, the pair PEO–PMMA should have the best chance of being compatible.

Flory Theory

As discussed earlier, the characteristic temperatures of the two polymers must be very close to one another ($\Delta T^* < 200°\text{K}$) for polymer compatibility. Larger ΔT^* differences can be tolerated if both $T_1^*/T_2^* > 1$ and $p_1^*/p_2^* > 1$; the combination $T_1^*/T_2^* > 1$ and $p_1^*/p_2^* < 1$ is very unfavorable. Unfortunately, the exact conditions on the T_1^*/T_2^* and p_1^*/p_2^* ratios have not yet been determined. Nevertheless, these general criteria can still be used.

Table III

Equation of State Parameters for the Flory and Lattice Fluid Theories and Solubility Parameters for Ten Selected Polymers[a]

Polymer	t(°C)	T^*(°K) Flory	T^*(°K) Lattice	p^*(bars) Flory	p^*(bars) Lattice	ρ^*(gm/cm³) Flory	ρ^*(gm/cm³) Lattice	v^*(cm³/mole) Lattice	ζ (cal/cm³)$^{1/2}$ Lattice
Poly(dimethyl siloxane) (PDMS) [28]	100	6050	569	3000	2440	1.159	1.046	19.3	6.6
Poly(ethylene oxide) (PEO) [35]	58	6460	592	6700	5550	1.328	1.208	8.87	10.4
Poly(vinyl acetate) (PVA) [36]	70	6720	615	6400	5290	1.397	1.270	9.67	10.2
Poly(n-butyl methacrylate) (PnBMA) [37]	106	7000	645	6210	5120	1.232	1.116	10.5	9.90
Poly(ethylene)(linear) (PE) [26]	100	7010	650	4730	3890	0.987	0.895	13.9	8.69
Polystyrene (PS) [25]	100	7950	708	5060	4240	1.219	1.115	13.9	9.26
	200	8660	803	4950	4060	1.194	1.082	16.4	8.82
Poly(isobutylene) (PIB) [27]	100	8030	717	4310	3610	1.042	0.953	16.4	8.57
Poly(vinyl chloride) (PVC) [38]	91	8130	720	6750	5670	1.603	1.466	10.4	10.8
Poly(2,6-dimethylphenylene oxide) (PDMPO) [39]	244	8260	777	6350	5170	1.287	1.161	12.5	9.57
Poly(methyl methacrylate) (PMMA) [37]	125	8440	756	6070	5070	1.362	1.245	12.4	10.1

[a] Literature sources for density, thermal expansion coefficient, and isothermal compressibilities (or thermal pressure coefficients) are indicated by a reference number for each polymer.

Notice that for polystyrene $T^* = 7950°$K and $p^* = 5060$ bars at $100°$C, and $T^* = 8660°$K and $p^* = 4950$ bars at $200°$C. This is a general characteristic of the Flory theory and of the lattice fluid theory; that is, the calculated characteristic temperatures tend to increase as the temperature at which they are calculated increases, whereas the characteristic pressure only decreases slightly. For polystyrene [25], polyethylene [26], polyisobutylene [27], and poly(dimethyl siloxane) [28] T^* increases at the rate of about $70°/10°$; that is, $dT^*/dT \simeq 7°$.

With the temperature dependence of T^* in mind, Table III indicates that there are five pairs with ΔT^* values of less than $200°$: PVAC–PEO, PVA/PnBMA, PVC–PIB, PE–PnBMA, and PS–PIB. The latter two pairs badly violate the criteria $p_1^*/p_2^* > 1$ when $T_1^*/T_2^* > 1$.

Lattice Fluid Theory

Published studies on low molecular weight solutions [14] and unpublished studies of several polymer solutions [29] have shown that ζ tends to fall in the general range

$$0.95 < \zeta < 1.0$$

for solutions that contain at least one nonpolar component. These results suggest that most nonpolar–nonpolar and nonpolar–polar polymer mixtures have ζ interaction parameter values in this range.

As in the Flory theory, the characteristic temperature T^* tends to increase as the temperature T at which T^* is determined increases, but p^* only slightly decreases. For polystyrene $T^* = 708°$K at $100°$C and $803°$K at $200°$C. The mer volume v^* also increases with the measurement temperature. Rough estimates of the temperature dependences of T^* and v^* are

$$\partial \ln T^*/\partial T \simeq .0012/°\text{K} \qquad \partial \ln v^*/\partial T \simeq .0017/°\text{K}$$

The most favorable conditions for compatibility are those denoted as Case I $(\tau > 1; v > 1)$. For a given value of τ and ζ, v must exceed the value given in Table II to achieve compatibility over the entire composition range. Notice that PDMS has the smallest T^* and the largest v^* in Table II. It should be the worst possible candidate for blending with any other polymer listed in the table since it always forms a Case II pair $(\tau > 1; v < 1)$.

Using the rough estimates given above for the temperature dependences of T^* and v^*, τ and v values can be calculated for most of possible pairs listed in Table III. However, it is not recommended that parameter extrapolations be extended over more than $50°$. Calculation of the τ and v ratios indicates that if $\zeta = 1.0$, about 20 pairs would have a good chance of being compatible. Notice from Table II that for $\zeta = 1.0$, v need only exceed unity by a small amount to achieve the conditions favorable for compatibility.

When $\zeta = 0.99$, v must exceed unity by relatively larger margin, and only six pairs have v values sufficiently large to exceed the v_{min} values. When $\zeta = 0.98$, none of the pairs have a sufficiently large v value. The six pairs that should be compatible even if $\zeta = 0.99$ are PEO–PS, PEO–PIB, PVA–PIB, PnBMA–PIB, PEO–PMMA, and PS–PDMPO.

Table IV is a summary of the predictions of the three theories. Notice that some common predictions are made between the theories, while others are unique to a given theory. The pair PVC–PnBMA has been stated to be

Table IV^a

Compatibility Predictions

Solubility parameter method	Flory theory	Lattice fluid theory
PEO–PMMA		PEO–PMMA
PE–PIB		
PVA–PEO	PVA–PEO	
PVA–PnBMA	PVA–PnBMA	
	PVC–PIB	
		PEO–PS
		PEO–PIB
		PVA–PIB
		PnBMA–PIB
		PS–PDMPO

^{a} Of the 45 possible pairs in Table III, the pairs listed above are those that each theory rates as the best candidates for forming a compatible blend.

compatible [30], but none of the three theories suggests compatibility for this pair. The pair PEO–PMMA, which the solubility parameter method and lattice fluid theory favor, does form a compatible blend over its entire composition range [31]. The only other blend in Table III for which there is evidence for compatibility is PS–PDMPO [32–34]. Only the lattice fluid theory favors compatibility for this pair.

These results and comparisons suggest that the Flory theory is not as good as either the solubility parameter method or the lattice fluid theory, but this may be an erroneous conclusion because the Flory theory has not as yet been completely analyzed. More exact criteria need to be developed for the Flory theory and more comparisons of the theory with experiment are required before we can conclude which theory, if any, is superior. At the present time, it is recommended that all three theoretical methods be applied to a given polymer pair to determine the possibility of compatibility.

REFERENCES

1. P. I. Freeman and J. S. Rowlinson, *Polymer* **1**, 20 (1960).
2. P. J. Flory, R. A. Orwoll, and A. Vrij, *J. Amer. Chem. Soc.* **86**, 3515 (1964).
3. P. J. Flory, *J. Amer. Chem. Soc.* **87**, 1833 (1965).
4. B. E. Eichinger and P. J. Flory, *Trans. Faraday Soc.* **64**, 2035 (1968).
5. P. J. Flory, *Discuss. Faraday Soc.* **49**, 7 (1970).
6. D. Patterson, *J. Polym. Sci. Part C* **16**, 3379 (1968).
7. D. Patterson, *Macromolecules* **2**, 672 (1969).
8. D. Patterson and G. Delmas, *Discuss. Faraday Soc*, **49**, 98 (1970).
9. J. Biros, L. Zeman, and D. Patterson, *Macromolecules* **4**, 30 (1971).
10. I. Prigogine, "The Molecular Theory of Solutions," Chapter XVI. North-Holland Publ., Amsterdam, 1957.
11. P. J. Flory, "Principles of Polymer Chemistry," Chapter 12. Cornell Univ. Press, Ithaca, New York, 1953.
12. I. C. Sanchez and R. H. Lacombe, *J. Phys. Chem.* **80**, 2352 (1976).
13. I. C. Sanchez and R. H. Lacombe, *J. Polym. Sci. Polym. Lett. Ed.* **15**, 71 (1977).
14. R. H. Lacombe and I. C. Sanchez, *J. Phys. Chem.* **80**, 2568 (1976).
15. R. Simha and T. Somcynsky, *Macromolecules* **2**, 342 (1969).
16. T. Somcynsky and R. Simha, *J. Appl. Phys.* **42**, 4545 (1971).
17. T. Nose, *Polym. J.* **2**, 124 (1971).
18. T. Nose, *Polym. J.* **2**, 427 (1971).
19. J. G. Curro, *J. Macromol. Sci. Rev. Macromol. Chem.* **C11**, 321 (1974).
20. S. Beret and J. M. Prausnitz, *Macromolecules* **8**, 878 (1975).
21. L. P. McMaster, *Macromolecules* **6**, 760 (1973).
22. A. Bondi, *J. Phys. Chem.* **68**, 441 (1964).
23. P. J. Flory and H. Hocker, *Trans. Faraday Soc.* **67**, 2258 (1971).
24. P. J. Flory and H. Shih, *Macromolecules* **5**, 761 (1972).
25. H. Hocker, G. J. Blake, and P. J. Flory, *Trans. Faraday Soc.* **67**, 2251 (1971).
26. R. A. Orwoll and P. J. Flory, *J. Amer. Chem. Soc.* **69**, 6814 (1967).
27. B. E. Eichinger and P. J. Flory, *Macromolecules* **1**, 285 (1968).
28. H. Shih and P. J. Flory, *Macromolecules* **5**, 758 (1972).
29. Unpublished data of F. E. Karasz and W. J. MacKnight, Univ. of Massachusetts.
30. S. Krause, *J. Macromol. Sci. Rev. Macromol. Chem.* **7**, 251 (1972).
31. D. M. Hoffman, unpublished data, University of Massachusetts.
32. W. J. MacKnight, F. E. Karasz, and J. Stoelting, *Polym. Eng. Sci.* **10**, 133 (1970).
33. W. J. MacKnight, F. E. Karasz, and J. Stoelting, *in* "Multicomponent Polymer Systems," Adv. Chem. Ser. 99. Amer. Chem. Soc., Washington, D.C., 1971.
34. A. R. Shultz and B. M. Beach, *Macromolecules* **1**, 902 (1974).
35. C. Booth and C. J. Devoy, *Polymer* **12**, 309 (1971).
36. J. E. McKinney and M. Goldstein, *J. Res. Nat. Bur. Stand.* **78A**, 331 (1974).
37. O. Olabisi and R. Simha, *Macromolecules* **8**, 206 (1975).
38. K. Hellvege, W. Knappe, and P. Lehmann, *Kolloid Z.Z. Polym.* **193**, 16 (1962).
39. At 244°C, $\rho = 1.00$ gm/cc and $\alpha = 6.9 \times 10^{-4}$ deg^{-1} obtained from unpublished data of F. E. Karasz. The compressibility at 244°C was estimated to be 9.3×10^{-5} bar^{-1} by using McGowan's parachor correlation [J. C. McGowan, *Polymer* **10**, 841 (1969)].

Chapter 4

Phase Separation Behavior of Polymer–Polymer Mixtures

T. K. Kwei and T. T. Wang

Bell Laboratories
Murray Hill, New Jersey

I. Introduction 141
II. Upper and Lower Critical Solution Temperatures 146
   A. Systems Exhibiting UCST 146
   B. Systems Exhibiting LCST 147
III. Phase Separation 152
   A. Phase Diagram and Stability 152
   B. Definitions of Nucleation and Growth Mechanism and Spinodal Decomposition Mechanism 152
   C. Kinetics of Phase Separation 154
   D. Morphology of Spinodally Decomposed Mixtures 158
IV. Experiment 159
   A. Identification of the Stable Region 159
   B. Phase Separation 161
V. Results. 165
   A. Interaction Parameters. 165
   B. Phase Boundary 167
   C. Morphology. 175
   D. Kinetics of Phase Separation 179
VI. Conclusion 183
   References 183

I. INTRODUCTION

In this chapter we consider the phase relationship in polymer mixtures. The stability of a compatible polymer mixture and its separation into

141

multiple phases are discussed with emphasis on phase boundaries, phase composition, and the kinetics of phase separation.

The thermodynamic criterion for two phases at equilibrium is stipulated by the equality of the chemical potential of each component in the two phases. For the sake of convenience, let us confine our discussion to binary mixtures. The two coexisting phases on the binodal curve must meet the requirement

$$\mu_1 = \mu_1'; \qquad \mu_2 = \mu_2' \tag{1}$$

where the prime designates the second phase. The loci of points on the free energy composition diagram that meet the condition

$$\partial^2 \Delta G / \partial \phi_1 \partial \phi_2 = 0 \tag{2}$$

form the spinodal. At the critical point for incipient phase separation, both the second and the third derivatives of the free energy of mixing vanish.

In the application of the thermodynamic criteria to the study of phase equilibrium in polymeric systems, two complicating factors are encountered, namely, a suitable expression for the free energy of mixing and the polydisperse nature of the specimens. Let us begin with the simplest case of a binary mixture of monodisperse polymers. It is assumed that the Flory–Huggins expression for the free energy of mixing is adequate and that the interaction parameter is independent of concentration,

$$\Delta G / RT = [\phi_1 x_1^{-1} \ln \phi_1 + \phi_2 x_2^{-1} \ln \phi_2 + \chi \phi_1 \phi_2] \tag{3}$$

where ΔG is the free energy of mixing per mole of lattice site, ϕ is the volume fraction, and x is the chain length, with each repeating unit defined as occupying a lattice site. The critical conditions are

$$\phi_{1c} = x_1^{1/2}/(x_1^{1/2} + x_2^{1/2}) \tag{4}$$

$$\chi_c = \tfrac{1}{2}(x_1^{-1/2} + x_2^{-1/2})^2 \tag{5}$$

and the equation for spinodal is

$$\chi_{sp} = \tfrac{1}{2}(x_1^{-1}\phi_1^{-1} + x_2^{-1}\phi_2^{-1}) \tag{6}$$

A massive amount of work has been carried out to analyze the effect of polydispersity. The interested reader should consult the papers by Koningsveld and co-workers [1–3]. When both polymers are polydisperse, the terms $(\phi_1 x_1^{-1} \ln \phi_1)$ and $(\phi_2 x_2^{-1} \ln \phi_2)$ in Eq. (3) are replaced by $\Sigma \phi_{1i} x_{1i}^{-1} \ln \phi_{1i}$ and $\Sigma \phi_{2i} x_{2i}^{-1} \ln \phi_{2i}$, respectively. The spinodal condition is

$$2\chi_{sp} = (\phi_1 x_{w1})^{-1} + (\phi_2 x_{w2})^{-1} \tag{7}$$

and the critical composition is given by

$$\phi_{1c} = 1/(1 + x_{w1}x_{z2}^{1/2}/x_{w2}x_{z1}^{1/2}) \qquad (8)$$

where x_w and x_z are the weight and z average chain lengths. Thus, a mixture of two polydisperse polymers can be considered to be a quasibinary system when the appropriate molecular weight averages are used.

If χ is dependent on composition, the above expressions become

$$2\chi - 2(1 - 2\varphi_1)\partial\chi/\partial\phi_1 = \phi_1\phi_2\,\partial^2\chi/\partial\phi_1{}^2 = (\phi_1 x_{w1})^{-1} + (\phi_2 x_{w2})^{-1} \qquad (9)$$

for the spinodal and

$$-6\,\partial\chi/\partial\phi_1 + 3(1 - 2\phi_1)\partial^2\chi/\partial\phi_1{}^2 + \phi_1\phi_2\partial^3\chi/\partial\phi_1{}^3$$
$$= x_{z1}/x_{w1}^2\phi_1{}^2 - x_{z2}/x_{w2}^2\phi_2{}^2 \qquad (10)$$

for the critical point.

Calculated cloud point curves and spinodals are given by Koningsveld for a variety of conditions (Fig. 1). (Note that the symbol g, instead of χ, is used in his papers.) The location of the critical point and the shapes of the cloud point curve and the spinodal vary markedly as the molecular weight distributions of the two polymers differ. These calculated features find qualitative verification in experimental results (Fig. 2). In addition, the (at least quadratic) dependence of χ on ϕ may lead to a two-peaked cloud point curve (Fig. 3).

A more fundamental problem concerns the proper expression for the free energy of mixing. The merit of the equation of state approach has been expounded in Chapter 3. The use of Patterson's [5] or Flory's [6] thermodynamic theory of polymer solutions allows a better understanding of the phenomenon of lower critical solution temperature recognized during the last decade, and therefore occupies an important place in the discussion of phase separation. Specifically, we refer to the analysis of McMaster [7] on binodal, spinodal, and critical conditions based on an adaptation of Flory's theory. In McMaster's calculation, a segment is defined as a real repeating unit of the polymer and consequently the mass per segment enters explicitly into his equation. Second, he allows for a change in the external degree of freedom upon mixing. The rest of his equation is identical to Flory's, with the use of X_{12} and Q_{12} as energy and entropy of interaction between unlike segments. There is a point of departure in the evaluation of the chemical potential of binary monodisperse systems,

$$\Delta\mu_1 = (\partial\Delta G/\partial N_1)_{T,V,N_2} + (\partial\Delta G/\partial\bar{v})_{T,N_1,N_2}(\partial\bar{v}/\partial N_2)_{T,V,N_1} \qquad (11)$$

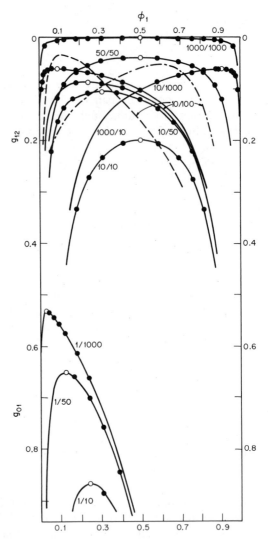

Fig. 1   Spinodals (——) and critical points (○) in quasi-binary systems showing the relative locations of the unstable regions in mixtures of two polydisperse polymers (upper set of curves, interaction parameter g_{12}) and solutions of a polydisperse polymer in a single, low molecular weight solvent (lower set of curves, interaction parameter g_{01}). The ratios of x_{w2}/x_{w1} are indicated. The two spinodals (---) and (-·-) refer to a concentration-dependent g_{12} ($\partial g_{12}/\partial \phi_1 = -0.1$ and $+0.1$, respectively). The critical points at the maxima of the spinodals refer to a_2/α_1 values of 1; those on the right-hand branches refer to α_2/α_1 values of 0.5, 0.2, 0.1, 0.05, 0.02, 0.01, 0.001 (from left to right, as far as they are indicated); those on the left-hand branches refer to α_2/α_1 values of 2, 5, 10, 20, 50, 100, 1000 (from right to left); α_2 and α_1 stand for $x_{z,2}/x_{w,2}$ and $x_{z,1}/x_{w,1}$.

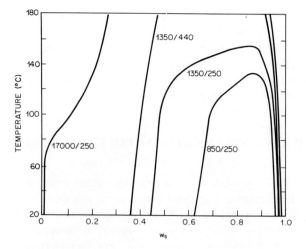

Fig. 2 Experimental cloud point curves of liquid mixtures of silicone (M values: 17000; 1350; 850) and polyisobutene (M: 440, 250); W_s: weight fraction of silicone.

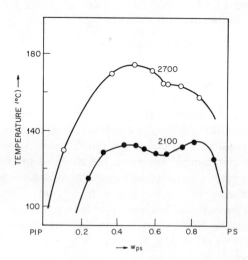

Fig. 3 Measured cloud point curves in the system polyisoprene ($M_n = 2700$–polystyrene (PS, M_n values indicated); w_{ps} = weight fraction polystyrene.

In Eq. (11), $\bar{v}$ is the reduced volume. The second term on the right-hand side of the above equation makes only a small contribution at low pressures and is neglected by Flory but is included in McMaster's derivation to evaluate the effect of pressure and thermal pressure coefficient. McMaster's equations are too cumbersome to be reproduced here. The results of his calculation are shown in Section II.

II. UPPER AND LOWER CRITICAL SOLUTION TEMPERATURES

When an initially homogeneous mixture undergoes phase separation upon lowering of temperature, the cloud point curve associated with the upper critical solution temperature (UCST) is convex upward and its maximum point shifts to a higher temperature with increasing molecular weight of the polymer. For many polymer–solvent and polymer–polymer mixtures, a decrease in mutual solubility is also observed as temperature is increased. The cloud point curve associated with phase separation at high temperatures is convex downward and the critical temperature is called lower critical solution temperature (LCST) although it occurs actually at a higher temperature than UCST. The LCST shifts to a lower temperature with increasing molecular weight. Studies by Patterson [8] and by Kaneko [9, 10] on LCST of polymer–solvent mixtures provide excellent background information about the subject.

A. Systems Exhibiting UCST

Most experimental studies in the past are focused on compatibility at room temperature. Once a compatible pair is found, there is little incentive to examine the possibility of phase separation below room temperature. From a purely experimental viewpoint, the study of phase separation at low temperatures may be impeded by kinetic reasons, especially if the glass transition temperature of the mixture is high. (One example of phase separation at low temperature has been reported—natural rubber, SBR [3].) For these reasons, the UCSTs reported in the literature are usually concerned with polymers of modest molecular weight for which macroscopic liquid phases separate into distinct layers, and the compositions of the conjugated phases can be determined by standard analytical methods. Examples in this category are polyisobutene–poly(dimethyl siloxane) [4], polystyrene–polyisoprene [11], and perhaps poly(ethylene oxide)–poly(propylene oxide) [12].

B. Systems Exhibiting LCST

Polystyrene–poly(vinyl methyl ether) [13] was the first pair for which thermally induced phase separation was observed. Subsequently polycapro-lactone–poly(styrene-*co*-acrylonitrile) [7] and poly(methyl methacrylate)–poly(styrene-*co*-acrylonitrile) [14] were also found to exhibit LCST behavior. In the following, phase diagrams and spinodals calculated by McMaster [7] for simulated binary monodisperse pairs are given to illustrate graphically the effect of molecular weight, interaction parameter, thermal expansion coefficient, and thermal pressure coefficient. The materials parameters used for simulation are listed in Table I. *An important conclusion is that LCST behavior should generally be anticipated for polymer–polymer systems.*

Table I

Base Case Parameters for Simulated Polymer–Polymer Phase Diagrams

Component	M	m	ρ (g/cm^3)	α ($^\circ$K^{-1})	γ (atm/$^\circ$K)	Temp. ($^\circ$C)
1	50,000	104	1.005	5.80×10^{-4}	8.47	135
2	4,000	58	1.055	7.23×10^{-4}	7.24	135
Mixture properties: $X_{12} = Q_{12} = 0$; $s_1/s_2 = 1.56$						
3	100,000	104	1.005	5.80×10^{-4}	8.47	135
4	100,000	104	1.005	6.15×10^{-4}	8.47	135
Mixture properties: $X_{12} = Q_{12} = 0$; $s_1/s_2 = 1.0$						

1. Effect of Molecular Weight

The mutual solubility of two polymers decreases as the molecular weight of either component increases. There is also a general skewness of the binodal and spinodal curves when the chain lengths differ considerably (Fig. 4).

2. Effect of Thermal Expansion Coefficient

The large difference in the thermal expansion coefficients α of the solvent and the polymer at high temperatures was recognized by Patterson in 1962 as a major cause for the LCST phenomena. According to McMaster, polymer–polymer solubility is sensitive to small differences in thermal expansion coefficients, as shown in Fig. 5. A difference of only 10% in the values of α can shift the LCST to a temperature below -100°C and lead to virtually complete immiscibility at ordinary conditions.

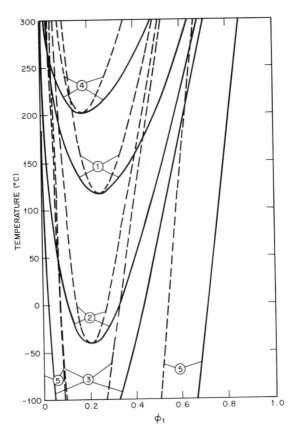

Fig. 4   Binodal and spinodal curves illustrating molecular weight effects (Components 1 and 2):

Curve no.	MW
1	$M_1 = 30,000$
2	$M_1 = 50,000$
3	$M_1 = 80,000$
4	$M_2 = 3,000$
5	$M_2 = 6,000$

3.   Effect of Thermal Pressure

The thermal pressure coefficient γ enters into consideration through its relation to the characteristic pressure parameter in the reduced variables. The effect of differences in γ values on phase diagrams is also considerable, as shown in Fig. 6.

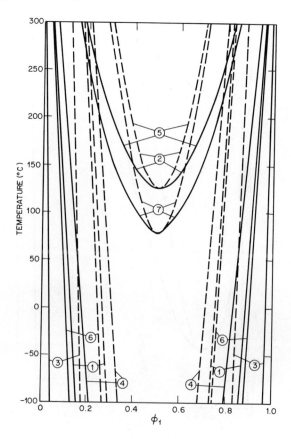

Fig. 5 Effects of molecular weight and the thermal expansion coefficient (Components 3 and 4), $\alpha_1 = 0.580 \times 10^{-3}\,°K^{-1}$:

Curve no.	MW	α_2 $(°K^{-1} \times 10^3)$	Rel. Diff. $(\alpha_2 - \alpha_1) \times$ $100/\alpha_1$ (%)
1	$M_1 = M_2 = 50,000$	0.640	10.3
2	$M_1 = M_2 = 50,000$	0.630	8.6
3	$M_1 = M_2 = 100,000$	0.630	8.6
4	$M_1 = M_2 = 100,000$	0.620	6.9
5	$M_1 = M_2 = 100,000$	0.615	6.0
6	$M_1 = M_2 = 200,000$	0.610	5.2
7	$M_1 = M_2 = 200,000$	0.605	4.3

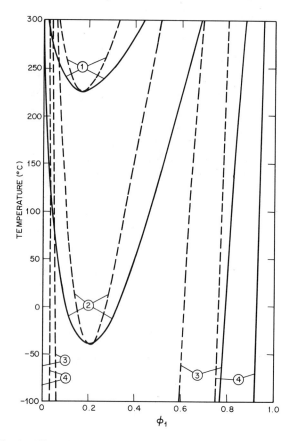

Fig. 6   Effect of thermal pressure coefficient (Components 1 and 2):

Curve no.	$\gamma_2 (\text{atm}/^\circ\text{K})$
1	6.50
2	7.24
3	8.47
4	10.00

4.   Effect of Interaction Parameters X_{12} and Q_{12}

A range of X_{12} values, from small negative values to small positive values, were examined (Fig. 7). As X_{12} decreases from zero to a negative value, the binodal and spinodal curves tend to flatten but the general features do not change markedly. Small positive values of X_{12} lead to both UCST

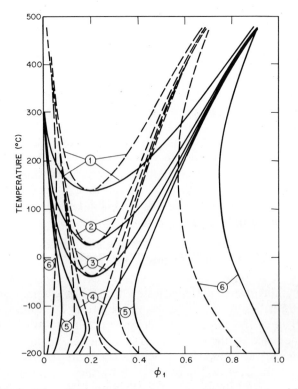

Fig. 7 Effect of interaction energy parameter (Components 1 and 2):

Curve no.	$X_{12}(\text{cal/cm}^3)$
1	−0.050
2	−0.100
3	0.0
4	0.005
5	0.010
6	0.100

and LCST. The two critical points may merge to result in hourglass-shaped phase diagrams sometimes seen in polymer–solvent systems.

The entropy interaction parameter Q_{12} appears in ΔG_m as a correction term for residual entropy. Positive values of Q_{12} enhances solubility.

In conclusion, McMaster's calculations demonstrate the influence of various parameters on LCST. The simulated phase diagrams, however, seem to be steeper than the experimental cloud point curves.

III. PHASE SEPARATION

In this section we consider the kinetic aspect of liquid–liquid phase transformation in binary mixtures. The study of the rate process is not only of basic importance but also of practical interest since, as we shall see, mixtures of different morphologies can result from different decomposition mechanisms. This affords a possibility of enhancing the properties of compatible systems by phase transformation. Despite these premises, however, the subject remains academic at least for the time being since there are so far only a handful of compatible polymer mixtures available for such a study. A concise review on the theoretical treatment based on Cahn's work [15] was given by McMaster [14] in his recent morphological study of phase separation in several polymeric systems.

A. Phase Diagram and Stability

We consider the phase diagram of a system containing a simple miscibility gap as shown in Fig. 8. At any given temperature T_0 inside the gap, we can expect the free energy of an undecomposed mixture to vary with composition as depicted by the top curve. The points of common tangency to this curve define the composition of two coexisting phases (binodal) at T_0. In addition, there are two inflection points (spinodal) at which $(\partial^2 G/\partial \phi^2)$ vanishes. Between these two points, the curvature $(\partial^2 G/\partial \phi^2)$ is negative, and the homogeneous mixture is said to be in an unstable state, because any small fluctuation in composition will cause a decrease in the free energy and hence phase separation. This can be seen by noting that any tangent to the curve lies above the curve between the spinodal. Outside the spinodal and up to the binodal the curvature is positive and the system is said to be metastable because the mixture is stable to small composition changes, but it is unstable to sufficiently large composition fluctuations. The spinodal is therefore the limit of metastability.

B. Definitions of Nucleation and Growth Mechanism
and Spinodal Decomposition Mechanism

Because of the different ways by which the energy state of the homogeneous mixture responds to composition changes, the rate processes that accompany phase transformation in the two regions are radically different. Inside the spinodal where the mixture is unstable to infinitesimal fluctuations, there is no thermodynamic barrier to phase growth, and thus, separation should occur by a continuous and spontaneous process. Since

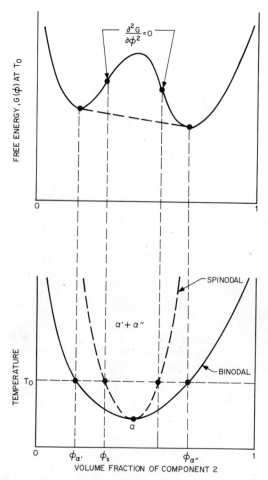

Fig. 8   The spinodal defined by the locus of temperatures on the phase diagram at which $(\partial^2 G/\partial\phi^2) = 0$. Inside the spinodal $(\partial^2 G/\partial\phi^2)$ is negative and outside it is positive.

the mixture is initially uniform in composition, this spontaneous reaction must occur by a diffusional flux against the concentration gradient, that is, by "uphill" diffusion with a negative diffusion coefficient. This process is called spinodal decomposition [15, 16].

In the metastable region, on the other hand, all small fluctuations tend to decay and hence separation can proceed only by overcoming the barrier with a large fluctuation in composition. This fluctuation is called a nucleus and, once such a nucleus is formed, it grows by a normal diffusion process. This is the nucleation and growth mechanism.

The major difference between the two mechanisms lies in the direction in

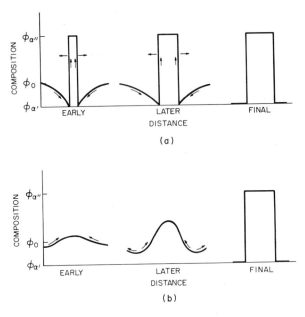

Fig. 9   Schematic drawing of time evolution of composition profiles for nucleation and growth and spinodal mechanism: (a) nucleation and growth; (b) spinodal mechanism. (From Cahn [15]. Reproduced with permission from SME–AIME Transactions, **242**, 171 (1968) in "Phase Separation Behavior of Polymer–Polymer Mixtures.")

which like molecules move in a concentration gradient. This is schematically shown in Fig. 9 [15]. In nucleation and growth, the new phase starts from small nuclei which then proceed to grow in extent. The molecules that feed the new phase follow the ordinary transport phenomenon by downhill diffusion (with a positive coefficient of diffusion). In spinodal decomposition, the new "phase" is formed by a continuous flux of like molecules against the direction of a concentration gradient created by small composition fluctuations. Since the growth rate of the new "phase" depends on the diffusion distance, decomposition should occur on a fine scale. In addition, there is no sharp interface between the phases until later stages of decomposition.

C. Kinetics of Phase Separation

1. Nucleation and Growth

Although several studies have been made on morphologies of polymer pairs decomposed by the nucleation and growth mechanism [14, 17] no attempt has been made to determine parameters governing the rate process.

The reason may be attributed, at least in part, to the lack of a kinetic theory appropriate for polymeric systems. Qualitatively, however, the rate process in polymer mixtures can be expected to obey the classical theory developed for systems of monatomic substances. For details of the classical treatment, the works by Turnbull [18] and Christian [19] may be consulted.

2. Spinodal Decomposition

Spinodal decomposition is one of the few transformations for which there is any plausible quantitative theory [20]. The reason for this is that the entire decomposition process can be treated as a purely diffusional problem and, further, many of the characteristics of decomposition, especially morphology, can be described by an approximate solution to the diffusion equation.

Cahn [16, 21] has derived a diffusion equation describing the kinetics of spinodal decomposition in isotropic binary mixtures. Basic to his theoretical treatment is the assumption that the free energy is a function of both concentration and concentration gradients [22]. This is simply a recognition of the fact that composition gradients can exist at interfaces in a phase-separated system. Cahn's approach is directly applicable to isotropic polymeric systems because the treatment is entirely based on phenomenological assumptions.

For the early stages of decomposition when changes in composition are small, Cahn's diffusion equation leads to the following linearized form:

$$\partial \phi / \partial t = M[(\partial^2 G / \partial \phi^2) \nabla^2 \phi - 2K \nabla^4 \phi] \tag{12}$$

where M is the mobility coefficient (positive [15]) and K is the energy-gradient coefficient arising from contributions of composition gradients to the free energy [22]. Alternatively, K can be regarded as a measure of energy contributions from incipient surfaces (surface tension) between regions differing in composition, and hence its sign must be positive [16, 21].

The product $M(\partial^2 G / \partial \phi^2)$ is defined as the diffusion coefficient $\tilde{D}$ since, in the absence of the second term in the brackets, Eq. (12) has the same form as Fick's diffusion equation. However, unlike Fick's equation, the sign of $\tilde{D}$ is dictated by $(\partial^2 G / \partial \phi^2)$, which can be either positive or negative, depending on the position in the phase diagram. The mathematical form of Eq. (12) is parabolic and, therefore, if $\tilde{D}$ is negative, its solution is inherently unstable for any boundary conditions. This is in accord with physical reality since $\tilde{D}$ is negative in the unstable region.

The general solution to Eq. (12) is

$$\phi - \phi_0 = \sum_{\text{all } \beta} \{\exp[R(\beta)t]\} [A(\beta)\cos(\boldsymbol{\beta} \cdot \mathbf{r}) + B(\beta)\sin(\boldsymbol{\beta} \cdot \mathbf{r})] \tag{13}$$

where ϕ_0 is the average composition, β is the wavenumber, and $R(\beta)$ is the kinetic amplification factor,

$$R(\beta) = -M\beta^2[(\partial^2 G/\partial\phi^2) + 2K\beta^2] \qquad (14)$$

Here A and B are constants to be determined from the existing composition fluctuation in the undecomposed system.

An important feature of the solution, Eq. (13), is the exponential dependence of the amplitude on $R(\beta)$. Because of the exponential dependence on $R(\beta)$, any compositional excursion from the average will decay or grow rapidly, depending on whether $R(\beta)$ is positive or negative. For the region outside the spinodal, $(\partial^2 G/\partial\phi^2) > 0$; consequently, $R(\beta) < 0$ and any existing fluctuation will diminish with time. Equation (12) therefore does not apply to nucleation expected in the metastable region.

Inside the spinodal, $(\partial^2 G/\partial\phi^2) < 0$; $R(\beta)$ is a nonmonotonic function of β and vanishes at $\beta = 0$ and at

$$\beta = \beta_c \equiv [-(\partial^2 G/\partial\phi^2)/2K]^{1/2} \qquad (15)$$

Rewriting $R(\beta)$ in terms of $\tilde{D}$ and β_c,

$$R(\beta) = -\tilde{D}(\beta_c^2 - \beta^2)(\beta/\beta_c)^2 \qquad (16)$$

we find, for $\beta > \beta_c$, $R(\beta)$ is negative, while for $0 < \beta < \beta_c$, $R(\beta)$ is positive. Moreover, $R(\beta)$ has a sharp maximum at $\beta = \beta_m$,

$$\beta_m \equiv \frac{\beta_c}{\sqrt{2}} = \frac{1}{2}\left[\frac{-(\partial^2 G/\partial\phi^2)}{K}\right]^{1/2} \qquad (17)$$

Thus, while all composition modulations with $0 < \beta < \beta_c$ will grow, growth will be fastest for those components with a wavelength β_m. Consequently the spinodal mechanism is dominated by growth of this particular modulation β_m.

For the convenience of our ensuing discussion on experimental results, Eq. (13) can be approximated, for the one dimensional case, by the following expression:

$$\phi - \phi_0 \approx \{\exp[R(\beta_m)t]\} A(\beta_m) \cos(\beta_m x) \qquad (18)$$

where

$$R(\beta_m) = -\tfrac{1}{2}\beta_m^2\tilde{D} \qquad (19)$$

The amplification factor $R(\beta)$ is a function of both β and $(\partial^2 G/\partial\phi^2)$. It is thus of interest to examine the dependence of $R(\beta)$ on $(\partial^2 G/\partial\phi^2)$ since the latter varies within the unstable region and vanishes at the spinodal. Equations (14) and (17) show that both $R(\beta)$ and β_m tend toward zero as the spinodal is approached from within the unstable region. Therefore, in the region sufficiently close to the limit of metastability, the spinodal mechanism may be so slow that it may not be able to compete with ordinary nucleation and growth resulting from finite fluctuation. This situation is most serious near the critical temperature since variation of $(\partial^2 G/\partial\phi^2)$ is extremely small there. A theory that accounts for nucleation and growth in the unstable region is not available at the moment [20].

The linearized diffusion equation, Eq. (12), applies only to the initial stages of decomposition when changes in the composition are small. As the amplitude of composition in the new "phase" increases, the nonlinear effect becomes important. Analyses of composition changes in the later stages, which involve sharpening of phase boundaries and flattening of composition profile of the new phase, have been attempted [23, 24] and found to be in qualitative agreement with experiment.

3. Determination of Kinetic Parameters Governing Spinodal Decomposition

In Cahn's theory, the kinetics of the initial stages of spinodal decomposition is characterized by two parameters $\tilde{D}$ and K. Experimental determination of β_m and $R(\beta_m)$ has been carried out for polymeric systems using light-scattering methods [25] and pulsed nuclear magnetic resonance (NMR) [18]. For the computation of the gradient–energy coefficient, additional relationships have to be invoked. The analogy between spinodal decomposition and critical opalescence of polymer solution was noted by van Aartsen [26]. In Debye's derivation of critical opalescence [27], the intensity of the scattered light is proportional to the mean square concentration fluctuation $\overline{(\phi - \phi_0)^2}(\gamma)$ and is scattered at an angle θ such that

$$\gamma = (4\pi/\lambda)\sin(\theta/2) = (2\pi/\lambda)S$$

The mean square concentration fluctuation is given by

$$\overline{(\phi - \phi_0)^2}(\gamma) = 2RT[(\partial^2 G/\partial\phi_1{}^2)_{\phi_{10}} + 2K(4\pi^2/\lambda^2)S^2]^{-1} \qquad (20)$$

The second derivative of the free energy of mixing can be estimated from the Flory–Huggins expression

$$(\partial^2 G/\partial\phi_1{}^2)_{\phi_{10}} = RT[(1/\overline{V}_1\phi_{10}) + (1/\overline{V}_2(1 - \phi_{10})) - (2\chi/\overline{V}_1)] \qquad (21)$$

In the above equation $\overline{V}_1$ and $\overline{V}_2$ are molar volumes of the constituents, and the derivative is evaluated at the mean volume fraction of Component 1 indicated by the second subscript 0.

The second derivative of free energy is, by definition, zero at the spinodal temperature T_s. In keeping with the conventional expression for the temperature dependence of χ, that is, $\chi = A + B\overline{V}_1/RT$, one obtains upon substitution,

$$(\partial^2 G/\partial\phi_1^2)_{\phi_{10}} = (2RT/\overline{V}_1)(\chi_{T_s} - \chi_T) = 2B(T/T_s - 1) \qquad (22)$$

The mean square concentration fluctuation is then found to be

$$\overline{(\phi - \phi_0)^2(\gamma)} = (RT/B)[(T/T_s - 1) + (K/B)(4\pi^2/\lambda^2)S^2]^{-1} \qquad (23)$$

When the above equation is compared with Debye's derivation of critical opalescence, it can be seen that

$$K = Bl^2/6 \qquad (24)$$

where l is the range of molecular interaction. Combination of Eqs. (17), (22), and (24) results in the following:

$$\lambda_m = 2\pi\beta_m^{-1} = 2\pi l[3(1 - T/T_s)]^{-1/2} \qquad (25)$$

In the application of van Aartsen's derivation to LCST phenomena, it should be noted that $\chi = A - B\overline{V}_1/RT$ in the temperature range of interest and the proper expression for the fluctuation wavelength is

$$\lambda_m = 2\pi l\{3[(T/T_s) - 1]\}^{-1/2} \qquad (26)$$

D. Morphology of Spinodally Decomposed Mixtures

In Section III, C it was shown that because of the gradient energy term, the spinodal mechanism tends to promote preferentially continuous but rapid growth of sinusoidal composition modulation of a certain wavelength β_m. Because of the exponential nature of the composition amplitude, Eq. (13), the basic morphological features developed during the earliest stage can be expected to persist through the rest of the decomposition period. For isotropic system's Cahn's linear theory predicts a two-phase structure that is characteristically uniform and highly interconnected [21]. In addition, Cahn has also shown that connectivity of phases is maintained as long as the volume fraction of the minor phase exceeds about 15%. For volume fractions less than 15%, the minor phase transforms into isolated droplets approximately one wavelength apart. Although connectivity of phases is an important feature of the spinodal mechanism, it has been shown that such a morphology can also arise from the coalescence of new phase particles formed by nucleation and growth [28]. Therefore, for positive identification of the spinodal mechanism, one must establish that either the diffusion coefficient is negative or the composition of the new phase changes continuously during phase separation.

The morphology of a spinodal structure can be affected by such factors as temperature and heating rate, as well as coherency strain energy and external fields (magnetic, electric, shearing, etc.) [15, 20]. The effect of an applied field is of particular interest because it could cause the molecules to orient and to flow in the direction of the field. An anisotropic spinodal structure can thus be expected. An example of magnetically aged spinodal decomposition was given by Cahn [29]. The many different means by which the texture of a two-phase mixture can be altered are of practical interest since they provide a way for intelligent manipulation of structure and properties in compatible systems.

IV. EXPERIMENT

Although the thermodynamic criterion is universal for molecules of all sizes, many standard techniques of measuring the physical properties of mixtures of simple liquids become insensitive for high molecular weight, solid polymers. Therefore, demonstration of the thermodynamic stability of a binary polymer mixture often requires a modification of conventional experiment methods.

A. Identification of the Stable Region

For the purpose of our discussion, the original Flory–Huggins formulation, rather than the equation of state approach, will be used in the following. It can be shown that, in the framework of the original equation by Flory, a negative value of the binary interaction parameter is a sufficient condition for thermodynamic stability, that is, $\partial^2 G/\partial \phi^2 > 0$. Therefore, the measurement of the interaction parameter occupies an important role in the study of compatibility.

Since there is no feasible experimental method to measure the interaction parameter of two solid polymers directly, use has been made of polymer solution theory applied to ternary systems consisting of one solvent and two polymers. The activity of the solvent in a ternary mixture is

$$\ln a_1 = \ln \phi_1 + (1 - \phi_1) - (x_1/x_2)\phi_2 - (x_1/x_3)\phi_3$$
$$+ (\chi_{12}\phi_2 + \chi_{13}\phi_3)(1 - \phi_1) - \chi'_{23}\phi_2\phi_3 \qquad (27)$$

where the subscript 1 refers to the solvent and subscripts 2 and 3 refer to the two polymers. The volume fraction of each component is represented by ϕ and the binary interaction parameter is given by χ. The symbol χ'_{23} is the interaction per segment of Polymer 2, that is, χ_{23}/x_2, where x is the number of segments in the macromolecule. The solvent–polymer interaction parameters χ_{12} and χ_{13} can be determined from separate experiments for binary solutions:

$$\ln a_1 = \ln \phi_1 + \phi_2 + \chi_{12}\phi_2^2 \qquad (28)$$

With a knowledge of χ_{12} and χ_{13}, the parameter χ'_{23} can be readily calculated from Eq. (27).

In many solvent–polymer mixtures the interaction parameters have been found to be dependent on concentration. It is difficult to judge a priori whether χ'_{23} is dependent on ϕ_1. For this reason, osmotic pressure measurement of dilute solutions does not necessarily provide the pertinent information. Instead, it appears desirable to study concentrated solutions and extrapolate the χ'_{23} values to $\phi_1 = 0$, so that the parameter may represent the interaction of a binary polymer mixture. In this regard the vapor sorption measurement offers advantages in experimental simplicity as well as flexibility. A polymer specimen is kept in an atmosphere of constant solvent vapor pressure, and the gain in weight due to vapor sorption is measured by a calibrated quartz spring or an electrobalance. The volume fractions of the components are computed from the weight gain at equilibrium with the assumption of no volume change upon mixing.

An alternate method using solvent as a probe is gas–liquid chromatography (GLC) in which the polymer is the stationary phase and the solvent probe is the moving phase. Both packed and capillary columns have been used. The specific retention time is a direct measure of the activity of the solvent from which the binary interaction parameter can be calculated. The use of GLC in measuring polymer–solvent interaction has been adequately described, and its use in measuring polymer–polymer interactions was demonstrated by Desphande et al. [32] and by Olabisi [33].

The approach outlined above is applicable to a mixture of two amorphous polymers. If one of the polymers is crystalline, the vapor sorption method no longer offers the same advantage. The uncertainty in the degree of crystallinity of the crystallizable component in the mixture often invalidates the calculation (see Chapter 10). Two methods have been devised to cope with this situation, both based on the depression of melting temperature of the crystallizable component. When the change in the chemical potential of the crystallizable segment in a binary mixture relative to its pure liquid state was equated to the difference in the chemical potential between a crystalline segment and the same segment in the pure liquid state, the following equation was obtained by Nishi and Wang [34]:

$$\Delta T_m / T_m^0 = -(\bar{V}_{2u} B_{12} / \Delta H_{2u}) \phi_1^2 \tag{29}$$

In the above equation, T_m^0 is the melting temperature of the undiluted crystalline polymer (Component 2), ΔT_m the lowering of the melting point $(T_m^0 - T_m)$, $\bar{V}_{2u}$ the molar volume per segment, ΔH_{2u} the heat of fusion per crystalline segment, and B_{12} is given by $\chi'_{12} = B_{12} \bar{V}_{1u}/RT$. It is seen from Eq. (29) that B_{12} plays a decisive role on the melting behavior of crystalline–amorphous polymer mixtures. A depression of the melting point can be realized only if B_{12} is negative. At the same time, since B_{12} and ΔH_{2u} enter Eq. (29) in the form of a ratio, it is no longer possible to deduce these two quantities simultaneously from T_m measurements alone, a feat that is possible in the case of polymer–solvent systems. When ΔH_{2u} is known from independent measurements, B_{12} can be readily computed from ΔT_m.

If ΔH_{2u} has not been determined previously, additional experiments are required. Shultz and McCullough [35] have combined data from binary (solvent–crystalline polymer) and ternary mixtures to obtain ΔH_{2u} and three χ_{ij}.

B. Phase Separation

1. Standard Methods

There are many well-established experimental methods to identify phase domains in polymers, namely, light and electron microscopy, light and x-ray scattering, thermal expansion, heat capacity, and dielectric or mechanical measurements. These methods are described extensively in other chapters.

2. Light Scattering of Polymer Solutions

Although light scattering of polymer solutions is a technique of long standing, the work by van Aartsen and Smolders [25] is described below because it pertains directly to the kinetics of spinodal decomposition and the application of Cahn's theory. In their paper, the authors used Cahn's equation for concentration fluctuations to calculate polarizability and obtained the following expression:

$$(1/h^2)(\partial/\partial t)\{\ln(\partial I_{vv}/\partial t)_{h^2}\} = 4MK(\beta_c^2 - h^2) \tag{30}$$

where I_{vv} is the intensity of the scattered light (vertically polarized), t time, M the mobility of the solvent, K is assumed to be constant for a given polymer–solvent pair, β_c the critical wavenumber in Cahn's theory,

$$h^2 = [(4\pi/\lambda)\sin(\theta/2)]^2$$

λ is the wavelength of the light, and θ is the scattering angle. Equation (30) predicts a linear variation of $\ln(\partial I_{vv}/\partial t)$ with time at the initial stages of spinodal decomposition. Furthermore, by measuring the intensity of the scattered light at different values of h, that is, different angles, a plot of the right-hand side of Eq. (30) versus h^2 should result in a straight line from which MK and β_c^2 can be computed.

3. Pulsed Nuclear Magnetic Resonance [36]

The magnetic energy of nuclei having nuclear spins is determined by the orientation of the nuclear magnetic dipoles relative to an applied magnetic field. In the applied field, the nuclei behave as magnetic dipoles precessing around the field axis. The allowed values of the vector component of the magnet moment in the direction of the field are dictated by magnetic quantum numbers, and the energy difference between the magnetic states is proportional to the field strength H_0 and the characteristic magnetic moment μ of the nuclei being studied. In a field of 10^4 G, the separation of energy levels is only about 10^{-2} cal for proton, which is small compared to the thermal energy kT. According to the Boltzmann distribution, the population of nuclei in the lower energy level exceeds that in the upper level by only a few parts per million. Therefore, the system may be considered to be at thermal equilibrium even in a strong magnetic field.

The transition from one energy state to the next can be brought about by the adsorption of energy $h\nu_0$, where h is Planck's constant and ν_0 is the nuclear precession frequency at which resonance occurs. For a proton in a field of 7050 G, the resonant frequency, about 30 MHz, is in the radio-frequency (rf) range.

In a pulsed NMR experiment, the sample is surrounded by an rf coil and placed in a permanent magnetic field. The rf coil functions both as a pulse generator and a detector. The arrangement is shown schematically in

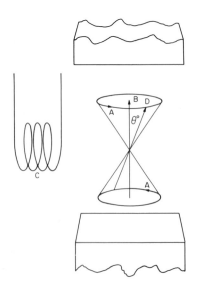

Fig. 10   Nuclear magnetic resonance: (A) direction of precession of magnetization; (B) equilibrium magnetization vector $\mathbf{M}_0$; (C) radiofrequency pulse and detector coil; (D) vector of magnetization following a radiofrequency $\theta°$ pulse.

Fig. 10. The equilibrium magnetization M_0 of the material in the magnetic field is parallel to the field H_0 in the absence of any disturbing force. Upon application of a rf pulse at the resonant frequency, an alternating magnetic field H_1 is produced, and the magnetization vector is tilted from its alignment along H_0 by an angle θ. A nonequilibrium population of nuclear spins in the high-energy state is generated, and it is said that the spin temperature of the nuclear ensemble is high. For thermal equilibrium to be restored, the excited nuclei must return to the lower energy state until the Boltzmann distribution is reestablished. The energy transfer occurs through two mechanisms. In the first process the reorientation of the nuclear dipoles takes place by an exchange of energy with surrounding molecules. The random motion of neighboring molecules containing magnetic nuclei generates fluctuating magnetic fields. If one of the component frequencies of the spectrum of molecular motions is the same as the precession frequency of the proton, this particular magnetic field can serve as a medium for energy transfer. This type of relaxation is variously termed thermal relaxation because the spin temperature is cooled to the equilibrium temperature, or longitudinal relaxation because the net magnetic moment along the field axis is returned to normal, or spin–lattice relaxation because the energy transfer occurs between nuclear spins and neighboring molecules or "lattice." The characteristic relaxation time is designated T_1.

The second relaxation process is concerned with the randomization of nuclear spins. The precessing dipoles gradually get out of phase with one another and a decay in the transverse component of the nuclear magnetization is observed. This occurs with a characteristic time T_2 known as the transverse or spin–spin relaxation time.

The angle θ through which M_0 is turned depends on the duration T of the rf pulse. If the value of T is chosen so that

$$\pi = (H_1/H_0)\omega_0 T \tag{31}$$

where ω_0 is the angular precession frequency $(= 2\pi\nu_0)$ and H_1 is strength of the alternating field produced by the rf; M_0 is turned 180 degrees. With a shorter pulse duration, $T/2$, the net magnetization is rotated 90 degress to the transverse plane and can be detected by the voltage it induces in the rf coil.

The spin–spin relaxation time can be obtained by measurement of the free induction decay following a 90-degree pulse. For signal decay obeying a simple exponential function of time t, $\exp(-t/T_2)$, T_2 is computed in a straightforward manner. Free-induction decay signals (S) from materials that contain two phases will generally exhibit two exponentials,

$$S = A\exp(-t/T_{2a}) + B\exp(-t/T_{2b}) \tag{32}$$

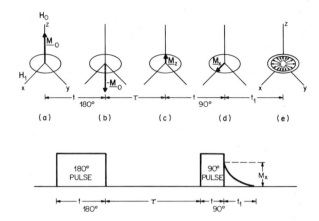

Fig. 11   Pulse sequence in a pulsed NMR experiment.

where A and B are the volume fractions of the two phases and T_{2a} and T_{2b} are the corresponding relaxation times.

The spin–lattice relaxation experiment is performed with a $180°–\tau–90°$ sequence, where τ is the time interval between two pulses. The pulse sequence and the state of magnetization vector $\mathbf{M}$ during the experiment are depicted schematically in Fig. 11. The return of the magnetization vector from $-\mathbf{M}_0$ after the 180-degree pulse to equilibrium is governed by time constant T_1. At time τ after the 180-degree pulse the vector $\mathbf{M}$ has changed from $-\mathbf{M}_0$ to $\mathbf{M}_z$. In order to monitor the magnitude of $\mathbf{M}_z$ at time τ, a 90-degree pulse is applied to tilt $\mathbf{M}_z$ into the transverse plane so that the detector coil can receive a signal proportional to the magnitude of $\mathbf{M}_z$ at τ. The nuclear spin system will now randomize in the x–y plane while returning to equilibrium magnetization $\mathbf{M}_0$.

The spin–lattice relaxation process is described by

$$(|\mathbf{M}_0| - |\mathbf{M}_z|)/2|\mathbf{M}_0| = \exp(-\tau/T_1) \tag{33}$$

A series of $180°–\tau–90°$ experiments conducted at different intervals of τ then permit the calculation of T_1. Where a single relaxation mechanism is operative, the null method of computing T_1 has been found to be convenient. In the null method, the time τ at which $\mathbf{M}_z$ is zero is identified from the oscilloscope traces, and T_1 is simply $(\tau)_{\text{null}}/(\ln 2)$.

The spin–lattice relaxation process for a two-phase material can be approximated by two exponential functions, similar to Eq. (32).

A third method of relaxation measurement makes use of the rotating frame. The longitudinal relaxation time in the rotating frame $T_{1\rho}$ is measured by observing the amplitude of the free induction decay that follows the

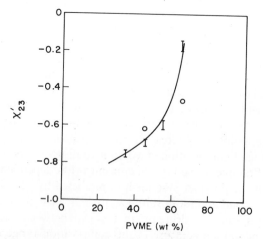

Fig. 12   Interaction parameter of PS–PVME: (I) 30°C; (○) 50°C.

termination of a resonant rf pulse, the nuclear magnetization having been aligned with the rotating field **H**, at the beginning of the pulse. The field **H$_1$** and the magnetization vector **M** are in the same direction during the decay period. The rotating frame method extends the spin–lattice relaxation measurement to shorter relaxation times.

V.   RESULTS

A.   Interaction Parameters

1.   Amorphous Mixtures

As early as 1950, Stockmayer and Stanley [37] applied Eq. (27) in their attempt to determine the interaction parameter of polystyrene with poly-(methyl methacrylate) in a dilute solution of butanone from light-scattering measurements. The value of χ_{23} was estimated to be 0.014 ± 0.01. The results of GLC studies were described fully by Desphande *et al.* [32] and by Olabisi [33] for many mixtures.

A recent study [38] concerns the system polystyrene (Component 2) and poly(vinyl methyl ether) (PVME) (Component 3). Compatible films were cast from toluene solution and the solvent chosen for the vapor sorption study was benzene (Component 1). The quantity of benzene absorbed by a mixture was found to be always smaller than the amount predicted from the simple rule of additivity. This observation suggests favorable interaction between the two polymers (see Chapter 10). In the application of Eq. (27) to determine χ'_{23} it is important to use the appropriate value of χ'_{12} at a given vapor pressure, which can be obtained from a plot of χ'_{12} versus vapor activity for the binary systems. Several interesting features are noted in the χ'_{23} results (Fig. 12). First, the quantity χ'_{23} does not vary significantly with ϕ_1 in the range studied. The same χ'_{23} value therefore applies to a solid film containing no solvent. Second, the negative value of χ'_{23}, which is a sufficient condition for thermodynamic stability, confirms the compatibility of this polymer pair. The third point of interest is the dependence of the interaction parameter on blend composition. Two contributing factors may come into play here. The first one is the type of concentration dependence commonly observed in solvent–polymer systems and predicted by Flory's equation of state theory. A second and more subtle factor may arise from the possibility that there is a systematic decrease in the gap between the cloud point and the experimental temperature in the composition range of interest. In

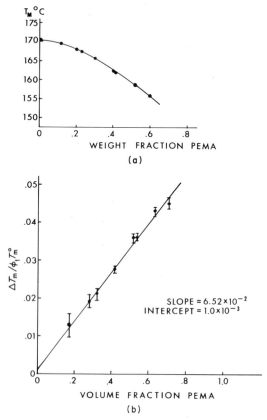

Fig. 13   Melting point depression of PVF$_2$.

any case, extrapolation of χ'_{23} values to higher PVME content indicates the likelihood that the interaction parameter would become positive at a PVME weight fraction of about 0.7. Indeed a film of this composition was initially transparent but turned opaque after standing at room temperature for several months. A 75% PVME film became opaque after three weeks. The latter opaque film, after being heated at 47°C for about 40 hr, turned transparent again. Thus, the phase boundary associated with the UCST is near room temperature for the 75% PVME film, and the experimental observation of phase separation is in full accord with the prediction from thermodynamic data.

2. Mixture Containing One Crystallizable Component

The melting temperatures of poly(vinylidene fluoride), a crystalline polymer, in the presence of poly(methyl methacrylate) (PMMA) [34] or poly(ethyl methacrylate) (PEMA) [39] affords a good test of the validity of Eq. (29). The graphs shown in Fig. 13 yield B_{12} values of -2.98 cal/cm³ (PMMA) and -2.86 cal/cm³ (PEMA). The corresponding χ/x_2 values are -0.30 and -0.34, respectively, at 160°C.

In the study by Schultz and McCullough [35] on the crystallization and melting of poly(2,6-dimethyl-1,4-phenylene oxide) in ternary mixtures with polystyrene and toluene, experimental data were found to be consistent with $\chi_{23} = 0$.

B. Phase Boundary

The determination of phase compositions in a solid polymer mixture poses a challenge to analytical skills. Two methods of estimating phase compositions will be discussed, namely, T_g and NMR measurements.

1. Phase Composition by Glass Transition Temperature Measurements

Where two sharp glass transitions are observed, phase compositions can be estimated from T_g–concentration relationships (see Chapter 5). Although elegant studies using T_g measurements have appeared in the literature, blends of poly(vinylidene fluoride) and poly(methyl or ethyl methacrylate) offer good examples of the high degree of complexity encountered when one component is crystallizable.

Poly(vinylidene fluoride) (PVF$_2$) crystallizes readily. In the presence of methacrylate polymers, crystallinity can still be detected, both by calorimetry and by microscopy, in solvent cast films containing minor quantities of PVF$_2$. When quenched from the melt, crystallization of PVF$_2$ can be suppressed completely in PVF$_2$–PMMA mixtures, and the T_g of the blend

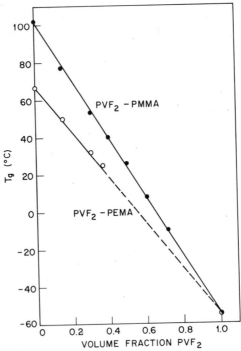

Fig. 14 Glass transition temperatures of quenched PVF_2–poly(methyl or ethyl methacrylate) mixtures.

is simply the volume fraction average of the component T_g's (Fig. 14). In PVF_2–PEMA mixtures, however, crystallization of PVF_2 is observed even with rapid quenching when the PEMA content is low. In each of these crystallized specimens two glass transition temperatures are revealed. A low-temperature transition occurs always at about $-50°C$ and a high-temperature transition at $27°C$. When compared with the T_g of PVF_2, namely, $-54°C$, the low-temperature transition near $-50°C$ in each mixture suggests the presence of an amorphous phase comprised nearly of pure PVF_2. The upper T_g corresponds to a phase composition of about 45% PVF_2. Therefore, the microstructure of these films can be depicted as a composite of crystalline regions coexistent with two conjugate amorphous phases.

2. Phase Composition by NMR Measurement [38]

In this section the use of pulsed NMR in the study of phase separation is discussed. The method offers the advantage that two relaxation times appear as phase separation occurs.

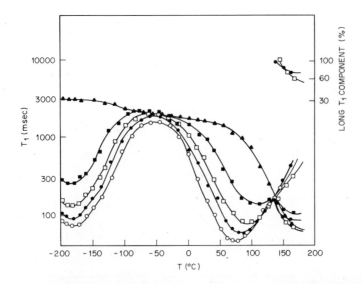

Fig. 15   Spin-lattice relaxation times of PS–PVME mixtures: (○) PVME; (●) PVME/PS =
75/25; (□) PVME/PS = 50/50; (■) PVME/PS = 25/75; (▲) PS.

The pair PS–PVME is particularly attractive because the relaxation
spectra of the pure components are relatively simple and the nature of
molecular motion can be identified with confidence. The temperature
dependence of spin–lattice relaxation time T_1 and fraction of long T_1 component for polystyrene (PS), PVME, and three compatible mixtures are
given in Fig. 15. Only one T_1 was observed for each sample except for
PVME/PS = 75/25 and 50/50 above 140°C at which thermally induced
phase separation associated with LCST began to occur and two T_1's were
found.

In the low-temperature region near -180°C, there is almost no change in
molecular mobility in PS. The T_1 minimum for PVME may be ascribed
to methyl group rotation. As PS is added to PVME, the position of the
T_1 minimum remains unchanged, but the value of T_1 at the minimum
increases progressively with PS content. Figure 16 shows a linear dependence
of the methyl T_1 minimum values on the relative number of methyl protons.
The same relationship found in mixtures of n-alkanes has been interpreted
in terms of the spin diffusion process in which the energy of the methylenic
spin system is conveyed to methyl protons where dissipation to the molecular
lattice is effective. The distance of effective spin diffusion is estimated from
the relaxation time to be about 100 Å and can be considered as the upper
limit of the size of the heterogeneity in the mixture.

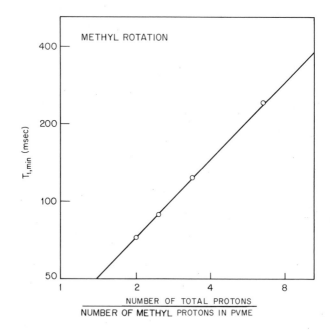

Fig. 16 The methyl T_1 minimum value as a function of the relative number of methyl protons in the system.

The high-temperature T_1 minimum at 75°C for PVME is associated with a glass transition at 30 MHz. The increase in the temperature of the T_1 minimum with PS content in the mixture also parallels the trend observed for glass transitions determined by the DSC method. While the single T_g manifested for each mixture is consistent with the earlier conclusion of a high degree of homogeneity, the increase in the T_1 minimum value from 40 to 140 msec serves as a caution to premature judgment. The T_1 minimum (at about 200°C) value of PS is similar to that of pure PVME. If the mixture were completely homogeneous, the increase in the value of T_1 minimum would not be expected to occur. The possible presence of microheterogeneous regions appears to be supported by the spin–spin relaxation data shown in Fig. 17. Two T_2's are evident for a 50/50 mixture above 25°C, indicative of dual mechanisms of segmental motion.

Aside from the above reservations about the fine scale of mixing, the PS–PVME films cast from aromatic solvents can be regarded as compatible.

There is no doubt, however, that phase separation occurred when samples were heated above 150°C. After NMR measurements were completed at 170°C, the samples were quenched in ice water, and measurements were

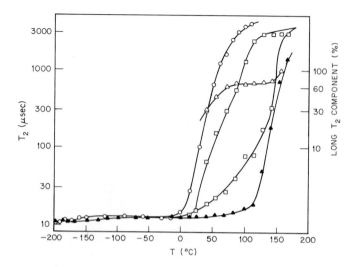

Fig. 17   Spin–spin relaxation times of PS–PVME: (◯) PVME; (▢) PVME/PS = 50/50; (▲) PS; (△) long T_2 component.

carried out again. Two T_1's can be resolved readily for the phase-separated 50/50 sample, as shown in Fig. 18. The phase compositions, calculated from the linear correlation of T_1 minimum value at $-180°C$ and the number of methyl protons, are 10 and 71% PVME, respectively. The weight percent of the PS-rich phase is estimated from a material balance to be 34%. This value predicts about 30% signal intensity for the long T_1 component, in reasonable agreement with experimental results. Further inspection of the experimental data reveals that measurement at 50°C also yields the correct value of the long T_1 component.

3.   Effect of Molecular Weight on Cloud Point

Monodisperse polystyrene with molecular weights ranging from 2.1×10^3 to 2×10^6 were mixed with PVME which is polydisperse with a $\overline{M}_v$ of 5.15×10^3 calculated from intrinsic viscosity. The cloud point curves associated with LCST are shown in Fig. 19. When the molecular weight of PS is below 5.1×10^4, the cloud point curves shift strongly to lower temperatures as molecular weight increases. At higher molecular weights, the shift in the position of the cloud point curve levels off and in fact reverses for PS having molecular weight above 1.1×10^5. The reversal phenomenon is not well understood and will not be discussed here.

In dealing with solvent–polydisperse polymer systems, an important point of distinction is that the extremum of the cloud point curve does not

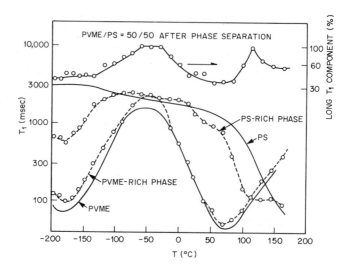

Fig. 18   Temperature dependence of T_1 and fraction of long T_1 component for phase separated PVME–PS (50/50).

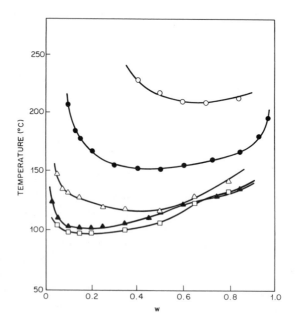

Fig. 19   Cloud point curves for PVME and monodisperse PS mixture where molecular weight $(\overline{M}_w)$ of PS is changed: (○) 10,000; (●) 20,400; (△) 51,000; (▲) 110,000; (□) 200,000. W is the weight fraction of PS.

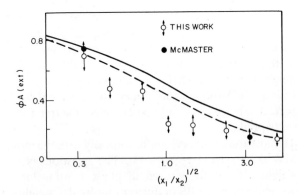

Fig. 20 Composition of the extremum point $\phi_{A(\text{ext})}$ and reduced variables x_A and x_B, where x_A and x_B are the mean degree of polymerization of PS and PVME, respectively: (———) Eq. (4); (– – –) Eq. (8).

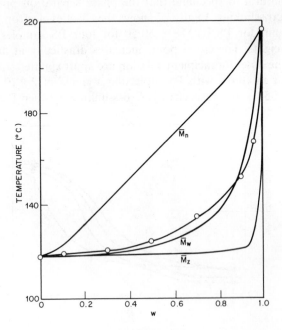

Fig. 21 Cloud point temperatures for PS/PVME = 50/50 mixture, where PS is a mixture of two monodisperse PS (1) and PS (2). The weight fraction of PS (1) is W and the molecular weights of PS (1) and PS (2) are 10,000 and 110,000, respectively.

coincide with the critical point. The latter tends to shift toward the right-hand branch of the cloud point curve as polydispersity increases. The following discussion regarding the position of the extremum is therefore qualitative in nature. With the assumption of a most probable distribution for the molecular weight (MW) of PVME, one can compare $\phi_{A(\text{extremum})}$ with $\phi_{A(\text{crit})}$ prescribed by Eq. (8). Although the numerical agreement between the two is far from perfect (Fig. 20), as would be expected, the trend of experimental data is correctly predicted.

The cloud temperature decreases monotonically with the molecular weight of PS for $MW \leqslant 1.1 \times 10^5$ (see Fig. 19). The question as to the correct molecular weight average to be used in phase relationships is answered by mixing two monodisperse PS of different molecular weights. Polystyrene samples with $MW = 1.0 \times 10^4$ and 1.1×10^5 were chosen, and the blend compositions were always kept at PS/PVME = 50/50. When the variation in the cloud temperature is examined as a function of the volume fraction of the PS component with $MW = 1.0 \times 10^4$ in the total PS (Fig. 21), it is evident that $\overline{M}_w$ is a good measure for correlating the cloud point of poly-disperse polymers.

It seems natural to speculate that the phase separation process will be retarded by crosslinking. Figure 22 shows the effect of electron radiation on the cloud point for PS/PVME = 50/50 for four PS samples of different molecular weights. The cloud point increases drastically at around 50 to 100 Mrad. The doses for incipient gelation are approximately 10, 20, 50, and 100 Mrad for mixtures with PS molecular weights of 2×10^6, 4.98×10^5, 1.1×10^5, and 5.1×10^4, respectively. Cross-linking between PS and PVME

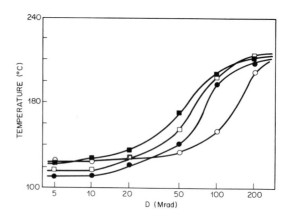

Fig. 22 Electron irradiation dose D dependence of cloud point temperature for PS/PVME = 50/50, where the molecular weight of PS has been changed: (○) 51,000; (●) 110,000; (□) 498,000; (■) 2,000,000.

has probably taken place to form an interpenetrating network (see Volume 2, Chapter 11) and the temperature of phase separation increases as a result of the restricted mobility of the chains.

C. Morphology

Connectivity of phases is an important morphological feature of spinodal decomposition. The presence of such morphological characteristics was observed in polystyrene–poly(vinyl methyl ether) [18] and poly(styrene-*co*-acrylonitrile)–poly(methyl methacrylate) (SAN–PMMA) [14] mixtures. The former system was examined under a phase contrast microscope under a variety of conditions indicated by open and filled circles in Fig. 23. The samples represented by open circles have the following characteristics (Fig. 24): (a) phase separation proceeds rapidly and the precipitated phases are interconnected with each other; (b) the spacing of the pattern is almost uniform; (c) phase pattern is finer at higher decomposition temperature; (d) at the later stage of phase separation, the phase domains seem to grow

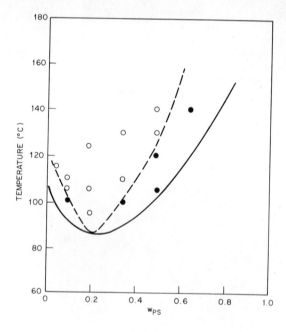

Fig. 23 Several compositions and temperatures where phase separation by what appears to be spinodal mechanism (○) and nucleation and growth mechanism (●) are observed under a microscope. The dotted curve represents the line of demarcation of the two morphologies.

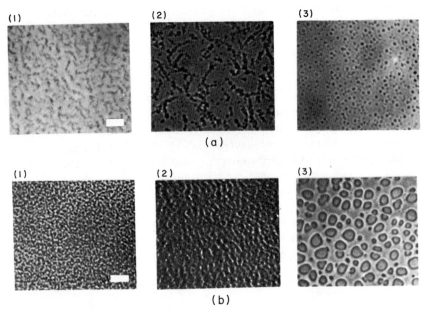

(a)

(b)

Fig. 24   Phase separation behavior after the sample has been heated rapidly to and maintained at high temperatures (open circle condition). The white bar in the figure corresponds to 10 μm. Phase patterns at different stages of decomposition are shown. (a) PS/PVME = 5/95, decomposed at 115°C: (1) 70 sec; (2) 600 sec; (3) 3800 sec. (b) PS/PVME = 20/80: (1) 300 sec; (2) 1800 sec; (3) 3800 sec.

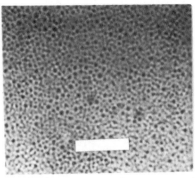

Fig. 25   Phase separation by what appears to be nucleation and growth mechanism for the PS/PVME = 65/35 mixture, decomposed at 140°C for 10,700 sec. The white bar in the figure corresponds to 10 μm.

(a)   **(b)**   **(c)**

Fig. 26 Initial phase separation pattern for **PS–PVME** = 50/50: (a) decomposed at 120°C for 3800 sec; (b) decomposed at 130°C for 220 sec; (c) decomposed at 140°C for 80 sec. The white bar in the figure corresponds to 10 μm.

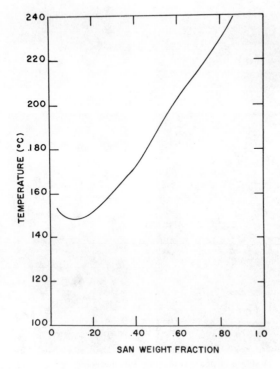

Fig. 27 Experimental cloud point curve for SAN–PMMA. (From L. P. McMaster, "Copolymers, Polyblends, and Composites," Advances in Chemistry Series No. 142, 1975. Copyright by the American Chemical Society.)

in size while maintaining their connectivity and eventually either break up into small spheres or merge into macrospheres. These characteristics agree with Cahn's prediction of a spinodal mechanism.

Samples prepared under conditions represented by filled circles have entirely different morphologies (Fig. 25): (a) phase separation proceeds

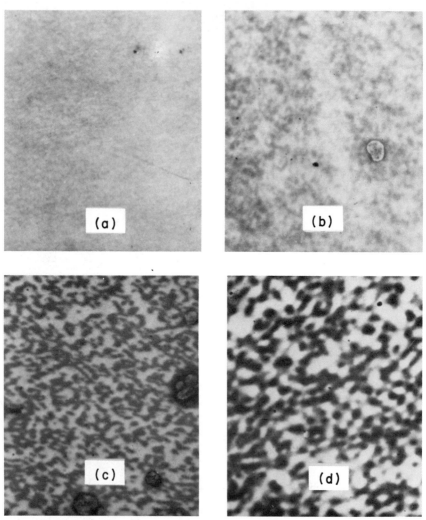

Fig. 28 TEM of stages of phase separation for interconnected structures 25% SAN–75% PMMA; $T = 180°C$. Time (min): (a) 3.4; (b) 4.4; (c) 13.4; (d) 32.4. The white bar in the figure corresponds to 0.3 μm. (From L. P. McMaster, "Co-polymers, Polyblends, and Composites," Advances in Chemistry Series No. 142, 1975. Copyright by the American Chemical Society.)

slowly and the precipitated phase is spherical in shape; (b) the dimension of these spheres is small in comparison with the average spacing of the phase pattern observed under open circle conditions. These characteristics are apparently representative of the mechanism of nucleation and growth.

Further evidence for the two mechanisms is shown in Fig. 26 for a 50/50 mixture decomposed at different temperatures. The line of demarcation between the two types of morphologies is the approximate location of the spinodal curve (see Fig. 23).

Morphological studies of phase-separated mixtures of poly(methyl methacrylate) and poly(styrene-*co*-acrylonitrile) (28% acrylonitrile) were conducted by McMaster [14]. The cloud point curve of the mixtures is shown in Fig. 27. When a 75% SAN specimen is held at 265°C for several minutes, spherical domains of PMMA-rich phase are clearly observable. After 2 hr, the domain size increases by only an approximate factor of 2. The mechanism of nucleation and growth appears to dominate in this case.

A 25% SAN composition held at 180°C develops interconnecting domains whose size does not change appreciably during the early stages of phase separation (Fig. 28). The compositions of the two phases, as deduced from difference in contrast, change only gradually. These observations corroborate the spinodal mechanism. At very long times, the interconnecting domains break into droplets approximately 500 μm in size.

D. Kinetics of Phase Separation

1. Light Scattering of Polymer Solutions [25]

Upon cooling a homogeneous solution of poly(2,6-dimethyl-1,4-phenylene oxide) (PPO) in caprolactam (CL), turbidity is detected at temperatures of about 110–120°C. The increase in the intensity of scattered light with time is represented schematically in Fig. 29. In the figure, t_0 is the time when the homogeneous solution at 180°C is transferred to a bath kept at a lower temperature, t_b is the point at which an increase in turbidity begins to occur, and t_i is the time to reach half of the maximum intensity. Neither t_b nor t_i depends on the angle of measurement; both depend only on the temperature of phase separation. A mechanism of nucleation and growth is highly improbable in view of this experimental observation.

In line with Cahn's theory adapted by van Aartsen and Smolders to light-scattering measurements, plots of $\ln(dI_{vv}/dt)$ as a function of time yield rather good straight lines, in agreement with Eq. (30), although results obtained at high temperatures deviated from linearity at long times. The computed values of β_c and λ_m are summarized in Table II.

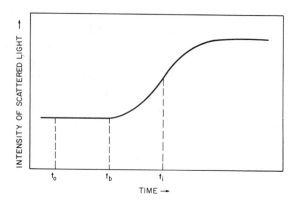

Fig. 29   Schematic representation of the intensity of scattered light as a function of time during the phase-separation process. t_0 corresponds to the moment the thermostats are changed.

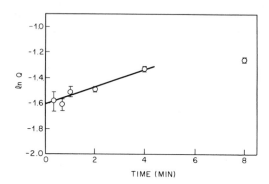

Fig. 30   Logarithm of total change of PVME concentration in PS-rich phase versus decomposition time. The data are based on the results listed in the last column of Table III.

Table II

Phase Separation Parameters for PPO–CL

$T(°C)$	55	103	112.5	119	128
$\beta_c^2(\mu^{-2})$	140 ± 15	107 ± 5	100 ± 10	76 ± 4	75 ± 6
$\lambda_m(\mu)$	0.75	0.86	0.89	1.02	1.03

2. NMR Studies

Because the composition of phases changes continuously with time while the spacing of the phase pattern (that is, phase volume) remains nearly constant in the early stage of spinodal decomposition, the verification of this crucial conclusion will serve as an important step to demonstrate the validity of Cahn's theory. Pulsed NMR is a particularly powerful tool for this purpose. In Table III, NMR data are given for the 50/50 PS/PVME mixture

Table III

Spin–Lattice Relaxation Time T_1 Analysis at 50°C for PS–PVME = 50/50 Compatible Polymer Mixtures Decomposed at 130°C for Various Lengths of Time

Decomposition time	Long T_1 (msec)	Short T_1 (msec)	Long T_1 component (%)	PVME (% in PS-rich phase)
Original		195		
20 sec	390 ± 30	115 ± 15	40 ± 8	29.5 ± 1.5
40 sec	375 ± 20	130 ± 10	35 ± 6	30 ± 1
1 min	420 ± 20	120 ± 10	37 ± 6	28 ± 1
2 min	410 ± 10	115 ± 5	39 ± 4	27.5 ± 0.5
4 min	530 ± 10	112 ± 5	38 ± 4	23.5 ± 0.5
8 min	610 ± 15	120 ± 5	31 ± 4	21.5 ± 0.5
32 min	680 ± 15	115 ± 5	33 ± 4	19.5 ± 0.5

conditioned at 130°C for various lengths of time. The volume fraction of the PS-rich phase as indicated by the long T_1 component remains nearly constant at 35% up to about 4 min, while the composition in each phase as indicated by the two T_1 relaxation times, changes continuously in the same period. These results offer convincing proof that phase separation indeed proceeds by the spinodal mechanism under the experimental conditions.

3. Kinetic Parameters

A combination of NMR, morphological, and thermodynamic data enables the computation of all the kinetic parameters in Cahn's theory. The composition ϕ in Eq. (18) is identified with the concentration of PVME in the PS-rich phase. If we denote by Q the total change in the amount of PVME in the PS-rich phase during spinodal decomposition, then Q may be obtained approximately by the following form,

$$Q \approx \left[\left| \int_{3\pi/2\beta_m}^{3\pi/2\beta_m} (\phi - \phi_0)\,dx \right| \right]^3 = [(2/\beta_m)\,A(\beta_m)]^3 \exp[3R(\beta_m)\,t]$$

so that

$$\ln Q = 3 \ln \left[(2/\beta_m) A(\beta_m) \right] + 3R(\beta_m) t \tag{34}$$

According to the above equation, a plot of $\ln Q$ against t should yield a straight line with a slope of $3R(\beta_m)$. Experimental values of $\ln Q$ are indeed approximately linear with time up to about 4 min, beyond which the curve levels off (Fig. 30). The initial slope of the curve yields

$$R(\beta_m) = 3.8 \times 10^{-4} \quad \sec^{-1} \tag{35}$$

The parameter β_m can be estimated from the phase pattern developed at 130°C. With the assumption that the average spacing, about $0.6\,\mu m$, corresponds to the half-wavelength of the fastest-growing wave, β_m can be computed as

$$\beta_m = 2\pi/\lambda_m = 5.2 \times 10^4 \quad cm^{-1} \tag{36}$$

Upon substitution of the values of $R(\beta_m)$ and β_m into Eq. (19), the diffusion coefficient $\tilde{D}$ is found to be

$$\tilde{D} = -2.8 \times 10^{-13} \quad cm^2/\sec \tag{37}$$

Although the magnitude of the diffusion coefficient is only approximate, the negative sign, which is beyond question, constitutes additional proof of spinodal decomposition.

The next step in the computation of kinetic parameters is the calculation of the gradient energy coefficient K. From a knowledge of the wavelengths of the dominant composition fluctuations at two temperatures, namely, $1.2\,\mu m$ at 130°C and $0.9\,\mu m$ at 140°C, it is found with the use of Eq. (26) that

$$l = 580 \quad \text{Å}$$

and

$$T_s = 391°K \quad (118°C)$$

The computed value of T_s is slightly lower than 120°C, at which temperature the 50/50 mixture undergoes phase separation by the mechanism of nucleation and growth. This slight discrepancy is not surprising because, as T_s is approached from the unstable region, the spinodal mechanism will slow down and become less important whereas the rate of nucleation and growth does not vanish there.

The magnitude of B is estimated from the temperature coefficient of χ to be 15.8 cal/cm^3 of PVME. The gradient energy coefficient K is then calculated to be 3.7×10^{-3} erg cm^2/cm^3 PVME and the second derivative of free energy with respect to concentration, $\partial^2 G/\partial \phi^2 (= -4K\beta_m^2)$, is -4.1×10^7 erg/cm^3 PVME.

Throughout the above calculations, many approximations have been made. It is difficult to assess the combined errors introduced by these approximations but the experimental results appear to be internally consistent and the kinetic parameters have at least the correct signs.

VI. CONCLUSION

Although a theoretical framework has been built during the last decade to understand phase relationships in polymer mixtures, its application to real systems is still fraught with experimental difficulties. In particular, new methods are needed for the determination of domain sizes and phase compositions. The understanding of the mechanisms of phase separation, via nucleation and growth or spinodal decomposition, is a formidable challenge, and much work remains to be done to control phase transformation in polymer mixtures.

REFERENCES

1. H. A. G. Chermin and R. Koningsveld, *Macromolecules* **2**, 207 (1969).
2. R. Koningsveld and H. A. G. Chermin, *Proc. Roy. Soc. London* **A319**, 331 (1970).
3. R. Konningsveld, L. A. Kleintjens, and H. M. Schoeffelers, *Pure Appl. Chem.* **39**, 1 (1974).
4. G. Allen, G. Gee, and J. P. Nicholson, *Polymer* **2**, 8 (1961).
5. D. Patterson and G. Delmas, *Trans. Faraday Soc.* **65**, 708 (1969).
6. B. E. Eichinger and P. J. Flory, *Trans. Faraday Soc.* **64**, 2035 (1968).
7. L. P. McMaster, *Macromolecules* **6**, 760 (1973).
8. G. Delmas and D. Patterson, *Int. Symp. Macromol. Chem., Toronto* (1968).
9. S. Saeki, N. Kuwahara, S. Konno, and M. Kaneko, *Macromolecules* **6**, 589 (1973).
10. S. Konno, S. Saeki, N. Kuwahara, M. Nakata, and M. Kaneko, *Macromolecules* **8**, 799 (1975).
11. D. McIntyre, N. Rounds, and E. Campos-Lopez, *Amer. Chem. Soc. Polym. Prepr.* **10**, 531 (1969).
12. C. Booth and C. J. Pickles, *J. Polym. Sci. Part A-2* **11**, 595 (1973).
13. M. Bank, J. Leffingwell, and C. Thies, *J. Polym. Sci. Part A-2* **10**, 1097 (1972).
14. L. P. McMaster, "Aspects of Liquid–Liquid Phase Transition Phenomena in Multicomponent Polymeric Systems," Adv. Chem. Ser. No. 142. Amer. Chem. Soc., Washington, D.C., 1975.
15. J. W. Cahn, *Trans. Met. Soc. AIME* **242**, 166 (1968).
16. J. W. Cahn, *Acta Met.* **9**, 795 (1961).
17. T. Nishi, T. T. Wang, and T. K. Kwei, *Macromolecules* **8**, 227 (1975).
18. D. Turnbull, *Solid State Phys.* **3**, (1966).
19. J. W. Christian, "The Theory of Transformations in Metals and Alloys." Pergamon, Oxford, 1965.

20. J. E. Hilliard, "Phase Transformation," Chapter 12. Amer. Soc. Metals, Metals Part, Ohio, 1970.
21. J. W. Cahn, *J. Chem. Phys.* **42**, 93 (1963).
22. J. W. Cahn and J. E. Hilliard, *J. Chem. Phys.* **28**, 258 (1958).
23. J. W. Cahn, *Acta Met.* **14**, 1685 (1966).
24. M. Hillert, *Acta Met.* **9**, 179 (1961).
25. J. J. van Aartsen and C. A. Smolders, *Eur. Polym. J.* **6**, 1105 (1970).
26. J. J. van Aartsen, *Eur. Polym. J.* **6**, 919 (1970).
27. P. Debye, *J. Chem. Phys.* **31**, 680 (1959).
28. T. P. Seward, III, D. R. Uhlmann, and D. Turnbull, *J. Amer. Ceram. Soc.* **51**, 634 (1968).
29. J. W. Cahn, *J. Appl. Phys.* **34**, 3581 (1963).
30. D. Patterson, Y. B. Tewari, H. P. Schreiber, and J. E. Guillet, *Macromolecules* **4**, 356 (1971).
31. R. D. Newmann and J. M. Prausnitz, *J. Phys. Chem.* **76**, 1492 (1972).
32. D. D. Desphande, D. Patterson, H. P. Schreiber, and R. S. Su, *Macromolecules* **7**, 530 (1974).
33. O. Olabisi, *Macromolecules* **8**, 316 (1975).
34. T. Nishi and T. T. Wang, *Macromolecules* **8**, 909 (1975).
35. A. R. Shultz and C. R. McCullough, *J. Polym. Sc. Part A-2* **10**, 307 (1972).
36. G. P. Jones, D. C. Douglass, and D. W. McCall, *Rev. Sci. Instrum.* **36**, 1460 (1965).
37. W. H. Stockmayer and H. E. Stanley, *J. Chem. Phys.* **18**, 153 (1950).
38. T. K. Kwei, T. Nishi, and R. F. Roberts, *Macromolecules* **7**, 667 (1974).
39. T. K. Kwei, G. D. Patterson, and T. T. Wang, *Macromolecules* **9**, 780 (1976).
40. T. Nishi and T. K. Kwei, *Polymer* **16**, 285 (1975).

Chapter 5

Solid State Transition Behavior of Blends

W. J. MacKnight and F. E. Karasz

Department of Polymer Science and Engineering
University of Massachusetts
Amherst, Massachusetts

J. R. Fried

Corporate Research and Development Staff
Monsanto Company
St. Louis, Missouri

I.	Criteria for Miscibility	186
	A. Optical Clarity, Mechanical Properties	186
	B. Transitions	188
II.	Measurements of the Glass Transition Temperature in Polyblends . . .	193
	A. Differential Scanning Calorimetry	193
	B. Dynamic Mechanical Relaxation	194
	C. Dielectric Relaxation	196
	D. Dilatometry	196
III.	Limitations on the Use of Glass Transition Temperatures as a Criterion for Blend Compatibility	197
	A. Differences in Glass Transition Temperatures of Homopolymers . .	197
	B. Magnitude Dependence on Concentration of Components. . . .	198
	C. Interference by Crystallinity	200
	D. Differences of Results Depending on Measurement Technique . . .	200
IV.	Solid State Transition Behavior of Poly(2,6-Dimethylphenylene Oxide)–Polystyrene Blends	201
	A. Glass Transition Temperature	202
	B. Crystalline Melt Temperature	209
V.	Solid State Transition Behavior of PPO–Polychlorostyrene Blends . . .	218
	A. Compatibility versus Incompatibility: Chlorostyrene–Styrene Copolymers	218
	B. Glass Transition Temperature	219
	C. Analysis of Dielectric Results	222
	D. Compatibility–Incompatibility Transition	224

185

VI. Unsolved Problems 229
   A. Effect of Size of Dispersed Phase on the Glass Transition Temperature . 229
   B. Kinetics versus Thermodynamics in Melt Temperature Depression . 231
   C. Differences of Transition Behavior with Different Techniques. . . 232
VII. Commercial Blends of PPO–Polystyrene: Noryl 234
   A. Composition 236
   B. Properties 236
   References 238

I. CRITERIA FOR MISCIBILITY

A. Optical Clarity, Mechanical Properties

Films or molded objects made from two mutually miscible or compatible polymers are optically clear and have good mechanical integrity, whereas those made from incompatible polymers are usually translucent or opaque and weak. For example, Yuen and Kinsinger [1] have shown that only a small amount of a second incompatible polymer need be added to cause a film to appear opaque; they showed that for poly(methyl methacrylate) (PMMA) polymerized in the presence of polystyrene (PS), as little as 0.01 wt % PS causes the resulting film to appear opaque. Although films of any two amorphous, compatible polymers are always clear, absolute judgment on the compatibility of the two polymers cannot be made on this basis. Under special circumstances, films made from blends of two incompatible or semicompatible polymers can be optically clear. For example, films that are very thin, so that light encounters only one of the two phases in passing through the material, can appear to be clear [2]. In addition, films consisting of two layers as a result of phase separation during casting may be transparent [3]; however, the majority of cases reported in which clear films can be obtained in a heterogeneous polymer system are a result of either the two polymers having equal refractive indices [2–4] or as a result of the dispersed phase having dimensions smaller than the wavelength of visible light [2, 4].

Rosen [4] indicated that the critical domain size required for film transparency in a microheterogeneous blend is approximately 0.1 μm or 1000 Å. Jenckel and Herwig [5] have suggested that the clarity observed in heterogeneous blends of poly(vinyl acetate) (PVA) and poly(methyl methacrylate (PMMA) was probably due to the small size of the dispersed domains of PVA in the PMMA matrix. The effect of domain size on film clarity was observed by Bank *et al.* [6] for phase separation at the lower critical solution temperature (LCST) of the blend polystyrene–poly(vinyl methyl ether). As the two polymers phase separated and the domain size of the dispersed phase enlarged, the film changed from complete clarity to a bluish appearance as a result of Rayleigh scattering of incident light at the critical domain size. Finally, with prolonged heating, the domains enlarged sufficiently to give the film an opalescent or white appearance.

It is also possible for block copolymers whose corresponding homopolymer components are incompatible to form clear films if the domain sizes are below the critical size. For example, Perry [7] has shown that block copolymers of styrene and acrylonitrile give transparent films when the acrylonitrile domains are smaller than 560 Å. More recently, Kenny [8] has demonstrated film clarity for block copolymers of styrene and butadiene for which the butadiene domains are approximately 300 Å in size.

As mentioned above, films prepared from two incompatible polymers can be transparent if the two component polymers have equal refractive indices. Bohn [9] has indicated that for transparency, the critical difference between refractive indices can be no greater than 0.01. For example, incompatible blends of polystyrene and bisphenol A polycarbonate, with refractive indices 1.585 and 1.590, respectively, yield films that are clear [10]. Bauer et al. [11] have shown that it is possible to produce transparent high impact thermoplastic elastomer blends by selective copolymerization of one component to match the refractive index of the other. This is an area of significant commercial interest. For example, Platzer [12] reports rigid vinyl bottle compounds prepared from blends of poly(vinyl chloride) (PVC), and methyl methacrylate–butadiene–styrene graft copolymers can have matched refractive indices. Although at a given temperature such heterogeneous blends may appear to be clear, they may be translucent or opaque at higher or lower temperatures if the temperature coefficients of the refractive indices of the two components do not match [4, 13].

In addition to film clarity, blends of compatible polymers exhibit good mechanical properties, especially tensile strength. Compatible systems exhibit tensile strengths as a function of blend composition that are at least a weighted average of the values corresponding to the two components, for example, a near linear dependence. Often compatible blends have been reported to show a small maximum in tensile strength over certain blend compositions [8, 15, 16]. This synergism in tensile strength has not been adequately explained but is probably a result of strong specific interactions that lead to better molecular packing, that is, greater densities than can be calculated from a simple weighted average of specific volumes, as has been reported for several compatible blends [17–19].

In contrast, blends of incompatible polymers have been reported to exhibit a broad minimum in tensile strength as a function of composition [7, 15, 20, 21]. Based upon these observations, it is probable that plots of tensile strength versus composition can be used as a relative indication of blend homogeneity. Recently, Fried [15] has shown that such measurements are very sensitive to blend compatibility. Vinogradov [22] has suggested using a parameter ε defined by

$$\varepsilon = (A - A_1)/(A_2 - A_1) \tag{1}$$

as a quantitative measure of homogeneity. In Eq. (1), A is some measured quantity such as tensile strength, whereas A_1 and A_2 are corresponding limiting values calculated on the basis of a two-phase and hypothetical compatible state, respectively.

Improved tensile strength as a result of compatibility is generally realized at the expense of impact strength [20]. It is well known that high-impact thermoplastics such as rubber filled materials require a dispersed phase with good boundary adhesion with the matrix [4, 23] (see Volume 2, Chapters 13 and 14). The size of this dispersed phase is also important. Wetton et al. [24] have shown that in rubber filled thermoplastics, impact strength increases with particle size within a range between 0.1 and 0.4 μm (above the critical phase size for transparency). Razinskaya et al. [25] have shown that optimum impact strength is obtained when the area per unit volume of the dispersed phase is between 1.5 and 25 μm^{-1}. The impact strength of compatible polymer blends such as those of poly(2,6-dimethyl-1,4-phenylene oxide) (PPO) and polystyrene may be improved by adding a dispersed rubbery phase, which is present in the form of high-impact polystyrene in this particular system [26].

B. Transitions

1. Single, Composition-Dependent Glass Transition Temperature

Perhaps the most unambiguous criterion of polymer compatibility is the detection of a single glass transition whose temperature is intermediate between those corresponding to the two component polymers. This has been taken to imply that within the limits of detection of the technique used to measure the glass transition temperature T_g, the blend is molecularly homogeneous. Krause and Roman [27] have demonstrated that compatible blends of poly(isopropyl methacrylate) and poly(isopropyl acrylate) exhibit the same T_g by dilatometry as random copolymers of the corresponding two monomers at the same composition. In this particular case, this can be adduced as evidence that the local molecular environment between the two homopolymers in the blend approximates that experienced by different adjacent monomer units on the same chain.

At the other extreme, blends of incompatible polymers that segregate into distinct phases exhibit glass transitions identical in temperature and width to those of the unblended components. In intermediate cases where there is partial mixing between components or if the dispersed phase is very small, the T_g's of the individual components may be shifted [15, 28, 29]. In addition, broadening in the calorimetrically measured width of the glass transition has been observed in both compatible blends [6, 15] and in blends of partial miscibility [15, 30]. Such broadening may be indicative of microheterogeneity where local composition fluctuations are in excess of normal density and temperature fluctuations [31]; similar broadening has been reported in the glass transitions of heterogeneous copolymers [32]. The combination of transition broadening and elevation in temperature of the low-T_g transition and depression of the high-T_g transition in two-phase polymer blends may indicate that the system is very close to being compatible [15]. On the other hand, sharp glass transitions and coincidence of the component T_g's with those of the unblended polymers is indicative of macrophase separation.

For compatible blends, many attempts have been made to correlate the observed T_g with blend composition as is frequently done with random copolymers. Several compatible blends [17, 33, 34] exhibit T_g–composition dependences, which can be predicted by the simple Fox relationship [35]

$$1/T_{g, b} = w_1/T_{g, 1} + w_2/T_{g, 2} \qquad (2)$$

where $w_{1, 2}$ represents the mass fraction of the respective components and $T_{g, b}$, $T_{g, 1}$, and $T_{g, 2}$ are the T_g's of the blend, Component 1, and Component 2. Other compatible blends have been reported to follow the more detailed Kelley–Bueche [36] expression [37, 38]

$$T_{g, b} = [T_{g, 1} + (KT_{g, 2} - T_{g, 1})\phi_2]/[1 + (K - 1)\phi_2] \qquad (3)$$

where the parameter K represents the ratio of the thermal expansion coefficient differences above and below the glass transition for Components 2 and 1 ($K = \Delta\alpha_2/\Delta\alpha_1$) and ϕ is the volume fraction of each component. Still other compatible blends [27] exhibit at T_g–composition dependence which cannot be approximated by any of the well-known equations such as the Fox, Kelley–Bueche, Gordon–Taylor [39], Gibbs–Dimarzio [40], or Känig [41] expressions.

Similar difficulties in obtaining widely applicable expressions relating T_g to composition have been reported for certain random copolymers [27, 42, 43]. A more promising approach for copolymers has been the use of equations such as the Gibbs–Dimarzio expression modified to take into consideration the effect of dyad sequences on chain stiffness [42–44]. It may be possible to obtain information on the statistical nature of intermolecular contacts in compatible blends from the form of the T_g dependence using an approach of this type, but at present there has been no effort in this regard.

2. Melting Point Phenomena

In an early review of polymer blends, Bohn [45] observed that, based upon enthalpy considerations, compatible blends of one crystalline polymer with any other polymer were unlikely except for the remote possibility of mixed crystal formation. However, soon after 1968, several groups showed that compatible polymer blends could be obtained in systems in which one component was a crystalline polymer. Examples of four such systems are blends of poly(vinylidene fluoride) (PVF$_2$) with poly(methyl methacrylate) (PMMA)[37, 46, 47] and with poly(ethyl methacrylate)(PEMA) [37, 48, 49]; blends of poly(vinyl chloride) (PVC) with poly(ε-caprolactone) (PCL) [50–52]; and blends of poly(2,6-dimethyl-1,4-phenylene oxide) (PPO) with atactic or isotactic polystyrene (PS) [53, 54, 56]. In the latter system, it is possible to crystallize isotactic PS alone in blends of PPO and isotactic PS, or to crystallize both PPO and isotactic PS simultaneously (see Section IV.B) by appropriate solvent and/or thermal treatment.

All these crystalline–amorphous polymer blends share three characteristics. First, the two polymers are thought to be compatible in the molten state. Second, as the blends are cooled from the melt, crystallization of the crystallizable component occurs, but the total degree of crystallinity of the blend decreases rapidly with increasing content of the second, amorphous, component. The latter behavior is illustrated in Fig. 1 for the PVF$_2$–PMMA

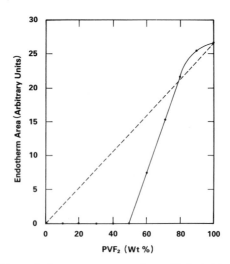

Fig. 1   Area of melting endotherm as a measure of crystallinity in blends of poly(vinylidene fluoride) (PVF$_2$) and poly(methyl methacrylate) (PMMA); determined by differential thermal analysis (DTA). (Reproduced with permission from Paul and Altamirano [47] in "Copolymers, Polymer Blends, and Composites," Advances in Chemistry Series 142, pp. 371–385, 1975. Copyright by the American Chemical Society.)

system. Here the area of the melting endotherm (normalized for sample mass from differential thermal analysis) is plotted as a function of weight composition of the crystalline component. The broken line represents the expected endotherm if PMMA were merely to dilute the PVF_2 and did not interfere with its crystallization. As indicated by the rapidly decreasing endotherm line, crystallization of the PVF_2 is inhibited below 80 wt % PVF_2 and, for the particular thermal history imposed upon these samples, no level of crystallinity is detected in blends with more than 50 wt % of the amorphous, PMMA component.

A third characteristic feature of crystalline–amorphous blends is a substantial depression of the crystalline melt temperature (T_m) as a result of the diluent effect of the amorphous component. In the upper half of Fig. 2 T_m is plotted as a function of wt % PVF_2 for the PVF_2–PMMA system. For this blend, T_m is depressed by approximately 10° from the equilibrium melt temperature (T_m^0) of unblended PVF_2 for a blend composition of 40 wt % PMMA. Similar magnitudes in melting point depression have been observed in other blends.

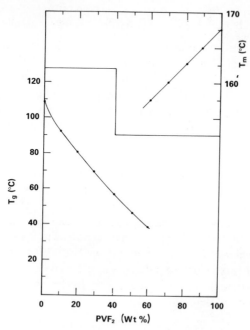

Fig. 2   Glass transition temperature (T_g) and crystalline melting temperature (T_m) as a function of composition for blends of poly(vinylidene fluoride) (PVF_2) and poly(methyl methacrylate) (PMMA). (Reproduced with permission from Paul and Altamirano [47] in "Copolymers, Polymer Blends, and Composites," Advances in Chemistry Series 142, pp. 371–385, 1975. Copyright by the American Chemical Society.)

Such a drop in T_m in part explains the reluctance of blends with a high content of the amorphous component to crystallize. In addition, in all blends considered this drop in T_m is accompanied by an elevation in T_g as the low-T_g component crystallizes out from solid state solution and as the content of the amorphous, high-T_g component increases (Fig. 2). For these reasons, the T_m–T_g interval decreases; this places a severe kinetic restriction on the crystallization process. An additional cause for the inability of blends with high amorphous content to crystallize is the growing isolation of the individual chains of the crystalline polymer with increasing amorphous polymer content due to the random placement of different polymer molecules in the homogeneous compatible state and the above restriction on molecular mobility during the crystallization process. As a result, at high amorphous component concentrations, the domain of the crystalline polymer is likely to be smaller than the critical nucleus size for crystallization [52].

Early attempts have been made to fit the observed melting point depression to the well-known equation of Flory [57–59] for polymer–diluent systems. For high molecular weight polymers and low molecular weight diluents, this equation is

$$1/T_m - 1/T_m^0 = (R/\Delta H_u)(V_u/V_1)(\phi_1 - \chi\phi_1^2) \tag{4}$$

where R is the ideal gas constant, ΔH_u is the heat of fusion of the unblended crystalline polymer unit, ϕ_1 is the volume fraction of the diluent, V_u and V_1 are the molar volumes of the polymer unit and diluent, respectively. χ is the polymer–diluent interaction parameter, and T_m and T_m^0 are the observed and equilibrium ($T_m^0 = \Delta H_u/\Delta S_u$) melting temperatures. To explain the magnitude of the T_m depression observed in the above polymer–polymer systems, the ratio of the molar volumes of the crystalline polymer unit and the diluent, in this case the amorphous polymer, would have to be in the order of 10^{-1} to 10^1 [60]. This means that the effective molar volume of the polymer diluent is substantially less than that of an entire polymer chain length and has been cited as thermodynamic evidence for mixing at the intimate segmental level in compatible blends [53].

Recently, Nishi and Wang [46] have derived an equation more appropriate for polymeric diluent systems from Scott's equation [61] for the thermodynamic mixing of two polymers based on an extension of the Flory–Huggins lattice theory. In the form given by Imken et al. [48], this equation for the melting point depression is

$$T_m^0 - T_m = -T_m^0(V_{2u}/\Delta H_{2u})B\phi_1^2 \tag{5}$$

where Component 2 is the crystalline polymer and B is related to the interaction parameter χ by the expression

$$B = RT(\chi/V_1) \tag{6}$$

A plot of $T_m^0 - T_m$ versus ϕ_1^2 should thus be linear with a zero intercept. The interaction parameter for the polymer–polymer blend can be obtained from the slope of the plot.

Nishi and Wang [46] found that for PVF_2–PMMA, Eq. (5) fitted the observed melting point data well (linear extrapolation to a near zero intercept) and gave an interaction parameter χ of -0.295 at 160°C. This is a reasonable value of χ because of the thermodynamic restrictions on mixing two high molecular weight polymers (see Chapters 2–4). Similar results were found by Kwei *et al.* [49] and by Imken *et al.* [48] for the blend PVF_2–PEMA.

II. MEASUREMENTS OF THE GLASS TRANSITION TEMPERATURE IN POLYBLENDS

As indicated in the previous discussion (Section I.B.1), a more reliable criterion of polymer compatibility than film clarity is the detection of a single, composition-dependent T_g. There are many methods by which T_g's of polymers may be determined. Examples are differential thermal analysis (DTA), thermal optical analysis (TOA), probe penetration thermomechanical analysis (TMS), differential scanning calorimetry (DSC), dynamic mechanical and dielectric measurements, and dilatometry. Some of the more widely used methods of T_g measurement in polymer blends are reviewed briefly below with respect to their advantages and limitations in the study of polymer compatibility.

A. Differential Scanning Calorimetry

Differential scanning calorimetry (DSC) has been widely used in the study of blend compatibility as well as in the determination of crystallinity in crystalline–amorphous polymer blends [46, 49, 53]. For example, de Boer and Challa [62] have used DSC to study the compatibility of isotactic and syndiotactic PMMA; Schneier [63] to study the effects of mixing conditions on the compatibility of PMMA and poly(vinyl acetate) (PVA); Frisch *et al.* [64] to study interpenetrating networks of polyacrylates and polyurethanes; Bank *et al.* [6] to study the influence of the choice of solvent on the compatibility of polystyrene and poly(vinyl methyl ether) (PVME); and Fried [15] to study the compatibility–incompatibility transition as a function of the composition of the copolymer in the blend PPO–poly[(*p*-chloro-styrene)-*co*-styrene] (e.g., Section V.D.1).

Differential scanning calorimetry (DSC) thermograms are plots of heat capacity (C_p) versus temperature from which T_g can be defined either as the inflection point at the discontinuity in C_p at the glass transition, the intersection point of the projection of the baseline with the tangent to the discontinuity (intersection method), or the temperature corresponding to the midpoint of the discontinuity ($\frac{1}{2}\Delta C_p$). Griffiths and Maisey [65] have shown that for unblended polymers optimum consistency was obtained using a heating rate of 8°/min and with T_g defined by the intersection technique. Richardson and Savill [66] claim that by converting DSC thermograms directly into enthalpy curves, T_g can be determined to an accuracy of 1°. General uses of DSC in the study of polymer thermodynamics have been reviewed by Flynn [67].

Besides rapidity of measurement, the principal advantages of DSC in the study of polymer blends are that only small (typically 5–30 mg) samples are required and that the available rapid temperature scan permits the study of heat sensitive polymers such as PVC [30] without the addition of stabilizers. In addition, coprecipitated blends in powder form can be studied; thereby, accurate control of thermal history can be maintained through programmed heating and cooling cycles. An additional advantage is that from the measured T_g's and heights (ΔC_p) of the glass transition, a quantitative determination of the composition of unknown multi-phase or compatible blends can be made [68].

B. Dynamic Mechanical Relaxation

Dynamic mechanical measurements have been widely used in the study of polymer compatibility. With this technique, a dynamic modulus can be measured as a function of temperature over a range of frequencies [69]. From traditional torsion pendulum measurements, the dynamic shear loss modulus (G'') and the dynamic shear storage modulus (G') may be obtained as a function of temperature at a nominal frequency in the vicinity of 1 Hz. The T_g (or T_g's) of the blend is defined as the temperature corresponding to the maximum in G'' or $\tan\delta$ ($\tan\delta = G''/G'$) at the main relaxation, which marks the onset of main chain segmental mobility corresponding to the glass transition (often designated the α relaxation in amorphous polymers). Commercially available instruments such as the Rheovibron, used in the standard sample configuration, permit the dynamic tensile loss modulus (E'') to be determined over a wider range of frequencies; in the latter case, $\tan\delta$ is defined as the ratio E''/E'. By appropriate modification of the sample holder geometry, either the dynamic shear moduli [70] or bulk moduli [71] can be measured with the Rheovibron.

Considerable success has been achieved in modeling the dynamic mechanical spectra of two-phase systems by combinations of series and parallel elements [72–74]. In addition, Bohn [75] has been able to obtain quantitative information about phase volumes and structure in two-phase systems from analysis of the dynamic mechanical measurements using existing theories of the elastic moduli of composites (see Chapter 8). Dynamic mechanical measurements have been used by Nishi *et al.* [76] to study phase separation in blends of PVC and Hytrel thermoplastic elastomers by resolution of the broad, overlapping tan δ peaks observed in these blends using a Fuoss–Kirkwood distribution of relaxation times. Recently, Akiyama *et al.* [77] have used the dynamic mechanical technique to study phase separation at the lower critical solution (LCST) of blends of poly(vinyl nitrate and poly(methyl acrylate). The dynamic mechanical technique has been used by Marcincin *et al.* [78] to study the incompatibility–compatibility transition in blends of PVC and chlorinated poly[ethylene-*co*-(vinyl acetate)] (EVA) as the chlorine content of the EVA component was raised above 38 wt %. The above are typical studies; the total number of references to the use of dynamic mechanical studies in the investigation of polymer compatibility is too large for complete coverage here.

One disadvantage of the dynamic mechanical method compared to the DSC technique is that only films or fibers (i.e., not powders) made from the blends can be studied. This means that particular attention must be paid to the thermal history of the test specimen and that in the cases of cast films or spun fibers the choice of solvents may be important in the interpretation of results. A principal advantage over DSC is that the effect of blending on the low-temperature secondary relaxations of both component polymers may be studied. Such secondary relaxations may arise from small scale, limited segmental motions such as the Schatzki [79] crankshaft motion or torsional and rotational movements of substituent groups. O'Reilly and Karasz [80] have concluded that heat capacity measurements such as those by DSC do not show such relaxations because they are not sufficiently cooperative to occur over a narrow temperature range. These secondary relaxations are more sensitive to small-scale local environment and as such may be more revealing measures of the intimacy of mixing between different polymer molecules in the compatible state.

Often, secondary relaxations in compatible blends have been reported to be broadened, and the maxima of component polymers may overlap with consequent interpretative difficulties [34]. For some compatible blends, for example, PVC and EVA (in the compatible copolymer composition range lying between 65 and 70 wt % vinyl acetate), no effect upon the secondary relaxations of either component has been reported although the main relaxations marking the glass transitions of the two component polymers are merged and shifted in temperature [33]. In other blends where specific interchain interactions may be relatively greater, a shift in temperature and/ or diminution of intensity of the secondary relaxation maxima may be noted [16, 17, 50, 81].

C. Dielectric Relaxation

Compared with DSC and dynamic mechanical measurements, the use of dielectric techniques in the study of polymer blends has been limited. This is in part due to the greater experimental difficulties inherent in this method and, in part, due to the lack of sensitivity of dielectric measurements to blends of nonpolar polymers. Analogous to the case of dynamic mechanical measurements in which the components of the complex modulus are measured over a temperature range, the components of the complex dielectric constant (ε^*) are determined as a function of temperature and frequency. Unlike the dynamic mechanical method, a wide and continuous range of frequencies can be readily used. This permits the dynamic dielectric constants ε'' and ε', or $\tan \delta$ ($\tan \delta = \varepsilon''/\varepsilon'$), to be determined either as a function of temperature (temperature–plane measurements) over many frequencies or as a function of frequency (frequency–plane measurements) at selected temperatures. Frequency–plane measurements are particularly important because blend compatibility may be studied at temperatures above or below T_g permitting study of phase separation at lower or upper critical solution temperatures [82]. Full advantage of this opportunity has not as yet been taken.

Mehra [83] has used the dielectric technique in the study of phase separation in block copolymers and blends of various homopolymers; Bank et al. [84] to study the influence of the choice of solvent on the compatibility of PS and PVME in cast films; and Feldman and Rusu [21] to study compatibility in PVC–EVA blends. Recently, Akiyama et al. [85] have shown how differential thermal analysis (DTA) instrumentation can be modified to obtain dielectric measurements from the generation of dielectric heat of a polymer sample in an alternating electric field. They also showed that the width of the normalized dielectric curves ($\varepsilon''/\varepsilon'_{max}$ versus f/f_{max}) may be used as a qualitative measure of blend homogeneity.

D. Dilatometry

The earliest used method of determining T_g in polymers, dilatometry, has been infrequently employed in blend studies because of the greater speed and versatility of modern thermal analysis instrumentation. The glass transition temperature determined by dilatometry corresponds to that temperature marking the discontinuity in the plot of specific volume as a function of temperature. In contrast to DSC, dilatometry requires larger samples and more time and care in sample preparation and measurement. Probably the earliest reference to the use of dilatometry in the study of blends is that of Boyer and Spencer [86]. More recently, Bartenev and Kongarov [87] have used dilatometric measurements in the study of compatible blends of elastomers; Krause and Roman [27] in the study of compatible blends of poly(isopropyl acrylate) and poly(isopropyl methacrylate); Sperling *et al.* [88] in the study of interpenetrating networks; and Noland *et al.* [37] in the investigation of compatible crystalline–amorphous blends of PVF_2 and PMMA and PEMA (e.g., Section I.B.2).

III. LIMITATIONS ON THE USE OF GLASS TRANSITION TEMPERATURES AS A CRITERION FOR BLEND COMPATIBILITY

Methods of detection of T_g in blends have been discussed above. In systems for which compatibility determination on the basis of film clarity cannot be made because of refractive index or domain size considerations (e.g., Section I.A), the presence of a single T_g may be a definitive test of blend compatibility (e.g., Section I.B.1); however, there are certain circumstances under which the T_g criterion may be inapplicable or even misleading as outlined below.

A. Differences in Glass Transition Temperatures of Homopolymers

Application of the single T_g criterion requires that the T_g's of the components be sufficiently displaced from each other so that resolution is possible. The limits of resolution of the various techniques outlined in Section II vary somewhat but in all cases resolution of T_g's less than about 20° apart is poor. For example, Hughes and Brown [89] have reported difficulty in resolving the two T_g's of incompatible blends of poly(methyl acrylate) and poly(vinyl acetate) of which the T_g's of the components are only 22° apart as determined by torsion modulus measurements. In dynamic mechanical measurements, the loss maxima of two incompatible polymers whose peak (T_g) temperatures are less than 30–40° apart will overlap with the minor component appearing as a shoulder on the larger peak, but often temperature assignments of the individual relaxations are possible. For example, Seefried and Koleske [90] reported that they were able to identify the separate shear loss (G'') peaks of quenched, amorphous blends of poly(ε-caprolactone) and poly(ε-methyl-ε-caprolactone) whose peak temperatures appear at -60 and $-40°C$, respectively. Buchdahl and Nielsen [31] were able to resolve the loss maxima of PVC and PS separated by only $25°$. At much narrower separations in temperature, identification of separate peaks of equal intensity may not be possible and only a broad, single peak may be observed.

The closeness of blend component T_g's may be even a more severe problem for detection by DSC and dilatometry. The difficulty of extrapolating straight lines for multiple discontinuities in specific volume–temperature plots has been commented upon [27]. By DSC, detection of T_g's separated by less than about $30°$ by standard techniques involves a high degree of uncertainty. Landi [91] has shown in the case of heterogeneous copolymers of butadiene and acrylonitrile, for which phase separation would be predicted theoretically, that the separate T_g's corresponding to the two mixed composition domains could be resolved by a graphical treatment of the slope of the DSC thermogram corrected for deviations from baseline linearity. The as-recorded thermogram indicated only a single, although broadened, glass transition. The derivative of the heat capacity–temperature curve can be electronically calculated and plotted automatically together with the standard thermogram in certain commercial instrumentation. Such improved techniques allow greater resolution than previously possible.

B.   Magnitude Dependence on Concentration of Components

In two-phase systems, if one component is present in a very small quantity, it may not be detected depending on the sensitivity of the technique used. This may lead to an erroneous assumption of compatibility; however, based upon the T_g assignment of the single transition in relation to the T_g's of the two components and with additional information such as film clarity, further judgment of the compatibility of the two polymers can be made. If the two components have different refractive indices, the film clarity criterion may be a very sensitive means of confirming suspected compatibility under these circumstances. As previously mentioned (Section I.A), as little as 0.01 wt % of an incompatible polymer may be sufficient to cause a film to appear opaque. Such small concentrations are far beyond the sensitivity of the T_g techniques outlined in Section II. Of these, the dynamic relaxation techniques (mechanical or dielectric) may be the most sensitive. For example, no difficulties have been reported in the detection of a minor component, 10 wt % of the total, by dynamic mechanical measurements [90]. In cases in which the minor component cannot be detected by DSC because of small samples and low signal-to-noise ratio, dynamic mechanical measurements may be more revealing [92]. Improved sensitivity in the detection of very small amounts of a minor component in an incompatible blend may be achieved with other techniques such as light scattering [1], NMR [163], or gas permeation studies [18] (see Chapter 10).

An additional complication may arise in the detection of a minor component in blends that are nearly compatible, that is, those in which broadening of the transitions and temperature shifts are observed (e.g., Section I.B.1). In such systems there is substantial evidence to suggest that significant amounts of both component molecules may intermix in interfacial regions between the dispersed phase and the matrix (see Volume 2, Chapter 21). It is possible that this interfacial mixing may involve the lower molecular weight molecules of both components for which the thermodynamic driving force for compatibility is the greatest [15], but no firm evidence to support this hypothesis is available at present. There is substantial evidence indicating that the size of this interfacial area is different for different polymer pairs and may be a relative indication of an approach to compatibility in these systems. Helfand [93] has calculated the concentration profiles of two immiscible polymers in the interfacial region and has concluded that it may be possible to obtain information on the interaction parameter χ from the measurement of the size and the distribution of polymers in the interfacial region. Evidence for mutual penetration of components in blends of PVC and PMMA and blends of poly(butyl methacrylate) and PVC has been obtained by Wojuzkij [94] from visual, ultraviolet, and electron microscopy. It was claimed that in the former blend penetration was greater due to the greater polarity and possibly smaller interaction parameter of the components. While Avgeropoulos *et al.* [95] have observed sharp boundaries in blends of polybutadiene and ethylene–propylene–diene rubbers, Lebedev *et al.* [96] have detected interfacial regions approximately 600 Å in width for blends of incompatible polyethylene and polyoxymethylene. Letz [97] using phase contrast microscopy studies, has measured interfacial regions as large as 28,000 Å (2.8 μm) for isotactic polypropylene–polyethylene mixtures and 89,600 Å (8.96 μm) in PVC–polyethylene mixtures. These results suggest that for blends in which two phases are present but in which there is a thermodynamic driving force for compatibility, minor component transitions may be significantly reduced in intensity as a result of molecular migration from well-defined phases into broad variable composition interfacial regions whose detection, because of the broad dispersion of the transition, may not be evident. From quantitative DSC studies of incompatible blends of PPO and random copolymers of styrene and *p*-chlorostyrene with compositions around the compatibility–incompatibility transition (e.g., Section V.D.1), Fried [15] has shown that as much as 40% of the total sample is undetected in terms of C_p discontinuities; whether this "absent" material can be accounted for in terms of large, diffuse interfacial regions which are undetected by DSC or because some dispersed phases may be too small for detection by DSC (e.g., Section III.D) as has been suggested in other studies of blends (Schurer et al.) [92] or a combination of both effects is impossible to state at present.

C. Interference by Crystallinity

Another complication to the use of the T_g criterion in compatibility determinations may be a suppression or apparent absence of any glass transition in blends of one or more crystalline polymers. For certain polymers with well-defined crystalline–amorphous regions, it is possible to detect both a melting endotherm and a glass transition by calorimetric methods. In these cases, the degree of crystallinity x may be obtained either from the area of the endotherm or from the magnitude (ΔC_p) of the glass transition by the relation [98]

$$x = (\Delta Q_f / \Delta H_f) \approx 1 - (\Delta C_p^{obsd} / \Delta C_p^{a}) \qquad (7)$$

where ΔH_f is the heat of fusion of the completely crystalline polymer, ΔQ_f is the heat of fusion measured for the semicrystalline sample, ΔC_p^{obsd} is the observed height of the glass transition, and ΔC_p^{a} is the change in heat capacity at the glass transition for a totally amorphous sample. Isotactic polystyrene, for example, has been reported to obey this strict two-phase model rather closely [98]; this implies that the boundaries between amorphous and crystalline regions are sharp and distinct [99]. Most other semicrystalline polymers, for example PPO, display a different phenomenon [53, 100]; as crystallinity in the sample is increased, a disproportionate reduction in the magnitude of the glass transition (ΔC_p) is observed. This effect renders the detection of T_g difficult, especially by calorimetric techniques. Similarly, for blends of amorphous and crystalline polymers, the glass transition of the amorphous region(s) may be obscured at high blend crystallinity [53]. In such cases, dynamic mechanical measurements may be preferable to those of DSC [48].

D. Differences of Results Depending on Measurement Technique

If all compatible blends are molecularly homogeneous (i.e., segmental mixing occurs) or if all measurement techniques for T_g are responsive to motions occurring within equal domain sizes, the detection of one or two glass transitions and the temperature assignment of these relaxations would not be expected to depend on the particular measurement technique selected. Whether or not an apparent compatible blend has reached its thermodynamic equilibrium state will depend upon thermal history and mixing conditions. It is readily possible that a given compatible system may not be molecularly homogeneous under a particular set of conditions and that individual molecules of each component may cluster in very small domains (i.e., microheterogeneity). Then if different techniques are responsive to cooperative molecular motions occurring over different region sizes, one may be led to different conclusions concerning the compatibility of the blend or see differences in the T_g–composition dependence. For example, if the molecular process reponsible for the discontinuity in heat capacity observed by DSC involves longer range motions than the segmental microbrownian motions responsible for the dynamic mechanical loss peaks, a particular blend may be judged compatible by DSC but heterogeneous by dynamic mechanical measurements [101]. There is at present substantial evidence to support these conclusions. Some of the existing results will be reviewed at a later point (Section VI.C). It is clear that a full assessment of blend homogeneity may require the use of several techniques for comparison, as well as supplementary information concerning, for example, domain size by electron microscopy.

IV. SOLID STATE TRANSITION BEHAVIOR OF POLY(2,6-DIMETHYLPHENYLENE OXIDE)– POLYSTYRENE BLENDS

In the previous sections, criteria for the compatibility of polymer blends and the techniques for measurement of T_g have been reviewed. Of the large number of polymer blends that have been investigated by these methods over the years, only a few can be judged compatible by the criteria of film clarity and a single, composition-dependent T_g. Most of the known compatible blends have been included in reviews by Bohn [45], by Petersen *et al.* [102], by Krause [3], and in Chapter 2. Some of these blends are compatible over a limited range of compositions. One of the most studied compatible blends is that of poly(2,6-dimethyl-1,4-phenylene oxide) (PPO) and polystyrene (PS) (see also Chapter 10). PPO and PS are compatible in all proportions and films made from their blends are clear; the refractive index has been reported to increase linearly with increasing PPO content [19]. In addition, only single glass transitions intermediate in temperature between the T_g's of PPO and PS have been reported and good mechanical properties (e.g., Section VII.B) are obtained at all compositions. These blends also have significant commercial application (Section VII.A). An interesting review of PPO and PPO–PS blends has been given by Hay [103].

Even among related polyphenylene oxides, the compatibility of PPO and

PS appears unique. For example, although PPO is compatible with poly(2-methyl-6-phenyl-1,4-phenylene oxide) [28, 104] and with poly(2-methyl-6-benzyl-1,4-phenylene oxide) [105], neither poly(2-methyl-6-phenyl-1,4-phenylene oxide) or poly(2,6-diphenyl-1,4-phenylene oxide) are compatible with PS.

The dependence of PPO–PS blend T_g on composition has been studied by DSC, probe penetration thermomechanical analysis (TMS), thermal optical analysis (TOA), and dynamic mechanical and dielectric relaxation techniques. In several cases, different T_g methods gave different T_g–composition relationships. The results of these studies and their implications as to the intimacy of PPO–PS mixing are discussed below.

A. Glass Transition Temperature

1. *Differential Scanning Calorimetry*

The T_g's of unblended PPO and PS as determined from DSC and the related techniques of TOA, DTA, dilatometry, and adiabatic calorimetry are listed in order of increasing heating rates in Table I. Depending on technique, heating rate, and sample history, T_g's reported for PS range from 89 to 113°C, while those of PPO range from 207 to 234°C.

The T_g's of PPO–PS blends as measured by DSC, or by TOA, DTA, and TMS techniques follow the same dependence on composition. In each case, the curve relating blend T_g to composition falls below the tie line connecting the T_g's of the pure components, that is, a concave relation (Fig. 3). Jacques

Table I

Glass Transition Temperatures of PS and PPO

T_g(°C)		Technique	Heating rate (°C/min)	Reference
PS	PPO			
89	—	Adiabatic calorimetry	~0.1	98
—	207	Adiabatic calorimetry	~0.1	106, 107
100	—	Dilatometry	~1	108
101	219	DSC	10	38
108	225	DSC	10	101
113	222	TOA	10	109
—	230	DSC	16	110
—	234	DTA	20	110
106	220	DSC	40	68
—	225	DSC	40	110

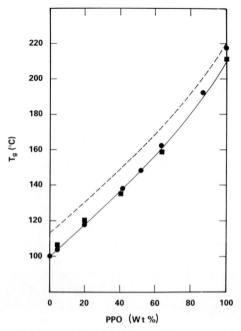

Fig. 3 Glass transition temperatures (T_g) as a function of composition for blends of poly(2,6-dimethyl-1,4-phenylene oxide) (PPO) and polystyrene as determined from DSC and probe penetration (TMS) studies at 10°/min (Prest and Porter [38]); circles indicate DSC data, squares probe data. Solid and broken curves were drawn from equations given by Shultz and Gendron [109] for their DSC (20°/min) and thermal optical analysis (TOA) data (10°/min), respectively. (Reproduced with permission from Fried [15].)

et al. [111] have shown that this curve retains the same shape but is shifted in temperature if the DSC T_g is defined alternately as the temperature at the onset, at the midpoint ($\frac{1}{2} \Delta C_p$), or at the maximum of the transition.

Prest and Porter [38] have claimed that the experimental data for blend T_g can be fitted closely by the relation derived by Kelley and Bueche [36] for polymer–diluent systems (Eq. 3, Section I.B.1). More recently, Fried [15] has shown that, when a more reliable set of thermal expansion coefficients for PPO and PS are used in the Kelley–Bueche expression, the K parameter is calculated to be 1.73 as compared with 0.68 in the Prest and Porter study. This value results in a calculated convex dependence of blend T_g on composition in contrast to the observed concave relationship (Fig. 3). In addition, the related Gordon–Taylor and Känig equations (Section I.B.1), which also employ a K parameter defined in the same way, were shown to be similarly inadequate. This is not surprising because the free volume arguments upon which these equations depend are probably not applicable in compatible polymer systems. Better agreement with experimental T_g values was found by using the Gibbs–Dimarzio and especially the Fox equation (Eq. 2, Section I.B.1).

Shultz and Gendron [109] have also shown that the Fox equation predicts the actual T_g–composition dependence accurately for their DSC and TOA data. Offered for comparison with the DSC and TMS data of Prest and Porter in Fig. 3, two curves have been drawn from T_g's calculated from the Fox equations given by Shultz and Gendron for T_g expressed in °K

$$1/T_g(\text{TOA}) = w_1/386.2 + w_2/495.2 \tag{8}$$

$$1/T_g(\text{DSC}) = w_1/372.2 + w_2/485.2 \tag{9}$$

where w_1 and w_2 correspond to mass fractions of PS and PPO, respectively. The upper curve in Fig. 3 corresponds to the $T_g(\text{TOA})$ values while the lower curve corresponds to the $T_g(\text{DSC})$ values. Both curves follow the same concave composition relationships, as do the TMS- and DTA-determined T_g's, but the $T_g(\text{TOA})$ values are consistently about 13° higher in temperature. Agreement between the TMS and DTA data and the Fox DSC curve, Eq. (9), is excellent considering the differences in heating rate and T_g technique.

2. Dynamic Mechanical Method

Although the relaxation processes of PPO have been investigated by electron spin resonance [112] and by NMR techniques [113], the most extensive studies have been made by dynamic mechanical methods. Although all sources agree to the presence of a well-defined α relaxation associated with the thermal excitation of cooperative motions in the chain above T_g at approximately 480°K [114], there remains some controversy as to the number (variously claimed to be from one to four), the location, and the assignment of specific molecular motions. There is evidence to suggest that some of the secondary relaxations may be extremely sensitive to trapped impurities such as diphenoquinone or copper salts [115] and to the presence of water [115–117]. It is therefore reasonable to expect that some of this controversy may be due to differences in the polymerization, purification, thermal history, or drying procedures used in different studies. The effects of different sample preparations and drying procedures have been studied by Eisenberg and Cayrol [117]. The controversial nature of the mechanical spectrum of pure PPO makes difficult an unambiguous interpretation of the dynamic mechanical results for the PPO–PS blends. Some of the conclusions reached by different authors are reviewed briefly below.

Stoelting *et al.* [101] have studied blends of PPO and PS by DSC and by dynamic mechanical measurements (Rheovibron) of films that were compression molded from a physical mixture of the two powders. Their results for unblended PPO agree with those of Heijboer [116] who found a small, low-temperature ($-115°C$) γ peak and a broad, poorly defined peak near 5°C in addition to the main α glass transition above 200°C. The previous identification of the γ peak with the presence of water was questioned because Stoelting *et al.* found that this peak persisted when the compression molded films of PPO were annealed at elevated temperatures in vacuum. No β peak was found at 40°C and 110 Hz as would have been expected from the previous results of de Petris *et al.* [114]. This β peak has been attributed to hindered torsional oscillations of the phenylene units in the backbone around the carbon–oxygen–carbon axis [114, 117]. For pure (unblended) PS, Stoelting *et al.* identified three peaks designated as α(106°C), β(52°C), and γ(−140°C).

Stoelting *et al.* obtained dynamic mechanical spectra for three PPO–PS compositions (25, 50, and 75 wt % PPO) for which films had been compression molded at different temperatures (290 and 330°C) for different times (2–10 min), slow cooled, and annealed at 180°C for 0–12 hr. The resulting spectra showed a small γ maximum in the dynamic tensile loss modulus (E'') at 110 Hz located between the corresponding γ maxima of unblended PS and PPO (−104°C). In addition, the α relaxations of the blends appeared at intermediate temperatures between those of unblended PPO (224°C) and PS (114°C) but were very broad with a distinct shoulder, suggesting two overlapping peaks. This latter result was interpreted as evidence for the existence of two mixed PPO–PS phases, one PS rich and the other PPO rich in composition, rather than a single molecularly homogeneous PPO–PS region. The temperatures corresponding to the maxima of each of these two overlapping peaks is plotted versus wt % PPO in Fig. 4. The low-temperature (PS-rich) peak maxima follow a similar concave dependence on composition as did the DSC, DTA, TOA, and TMS data cited previously (Section IV.A.1). In contrast, the high-temperature or PPO-rich maxima follow a convex dependence on composition (Fig. 4). These same blends appeared homogeneous by DSC measurement in which only a single step in heat capacity was observed.

Stoelting *et al.* calculated the compositions of the PS-rich and PPO-rich phases apparent in the dynamic mechanical spectra from the maximum peak temperatures and the equation

$$T_g = (1 - w_2) T_{g,1} + w_2 T_{g,2} \tag{10}$$

which is simply the equation of the tie line joining the T_g's of PS and PPO (Component 2). The apparent blend compositions calculated from the substitution of DSC T_g values in Eq. (10) were roughly comparable with

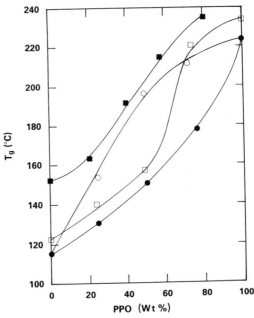

Fig. 4   Glass transition temperatures (T_g) measured by dynamic and dielectric techniques as a function of composition for blends of PPO and polystyrene. The filled and open circles represent temperatures at the maxima of the low- and high-temperature shoulders, respectively, observed in the dynamic mechanical α peak by Stoelting *et al.* [101]. Filled squares indicate temperatures at the maximum of the dynamic mechanical tan δ from the data of Shultz and Beach [119]. Open squares represent temperatures at the maxima of the dielectric loss from the data of MacKnight *et al.* [122]. (Reproduced with permission from Fried [15].)

those calculated for the dynamic mechanical PS-rich phases, suggesting that the observed DSC glass transition reflected only relaxations in the PS-rich phase. As previously mentioned, both the DSC T_g and the T_g corresponding to the PS-rich maxima exhibit similar concave composition relationships.

Following these early studies, Shultz and Beach [119] and more recently Yee [16] have performed dynamic mechanical measurements (also Rheo-vibron) on PPO–PS blends but for blending procedures different from those used by Stoelting *et al.* Shultz and Beach coprecipitated their blends from 10% toluene solutions into large excesses (20/1) of a nonsolvent, methanol. The dried precipitates (80°C for 16 hr in vacuum) were then compression molded into thin films at temperatures ranging from 180°C for PS to 270°C for PPO. Yee studied blends made by coextruding PS pellets and PPO powder in a master batch of 50/50 composition. The extrudate was subsequently chopped and coextruded with additional PS or PPO to give final compositions of 10, 30, 50, 70, and 90% PPO. This final extrudate was then chopped and compression molded into films at the same temperatures employed by Shultz and Beach.

In their dynamic mechanical studies, Shultz and Beach reported that each PPO–PS blend (20, 40, 60, and 80% PPO) exhibited a single dynamic tensile storage modulus (E') drop at T_g and a single, well-defined loss tangent with no apparent shoulder in contrast to the findings of Stoelting *et al.* [101]. The temperatures corresponding to maximum tan δ for each blend showed a smooth, nearly linear dependence on blend composition, as indicated in Fig. 4.

Yee using the same frequency of 110 Hz as in the other studies, observed a broad shoulder in the β transition for PS which was discernible in PPO–PS blends with high PS content (10–30% PPO). For those blends containing 50, 70, and 90% PPO, there was a suppression of the broad β peak of pure PPO with a new, small, unassigned peak centered around 80–90°C. The suggestion was made that the latter small peak may be that responsible for the shoulder in unblended PS and may be apparent in the blends as a result of trace amounts of PS not perfectly mixed.

Wellinghoff and Baer [81] have investigated the dynamic mechanical spectra of blends of PPO and isotactic polystyrene (iPS) by torsion pendulum measurements at 1 Hz. When quenched to the amorphous state, iPS has nearly identical thermal and mechanical properties as those of atactic PS [53, 98]. In agreement with the results of Yee, they found a suppression of the broad β peak of PPO in blends of 15 wt % iPS. They remarked that similar suppression of the β relaxations of bisphenol A polycarbonate or PVC occurs at equivalent amounts of antiplasticizers [120] for these polymers. Further indication that PPO and iPS or atactic PS mix at the segmental level comes from the analysis of the Fourier transform infrared (FTIR) spectra of PPO–iPS. They noted an increase in band intensity and a frequency shift to higher wave numbers for the C–H out-of-plane bending mode for PPO when a solvent such as ethyl benzene was added. A similar change in band intensity and an even larger increase in intensity was observed in the spectrum of PPO–iPS.

The single, well-defined tan δ peaks found by Shultz and Beach, the β suppression observed by Yee and by Wellinghoff and Baer, and the shift in the C–H bending mode are strong evidence that PPO and PS approach segmental compatibility. This conclusion suggests a reevaluation of the dynamic mechanical results of Stoelting *et al.* The fact that PPO and PS may be thermodynamically compatible [which can be stated conclusively only when the heat of mixing (ΔH_m) is measured for this polymer pair] does not imply that the kinetics of the mixing process need not be considered. It is clear that the interdiffusion of large polymer molecules at temperatures not far above the melt temperature of one of the components, as is the case for the physically mixed powder of PPO and PS in the Stoelting study, is an exceedingly slow process. Bohn [45] has cited studies indicating that in the melt state, diffusion constants on the order of 10^{-12} to 10^{-13} cm^2/sec are probable; these values are smaller by three or four orders of magnitude than for the diffusion of high viscosity plasticizers in PVC, for example. A nonequilibrium thermodynamic state as a result of the kinetic restrictions on the mixing process may therefore be the explanation for the small, mixed-composition domains evident from the dynamic mechanical studies of Stoelting *et al.* In this study, the DSC technique and, as subsequently shown (Section IV.A.3), the dielectric method may not be able to resolve two phases whose T_g's are close or to detect domains smaller than a critical dimension (Section VI.A). The mixing techniques of Shultz and Beach and of Yee allowed greater opportunity for an equilibrium state to be reached by effectively reducing viscosity through coprecipitation from dilute solution in the former case and by greater residence time and good mechanical mixing in the latter case.

3. Dielectric Method

The dielectric properties of unblended PPO have been investigated by Karasz *et al.* [121]. As in the dynamic mechanical spectrum, PPO is characterized by a dielectric α peak denoting the glass transition and a low-temperature γ peak presumably attributable to a localized vibrational or librational mode of a dipolar moiety. The temperatures at the maxima of the α and γ dielectric peaks are comparable to those observed in the dynamic mechanical spectrum when compared at equivalent frequencies. The polymer PPO is only weakly polar, the only dipolar contribution arising from the cosine projection of the two *ortho*-methylphenylene dipoles along the main chain axis. For this reason, the intensities of both the α and γ peaks are weak and, unlike the dynamic mechanical case, both peaks are nearly equal in intensity. Reed [118] has shown that the dielectric properties of PPO may be strongly influenced by the presence of ionic impurities in the form of amines, both free and incorporated in the chain, attributable to the amine catalyst used in the polymerization of PPO. Activation energies have been given as 150 kcal/mole for the α relaxation and 8.7 kcal/mole for the γ relaxation.

The dielectric properties of PPO–PS blends have been described by MacKnight *et al.* [122]. These blends were prepared by the same technique of compression molding physically mixed powders of PPO and PS as used in their dynamic mechanical studies [101] with similar thermal treatment. Unlike the dynamic mechanical results, each blend composition displayed a single, broadened α relaxation with no evidence of shoulders indicative of

PPO-rich or PS-rich phases. The temperatures corresponding to the maxima of the dielectric α peaks of the PPO–PS blends are plotted versus PPO content (wt % PPO) in Fig. 4. There is a pronounced increase in T_g above 50% PPO to give a nearly sigmoidal relationship between T_g and composition. The authors noted that if only the stronger of the overlapping peaks in the dynamic mechanical data are considered, then a similar sigmoidal curve would result. Shultz and Beach [119] in studying blends of PPO and random copolymers of styrene and *p*-chlorostyrene (e.g., Section V) which are at the verge of incompatibility have found a similar sigmoidal relationship of T_g by TOA to blend composition. This comparison is interesting because one might expect similar mixed-composition domains in the PPO–copolymer blends that are near thermodynamic incompatibility as in the thermodynamically compatible PPO–PS blends that probably are only partially mixed due to the blending procedure.

4. Comparison among the Different Methods

As shown previously (Section IV.A.1), DSC and the related techniques of DTA, TOA, and TMS indicate that the form of the T_g–composition curve for PPO–PS is concave and can be fitted reasonably well by the simple Fox equation. Differences in the conclusions reached in the dynamic mechanical studies of Stoelting *et al.* and those of Shultz and Beach, Yee, and Wellinghoff and Baer may be the result of incomplete mixing during blending in the former case. An interesting result of the Stoelting studies was that under the same mixing conditions and thermal history of the PPO–PS blends, DSC and dielectric techniques [122] indicated homogeneity, whereas the dynamic mechanical measurements detected the presence of two phases of different PPO and PS compositions. In addition, the relation of blend T_g to composition differed among the three techniques. These observations would suggest that different techniques of T_g measurement may be sensitive to different domain sizes and that dynamic mechanical measurements may be the most sensitive to small phases. Additional evidence for varying effective "probe" sizes of different T_g techniques and some conclusions are offered in Section VI.C.

B. Crystalline Melt Temperature

1. Morphology of Crystalline Isotactic Polystyrene

a. Small-Angle x-Ray Scattering (SAXS)

The morphology of crystalline isotactic polystyrene (iPS) has been investigated by a variety of techniques including light scattering, x-ray scattering, optical microscopy, and electron microscopy. The results of these investigations have led to the conclusion that iPS normally crystallizes as stacks of folded chain lamellae, which are arranged in a volume-filling spherulitic superstructure. In this section we shall focus on information provided from small angle x-ray scattering studies (SAXS). As is well known, the interpretation of SAXS data rests on the assumption of a model that accurately reflects the structural elements of which the polymer super-structure is composed. In the case of lamellar semicrystalline polymers, the models most frequently employed are those of Tsvankin and Buchanan [123], Kortleve [124], and Hosemann [125]. All three models are based on a stack of parallel alternating layers of crystalline and amorphous regions. As it is not our intention to review here methods of analysis of SAXS, the details of these models will not be discussed. Suffice it to say that it has been shown that the Hosemann model is the only one that provides a reasonable fit to the SAXS curves from crystalline iPS and yields values for the degree of crystallinity that agree with those obtained from other techniques, such as thermal analysis [56, 126]. The principle parameters obtained from the Hosemann analysis are the average thickness of the amorphous layer ($\bar{X}_a$), the average thickness of the crystal lamellae ($\bar{X}_c$), the linear fractional crystallinity (ϕ_c), and an order parameter (N). The parameter N refers to the number of lamellae oriented parallel to each other. However, there is very likely to be considerable lamellar branching within a fibril in iPS, and this phenomenon would have the effect of decreasing the order in a stack of lamellae without decreasing the number of lamellae within the stack [126]. Table II is a collection of these parameters for iPS samples crystallized at three different temperatures. The results are in accord with the generally observed trends in other semicrystalline polymers. As the crystallization temperature increases, so does the melting point and the lamellar thickness. The parameter N also shows a slight increase with increasing crystallization temperature, perhaps indicating better stacking order at the higher

Table II

SAXS Structural Parameters from the Hosemann Model for
iPS Crystallized at Different Temperatures

Crystallization temperature (°C)	Melting point[a] (°C)	ϕ_c	$\bar{X}_c$ (Å)	$\bar{X}_a$ (Å)	N
170	220	0.32	45	95	2
200	223	0.30	48	112	2.2
210	229	0.30	52	120	2.4

[a] Determined by differential scanning calorimetry (DSC).

temperatures. It is to be noted that ϕ_c remains essentially unchanged at all three crystallization temperatures.

b. Melting Point (T_m)

The melting behavior of iPS has been the subject of numerous investigations. Adiabatic calorimetry reveals an equilibrium T_m of 240°C [98]. In contrast, differential scanning calorimetry (DSC) measurements typically result in multiple melting peaks whose magnitudes and temperatures are dependent on crystallization conditions [127]. Thus, when samples of iPS that have been crystallized at 180°C or lower are scanned at 10° per minute, three melting endotherms appear [53]. It appears that the onset of the melting process increases linearly with crystallization temperature (T_c) and it lies approximately 4° or 5° above T_c. The temperature at which the melting process ends is essentially independent of T_c. If T_c is above 180°, only two peaks are discernible, and if T_c is in excess of 200°C, only one peak is seen. This behavior is summarized in Figs. 5 and 6. In general, there are two explanations of this phenomenon. One is that the multiple peaks reflect the presence of different crystal structures [128], and the other is that the

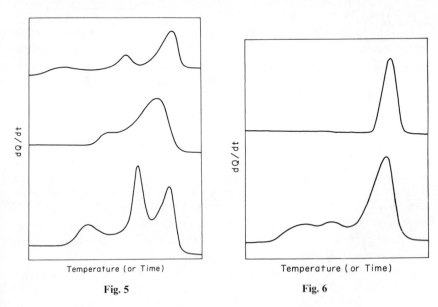

Temperature (or Time)   Temperature (or Time)

Fig. 5   Fig. 6

Fig. 5   DSC thermograms of iPS annealed at 150°C (upper), 190° (middle) and 170°C (lower) for 24 hr. (From Neira-Lemos [53].)

Fig. 6   DSC thermograms of iPS annealed at 210°C (upper) and 180°C (lower). (From Niera-Lemos [53].)

peaks are a consequence of reorganization occurring during the heating process, perhaps involving melting and recrystallization or merely processes such as lamellar thickening. The low angle x-ray data cited in the preceding section are consistent with the reorganization hypothesis since the lamellar thickness increases with increasing T_c.

c. Solvent versus Thermal Crystallization

The solvent induced crystallization of iPS appears to result from a lowering of both the T_g and T_m of the iPS so that crystal nucleation and growth can occur. In a study utilizing both acetone and methyl ethyl ketone (MEK), it was found that the solvents exerted no significant effect on the final crystal morphology or the crystallization kinetics [53]. As a result, the solvent treatment may be regarded simply as a device to lower T_c in the case of iPS. As will be discussed subsequently, the effect of solvents on PPO is much more complex.

2. Morphology of Crystalline PPO

a. Small Angle x-Ray Scattering (SAXS)

Bulk PPO cannot be thermally crystallized but crystallizes fairly readily in the presence of solvents. Horikiri and Kodera [129] have reported the growth of spherulitic superstructure in PPO when thin films are exposed to α-pinene or tetralin. It was noted by these workers that spherulites are not formed when the crystallization is carried out in tetralin vapor at temperatures greater than 50°C. This work demonstrates the dependence of the morphology as well as the crystal structure of PPO on the nature of the solvent and the conditions of crystallization. The SAXS of PPO films crystallized by exposure to MEK vapor at 75°C is discussed in this section and is compared with blend studies below. This method of crystallization results in films which, when dried, show no evidence of crazing. This is in contrast to the behavior observed when the solvent used is acetone, for example [53]. Figure 7 is a display of the SAXS curve for PPO [55]. It may be seen that the Lorentz corrected curve in Fig. 7 exhibits two maxima. The more prominent of the two occurring at the lower angle corresponds to a Bragg spacing of 160 Å, and will be referred to as the long period. The second maximum is presumably a second order of the long period maximum although its angular position is somewhat less than double that of the long period. The overall shape of the Lorentz corrected SAXS curve is very reminiscent of those of many other semicrystalline polymers having highly ordered superstructures. The same one—dimensional paracrystalline model used in the analysis of the iPS SAXS was employed to interpret the PPO results. The

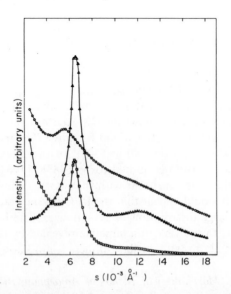

Fig. 7 SAXS curve of PPO (crystallized by exposure to MEK vapor). (○) measured intensity; (□) desmeared intensity; (△) Lorentz-corrected intensity ($l \times s^2$). (From Wenig *et al.* [55].)

lamellae were found to be 38 Å in thickness, while the amorphous layers were 114 Å in thickness.

b. Wide-Angle x-Ray Scattering (WAXS)

Figure 8 shows the x-ray intensity versus Bragg angle [55] for the crystalline PPO sample discussed above. The crystalline diffraction peaks are quite broad and ill defined. The scattering pattern is similar to that obtained by

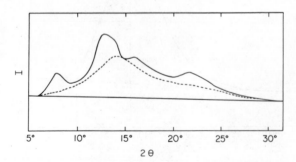

Fig. 8 WAXS curve of PPO (crystallized by exposure to MEK vapor). (From Wenig *et al.* [55].)

Barrales-Rienda and Fatou [130]. It is probable that solvent removal introduces lattice defects which result in a very imperfect crystal structure.

c. Melting Point Temperature

It has been reported [106] that the ratio of (T_g/T_m) in PPO is about 0.9, which is much higher than the "normal" values of 0.5–0.67 found in most other polymers. Figure 9 shows the DSC thermogram of the crystalline PPO discussed in Sections IV.B.2. a and b, above. The melting endotherm is rather broad, extending from 490 to 530°K, with a maximum at 513°K. The equilibrium T_m of PPO is quoted to be 533°K [106]. The breadth of the melting endotherm is consistent with the poorly defined WAXS profile described above. It would appear that the PPO crystals are quite small and imperfect with a large distribution of crystal sizes. The high (T_g/T_m) ratio may be due in part to the very thin lamellae revealed by the SAXS studies. The reason for the abnormal lamellar thinness is not known.

3. Morphology of Blends of Amorphous PPO and Crystalline Isotactic Polystyrene

a. Small-Angle x-Ray Scattering (SAXS)

Since iPS can be crystallized thermally and PPO cannot, it is possible to prepare blends of iPS and PPO that contain a crystalline iPS phase and an amorphous mixed iPS–PPO phase. Table III is a summary of the SAXS results for blends crystallized by thermal annealing at 170°C for 24 hr. It is seen that the crystal lamellar thickness decreases with increasing PPO

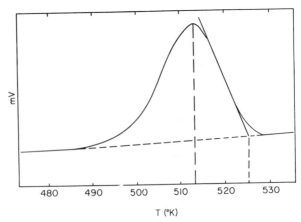

Fig. 9 DSC thermogram of PPO (crystallized by exposure to MEK vapor). (From Wenig *et al.* [55].)

Table III

SAXS Structural Parameters from the Hosemann Model
for iPS–PPO Blends Crystallized at 170°C

Blend composition (iPS/PPO) (%)	$T_m(°C)^a$	ϕ_c	$\bar{X}_c(\text{Å})$	$\bar{X}_a(\text{Å})$	N
100/0	220	0.32	45	95	2
90/10	216	0.31	41	89	1.7
80/20	212	0.20	28	112	1.4
70/30	209	0.18	21	124	1.2

^{a} Measured by DSC.

content, while the amorphous layer thickness increases. The Hosemann correlation parameter N also decreases with increasing PPO content. It appears that the mixed amorphous phase is concentrated between the lamellae of the spherulites. This conclusion was also reached by Berghmans and Overbergh on the basis of spherulitic growth rate studies [131].

b. Wide-Angle x-Ray Scattering (WAXS)

Figure 10 illustrates the WAXS profile for blends of iPS–PPO thermally crystallized at 170°C for 24 hr [56]. It may be seen that the crystal diffraction

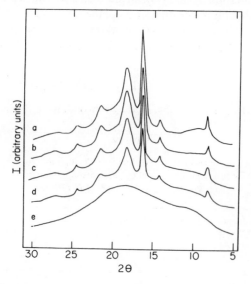

Fig. 10 WAXS curves of blends of iPS–PPO thermally crystallized at 170°C for 24 hr: (a) 100/0; (b) 90/10; (c) 80/20; (d) 70/30; (e) 65/35. (From Wenig *et al.* [55].)

peaks decrease in intensity with increasing PPO content until they become extremely ill defined in the 65/35 iPS/PPO blend. The degrees of crystallinity of the blends may be obtained from the plot shown by the following relationship:

$$X_{cr} = \frac{(\text{total area}) - (\text{amorphous area})}{(\text{total area})} \tag{11}$$

Results from such calculations are compared with the linear crystallinities obtained from SAXS in Table IV. The agreement is excellent, indicating that the model used to interpret the SAXS data is physically reasonable. It should also be noted that no thermal crystallization of iPS can be accomplished under any conditions when the blends contain less than about 60% by weight of iPS.

Line broadening analysis was carried out on the 110, 300, 220, and 211 iPS crystal diffraction peaks in order to obtain some idea of the dependence of crystallite size on blend composition. The 211 and 300 peaks were analyzed by the Scherrer [132] formula

$$I_{hkl}(\text{Å}) = K(\lambda/\beta)\cos\theta_0 \tag{12}$$

where K is a proportionality constant ($K = 1$ in this study), $\lambda = 1.45$ Å for Cu $K\alpha$ radiation, β is the integral peak breadth, and θ_0 is the scattering angle of the peak maximum.

The Scherrer relation gives a mean diameter for the crystallites in the direction perpendicular to the diffraction plane. The 110 and 220 peaks were analyzed by the Hosemann [133] method in an attempt to separate broadening caused by small crystallites from broadening caused by para-crystalline effects:

$$\beta_0{}^2 = \beta_c{}^2 + \beta_{II}^2 = (1/I_{hkl}^2) + [(\pi g_{II})^4/d_{hkl}^2] m^4 \tag{13}$$

where β_0, β_c, and β_{II} are the observed line broadening, broadening due to

Table IV

Comparison of Weight Fraction Crystallinity X_{cr}
with Linear Fractional Crystallinity ϕ_c Obtained
from WAXS and SAXS of iPS–PPO Blends

Blend composition (iPS–PPO) (%)	ϕ_c	X_{cr}
100/0	0.32	0.32
90/10	0.31	0.29
80/20	0.20	0.23
70/30	0.18	0.19

Table V

Crystallite Sizes and Paracrystallinity Parameters for Thermally Crystallized iPS–PPO Blends

Blend composition (iPS/PPO) (%)	I_{110} (Å)^a	g_{II} (Å)^a	I_{300} (Å)^b	I_{211} (Å)^b
100/0	542 ± 54	3.2 ± 0.1	300 ± 30	84 ± 8
90/10	408 ± 34	3.3 ± 0.1	298 ± 30	78 ± 8
80/20	447 ± 41	2.9 ± 0.1	314 ± 30	82 ± 8
70/30	542 ± 66	2.9 ± 0.1	349 ± 35	82 ± 8

^{a} By the Hosemann method, Eq. (13).
^{b} By the Scherrer method, Eq. (12).

crystallite size, and broadening caused by paracrystalline distortions of the second kind, respectively; g_{II} is the paracrystalline parameter, d is the Bragg spacing in angstroms, and m is the order of reflection.

Table V is a summary of the results of these calculations. It appears that crystallization of iPS from blends has no appreciable effect on the crystallite dimensions, in contrast to the considerable decrease in lamellar thickness that occurs when PPO content increases.

c. T_m Depression

A reasonable conclusion from the SAXS and WAXS studies of iPS and iPS–PPO blends is that the PPO component of the mixed amorphous phase affects the crystallization of the iPS in the same manner as crystallization of pure iPS at lower temperatures. That is, the PPO increases the viscosity of the amorphous phase as is manifested from its effect on T_g, and it would thus be necessary to crystallize the iPS in the blend at a higher temperature than pure iPS in order to achieve the same amorphous viscosity, the same growth rate, and the same lamellar thickness. The observed melting point depression is thus accounted for on a kinetic basis related to the thinning of the lamellae. It follows that the thermodynamic melting point (equilibrium melting point) T_m^0 is the same for both the blends and pure iPS, that is, 240°C. This conclusion is also reached by Berghmans and Overbergh [131]. From the discussion in Section I.B.2 above, and in the light of Eq. (5), it is clear that this result implies an interaction parameter χ between iPS and PPO of near zero. A direct confirmation of this result must again await measurement of the enthalpy of mixing of these blends.

4. Morphology of Blends of Crystalline PPO and Crystalline iPS

As already discussed, two crystalline phases and a mixed amorphous phase may be obtained by a combined solvent–thermal treatment of iPS–PPO blends [53]. A pseudo-phase diagram may be constructed for such a system and this is shown in Fig. 11. At temperatures below the T_g line, two crystalline phases and a mixed glassy amorphous phase exist at compositions between 0.2 and 0.5 weight fraction PPO. At less than 0.2 weight fraction PPO only a crystalline iPS phase and a mixed amorphous phase exist. At compositions greater than 0.5 weight fraction PPO, only a crystalline PPO phase and an amorphous mixed phase exist. At temperatures above the T_g line, the mixed amorphous phase is in a rubbery or liquidlike state, and above the melting lines there is only a simple, mixed liquidlike phase. The melting point depression of the PPO observed in Fig. 9 is presumed to arise from kinetic factors similar to those discussed for iPS above. Final proof of this proposition must await further experimental work.

V. SOLID STATE TRANSITION BEHAVIOR OF PPO–POLYCHLOROSTYRENE BLENDS

A. Compatibility versus Incompatibility: Chlorostyrene–Styrene Copolymers

In the previous sections, evidence of the compatibility of PPO and PS was presented from DSC, dynamic mechanical, and dielectric studies. Although PPO is compatible with PS, it is incompatible with chlorine substituted polystyrenes such as poly(p-chlorostyrene) (PpClS) [15, 70, 119, 134] and poly(o-chlorostyrene) (PoClS) [15]. Films made from blends of PPO and either PpClS or PoClS were reported to be opaque, and two T_g's, corresponding in temperature to those of the unblended component polymers, were evident from DSC, TOA, dynamic mechanical, and dielectric measurements. Transmission electron microscopy of fractured film surfaces of PPO–PpClS blends revealed macrophase separation, that is, dispersed domains in the order of 2 μm in size [70]. In agreement with results for other incompatible blends, the measured densities of PPO–PpClS and PPO–PoClS blends were equal to densities calculated from simple additivity of weighted specific volumes (e.g., Section V.D.2), that is, no excess volume of mixing [15].

In the same studies cited above, PPO was found to be either compatible or incompatible with random copolymers of styrene and p-chlorostyrene (pClS), depending only on the composition of the copolymer. Shultz and Beach [119] have shown that copolymers with $\leqslant 65.3$ mole % pClS were compatible with PPO, while those $\geqslant 68$ mole % pClS were not. Thus it was found that blends of PPO and p-chlorostyrene–styrene copolymers

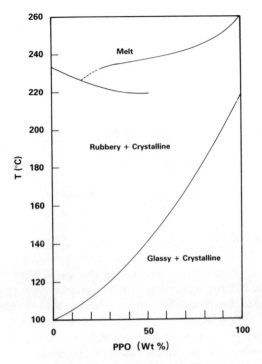

Fig. 11 Phase diagram for crystallized blends of PPO and iPS. (From Niera-Lemos [53].)

exhibited a compatibility–incompatibility transition between copolymer compositions of 65.3 and 68 mole % pClS. More recent studies by Fried [15] have indicated that this transition region may be even narrower, that is, between 67.1 and 67.8 mole % pClS (e.g., Section V.D.1).

B. Glass Transition Temperature

1. Differential Scanning Calorimetry

The compatibility of PPO and pClS copolymers of various compositions has been studied by DSC by Shultz and Beach [119], Tkacik [70] and Fried [15]. Tkacik showed that copolymers with compositions of 47 or 60 mole % pClS in blends of 25, 50 or 75 wt % PPO exhibited single, composition-dependent T_g's. By contrast, blends of PPO and a copolymer composed of 68 mole % pClS displayed two glass transitions corresponding in temperatures with those of the unblended component polymers although there was evidence for some elevation of the low T_g and a lowering of the high T_g.

In more extensive studies by Shultz and Beach covering a wider range of copolymer and blend compositions, single, composition-dependent T_g's were reported for all PPO–copolymer blends, provided copolymer composition was $\leqslant 65.3$ mole % pClS. They also reported that the minor component by weight in the incompatible blends was often difficult to detect or was not apparent at all. For example, for the incompatible blends whose compositions were $\leqslant 37.5$ wt % PPO, the glass transition of only the copolymer was evident. For one copolymer composition of 69.5 mole % pClS (near the compatibility–incompatibility region), blends with > 50 wt % PPO exhibited only the PPO transition. These observations were tentatively attributed to a lack of sensitivity of the DSC measurements. In addition, Shultz and Beach observed a similar elevation of the lower T_g (copolymer-rich phase) and a lowering of the upper T_g (PPO-rich phase) to that noted by Tkacik, particularly for blends with copolymer compositions within the compatibility–incompatibility region between 66.6 and 68.0 mole % pClS; however, specific trends in the T_g shifting could not be defined.

Using somewhat improved DSC techniques, Fried [15] has determined that the T_g of the PPO-rich phase in such transitional copolymer blends may be lowered by as much as 20°C (e.g., Section V.D.1). This reduced T_g corresponds to a phase containing about 14 wt % copolymer. Although the lower T_g of this PPO-rich phase was unaffected by blend composition, the T_g of the copolymer-rich phase increased linearly with overall PPO composition of the blend. These differences in phase T_g behavior were explained in terms of a shifting of the miscibility gap toward the lower molecular weight (PPO) component as discussed by Koningsveld [135, 136]. In addition, there was a significant reduction in ΔC_p of both transitions in the incompatible blends especially those with copolymers in the compatibility–incompatibility region. Only about 60% of the total expected ΔC_p could be detected in such blends even when larger samples and corrections for baseline linearity were employed for maximum sensitivity. Such results could be explained only on the basis of extensive interfacial mixing (see Section III.B).

The T_g-composition dependence of all the compatible PPO–PS and PPO–copolymer blends followed the same concave relation (see Section IV.A.1) which could be approximated by the Fox expression given by Eq. (2). Fried has shown that an even better fit may be obtained by using the generalized Wood [137] relationship given as

$$T_{g,b} = [T_{g,1} + (kT_{g,2} - T_{g,1})w_2]/[1 + (k-1)w_2] \qquad (14)$$

where k is an empirical constant and $T_{g,1}$ represents the T_g of PS or one of the compatible copolymers, that is, those with $\leqslant 67.1$ mole % pClS. For these blends, k was determined to be 0.679.

2. Dynamic Mechanical Method

In addition to DSC measurements, PPO–copolymer blends have been studied by dynamic mechanical techniques [70, 119]. Shultz and Beach used the Rheovibron to determine the dynamic tensile moduli and tan δ of films compression molded from coprecipitated blends. In the case of the incompatible blends, only the tan δ peak of the copolymer (or PpClS) unchanged by blending was evident; the tan δ of the PPO phase could not be reached as a result of sample softening near the temperature it would be expected to appear (>250°C). In comparison, the compatible PPO–copolymer blends displayed a single, temperature-shifted tan δ. Using similar techniques but employing the Rheovibron in the shear mode geometry, Tkacik [70] observed shoulders in the tan δ peaks of compatible PPO–copolymer blends reminiscent of the dynamic mechanical results of Stoelting *et al.* [101] for PPO–PS (e.g., Section IV.A.2). Shultz and Beach reported a shoulder in tan δ only in the case of one blend of PPO and a copolymer of 69.5 mole % pClS in the region of the compatibility–incompatibility transition. The shoulders observed by Tkacik were attributed to the presence of copolymer-rich phases whose compositions were calculated from Eq. (10) as being richer in copolymer composition than the corresponding PS-rich phases of the PPO–PS blends were rich in PS. There was no evidence of corresponding PPO-rich phases in these studies.

The only investigation of the dynamic mechanical T_g–composition dependence of the compatible PPO–copolymer blends was that of Tkacik [70]. He showed that the temperature at the maximum of G'' for the main α relaxation (excluding shoulders) followed a nearly linear dependence on blend composition. Similar results appear in the data of Shultz and Beach for the PPO–PS blends in which the temperature at the maximum of tan δ (tensile modulus measurements) appeared to be nearly linearly dependent on blend composition (see Fig. 4).

3. Dielectric Method

The dielectric relaxations of blends of PPO and the pClS copolymers have been investigated by Tkacik [70] and Karasz *et al.* [134]. As in the DSC and dynamic mechanical studies, blends of PPO and copolymers with 47 and 60 mole % pClS were compatible as indicated by single, although broadened, ε'' peaks, whereas blends of PPO and PpClS or a copolymer with 68 mole % pClS displayed sharp ε'' peaks corresponding in temperature to those of unblended PpClS or the copolymer. The ε'' peak of PPO was not evident in the incompatible blends because in comparison to the dielectric strength of the polar pClS molecules, the dielectric loss of PPO was approximately three orders of magnitude lower in intensity and could not be detected.

The temperatures corresponding to the maximum of the ε'' peaks of the compatible PPO–copolymer blends displayed a nearly linear dependence on blend PPO composition as did the dynamic mechanical G'' peaks cited above.

4. Comparison among the Different Techniques

As indicated by the preceding DSC, dynamic mechanical, and dielectric studies, copolymers of styrene and pClS are compatible with PPO, provided the composition of the copolymer is $\leqslant 67.1$ mole % pClS. The T_g–composition relation for the compatible PPO–copolymer blends by DSC was found to be identical to that of PPO–PS. Thermal optical analysis studies by Shultz and Beach [119] indicated a similar dependence of T_g on composition of the PPO–copolymer blends, but the form of the curve changed from a concave to a sigmoidal form for copolymer compositions near the compatibility–incompatibility transition. As previously noted (e.g., Section IV.A.3), a similar sigmoid relation was observed by MacKnight et al. [122] in the dependence of the temperature maximum of the dielectric loss for PPO–PS blends compression molded from physical mixtures of PPO and PS powders (see Fig. 4). This may indicate that TOA may be more sensitive than DSC in detecting the presence of small phases in blends of PPO and pClS copolymers near thermodynamic incompatibility, a situation thus analogous to the case of poorly mixed PPO–PS blends in which dynamic mechanical and dielectric techniques appeared to be more sensitive (e.g., Section IV.A).

The dielectric studies of Tkacik indicated that PPO and copolymers with $\leqslant 60$ mole % pClS form compatible blends, although the dynamic mechanical results for the same blends indicate the existence of a second highly copolymer-rich phase evident as a shoulder in the low-temperature side of the main peak. No such shoulders were apparent at equivalent blend compositions in the study of Shultz and Beach. In both cases, coprecipitation was the technique employed for blending and the only difference in the dynamic mechanical measurements was the choice of the tensile mode (at 110 Hz) by Shultz and Beach and the shear mode (at 3.5 Hz) by Tkacik. Reasons for the apparently microheterogeneous morphology in the latter study are as yet unclear.

C. Analysis of Dielectric Results

1. Width of Transition

The dielectric studies of Tkacik [70] and Karasz et al. [134] have indicated that blends of PPO and copolymers of styrene and pClS are compatible at $\leqslant 60$ mole % pClS composition of the copolymer. An observation made in that study was that the width of the ε'' peak was considerably broader for the compatible blends than for the copolymer ε'' peak in the incompatible blends. For example, at 50 wt % PPO the peak width at the half-height of the ε'' peak of a compatible blend of a copolymer containing 47 mole % pClS was 41°C compared with the copolymer peak width of 25°C in an incompatible blend of PPO and a copolymer containing 68 mole % pClS.

In a recent dielectric study of several of the copolymer blends studied by DSC by Fried, Wetton *et al.* [82] have shown that the greater widths of the dielectric peaks of the compatible blends can be attributed to large local concentration fluctuations of PPO and copolymer molecules within the blend. The range of apparent concentrations at a given blend composition could be approximated by fitting a Fuoss–Kirkwood distribution of relaxation times to the normalized dielectric curves ($\varepsilon''/\varepsilon''_{\text{max}}$ versus f/f_{max}) of the compatible blends. By this means, it was shown that in the case of a blend of PPO and a copolymer of 58.5 mole % pClS at an overall blend composition of 60 wt % PPO, the local concentrations of PPO in the blend averaged between 40 and 80 wt % as a conservative estimate. Such broadening of the glass transition has been noted in other compatible blends and for different methods of T_g measurements (e.g., Section I.B.1). Evidence for the broadening of the DSC transition for the PPO–copolymer blends at copolymer compositions on the edge of the compatibility–incompatibility transition is presented in Section V.D.1.

2. *g Parameter*

Tkacik [70] applied his dielectric data to the Fröhlich [138] equation, which relates a dipole orientation function g to the limiting dielectric constants at high and low frequencies, ε_U and ε_R respectively. This equation is

$$g = [9kT(2\varepsilon_R + \varepsilon_U)(\varepsilon_R - \varepsilon_U)]/[4\pi N\mu_0^2\varepsilon_R(\varepsilon_U + 2)^2] \tag{15}$$

where k is Boltzmann's constant, T absolute temperature, N number of dipoles per cubic centimeter, μ_0 dipole moment of an isolated dipole unit, ε_R relaxed (low frequency) dielectric constant, and ε_U unrelaxed (high frequency) dielectric constant.

In the absence of orientational correlation between molecules, g is unity. For most macromolecules, g is less than unity because of chain configuration and hindered rotation. As the effective dipole concentration is decreased, as it is by dilution with a low molecular weight solvent, g increases. The dipole orientation function also increases if the effective dipole concentration is reduced by copolymerization of the polar entity with a nonpolar one. For example, Mikhailov *et al.* [139] have observed an increase in g for copolymerization of pClS with styrene from dielectric measurements in solution. Tkacik showed that, while *g* was independent of blend composition for the incompatible blends, *g* increased with increasing PPO composition for the compatible blends from a value of about 0.40 to 0.80. This trend is illustrated in Fig. 12 for two compatible polymer blends. These findings were interpreted as indicating that mixing in the compatible PPO–copolymer blends was as intimate as if the copolymer was dissolved in a low molecular weight solvent. In the case of the blends, PPO could be viewed as the solvent, and therefore *g* increased with increasing dilution, that is, increasing PPO composition.

D. Compatibility–Incompatibility Transition

1. Differential Scanning Calorimetry T_g Study

In his DSC study, Fried [15, 140] has shown that copolymers with ≤67.1 mole % *p*ClS are compatible with PPO. At each blend composition of 20, 40, 60, and 80 wt % PPO, a single, composition-dependent T_g was

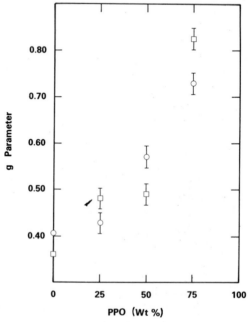

Fig. 12   Fröhlich *g* parameter versus composition for two compatible blends of PPO and copolymers of styrene and *p*-chlorostyrene. Circles and squares represent *g* values calculated for blends with copolymers containing 47 and 60 mole % *p*-chlorostyrene, respectively. (Reproduced with permission from Tkacik [70].)

observed by DSC. For copolymers with compositions of 68.6 and 67.8 mole %
pClS near the compatible compositions of 67.1 mole % pClS, the blends
exhibited two transitions characteristic of an incompatible blend at all
blend compositions except those for which the minor component was only
20 wt % of the total sample weight, in which case only the major component
T_g could be detected. These results indicate that the transition from
compatibility to incompatibility in blends of PPO and pClS copolymers
occurs in a very narrow range of copolymer compositions between 67.1 and
67.8 mole % pClS.

Several trends in the glass transition behavior of the blends were evident
as different copolymers were blended to pass through the incompatibility–
compatibility range. The first was that for the incompatible blends at the
edge of the transition region (copolymers with 68.6 and 67.8 mole % pClS),
the temperatures of the two transitions were shifted from those of the
corresponding unblended components (e.g., Section V.B.1). The trends in the
T_g–composition behavior are illustrated in Fig. 13 for four representative
blends. For the two compatible blends, PPO–PS and PPO–copolymer C
(67.1 mole % pClS), a single T_g was exhibited and followed a concave
dependence on composition that could be fitted by Eq. (14). For the incompatible blend of PPO–PpClS, two T_g's corresponding exactly in temperature with those of unblended PPO and PpClS were found as indicated
by the two horizontal lines. For an incompatible blend of PPO and copolymer E (68.6 mole % pClS) at the edge of compatibility, the T_g of the

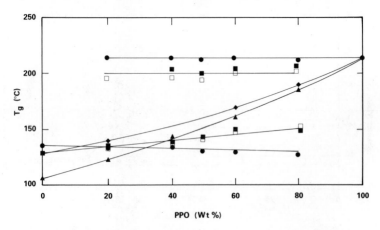

Fig. 13   Glass transition temperature (T_g) as determined by DSC for blends of PPO with
polystyrene, two copolymers of styrene and p-chlorostyrene, and poly(p-chlorostyrene) (PpClS):
Open symbols are for samples annealed at 20° below T_g: (▲) PS; (♦) Copolymer C; (■)
Copolymer E; (●) PpClS.

PPO phase was depressed, whereas the T_g of the apparent copolymer phase linearly increased with increasing PPO blend composition from that of the unblended copolymer. This result suggests that, near the edge of compatibility, the dispersed phase and matrix are no longer pure in each component, that is, a degree of mutual mixing occurs between components in each phase. The T_g of the copolymer-rich phase is elevated by the presence of PPO, and that of the PPO-rich phase is depressed as a result of some mixing with the lower T_g copolymer component. Almost identical results were obtained for blends of PPO and a copolymer with slightly less pClS content (67.8 mole % pClS). Annealing these edge of compatibility blends 20° below T_g had little effect on the temperature assignment of these transitions.

A second observation concerning the glass transitions of blends near the edge of compatibility was a significant increase in the width of the transition as the compatibility–incompatibility region is approached from either side, as illustrated in Fig. 14. In this case, the width of the DSC glass transition was defined as the difference between temperatures at the intersections of the discontinuity in C_p with the linear portions of the baseline above and below T_g. For the incompatible blends, an average of the two transitions was used for comparison with the compatible blends.

This observed trend in the transition width can be related to the extent of blend heterogeneity. The validity of this argument is supported by the early work of Nielsen [32] who showed that the width of the logarithmic decrement

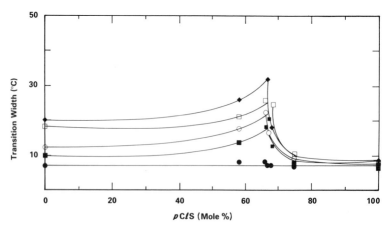

Fig. 14   Width of the glass transition from DSC thermograms of blends of PPO and styrene–p-chlorostyrene copolymers as a function of copolymer composition and blend composition: (●) 0 wt % PPO; (■) 20 wt % PPO; (○) 40 wt % PPO; (□) 60 wt % PPO; (♦) 80 wt % PPO. (Reproduced with permission from Fried [15].)

in the dynamic mechanical spectra of several vinyl copolymers of broad distributions in composition increased with copolymer heterogeneity. In addition, similar broadening of the loss peak in the dynamic mechanical spectrum of plasticized polymers has been observed [141]. For example, the width of the damping peak of PVC was observed to increase when plasticizers of decreasing solvating power were used. Buchdahl and Nielsen [31] have attributed such broadening in the dynamic mechanical spectrum to fluctuations in the interaction of chain segments with their nearest neighbors which are in excess of normal thermal fluctuations.

As shown by the analysis of dielectric broadening in PPO–copolymer blends [82] (Section V.C.1), compatible polymer blends experience local fluctuations in composition in addition to the thermal and density fluctuations experienced by molecularly homogeneous media. In the case of compatible PPO–copolymer blends, the local fluctuations in composition become larger and the glass transition broadens for copolymers with higher pClS composition until phase separation occurs. At this point, as indicated by the shifting of the T_g's, the phase separated blends have a matrix and dispersed phase that are not pure in each component. There is some PPO dissolved in the copolymer-rich phase and some copolymer in the PPO-rich phase. This means that there are compositional fluctuations in each phase, and therefore the individual transitions are broadened. As the pClS composition of the copolymer is raised (above 68.6 mole % pClS), the PPO and copolymer components reach a more extreme state of incompatibility (with probably a higher interaction parameter). The individual phases are then pure in each component as evidenced by the observed equivalency of the individual phase T_g's with the corresponding unblended components and therefore composition fluctuations are minimized, that is, the width of each glass transition begins to narrow. Similar trends in the broadening of the dynamic mechanical spectrum (torsion pendulum) of blends have been observed by Robeson *et al.* [29] for incompatible blends of PS and poly(α-methylstyrene). Decreasing the molecular weight (i.e., lowering the interaction parameter) resulted in broader spectra. Similarly, block copolymers of PS and poly(α-methyl styrene) exhibited a single loss peak, which broadened as the molecular weight of the blocks was increased (i.e., increasing the interaction parameter).

Broadening of the transitions in the incompatible PPO–copolymer blends is more pronounced for the minor components. Similar results were noted by Schurer *et al.* [92] for blends of isotactic PMMA and PVC for which some T_g shifting was also observed. A probable explanation is that the minor component (dispersed phase) is surrounded by an interfacial mixed-composition zone (Section III.B) which for small phases becomes appreciable. The effective probe size (see Section VI.C) of the measurement technique is responsive to both the molecular composition of the inner phase and the diffuse interfacial layer between the matrix and dispersed phase, leading to an apparent broadened transition. In addition, as copolymers with compositions approaching the incompatibility–compatibility region are blended, the interfacial region enlarges as suggested from the results of the other studies [93, 94, 97] cited in Section III.B. Due to the concentration gradient in the interfacial region, a continuous distribution of T_g's would result and, therefore, be undetected. The dispersed phase and matrix as a result of depleted material involved in the interfacial region would appear smaller in magnitude. From quantitative measurement of the DSC transition height, Fried [15] showed that the magnitude of the transitions of the incompatible blends decreased with decreasing pClS composition of the copolymer, that is, toward increasing compatibility. At the edge of compatibility, as much as 40% of the total expected transition heights were undetected.

2. Density Behavior

The density of a polymer blend (ρ_b) may be calculated from the densities of the components by assuming a two-phase model [17, 34] or an ideal solution for which

$$1/\rho_b = w_1/\rho_1 + w_2/\rho_2 \tag{16}$$

Densities measured for incompatible blends have been reported to agree with values calculated from the simple additivity relation given by Eq. (16) [14, 18, 143]. Those of compatible blends may be up to 5% larger than the additive values [17, 18, 34, 144–146]. The increase in density or negative excess volume of mixing observed for compatible blends is indicative of strong intermolecular interactions favoring better packing between molecules. The largest increases in density have been observed in blends of polar polymers such as PVC and butadiene–acrylonitrile copolymers [18, 144] for which intermolecular interactions would be expected to be greater. In the case of the PPO–PS blends, densities about 1% larger than additive values have been reported [15, 19, 147]. This comparatively small increase in density suggests that only weak intermolecular interactions are present as substantiated by the observed near zero value for the interaction parameter [148]; stronger interactions between blended polymers would be reflected in more negative interaction parameters.

In the case of the incompatible blends of PPO and PpClS or PoClS, densities were found to be additive [15]. Blends of PPO and the pClS copolymers in the compatible and the edge of compatibility composition range exhibited densities which like those for PPO–PS were larger than additive. Densities measured for PPO–PS, PPO–PpClS, and one compatible blend of PPO and a copolymer (58.5 mole % *p*ClS) are plotted against wt % PPO in Fig. 15. Compared with T_g and tensile strength measurements [15], measurements of blend density appear to be less sensitive indicators of the compatibility–incompatibility transition. One incompatible blend of PPO and a copolymer of 75.4 mole % *p*ClS exhibited a small but experimentally significant density increase at all blend compositions. It is possible that the observed excess density in such incompatible blends may be the result of contributions from the mixed-composition interfacial region. Such an effect has been observed by Wojuzkij *et al.* [94] who noted an increase in density in the interfacial regions in incompatible blends of PMMA–PVC and PBMA–PVC.

VI. UNSOLVED PROBLEMS

A. Effect of Size of Dispersed Phase on the Glass Transition Temperature

Incompatible blends are characterized by two distinct phases. Depending on composition, melt viscosity, and mixing conditions, one component may

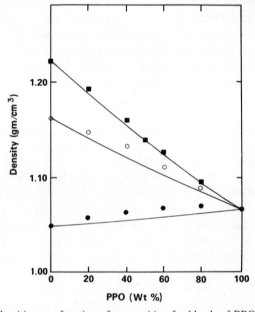

Fig. 15 Blend densities as a function of composition for blends of PPO and polystyrene, one compatible copolymer composition near the compatibility–incompatibility transition, and poly(*p*-chlorostyrene) (P*p*ClS): (●) PS–PPO; (○) Copolymer B–PPO; (■) P*p*ClS–PPO. (Reproduced with permission from Fried [15].)

be present as a dispersed phase in a matrix of the other component. The size of the dispersed phase will depend upon mixing variables (see Chapter 7) and to an extent on the thermodynamics of mixing which will also determine such features as the sharpness of the boundary between the dispersed phase and the matrix.

There is some evidence suggesting that the size of the dispersed phase can affect the location of the observed glass transition, the width of the transition, transition height, and even whether one or two glass transitions will be apparent. Most of these conclusions have been based upon a comparison between the observed dynamic mechanical spectrum and the actual blend morphology as determined from electron micrographs. Provided the blend can be stained preferentially by such techniques as osmium tetroxide staining [149], domains as small as 100 Å (0.01 μm) can apparently be detected in electron micrographs of ultrathin sections. Phase contrast microscopy is useful for identifying phases only in the 0.2–10-μm range [150].

Wetton et al. [24] have shown that the mechanical loss peaks decrease in amplitude with decreasing phase size because of the increasing mechanical isolation of the smaller phases. In addition, Bares [151] has concluded that, if the dispersed phase is in a matrix of lower T_g and is sufficiently small, then, instead of the unchanged T_g of the microphase and an intermediate transition of the intermixed zone, an average lower T_g will be apparent. In the case of styrene–butadiene–styrene triblock copolymers with an average styrene domain size of 120 Å, the observed T_g was depressed by nearly 20°. The dependence of phase T_g upon size was given in an extended form of the well-known Flory–Fox equation

$$T_g = T_{g_\infty} - K_m/M - K_s(S/V) \tag{17}$$

where T_{g_∞} is the T_g of very high molecular weight polymer, K_m is the Flory–Fox constant, M is molecular weight, K_s is a phase constant, S is the surface area, and V is the volume of that phase.

Razinskaya et al. [25] report that, when the relative specific surface (total area of the surface per unit volume of the dispersed phase) was less than 25 μm^{-1}, two loss maxima were detected, but when the relative specific surface was greater than 25 μm^{-1}, only one loss maximum was apparent. Similarly, Matsuo et al. [152] observed only one E'' peak for blends of PVC and a butadiene–acrylonitrile copolymer, although evidence of domains ~100 Å in size was apparent in electron micrographs of osmium tetroxide stained ultrathin sections (see Section VI.C). Kollinsky and Markert [153] have also observed that in the case of blends of methyl methacrylate–butyl acrylate copolymers only single, well-defined damping peaks (torsion pendulum) were evident even though electron micrographs revealed micro-heterogeneous structure in the order of 30 Å.

B. Kinetics versus Thermodynamics in Melt Temperature Depression

In Section I.B.2, evidence from some recent studies was presented indicating that the observed T_m depression in crystalline–amorphous polymer blends may be explained by simple thermodynamic arguments based on the Flory–Huggins lattice theory. The substantiation came from the observations that T_m data for PVF$_2$–PMMA or PVF$_2$–PEMA blends fitted to Eq. (5) gave linear plots ($T_m^0 - T_m$ versus ϕ_1^2) with a near zero intercept. The very small intercept ($\sim 1°C$) was attributed to a small entropic effect. The values of the interaction parameter obtained from the slope of the linear plot were reasonable for compatible blends, that is, small and negative, suggesting the adequacy of the thermodynamic arguments. In these studies [46, 48, 49], the observed T_m depression was not attributed to such changes in the crystal morphology as reduction in crystal size or lamellae thickness as has been suggested to explain the observed T_m depression in other studies [47, 52–56]. Nishi and Wang [46] supported their thermodynamic argument with two observations which suggested that at least for PVF$_2$–PMMA blends, morphological effects were not significant in T_m depression. First, in their isothermal crystallization experiments, they showed that their data could be fitted by the equation

$$T_m^0 - T_m = \Phi(T_m^0 - T_c) \tag{18a}$$

or by rearranging terms

$$T_m = \Phi T_c + (1 - \Phi) T_m^0 \tag{18b}$$

where Φ is a stability parameter and T_c is the crystallization temperature. Plots of T_m versus T_c were linear with constant slope independent of blend composition, that is, constant Φ. If T_m depression is mainly due to morphological effects, then Φ, which is in fact a morphological parameter, would not be independent of composition, and the plots of T_m versus T_c would have different slopes extrapolating to a single equilibrium melting point. In addition, they found that the T_m depression was affected less by heating rates (DSC study) than would be expected if morphological effects were significant.

These observations were in contrast to direct morphological studies of other crystalline–amorphous blends by small- and wide-angle x-ray and light scattering experiments [52, 54–56] in which significant changes in lamellae thickness and spherulite size followed changes in blend composition (e.g., Section IV.B). This would suggest that the significance of morphological effects on T_m depression may be different in different blends or that the thermodynamic arguments may be too simplistic. The inadequacies of the Flory–Huggins theory for polymer blends especially in the consideration of volume changes in mixing (see Chapter 3) are well known. There is also some evidence [154] that melting point depressions may occur in incompatible blends as well. If this is confirmed, then T_m depression in such blends must be attributed to morphological effects of the polymer diluent on reducing crystal sizes or inducing crystal defects in the crystalline lattice. What is needed is the application of a more exact thermodynamic treatment (e.g., corresponding states theory) accompanied by direct morphological studies to determine the relative importance of thermodynamic versus morphological effects.

C. Differences of Transition Behavior with Different Techniques

From the previous discussion (Section VI.A), phase size below a critical dimension was shown to have a significant effect on T_g. In heterogeneous blends, as the size of the dispersed phase decreases, both T_g and the magnitude of the dynamic mechanical loss peaks may be affected. In compatible blends in which only a single glass transition is evident, micro-heterogeneous structure may still persist as alluded to previously. This was shown by the dynamic mechanical studies of Matsuo [152] for blends of PVC and butadiene–acrylonitrile rubbers. For NBR rubbers with 40 wt % acrylonitrile, these blends exhibit only one E'' peak intermediate in temperature between those of the two components, as shown in Fig. 16, whereas

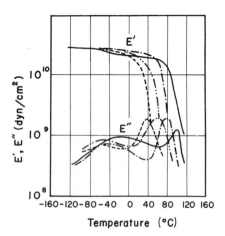

Fig. 16   Temperature dependence of dynamic storage and loss moduli E' and E'', respectively, for compatible blends of PVC and butadiene–acrylonitrile (NBR) copolymer: (———) 100/0; (—·—) 100/10; (—··—) 100/25; (———) 100/50, all PVC–NBR-40. (Reproduced with permission from Matsuo [152].)

electron micrographs of ultrathin sections stained with osmium tetroxide (Fig. 17) reveal rubber domains less than ~100 Å in size.

Recently, Kaplan [156] has reviewed several studies of polymer blends including that of Matsuo in which both dynamic mechanical and electron microscopy information was available. His conclusions from this evidence was that two glass transitions will be apparent in blends where the level of heterogeneity is greater than ~1000 Å (0.1 μm). Between 200 and 1000 Å, two transitions will appear but will be broadened, while for less than 150 Å the blend will exhibit only one glass transition. This means that molecular homogeneity and the detection of a single, composition-dependent glass transition are not necessarily synonymous. Vinogradov *et al.* [22] have suggested that as a relative measure of homogeneity, a parameter given by

$$\mu = 1/B \tag{19}$$

may be used, where B is the average linear dimension of the micro-heterogeneous regions.

The above results indicate that the minimum domain size required for detection as a main T_g peak in dynamic mechanical measurements is approximately 150 Å or between 100 and 5000 carbon–carbon bonds,

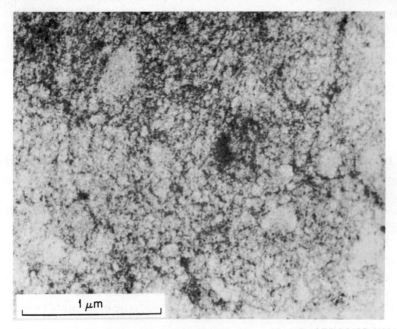

Fig. 17 Transmission electron micrograph of osmium tetroxide stained PVC/NBR (100/15) blend. (Reproduced with permission from Matsuo [152].)

depending on whether the polymer conformation is highly restrictive or freely rotating [156]. This is appreciably larger than the 10–30 carbon atom segments suggested by Hammer [33] or the 30–40 main chain carbon atoms postulated by Stoelting et al. [101]. The principal unanswered question is: What is the effective probe size of the other T_g techniques such as calorimetric (DSC, etc.) and dielectric? Stoelting et al. have proposed that the molecular process responsible for the discontinuity in heat capacity observed by DSC may involve longer-range motions than the segmental microbrownian motions responsible for the dynamic mechanical loss peaks. Their studies with PPO–PS, as well as additional evidence from other studies, suggest that DSC and dielectric measurements are less sensitive to small-phase structure, that is, they have larger effective probe sizes than dynamic mechanical techniques. For example, Feldman and Rusu [21] have concluded that blends of PVC and EVA (45 wt % vinyl acetate) are compatible, based on the detection of a single dielectric tan δ and ε'' peaks, while for identical blends Marcincin et al. [78] have observed two dynamic mechanical loss peaks.

A principal difficulty in comparisons of this type is that of resolution and the subjectivity of individual observations. What may appear to be a single, broad loss peak may be actually two overlapping peaks. For example, Nishi et al. [76] have shown that in the case of blends of PVC and Hytrel rubber for which electron microscopy indicated microheterogeneity in the range of 100 Å, the broad, dynamic mechanical tan δ peak could be resolved by application of a Fuoss–Kirkwood distribution of relaxation times.

VII. COMMERCIAL BLENDS OF PPO–POLYSTYRENE: NORYL[†]

As illustrated by the results of the T_g studies cited in Section IV, PPO and PS are compatible in all proportions. PPO has the properties of a good engineering plastic as shown by the values given in Table VI, including good hydrolytic and dimensional stability, high heat distortion temperature (355°F at 66 psi), and considerable impact strength (1.2 ft-lb/in. notch at 73°F); however, its high melt viscosity and the occurrence of some degradative processes such as methyl group oxidation [103] at molding temperatures limit the use of melt processing. Blending with PS not only lowers the price of the final product but enables melt processing at lower temperatures without sacrificing many of the attractive mechanical properties

[†] Registered trademark of the General Electric Company, Plastics Department, Selkirk, New York 12158.

Table VI

Properties of PPO and a Noryl Resin[a]

Properties	ASTM method	PPO	Noryl
Physical			
Specific gravity, 73°F	D792	1.06	1.06
Specific volume, in.^{3}/lb		26.1	26.1
Water absorption, 24 hr, 73°F	D570	0.03	0.07
Thermal			
Heat deflection temperature (°F)	D648		
at 66 psi		355	279
at 264 psi		345	265
Mold shrinkage, in./in.	D1299	0.007–0.009	0.005–0.007
Mechanical			
Tensile strength (psi)	D638		
at 73°F		11,600	9,600
at 200°F		8,000	6,500
Elongation at break (%)	D638		
at 73°F		20–40	20–30
at 200°F		30–70	30–40
Tensile modulus (psi)	D638		
at 73°F		390,000	355,000
at 200°F		360,000	230,000
Flexural strength (psi)	D790		
at 0°F		19,100	15,900
at 73°F		16,500	13,500
at 200°F		12,600	7,300
Flexural modulus (psi)	D790		
at 0°F		385,000	380,000
at 73°F		375,000	360,000
at 200°F		360,000	260,000
Shear strength (psi)	D732	11,000	10,500
Deformation under load (%)			
at 122°F, 2000 psi	D621	0.1	0.3
Creep, 300 hr, 73°F, 2000 psi (%)	D674	0.5	0.75
Izod impact strength, notched $\frac{1}{4} \times \frac{1}{2}$-in. bar, ft-lb/in. notch	D256		
at −40°F		1.0	1.4
at 73°F		1.2	1.8
at 200°F		1.7	4.2

[a] From Hay [103].

of PPO itself. The first commercial PPO–PS blends were marketed by General Electric as the Noryl series of engineering resins in 1966 [103].

A. Composition

From quantitative DSC measurements, Bair [68] has deduced that commercial Noryl resins are actually blends of PPO and high-impact polystyrene (HIPS) with about 1% low density polyethylene added as a processing lubricant. Cizek [26] has disclosed that the preferred composition is between 40 and 85 wt % PPO although a wider composition range may be chosen to meet processing requirements or to satisfy product specifications such as heat distortion temperature. Other commercial Noryl products include glass-reinforced Noryl resins for greater rigidity and dimensional stability and a series of flame retardant products [157].

B. Properties

The physical and mechanical properties of unblended PPO and a typical Noryl resin are given for comparison in Table VI. Many of the good mechanical characteristics of PPO itself are retained in the blend, while at certain blend compositions, the blend property may be in fact

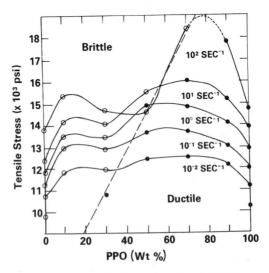

Fig. 18   Tensile stress of PPO–PS blends as a function of blend composition and strain rate. Broken line indicates region of brittle–ductile transition. (Reproduced with permission from Yee [16].)

superior to that of either PPO or PS (HIPS) alone. This is illustrated in Fig. 18 in which the tensile strength of PPO–PS blends have been plotted as a function of blend composition over four decades of strain rate [16]. Two maxima in tensile or yield strength are noted at both high and low PPO compositions. These maxima diminish in intensity with decreasing strain rate. Broad but singular maxima have also been observed in flexural and compressive strength, Rockwell hardness, and in modulus as functions of blend composition [26].

One of the most dramatic effects of blending is the transition from ductile to brittle modes of deformation with increasing PS composition of the blend. At temperatures below T_g, PPO yields and cold draws [81] as do other polymers with flexible oxygen bonds and main chain phenyl groups such as bisphenol A polycarbonate and poly(ethylene terephthalate). This is in contrast to PS and other vinyl polymers with bulky side groups, which fail in the brittle mode with extensive crazing at temperatures as low as $-90°C$. In contrast, PPO crazes only at temperatures approaching T_g or in the presence of crazing agents such as ethanol [158, 159].

In the case of PPO–PS blends of high PPO content the development of shear bands characteristic of PPO during deformation inhibits the growth of crazes beyond the critical defect size cited by Nicolais and DiBenedetto [160] for brittle failure. Similar behavior has been observed in the case of PPO–HIPS blends [161, 162]. This means that PPO–PS blends at such PPO compositions will yield and cold draw. Blends with lower PPO composition will be brittle although craze formation is retarded. This brittle–ductile transition is indicated in Fig. 18. As shown, decreasing the strain rate widens the composition range over which ductile deformation occurs. In their studies, Wellinghoff and Baer [81] showed that the ductile–brittle transition occurred at about 70–80 wt % PPO and was unaffected by the tacticity of the polystyrene component. It is interesting to note that at approximately the same composition, the broad β peak of PPO (e.g., Section IV.A.2) was suppressed. This suggests that strong intermolecular attractions resulting in an increase in blend density and/or the absence of low-temperature relaxation processes suppresses ductility in these blends. The location of the brittle–ductile transition is also affected by sample size and the temperature at deformation. Thinner films are more ductile than thick ones as a probable result of looser packing in the thin films due to the higher surface-to-volume ratio, which enables delocalized shear yielding to occur at lower stress levels and thereby prevents craze nucleation and growth.

The overall consequence of the ductile–brittle transition in PPO–PS blends is a loss of impact strength with increasing PS composition. Unlike tensile strength and other mechanical properties of PPO–PS blends cited above, impact strength at a given blend composition is below the weighted average of the impact strengths of unblended PPO and PS [26]. The use of HIPS in place of rigid polystyrene leads to a substantial increase in the overall impact strength of the Noryl resins. For example, the Izod impact strength of PPO–HIPS has been shown to exhibit a large, broad maximum near 50 wt % PPO that is nearly 200% larger in magnitude than that at the equivalent composition in PPO–PS blends.

REFERENCES

1. H. K. Yuen and J. B. Kinsinger, *Macromolecules* **7**(3), 329–336 (1974).
2. B. D. Gesner, *Encycl. Polym. Sci. Technol.* **10**, 694–709 (1969).
3. S. Krause, *J. Macromol. Sci. Rev. Macromol. Chem.* **7**(2), 251–314 (1972).
4. S. L. Rosen, *SPE Tech. Papers* **XII**, **XIV-4**, 1 (1966); *Polym. Eng. Sci.* **7**(2), 115–123 (1967).
5. E. Jenckel and H. U. Herwig, *Kolloid Z.* **148**, 57–66 (1956).
6. M. Bank, J. Leffingwell, and C. Thies, *J. Polym. Sci. Polym. Phys. Ed.* **10**, 1097–1109 (1972).
7. E. Perry, *J. Appl. Polym. Sci.* **8**, 2605–2618 (1964).
8. J. F. Kenney, *in* "Recent Advances in Polymer Blends, Grafts, and Blocks" (L. H. Sperling, ed.), pp. 117–139. Plenum, New York, 1974.
9. L. Bohn, *in* "Polymer Handbook" (J. Brandrup and E. H. Immergut, eds.), 2nd ed., pp. III–211. Wiley (Interscience), New York, 1975.
10. E. I. Nagumanova, R. Y. Sagitov, and V. A. Voskresenskii, *Kolloid Zh.* **37**(3), 571–573 (1975).
11. R. G. Bauer, R. M. Pierson, W. C. Mast, N. C. Bletso, and L. Shepherd, *Adv. Chem. Ser.* **99**, 237–250, 251–259 (1971).
12. N. Platzer, *Chem. Technol.*, pp. 56–60 (January 1976).
13. R. D. Deanin, A. A. Deanin, and T. Sjoblom, *in* "Recent Advances in Polymer Blends, Grafts, and Blocks" (L. H. Sperling, ed.), pp. 63–91. Plenum, New York, 1974.
14. O. F. Noel III and J. F. Carley, *Polym. Eng. Sci.* **15**(2), 117–126 (1975).
15. J. R. Fried, Ph.D. Thesis, Univ. of Massachusetts (1976).
16. A. F. Yee, *Polym. Prepr. Amer. Chem. Soc. Div. Polym. Chem.* **17**(1), 145–150 (1976).
17. J. J. Hickman and R. M. Ikeda, *J. Polym. Sci. Polym. Phys. Ed.* **11**(9), 1713–1721 (1973).
18. Y. J. Shur and B. Rånby, *J. Appl. Polym. Sci.* **19**, 1337–1346, 2143–2155 (1975).
19. C. H. M. Jacques and H. B. Hopfenberg, *Polym. Eng Sci.* **14**(6), 441–448 (1974).
20. G. Ajroldi, G. Gatta, P. D. Gugelmetto, R. Rettore, and G. P. Talamini, *Adv. Chem. Ser.* **99**, 119–137 (1971).
21. D. Feldman and M. Rusu, *Eur. Polym. J.* **10**, 41–44 (1974).
22. Ye. L. Vinogradov, M. A. Martynov, and G. A. Ol'shanik, *Polym. Sci. USSR* **17**(6), 1605–1610 (1975).
23. E. H. Merz, G. C. Claver, and M. Baer, *J. Polym. Sci.* **22**, 325–341 (1956).
24. R. E. Wetton, J. D. Moore, and P. Ingram, *Polymer* **14**, 161–166 (1973).
25. I. N. Razinskaya, L. I. Batueva, B. S. Tyves, L. A. Smirnova, and B. P. Shtarkman, *Kolloid Zh.* **37**(4), 672–677 (1975).
26. E. P. Cizek, U.S. Patent 3,383,435 (assigned to General Electric) (1968).
27. S. Krause and N. Roman, *J. Polym. Sci. Part A* **3**, 1631–1640 (1965).
28. A. R. Shultz and B. M. Gendron, *J. Macromol. Sci. Chem.* **8**(1), 175–189 (1974).
29. L. M. Robeson, M. Matzner, L. J. Fetters, and J. E. McGrath, *in* "Recent Advances

in Polymer Blends, Grafts, and Blocks" (L. H. Sperling, ed.), pp. 281–300. Plenum, New York, 1974.

30. C. Elmqvist and S. Svanson, *Eur. Polym. J.* **12**, 559–561 (1976).
31. R. Buchdahl and L. E. Nielsen, *J. Polym. Sci.* **15**, 1–8 (1955).
32. L. E. Nielsen, *J. Amer. Chem. Soc.* **75**, 1435–1439 (1953).
33. C. F. Hammer, *Macromolecules* **4**(1), 69–71 (1971).
34. G. A. Zakrzewski, *Polymer* **14**, 347–351 (1973).
35. T. G. Fox, *Bull. Amer. Phys. Soc.* **1**, 123 (1956).
36. F. N. Kelley and F. Bueche, *J. Polym. Sci.* **50**, 549–556 (1961).
37. J. S. Noland, N. N.-C. Hsu, R. Saxon, and J. M. Schmitt, *Adv. Chem. Ser.* **99**, 15–28 (1971).
38. W. M. Prest and R. S. Porter, *J. Polym. Sci. Polym. Phys. Ed.* **10**, 1639–1655 (1972).
39. M. Gordon and J. S. Taylor, *J. Appl. Chem.* **2**, 493–500 (1952).
40. J. Gibbs and E. Dimarzio, *J. Polym. Sci.* **40**, 121–131 (1959).
41. G. Känig, *Kolloid Z.* **190**, 1–16 (1963).
42. A. E. Tonelli, *Macromolecules* **7**(5), 632–634 (1974).
43. H. Daimon, H. Okitsu, and J. Kumanotani, *Polym. J.* **7**(4), 460–466 (1975).
44. N. W. Johnston, *J. Macromol. Sci. Rev. Macromol. Chem.* **14**(2), 215–250 (1976).
45. L. Bohn, *Rubber Chem. Technol.* **41**, 495–513 (1968).
46. T. Nishi and T. T. Wang, *Macromolecules* **8**(6), 909–915 (1975).
47. D. R. Paul and J. O. Altamirano, *Polym. Prepr. Amer. Chem. Soc. Div. Polym. Chem.* **15**(1), 409–414 (1974); *Adv. Chem. Ser.* **142**, 371–385 (1975).
48. R. L. Imken, D. R. Paul, and J. W. Barlow, *Polym. Eng. Sci.* **16**(9), 593–601 (1976).
49. T. K. Kwei, G. D. Patterson, and T. T. Wang, *Macromolecules* **9**(5), 780–784 (1976).
50. J. V. Koleske and R. D. Lundberg, *J. Polym. Sci. Polym. Phys. Ed.* **7**, 795–807 (1969).
51. L. M. Robeson, *J. Appl. Polym. Sci.* **17**, 3607–3617 (1973).
52. F. B. Khambatta, F. Warner, T. Russell, and R. S. Stein, *J. Polym. Sci. Polym. Phys. Ed.* **14**, 1391–1424 (1976).
53. R. A. Neira-Lemos, Ph.D. Thesis, Univ. of Massachusetts (1974).
54. W. Wenig, F. E. Karasz, and W. J. MacKnight, *J. Appl. Phys.* **46**(10), 4194–4198 (1975).
55. W. Wenig, R. Hammel, W. J. MacKnight, and F. E. Karasz, *Macromolecules* **9**(2), 253–257 (1976).
56. R. Hammel, W. J. MacKnight, and F. E. Karasz, *J. Appl. Phys.* **46**(10), 4199–4203 (1975).
57. P. J. Flory, *J. Chem. Phys.* **17**(3), 223–240 (1949).
58. L. Mandelkern, R. R. Garrett, and P. J. Flory, *J. Am. Chem. Soc.* **74**, 3939–3951 (1952).
59. L. Mandelkern, *Chem. Rev.* **56**, 903–958 (1956).
60. F. E. Karasz, U.S. Nat. Tech. Inform. Serv., AD Rep. 72, No. 735395 (1972).
61. R. L. Scott, *J. Chem. Phys.* **17**(3), 279–284 (1949).
62. A. de Boer and G. Challa, *Polymer* **17**, 633–637 (1976).
63. B. Schneier, *J. Appl. Polym. Sci.* **18**, 1999–2008 (1974).
64. K. C. Frisch, D. Klempner, S. Migdal, H. L. Frisch, and H. Ghiradella, *Polym. Eng. Sci.* **14**(1), 76–78 (1974).
65. M. D. Griffiths and L. J. Maisey, *Polymer* **17**, 869–874 (1976).
66. M. J. Richardson and N. G. Savill, *Polymer* **16**, 753–757 (1975).
67. J. H. Flynn, *Thermochim. Acta* **8**, 69–81 (1974).
68. H. E. Bair, *Polym. Sci. Eng.* **10**, 247–250 (1970).
69. L. E. Nielsen, *SPE J.* 525–533 (1960).
70. J. J. Tkacik, PhD Thesis, Univ. of Massachusetts (1975).
71. T. Murayama, *J. Appl. Polym. Sci.* **20**, 2593–2596 (1976).
72. M. Takayanagi and H. Harima, *Amer. Chem. Soc. Div. Org. Coatings Plast. Chem. Pap.* **23**(2), 75–82 (1963).

73. M. Takayanagi, H. Harima, and Y. Iwata, *Mem. Fac. Eng. Kyushu Univ.* **23**, 1–13 (1963).
74. T. Pakula, M. Krysezewski, J. Grebowicz, and A. Galeski, *Polym. J.* **6**(2), 94–102 (1974).
75. L. Bohn, *Adv. Chem. Ser.* **99**, 66–75 (1971).
76. T. Nishi, T. K. Kwei, T. T. Wang, *J. Appl. Phys.* **46**(10), 4157–4165 (1975).
77. S. Akiyama, I. Matsumoto, M. Nakata, and R. Kaneko, *Konbunshu* **33**(4), 238–241 (1976).
78. K. Marcincin, A. Ramanov, and V. Pollak, *J. Appl. Polym. Sci.* **16**, 2239–2247 (1972).
79. T. F. Schatzki, *J. Polym. Sci.* **57**, 496 (1962); *J. Polym. Sci. Polym. Symp. Ed.* **14**, 139–140 (1966).
80. J. M. O'Reilly and F. E. Karasz, *Polym. Prepr. Amer. Chem. Soc. Div. Polym. Chem.* **6**(2), 731–735 (1965).
81. S. Wellinghoff and E. Baer, *Amer. Chem. Soc. Div. Org. Coatings Plast. Chem. Pap.* **36**(1), 140–145 (1976).
82. R. E. Wetton, W. J. MacKnight, J. R. Fried, and F. E. Karasz, *Macromolecules* **11**(1), 158–165 (1978).
83. U. Mehra, L. Toy, K. Biliyar, and M. Shen, *Adv. Chem. Ser.* **142**, 399–408 (1975).
84. M. Bank, J. Leffingwell, and C. Thies, *Macromolecules* **4**(1), 43–46 (1971).
85. S. Akiyama, Y. Komatsu, and R. Kaneko, *Polym. J.* **7**(2), 172–180 (1975).
86. R. F. Boyer and R. S. Spencer, *J. Appl. Phys.* **15**, 398–405 (1944).
87. G. M. Bartenev and G. S. Kongarov, *Vysokomol. Soedin. Ser. A* **2**(11), 1692–1697 (1960).
88. L. H. Sperling, D. W. Taylor, M. L. Kirkpatrick, H. F. George, and D. R. Bardman, *J. Appl. Polym. Sci.* **14**, 73–78 (1970).
89. L. J. Hughes and G. L. Brown, *J. Appl. Polym. Sci.* **5**(17), 580–588 (1961).
90. C. G. Seefried Jr. and J. V. Koleske, *J. Polym. Sci. Polym. Phys. Ed.* **13**, 851–856 (1975).
91. V. R. Landi, *Rubber Chem. Technol.* **45**(1), 222–240 (1972).
92. J. W. Schurer, A. de Boer, and G. Challa, *Polymer* **16**, 201–204 (1975).
93. E. Helfand, *J. Chem. Phys.* **62**(4), 1327–1331 (1975).
94. S. S. Wojuzkij, A. N. Kamenskij, and N. M. Fodimann, *Kolloid Z. Polym.* **215**(1), 36–42 (1967).
95. G. N. Avgeropoulos, F. C. Weissert, P. H. Biddison, and G. G. A. Bohm, *Rubber Chem. Technol.* **49**, 93–104 (1976).
96. Ye. V. Lebedev, Yu. S. Lipatov, and V. P. Privalko, *Polym. Sci. USSR* **17**(1), 171–179 (1975).
97. J. Letz, *J. Polym. Sci. Polym. Phys. Ed.* **7**, 1987–1994 (1969).
98. F. E. Karasz, H. E. Bair, and J. M. O'Reilly, *J. Phys. Chem.* **69**, 2657–2667 (1965).
99. J. M. O'Reilly and F. E. Karasz, *J. Polym. Sci. Polym. Symp.* **14**, 49–68 (1966).
100. F. E. Karasz and J. M. O'Reilly, *J. Polym. Sci. Polym. Lett. Ed.* **3**, 561–563 (1965).
101. J. Stoelting, F. E. Karasz, and W. J. MacKnight, *Polym. Eng. Sci.* **10**(3), 133–138 (1970).
102. R. J. Petersen, R. D. Corneliussen, and L. T. Rozelle, *Polym. Prepr. Amer. Chem. Soc. Div. Polym. Chem.* **10**, 385–390 (1969).
103. A. S. Hay, *Polym. Eng. Sci.* **16**(1), 1–10 (1976).
104. A. R. Shultz and B. M. Gendron, *Polym. Prepr. Amer. Chem. Soc. Div. Polym. Chem.* **14**(1), 571–575 (1973).
105. A. R. Shultz and B. M. Gendron, *J. Polym. Sci. Polym. Symp. Ed.* **43**, 89–96 (1973).
106. F. E. Karasz, H. E. Bair, and J. M. O'Reilly, *J. Polym. Sci. Polym. Phys. Ed.* **6**, 1141–1148 (1968).
107. F. E. Karasz, J. M. O'Reilly, H. E. Bair, and R. A. Kluge, *Polym. Prepr. Amer. Chem. Soc. Div. Polym. Chem.* **9**, 822–824 (1968).
108. T. G. Fox and P. J. Flory, *J. Appl. Phys.* **21**, 581–591 (1950).
109. A. R. Shultz and B. M. Gendron, *J Appl. Polym. Sci.* **16**, 461–471 (1972).
110. L. Nicolais and R. F. Landel, *Polym. J.* **7**(2), 259–263 (1975).

111. C. H. M. Jacques, H. B. Hopfenberg, and V. Stannett, *Polym. Eng. Sci.* **13**, 81–87 (1973).
112. A. Savolainen and P. Tormala, *J. Polym. Sci. Polym. Phys. Ed.* **12**(6), 1251–1254 (1974).
113. G. Allen, M. W. Coville, R. M. John, and R. F. Warren, *Polymer* **11**, 492–495 (1970).
114. S. de Petris, V. Frosini, E. Butta, and M. Baccaredda, *Makromol. Chem.* **109**, 54–61 (1967).
115. T. Lim, V. Frosini, V. Zaleckas, D. Morrow, and J. A. Sauer, *Polym. Eng. Sci.* **13**(1), 51–58 (1973).
116. J. Heijboer, *J. Polym. Sci. Polym. Symp. Ed.* **16**, 3755–3763 (1968).
117. A. Eisenberg and B. Cayrol, *J. Polym. Sci. Polym. Symp. Ed.* **35**, 129–149 (1971).
118. C. W. Reed, *in* "Dielectric Properties of Polymers" (F. E. Karasz, ed.), pp. 343–369. Plenum, New York, 1971.
119. A. R. Shultz and B. M. Beach, *Macromolecules* **7**, 902–909 (1974).
120. W. J. Jackson Jr. and J. R. Caldwell, *Adv. Chem. Ser.* **48**, 185–195 (1965).
121. F. E. Karasz, W. J. MacKnight, and J. Stoelting, *J. Appl. Phys.* **41**(11), 4357–4360 (1970).
122. W. J. MacKnight, J. Stoelting, and F. E. Karasz, *Adv. Chem. Ser.* **99**, 29–41 (1971).
123. D. R. Buchanan, *J. Polym. Sci. Polym. Phys. Ed.* **9**, 645–658 (1971).
124. G. Kortleve and C. G. Vonk, *Kolloid Z.Z. Polym.* **225**, 124–131 (1968).
125. R. Hosemann and S. N. Bagchi, *in* "Direct Analysis of Diffraction by Matter." North-Holland Publ., Amsterdam, 1963.
126. F. Warner, R. S. Stein, and W. J. MacKnight, *J. Polym. Sci. Phys. Ed.* (1976) (in press).
127. J. P. Bell and J. H. Dumbleton, *J. Polym. Sci. Polym. Phys. Ed.* **7**, 1033–1057 (1969).
128. Z. Pelzbauer and R. St. John Manley, *J. Polym. Sci. Polym. Phys. Ed.* **8**, 649–652 (1970).
129. S. Horikiri and K. Kodera, *Polym. J.* **4**(2), 213–214 (1973).
130. J. M. Barrales-Rienda and J. M. G. Fatou, *Kolloid Z.Z. Polym.* **244**, 317–323 (1971).
131. H. Berghmans and N. Overbergh (1976) (in press).
132. P. Scherrer, *Gott. Nach.* **2**, 98 (1918).
133. R. Hosemann, *Z. Phys.* **128**, 464 (1950).
134. F. E. Karasz, W. J. MacKnight, and J. J. Tkacik, *Polym. Prepr. Amer. Chem. Soc. Div. Polym. Chem.* **15**(1), 415–420 (1974).
135. R. Koningsveld, *Chem. Z.* **26**, 263–287 (1972).
136. R. Koningsveld, L. A. Kleintjens, and H. M. Schoffeleers, *Pure Appl. Chem.* **39**, 1–32 (1974).
137. L. A. Wood, *J. Polym. Sci.* **28**, 319–330 (1958).
138. H. Frölich, "Theory of Dielectrics." Oxford Univ. Press, London and New York, 1958.
139. G. P. Mikhailov, A. M. Lobanov, and M. P. Platonov, *Polym. Sci. USSR* **9**, 2565–2574 (1967).
140. J. R. Fried, F. E. Karasz, and W. J. MacKnight, *Bull. Amer. Phys. Soc.* **21**(3), 237 (1976).
141. I. M. Ward, "Mechanical Properties of Solid Polymers," p. 17. Wiley (Interscience), New York, 1971.
142. R. Buchdahl and L. E. Nielsen, *J. Appl. Phys.* **21**, 482–487 (1950).
143. J. L. Work, *Polym. Eng. Sci.* **13**(1), 46–50 (1973).
144. B. G. Rånby, *J. Polym. Sci. Polym. Symp. Ed.* **51**, 89–104 (1975).
145. T. K. Kwei, T. Nishi, and R. F. Roberts, *Macromolecules* **7**(5), 667–674 (1974).
146. T. K. Kwei, T. Nishi, and R. F. Roberts, *Rubber Chem. Technol.* **48**, 218–235 (1975).
147. H. B. Hopfenberg, V. T. Stannett, and G. M. Folk, *Polym. Eng. Sci.* **15**(4), 261–267 (1975).
148. A. R. Shultz and C. R. McCullough, *J. Polym. Sci. Polym. Phys. Ed.* **10**, 307–316 (1972).
149. K. Kato, *Polym. Eng. Sci.* **7**, 38–39 (1967).
150. A. J. Yu, *Adv. Chem. Ser.* **99**, 2–14 (1971).
151. J. Bares, *Macromolecules* **8**(2), 244–246 (1975).
152. M. Matsuo, C. Nozaki, and T. Jyo, *Polym. Eng. Sci.* **9**(3), 197–205 (1969).
153. F. von Kollinsky and G. Markert, *Die Makromol. Chem.* **121**, 117–128 (1969).

154. M. Natov, L. Peeva, and E. Djagarova, *J. Polym. Sci. Polym. Symp. Ed.* **16**, 4197–4206 (1968).

155. D. Kaplan and N. W. Tschoegl, *in* "Recent Advances in Polymer Blends, Grafts, and Blocks" (L. H. Sperling, ed.), pp. 415–430. Plenum, New York, 1974.

156. D. S. Kaplan, *J. Appl. Polym. Sci.* **20**, 2615–2629 (1976).

157. M. Kramer, *Appl. Polym. Symp.* **15**, 227–237 (1971).

158. G. A. Berniev, and R. P. Kambour, *Macromolecules* **1**(15), 393–400 (1968).

159. R. P. Kambour and A. S. Holik, *J. Polym. Sci. Polym. Phys. Ed.* **7**, 1393–1403 (1969).

160. L. Nicolais and A. T. DiBenedetto, *J. Appl. Polym. Sci.* **15**, 1585–1598 (1971).

161. C. B. Bucknall, D. Clayton, and W. E. Keast, *J. Mater. Sci.* **7**, 1443–1453 (1972).

162. C. B. Bucknall, D. Clayton, and W. Keast, *J. Mater. Sci.* **8**, 514–524 (1973).

163. C. Elmqvist and S. E. Svanson, *Eur. Polym. J.* **11**, 789–793 (1975).

Chapter 6

Interfacial Energy, Structure, and Adhesion between Polymers

Souheng Wu

Central Research and Development Department, Experimental Station
E. I. du Pont de Nemours & Company
Wilmington, Delaware

I.	Introduction	244
II.	Surface Tension	245
	A. Temperature Dependence	245
	B. Molecular Weight Dependence	247
	C. Effects of Glass and Phase Transitions	249
	D. Solid Polymers	252
	E. Copolymers, Blends, and Additives	252
	F. Effects of Mold Surfaces	256
	G. Methods of Estimation	258
	H. Tabulation of Surface Tension Data	260
III.	Interfacial Tension	263
	A. Works of Adhesion and Cohesion	263
	B. Theories of Interfacial Tension	263
	C. Temperature Dependence	268
	D. Effect of Polarity	269
	E. Molecular Weight Dependence	272
	F. Effects of Additives	272
	G. Methods of Estimation	274
	H. Tabulation of Interfacial Tension Data	277
IV.	Adhesion between Polymers	278
	A. Basic Concepts	278
	B. Fracture Theory of Adhesion	279
	C. Weak Boundary Layer Theory	281
	D. Wetting–Contact Theory of Adhesion	282
	E. Diffusion Theory of Adhesion	284
	F. Chemical Adhesion	287
V.	Summary	288
	References	288

I. INTRODUCTION

Most small-molecule organic liquids are mutually miscible and their mixtures do not form stable interfaces. Polymers are, however, usually immiscible and their mixtures form multiphase structures with stable interfaces. The dispersion, morphology, and adhesion of the component phases are greatly affected by the interfacial energies, which thereby play an important role in determining the mechanical properties of a multiphase polymer blend [1–7]. See also Chapters 4, 7, 8; Volume 2, Chapters 12–16.

Despite its importance, reliable measurements of the interfacial tension between polymers have not been reported until 1969 because of the experimental difficulty in handling highly viscous polymer melts [8,9]. Statistical mechanical theories of polymer interfaces have been reported since 1971 [10–14]. Considerable insights into the nature of polymer interfaces have been gained [15].

Interfacial energy, structure, and adhesion between polymers as related to the properties of polymer blends are discussed in this Chapter. Various thermodynamic functions of interfaces have been defined elsewhere [16].

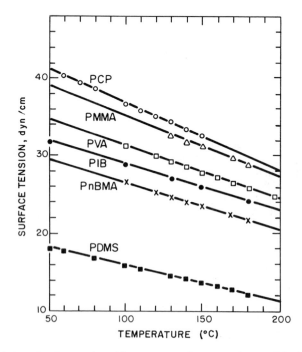

Fig. 1   Variation of surface tension with temperature for some polymers. (After Wu [8, 20, 22, 64].)

Methods for the measurement of surface and interfacial tensions of molten polymers have been discussed before [15].

II. SURFACE TENSION

A. Temperature Dependence

Surface tension of small-molecule liquids are known to vary linearly with temperature, at temperatures far below the critical temperature, with $-(d\gamma/dT)$ values of about 0.1 dyn/cm-degree [15, 17–19], where γ is the surface tension and T the temperature. For polymers, the $-(d\gamma/dT)$ values are also independent of temperature, but smaller in the range of 0.05 to 0.08 dyn/cm-degree [8, 20, 21]. Since $-(d\gamma/dT)$ is the surface entropy, the smaller values for polymers may thus be attributed to the conformational restrictions of the long chain molecules [8, 21]. The γ versus T plots for some polymers [8, 20–22] are shown in Figs. 1 and 2. Linearity is found for all the polymers

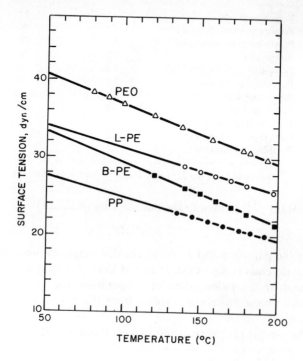

Fig. 2   Variation of surface tension with temperature for some polymers. (After Wu [8, 20, 22, 64].)

shown. One exception is perfluoroalkanes whose γ versus T plot appears to better fit a quadratic equation [23].

MacLeod's equation [24] has been shown to be applicable to polymers [8, 15, 21]:

$$\gamma = \gamma^0 \rho^n \tag{1}$$

where ρ is the density and γ^0 and n are positive constants independent of temperature. In Eq. (1), we have neglected the vapor density, which is negligible for polymers. This relation is accurately obeyed with $n = 4$ for unassociated, small-molecule liquids [24–26]. For polymers, the value of n has been found to vary from 3.0 to 4.4 and may be approximated by 4.0, just as for small-molecule liquids [8, 15, 20, 21]. Table I is a list of the MacLeod's exponents for some polymers [15].

Table I

MacLeod's Exponent for Some Polymers[a]

Polymer	MacLeod's exponent	References
Polychloroprene	4.2	22
Poly(methyl methacrylate)	4.2	20
Poly(n-butyl methacrylate)	4.2	20
Polystyrene	4.4	20
Poly(vinyl acetate)	3.4	8
Poly(ethylene oxide)	3.0	21
Polyisobutylene	4.1	8, 21
Polypropylene	3.2	21
Polyethylene, linear	3.2	8, 21
Polyethylene, branched	3.3	21
Poly(dimethylsiloxane)	3.5	21

[a] Reprinted from Wu [15] by courtesy of Marcel Dekker, Inc.

The constant γ^0 is related to the parachor by [8, 27, 28]

$$\gamma^0 = (P/M)^n \tag{2}$$

where P is the parachor and M the molecular weight of the molecule. If the effect of chain ends is neglected, P and M may be taken for a repeat unit. Since P and M are independent of temperature, the variation of surface tension with temperature arises solely from the variation of density with temperature.

According to Eq. (1), the temperature coefficient of surface tension is given by [15]

$$d\gamma/dT = -n\alpha\gamma \tag{3}$$

where α is the thermal expansion coefficient defined by $\alpha = (1/v_0)(dv/dT)$, where v is the specific volume. Equations (1) and (3) provide a basis for predicting the effects of primary and secondary transitions on the surface tension, as is discussed later.

Eötvös' equation [29, 30], although applicable to small-molecule liquids, is not applicable to polymers, since the equation will diverge when the molecular weight approaches infinity [15].

B. Molecular Weight Dependence

The surface tension of a homologous series increases with increasing molecular weight, while the surface entropy decreases with increasing molecular weight. Both the surface tension and the surface entropy are finite at infinite molecular weight.

Bulk properties of a homologous series are known to depend linearly on the reciprocal of the molecular weight [31–36],

$$X_B = X_{B\infty} - (k_B/M_n) \tag{4}$$

where X_B is a bulk property (such as glass transition temperature, heat capacity, specific volume, thermal expansion coefficient, refractive index, and tensile strength), $X_{B\infty}$ the bulk property at infinite molecular weight, k_B a constant and M_n the number-average molecular weight. Such a relation is, however, not applicable to surface properties. Plots of γ versus M^{-1} for n-alkanes [37], perfluoroalkanes [38], poly(α-methylstyrenes) [39], and polyisobutylenes [40] show various degrees of curvature.

Instead, the surface tensions of homologous series have been found empirically to depend linearly on the reciprocal two-thirds power of the molecular weight [41, 42], that is,

$$\gamma = \gamma_\infty - (k_e/M_n^{2/3}) \tag{5}$$

where γ_∞ is the surface tension at infinite molecular weight and K_e is a constant. Equation (5) can accurately fit the data for n-alkanes with standard deviations in γ of the order of 0.05 dyn/cm, and for perfluoroalkanes, polyisobutylenes, polydimethylsiloxanes, and polystyrenes with standard deviations in γ of the order of 0.2 dyn/cm, as shown in Table II.

A theoretical basis for Eq. (5) has been given by a simple lattice consideration [43]. Let γ_r be the surface tension of the repeat unit ($= \gamma_\infty$) and γ_e be that of the end group. Then, we have [43]

$$\gamma = \gamma_\infty - (\gamma_r - \gamma_e) f_e \tag{6}$$

where f_e is the fraction of the surface area occupied by the end groups and can be given by [43]

Table II

Constants for the Molecular Weight Dependence Relations[a]

Polymer	$T(°C)$	Eq. (11)			Eq. (5)		
		γ_∞	k_m	σ	γ_∞	k_e	σ
n-Alkanes	20	34.75	30.69	0.04	37.81	385.9	0.03
Polyisobutylenes	24	34.50	46.15	0.19	35.62	382.7	0.34
Poly(dimethylsiloxanes)	20	20.33	22.22	0.15	21.26	166.1	0.09
Perfluoroalkanes	20	23.94	12.21	0.29	25.85	682.8	0.30
Polystyrenes	176	29.50	75.42	0.01	29.97	372.7	0.08
Poly[(ethylene oxide) dimethyl ether]	24	41.50	28.13	0.38	44.35	342.8	0.44

[a] γ is dyn/cm; σ is the standard deviation in γ.

$$f_e \cong \{2m/[M + 2m(1-p)]\}^{2/3} \simeq (2m/M)^{2/3} \qquad (7)$$

where m is the formula weight of a repeat unit, M the molecular weight, and p the ratio of the formula weight of the end group to that of the repeat unit. Combination of Eqs. (6) and (7) gives Eq. (5), and the constant k_e is then given by [43]

$$k_e = (\gamma_\infty - \gamma_e)(2m)^{2/3} \qquad (8)$$

On the other hand, MacLeod's equation provides an alternative approach [15]. The densities of a homologous series are given by [33]

$$(1/\rho) = (V_r/m) + (V_e/M) \qquad (9)$$

where V_r is the molar volume of a repeat unit, and V_c that of the two chain ends. Combination of Eqs. (1), (2), and (9) gives

$$\gamma^{1/n} = (1/m) \left[\frac{P_r M + (mP_e - m_e P_r)}{(V_r/m) M + V_e} \right] \qquad (10)$$

where we have let $M = rm + m_e$, $P = rP_r + P_e$, r the number of repeat unit in the molecule, m_e the formula weight of the two chain ends, P the parachor of the molecule, P_r that of the repeat unit, and P_e that of the two chain ends.

Equation (10) can be recast in several functional forms [43]. A particularly useful equation is obtained by expanding and truncating the terms higher than $(1/M)$ and letting $n = 4$, that is,

$$\gamma^{1/4} = \gamma_\infty^{1/4} - (k_m/M_n) \qquad (11)$$

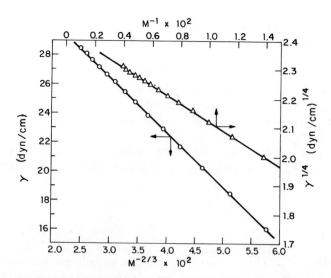

Fig. 3 Variation of surface tension with molecular weight for *n*-alkanes (orthobaric data at 20°C).

where k_m is a numerical constant given by

$$k_m = [(m_c P_r - m P_c)/V_r] + (m V_c P_r/V_r^2) \qquad (12)$$

Equation (11) can accurately fit the data for *n*-alkanes with standard deviations in γ of the order of 0.05 dyn/cm, and for perfluoroalkanes, polyisobutylenes, polydimethylsiloxanes, and polystyrenes with standard deviations in γ of the order of 0.2 dyn/cm, as shown in Table II. Thus, Eqs. (5) and (11) give comparable accuracy. Figure 3 gives the plots for *n*-alkanes and polydimethylsiloxanes. Note that Eq. (11) is different from and better than the previously proposed equation, $\gamma^{-1/4} = \gamma_\infty^{-1/4} + k_m/M$ [15]. On the other hand, the surface tensions of poly(ethylene glycols) have been found to be independent of molecular weight [44, 45]. This is due to the strong hydrogen bonding between the terminal hydroxyl groups. When the terminal hydroxyl groups are converted into non-hydrogen-bonding methyl ether groups or acetate groups; the surface tension then becomes dependent on the molecular weight, obeying Eq. (5) or Eq. (11).

C. Effects of Glass and Phase Transitions

MacLeod's equation also provides a basis for analyzing the effects of glass (secondary) and phase (primary) transitions. Based on Eq. (3), the temperature coefficient of surface tension in the glassy region (below the glass transition temperature, T_g) is related to that in the rubbery regions (above T_g) by

$$(d\gamma/dT)_g = (\alpha_g/\alpha_r)(d\gamma/dT)_r \tag{13}$$

where the subscripts g and r refer to glass and rubber, respectively, and α is the thermal expansion coefficient defined under Eq. (3) before. Thus, the γ versus T plot is continuous, but its slope changes at T_g. Since $\alpha_g < \alpha_r$, hence $-(d\gamma/dT)_g < -(d\gamma/dT)_r$.

On the other hand, at the crystal–melt transition, the surface tension of the crystalline phase γ^c is related to that of the amorphous melt phase γ^a by [15]

$$\gamma^c = (\rho_c/\rho_a)^n \gamma^a \tag{14}$$

where ρ_c is the density of the crystalline phase and ρ_a that of the amorphous phase. At the crystal–melt transition, the density changes discontinuously and so does the surface. Since ρ_c is usually greater than ρ_a [46, 47], the crystalline phase will usually have higher surface tension than the amorphous phase. For linear polyethylene, $n = 3.2$, and $\gamma^a = 35.7$ dyn/cm [8], $\rho_a = 0.855$, $\rho_c = 1.000$ gm/cc at 20°C. Using these values in Eq. (14) γ^c at 20°C is calculated to be 58.9 dyn/cm, which compares favorably with the experimental value of 53.6 dyn/cm reported for aggregates of linear polyethylene crystals [48].

Thermodynamic analysis [15] confirms that the surface tension changes discontinuously at the crystal–melt transition, but continuously with discontinuous temperature coefficient at the glass transition. Let ΔG_{cm} be the change in the Gibbs free energy of the entire system for the crystal–melt transition,

$$\Delta G_{cm} = G_c - G_m \tag{15}$$

where G_c is the total Gibbs free energy of the crystal phase and G_m that of the melt phase. Since the crystal–melt transition is a first-order transition, we have

$$\Delta G_{cm} = 0 \qquad \text{at constant } T, p \tag{16}$$

and

$$(\Delta \partial G_{cm}/\partial A)_{T,p} = (\partial G_c/\partial A)_{T,p} - (\partial G_m/\partial A)_{T,p} \tag{17}$$

$$= \gamma^c - \gamma^a \neq 0 \tag{18}$$

where A is the surface area. Thus, $\gamma^c \neq \gamma^a$.

On the other hand, let ΔG_{gr} be the change in the Gibbs free energy of the entire system for the glass–rubber transition,

$$\Delta G_{gr} = G_g - G_r \tag{19}$$

where G_g is the total Gibbs free energy for the glassy state, and G_r that for the rubbery state. Since glass–rubber transition is a second-order transition, we have

$$\Delta G_{gr} = 0 \quad \text{at constant } T, p \tag{20}$$

and

$$(\partial \Delta G_{gr}/\partial A)_{T,p} = \gamma^g - \gamma^r = 0 \tag{21}$$

$$(\partial^2 \Delta G_{gr}/\partial A \partial T) = (\partial \gamma^g/\partial T) - (\partial \gamma^r/\partial T) \neq 0 \tag{22}$$

where γ^g is the surface tension of the glassy state and γ^r that of the rubbery state. Thus, the surface tension is continuous, but its temperature coefficient is discontinuous at the glass–rubber transition.

Semicrystalline polymers tend to be covered with an amorphous surface layer because of the lower surface tension of the amorphous phase. Various degrees of crystallinity can, however, be induced on the surface by nucleating against certain contacting surfaces [49–58]. Table III compares the surface tensions of the amorphous and the crystalline phases for some polymers.

Table III

Effect of Surface Crystallinity on Surface Tension of Semicrystalline Polymers[a]

Polymer	Nucleation surface	Crystallinity on surface (%)	Surface tension (dyn/cm 20°C)
Polyethylene	Nitrogen	0	36.2
	Polytetrafluoroethylene	0	36.2
	Poly(ethylene terephthalate)	0	36.2
	Copper	5.1	37.4
	Nickel	53.3	51.3
	Tin	60.1	53.8
	Aluminum	63.2	54.9
	Glass	63.2	54.9
	Chromium	66.2	56.1
	Mercury	86.4	64.8
	Gold	93.6	69.6
Nylon 66	Gold	94.1	74.4
Polychlorotrifluoroethylene	Gold	100	58.9
Polypropylene (isotactic)	Gold	100	39.5
Polypropylene (atactic)	Gold	0	28.0
Poly(4-methyl-1-pentene)	Gold	100	22.0

[a] From Schornhorn [50].

D. Solid Polymers

The surface tensions of solid polymers can be obtained by two approaches: (a) extrapolation of γ versus T plot of the melt to the solid range; (b) extrapolation of γ versus M function of the low-molecular-weight liquids to the solid range. Table IV shows some results obtained. Equation (11) gives results in good agreement with those extrapolated from experimental $\gamma-T$ plots, but Eq. (5) consistently gives much higher values.

Table IV

Comparison of γ_s and γ_∞ [a]

| | | | γ_∞ | |
| | | γ_s (or γ) | | |
Polymer	T (°C)	from $\gamma-T$ plots	Eq. (11)	Eq. (5)
Polyethylene	20	35.7	34.75	37.81
Polyisobutylene	24	34.0	34.50	35.62
Poly(dimethylsiloxane)	20	19.8	20.33	21.26
Polytetrafluoroethylene	20	—	23.94	25.85
Poly(ethylene glycol)	24	42.5	41.50	44.35
Polystyrene	176	29.5	29.50	29.97

[a] γ_∞ is the surface tension at infinite molecular weight, extrapolated by Eq. (5) or Eq. (11); γ_s is the surface tension of the solid polymer, extrapolated from $\gamma-T$ plots. All γ data are in dyn/cm.

The surface tension of a solid polymer may also be evaluated indirectly from wettability data. Many methods have been proposed. Calculations are based on empirical or semiempirical relations between the surface tension and the contact angles. Critical surface tension [59] is a qualitative estimate of the surface tension and its values have been tabulated [60, 61]. On the other hand, the preferred method is based on the harmonic-mean equation, which can give reliable values of the surface tension and the polarity [22, 62, 63]. Alternatively, a geometric-mean equation may be used, which is mathematically simpler but gives less accurate results [22, 64–68].

E. Copolymers, Blends, and Additives

Low-energy components tend to adsorb preferentially on the surface and thus lower the surface tension of the mixtures of small-molecule liquids. Similar behavior is observed with polymers. The behavior with polymers is, however, more complex, and affected by the compatibility and the phase structure as well as the surface tension of the components.

Most polymer pairs are incompatible as a rule (see Chapter 2). Blends of two incompatible polymers will result in dispersion of one polymer in the matrix of another. Furthermore, graft and block copolymers tend to possess microphases. Systematic studies on the effects of compatibility and phase structure on the surface tension of polymer blends, grafts, and blocks have not been reported.

1. Random Copolymers

Random copolymers of ethylene oxide and propylene oxide have been found to show no surface excess behavior (surface activity) and approximately follow the linear relation [69]:

$$\gamma = x_1\gamma_1 + x_2\gamma_2 \qquad (23)$$

where γ is the surface tension of the random copolymer, γ_i that of component i, and x_i the mole fraction of component i. This is shown in Fig. 4. The lack of surface excess behavior in random copolymers may be due to conformational restrictions on the polymer chains, which preclude preferential accumulation of the low-energy segments on the surface.

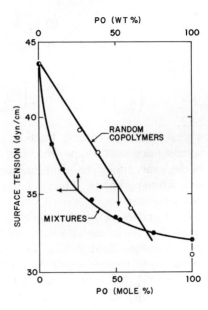

Fig. 4 Surface tension of random copolymers of ethylene oxide and propylene oxide, and physical mixtures of poly(ethylene oxide) and poly(propylene oxide). (After Rastogi and St. Pierre [69].)

2. Blends of Compatible Polymers

Blends of compatible homopolymers of ethylene oxide and propylene oxide, however, show pronounced surface excess behavior of the lower-energy component [69]. The surface excess behavior increases as the molecular weight increases, probably because of increased incompatibility, as show in Fig. 4 for the blends of poly(propylene glycol) and polyepichlorohydrin.

The equation of Belton and Evans [70] has been found to fit reasonably well the surface tension of the blends of poly(ethylene oxide) and poly(propylene oxide) and the blends of poly(propylene oxide) and polyepichlorohydrin [70].

$$\gamma = \gamma_1^0 + (kT/a)\ln\{c/[1 + (c-1)x_1]\}$$
$$c = \exp[a(\gamma_2^0 - \gamma_1^0)] \tag{24}$$

where γ_i^0 is the surface tension of component i, x_i the mole fraction of component i, a the molecular surface area, k the Boltzmann constant, and T the temperature.

For polymer solutions, the following equations obtained by the combination of Flory–Huggins lattice theory and the parallel layer theory of Prigogine and Marechal [71] have been found to give excellent results [72]:

$$\gamma = \gamma_i^0 + \frac{kT}{a}\left[\ln\frac{\phi_1^s}{\phi_1} + \frac{r-1}{r}(\phi_2^s - \phi_2)\right] - \frac{\beta}{a}(\phi_2)^2 \tag{25}$$

$$= \gamma_2^0 + \frac{kT}{ra}\left[\ln\frac{\phi_2^s}{\phi_2} + (r-1)(\phi_2^s - \phi_2)\right] - \frac{\beta}{a}(\phi_1)^2 \tag{26}$$

and

$$\left(\frac{\phi_1^s}{\phi_1}\right) = \left(\frac{\phi_2^s}{\phi_2}\right)^{1/r}\exp\left[\frac{(\gamma_2^0 - \gamma_1^0)a}{kT}\right]\exp\left[\frac{\beta}{kT}(1 - 2\phi_1)\right] \tag{27}$$

where ϕ_1 is the volume fraction of component 1 in the bulk, ϕ_1^s the volume fraction of component 1 in the surface layer, r the number of segment in the r-mer molecule, a the surface lattice spacing parameter, and β the interaction parameter. The parameter a may be taken to be the surface area of a solvent molecule, $a = (V_{m1}/N_0)^{2/3}$ where V_{m1} is the molar volume of the solvent and N_0 Avogadro's number. Also, r may be taken to be V_{m2}/V_{m1} where V_{m2} is the molar volume of the polymer. The equations have been used with excellent results with solutions of poly(dimethylsiloxane) in toluene and tetrachloroethylene [72], polyisobutylene in n-heptane and tetralin [73], three component mixtures of the above [74], and poly(dimethylsiloxane)–oligomers and polyisobutylene–oligomers [75].

3. Block Copolymers

Pronounced surface excess behavior of the lower-energy block is observed for ABA block copolymers of ethylene oxide (A block, higher energy) and propylene oxide (B block, lower energy) [69]. Figure 5 shows the γ versus % B plots at three levels of the degree of polymerization of the B block. The surface excess behavior increases as the length of B block increases. When the degree of polymerization of the B block is greater than about 56, the surface tension of the block copolymer becomes practically equal to that of the homopolymer of B block, even when the B block amounts to only 20% by weight of the composition. Evidently when the B block is sufficiently long, it can accumulate and orient on the surface independently of the rest of the molecule. For ABA block copolymers of polyether (A block, higher energy)

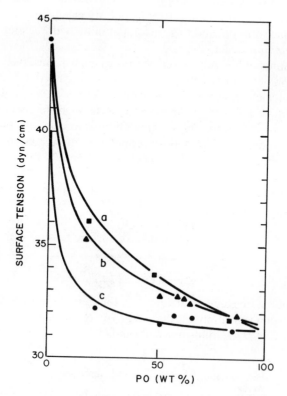

Fig. 5   Surface tension of ABA block copolymers of ethylene oxide (A block) and propylene oxide (B block) at 25°C. The degrees of polymerization of B blocks are (a) $X_n = 16$; (b) $X_n = 30$; (c) $X_n = 56$. (After Rastogi and St. Pierre [69].)

and dimethylsiloxane (B block, lower energy), a degree of polymerization above 20 is sufficient to reduce the surface tension of the block copolymer to that of poly(dimethylsiloxane) [76].

The behavior of graft copolymers has not been reported so far, but it is expected to be similar to the block copolymers with somewhat less surface activity. On the other hand, block and graft copolymers tend to show heterogeneous microphases. The effect of the microphases on surface tension has not been reported.

4. Additives

Additives with low surface tensions are known to modify the surface properties of polymers. Wettability and friction are lowered, for instance, with small amounts of fatty amides in polyethylene [77] and vinylidene chloride–acrylonitrile copolymers [78]; fluorocarbons in poly(methyl methacrylate) and poly(vinyl chloride) [79]; dimethylsiloxane block copolymers in bisphenol-A polycarbonate [80] and polystyrene [81]. These additives have low surface tension, and their incompatibility with the matrix polymer enhances their surface activity.

Small amounts of low surface tension polymers can effectively reduce the surface tension of the matrix polymer as already discussed in Section III.E.3 [69]. ABA block copolymers of polyether (A block) and poly(dimethylsiloxane) (B block at 0.1–1% will reduce the surface tension of a liquid polyol (32 dyn/cm at 25°C) to about 21 dyn/cm, resembling that of pure poly(dimethylsiloxane) [76]. A block copolymer of styrene and dimethylsiloxane can reduce the surface tension of polystyrene (27.7 dyn/cm at 200°C) to 14.7 dyn/cm (200°C) at 0.05% and to 11.9 dyn/cm (200°C) at 5%, as compared to the surface tension of pure poly(dimethylsiloxane) of 11.2 dyn/cm at 200°C [81].

The rate at which the additive migrates to the surface is diffusion controlled [78,81]. The rate of change of surface tension can be given by [81]

$$\gamma = \gamma_0 - 2RTc_0(Dt/\pi)^{1/2} \tag{28}$$

where γ is the surface tension at time t, γ_0 the initial (clean) surface tension, c_0 the bulk concentration of the additive, R the gas constant, T the temperature, and D the diffusion coefficient.

F. Effects of Mold Surfaces

Mold surfaces can greatly affect the surface properties of polymers by acting as a nucleating agent to change the crystallinity and morphology of the surface region [49–58] or as an adsorption site to change the polarity and chemical composition of the surface region [82–84].

1. Nucleation at the Mold Surface

While the bulk (interior) of most semicrystalline polymers is spherulitic, the surface region can be spherulitic, transcrystalline, or lamellar [49–58]. Transcrystalline surface layers are formed when the mold surface nucleates massive numbers of nuclei that are crowded, causing growth normal to the surface. To be effective, the nucleating efficiency of the mold surface should be equal to or greater than that of the adventitious nuclei present in the polymer; otherwise spherulites will be formed in the surface region as in the bulk. The transcrystalline regions as formed in polyamides [85, 86], polyurethanes [85], and polyolefins [49–58, 87] are often thicker than 10 μm. Such transcrystalline layers are thus effectively macroscopic and have different mechanical, transport, and surface properties from the bulk of the polymers [49–58].

Various degrees of surface crystallinity can be obtained using different mold surfaces, as shown in Table III. Polyethylene gives zero surface crystallinity when molded against poly(ethylene terephthalate) or polytetrafluoroethylene but gives highly crystalline surface when molded against gold or alkali halides [49–55]. On the contrary, polypropylene gives highly crystalline surface layer when molded against poly(ethylene terephthalate) or polytetrafluoroethylene [56–58]. Evidently, the nucleating activities of mold surfaces are quite specific and show an unusual dependence on the nuances of their chemical and physical properties. Table III lists the surface tensions of amorphous and crystalline surfaces for some polymers.

2. Adsorption at the Mold Surface

Certain chemical groups in a polymer molecule may adsorb preferentially onto the mold surface, giving surfaces with high concentrations of such chemical constituents. An acrylic copolymer containing both amide and acid groups was found to adsorb onto metals with its polar groups, so that the metal side of the film surface was more hydrophilic than the air side of the film surface [83].

Surface properties of an ethylene–acrylic acid copolymer containing 10% stearic acid, melt pressed against Mylar® poly(ethylene terephthalate) (PET) and Teflon® polytetrafluoroethylene (TFE), showed dependency on the substrate [82]. Stearic acid exudes rapidly to the surface of the Teflon TFE pressed sample which, initially clear, becomes opaque after 4 hr. On the other hand, stearic acid exudes slowly in the Mylar PET pressed sample, which remains clear even after 4 weeks [82].

Chemical compositions of the surface region of some epoxy formulations have been shown to vary with the nature of the mold surface and differ from those of the bulk region [84].

G. Methods of Estimation

1. Solubility Parameter Method

The surface tension of small-molecule liquids is related to the solubility parameter by [88, 89]

$$\gamma = 0.07147\,\delta^2\,V^{1/3} \tag{29}$$

where δ is the solubility parameter and V the molar volume. This equation has been derived theoretically [90].

In applying to polymers, V should not be identified as molar volume of the polymer molecule as a whole, nor as that of a repeat unit. Obviously, the surface tension of a polymer does not depend on the size of the molecule as such, nor on the size of the repeat unit. Rather, V should be identified as the molar volume of an interacting element, which is, however, not known a priori. An early approach is to take V as the average atomic volume of the molecule [61]. A recent improvement [91] is to let $V = M_i/\rho$, where M_i is the formula weight of the interacting element and ρ the density of the polymer. M_i should vary from polymer to polymer, but regression of experimental data shows that on average $M_i = 46.8$. Applying these in Eq. (29) gives [91]

$$\gamma = 0.2575\,\delta^2\,\rho^{-1/3} \tag{30}$$

which can predict the surface tension of polymers to within a few dyn/cm. The solubility parameter may be estimated from group additivity constants [92] by

$$\delta = (\rho \sum F_i)/M \tag{31}$$

where $\sum F_i$ is the sum of Small's molar attraction constants (see Chapter 2) for a repeat unit and M the formula weight of a repeat unit.

2. Parachor Method

The MacLeod–Sudgen parachor equation has been shown to be applicable to polymers [8, 15, 20, 21, 93]. Combination of Eqs. (1) and (2) gives

$$\gamma = (\rho P/M)^4 \tag{32}$$

Parachor values are independent of temperature and can be obtained from group contributions [28]. Reasonably good predictions of the surface tensions of polymers over a wide temperature range (20–200°C) have been reported, as shown in Table V.

3. Corresponding State Method

Prigogine's corresponding state theory [94], as extended for surfaces by Patterson and co-workers [95–97], gives the reduced surface tension $\tilde{\gamma}$ as

$$\tilde{\gamma} = \gamma/k^{1/3} P^{*\,2/3} T^{*\,1/3} \tag{33}$$

where k is the Boltzmann constant, P^* the Prigogine reduction parameter for pressure, and T^* the Prigogine reduction parameter for temperature. The reduced surface tension is a universal function of the reduced temperature $\tilde{T}(= T/T^*)$. Various models can be used to calculate the reduction parameters. The cell model gives

$$\alpha T = (1 - \tilde{V}^{-1/3})/[(\tfrac{4}{3})\tilde{V}^{-1/3} - 1] \tag{34}$$

where α is the thermal expansion coefficient and $\tilde{V}$ the reduced volume. When α is known, $\tilde{V}$ can be calculated by Eq. (34). Next, $\tilde{T}$ is calculated by

$$\tilde{T} = \tilde{V}^{-1}(1 - \tilde{V}^{-1/3}) \tag{35}$$

and P^* is found from

$$P^* = (\alpha/\beta) T \tilde{V}^2 \tag{36}$$

Table V

Comparison between Measured and Parachor-Calculated Surface Tension Values[a, b]

	Surface tension (dyn/cm)						
	Measured			Parachor calculated			
Polymer	20°C	140°C	180°C	20°C	140°C	180°C	Maximum difference
Polychloroprene	43.6	33.2	29.8	44.2	32.4	29.0	0.8
Polystyrene	40.7	32.1	29.2	37.5	30.9	28.5	1.2 (3.2)
Poly(methyl methacrylate)	41.4	32.0	28.9	41.1	32.0	28.9	0 (0.3)
Poly(n-butyl methacrylate)	31.2	24.1	21.7	38.1	28.6	26.0	4.5 (6.9)
Poly(vinyl acetate)	36.5	28.6	25.9	38.8	27.7	24.8	1.1 (2.3)
Polyethylene, linear	35.7	28.8	26.5	35.3	25.0	22.4	4.1

[a] Maximum difference between measured and calculated surface tensions. The measured surface tension values at 20°C were obtained by extrapolation of γ–T plots. The parenthetical maximum difference values refer to the case where the extrapolated γ at 20°C values are also considered. All data from Wu [8, 20, 64].

[b] From Wu [15].

where β is the isothermal compressibility. The reduced surface tension γ is calculated from the relation

$$\gamma \tilde{V}^{5/3} = 0.29 - (1 - \tilde{V}^{-1/3}) \ln [(\tilde{V}^{1/3} - 0.5)/(\tilde{V}^{1/3} - 1)] \qquad (37)$$

The surface tension is then obtained by inserting the values of P^*, T^*, and $\tilde{\gamma}$ in Eq. (33). The calculation requires the data for α and β, which are usually not readily available. The method gives results similar to those by the parachor method, which requires only density data.

Alternatively, the quantity $\gamma k^{-1/3} \alpha^{1/3} \beta^{2/3}$ has been shown to be a universal function of the quantity αT, that is,

$$\gamma k^{-1/3} \alpha^{1/3} \beta^{2/3} = f(\alpha T)$$

The universal function $f(\alpha T)$ has been given graphically [95–97]. Thus, if α and β are known, the values of γ can be obtained graphically [95–97].

H. Tabulation of Surface Tension Data

Table VI tabulates the available surface tension data.

Table VI

Surface Tension of Polymers

Polymers	Molecular weight	Surface tension (dyn/cm)			$-(d\gamma/dT)$ (dyn/ cm-degree)	References[a]
		20°C	140°C	180°C		
Hydrocarbon polymers						
Polyethylene, linear (L-PE)	$M_w = 67,000$	35.7	28.8	26.5	0.057	8 (21, 98)
Polyethylene, branched	$M_n = 7000$	35.3	27.3	24.6	0.067	22 (21, 98)
(B-PE)	$M_n = 2000$	33.7	26.5	24.1	0.060	98
Polyethylene, ideal (I-PE)	$MW = \infty$	36.8	30.0	27.7	0.056	15
Polypropylene, atactic (PP)	—	29.4	22.7	20.4	0.056	21
	$M_n = 3000$	28.3	23.5	21.9	0.040	99
Polypropylene, mixture of atactic and isotactic	—	30.1	23.1	20.8	0.058	100 (101)
Polyisobutylene (PIB)	$M_n = 2700$	33.6	25.9	23.4	0.064	8 (21, 40, 100, 102)
	$M_n = 2700$	34.0	26.1	23.4	0.066	21 (8, 40, 100, 102)
Polystyrene (PS)	$M_v = 44,000$	40.7	32.1	29.2	0.072	20 (100, 103–106)
	$M_n = 9290$	39.4	31.6	29.0	0.065	103 (20, 100, 104–1
	$M_n = 1680$	39.3	30.0	26.9	0.077	103 (20, 100, 104–1
Poly(α-methyl styrene) (PMS)	$M_n = 3000$	38.7	31.7	29.4	0.058	100 (39)

Table VI (*continued*)

Polymers	Molecular weight	Surface tension (dyn/cm) 20°C	140°C	180°C	$-(d\gamma/dT)$ (dyn/ cm-degree)	References[a]
ogenated hydrocarbon polymers						
olychloroprene (PCP)	$M_v = 30,000$	43.6	33.2	29.8	0.086	22 (64)
olychlorotrifluoroethylene (PCTFE)	$M_n = 1280$	30.9	22.9	20.2	0.067	107
olytetrafluoroethylene, ideal (PTFE)	$MW = \infty$	25.7	18.7	16.6	0.058	15
yl polymers						
oly(vinyl acetate) (PVA)	$M_w = 11,000$	36.5	28.6	25.9	0.066	8
ylic polymers						
oly(methyl methacrylate) (PMMA)	$M_v = 3000$	41.1	32.0	28.9	0.076	20
oly(*n*-butyl methacrylate) (P*n*BMA)	$M_v = 37,000$	31.2	24.1	21.7	0.059	20
oly(isobutyl methacrylate) (PiBMA)	$M_v = 35,000$	30.9	23.7	21.3	0.060	22
oly(*tert*-butyl methacrylate) (P*t*BMA)	$M_v = 6000$	30.4	23.3	21.0	0.059	22
yethers						
oly[(ethylene glycol)diol] (PEG-DO)	$M_w = 6000$	42.8	33.7	30.6	0.076	21 (44, 45, 69)
	$MW = 86–17,000$	42.9	31.1	27.2	0.098	44 (21, 45, 69)
oly[(ethylene glycol) dimethyl ether] (PEG-DME)	$MW = 114$	28.1 (25°C)	—	—		45 (44)
	$MW = 148$	30.6 (25°C)	—	—		45 (44)
	$MW = 182$	32.4 (25°C)	—	—		45 (44)
	$MW = 600$	37.1 (25°C)	—	—		45 (44)
oly[(propylene glycol)diol (PPG-DO)	$M_n = 3000$	30.9	21.4	18.3	0.079	45 (69)
	$MW = 400–4100$	30.4	20.8	17.6	0.080	45 (69)
oly[(propylene glycol) dimethyl ether] (PPG-DME)	$MW = 3000$	30.2 (25°C)	—	—		45 (69)
oly[(tetramethylene glycol)diol] (PTMG-DO)	$M_n = 43,000$	31.9	24.6	22.2	0.061	22 (9)
olyeplchlorohydrin (PECH)	$MW = 1500$	43.2 (25°C)	—	—		69

(*continued*)

Table VI (*continued*)

Polymers	Molecular weight	Surface tension (dyn/cm) 20°C	140°C	180°C	$-(d\gamma/dT)$ (dyn/ cm-degree)	References[a]
Polyesters						
Poly(ethylene	—	42.1	—	—	—	22
terephthalate) (PET)	—	27 ± 3 (290°C)	—	—	—	108
Polyamides						
Nylon–polyamide mixture	—	29.0 (290°C)	—	—	—	108
Nylon 66	—	47.9	—	—	—	22
	—	35.1 (285°C)	—	—	—	109
Nylon 6	—	36.1 (265°C)	—	—	—	109
Nylon 610	—	37.0 (265°C)	—	—	—	109
Nylon 11	—	22.6 (225°C)	—	—	—	109
Polycapramide	—	Data not reliable				110
Polyorganosiloxanes						
Poly(dimethylsiloxane)	60,000 cS	19.8	14.0	12.1	0.048	22 (21, 72, 75, 111)
	$M_n = 1274$	19.9	—	—	—	75
	$M_n = 607$	18.8	—	—	—	75
	$M_n = 310$	17.6	—	—	—	75
	$M_n = 162$	15.7	—	—	—	75
Poly(diethylsiloxane)	158 cS	25.7	16.9	14.0	0.073	111
Poly(methylphenylsiloxane)	102 cS	26.1	12.9	8.5	0.11	111
Random copolymers						
Poly[ethylene-*co*-(vinyl acetate)], 25% by weight of vinyl acetate (PEVA)	$M_n = 17,000$	35.5	27.5	24.8	0.067	9 (39, 112)
17.7% by weight VA	—	34.1	27.6	25.5	0.054	39 (112)
26.6% by weight VA	—	31.3	26.9	25.4	0.037	39 (112)
30.9% by weight VA	—	30.7	26.7	25.4	0.033	39 (112)
38.7% by weight VA	—	33.0	27.4	25.5	0.047	39 (112)
Poly[(ethylene oxide)-*co*-(propylene oxide)] (PEOPO)	—	Linear additivity with mole fraction of components, no surface excess (see Section II.E.1)				69
Block copolymers						
ABA block copolymers of ethylene oxide (A block), and propylene oxide (B block)	—	Pronounced surface excess behavior observed when the low-energy block is sufficiently long, i.e., the degree of polymerization greater than 20–50 (see Section II.E.3)				69

Table VI (*continued*)

Polymers	Molecular weight	Surface tension (dyn/cm)			$-(d\gamma/dT)$ (dyn/ cm-degree)	References[a]
		20°C	140°C	180°C		
ABA block copolymers of polyether (A block) and dimethylsiloxane (B block)	—	Same as above.			76	
Mixtures of homopolymers						
Mixtures of poly(ethylene glycol) and poly(propylene glycol)	—	Pronounced surface excess behavior observed (see Sections II.E.2 and II.E.4)				69 (70, 81)

[a] The first reference is the source from which the data listed are taken. The reference numbers in parentheses are additional sources.

III. INTERFACIAL TENSION

A. Works of Adhesion and Cohesion

The energy required to separate reversibly an interface is the work of adhesion, W_a, given by

$$W_a = \gamma_1 + \gamma_2 - \gamma_{12} \tag{38}$$

where γ_{12} is the interfacial tension between phases 1 and 2. If the two phases are identical, then

$$W_{cj} = 2\gamma_j \tag{39}$$

where W_{cj} is the work of cohesion of phase j.

If W_a can be expressed as a function of W_{c1} and W_{c2}, then Eq. (38) provides a basis for relating the interfacial tension to the surface tensions of the two phases. Such methods are discussed next.

B. Theories of Interfacial Tension

1. Antonoff's Rule [113–116]

Antonoff [113–116] proposed an empirical rule which states that the interfacial tension is equal to the difference between the two surface tensions, i.e., $\gamma_{12} = \gamma_1 - \gamma_2$, where $\gamma_1 \geq \gamma_2$. This can be correct only when phase 2 spreads on phase 1, and phase 2 must be a small-molecule liquid. This empirical rule is not applicable to polymer systems [15].

2. Theory of Good and Girifalco [117–121]

The ratio of the work of adhesion to the geometric mean of the works of cohesion for the two phases is defined as the interaction parameter ϕ, which can be given in terms of the molecular constants of the two individual phases by [117–121]

$$\phi = W_a/(W_{c1} W_{c2})^{0.5}$$
$$= 4(V_1 V_2)^{1/3}(V_1^{1/3} + V_2^{1/3})^{-2} \tag{40}$$
$$\times \frac{\frac{3}{4}(\alpha_1 \alpha_2)[2I_1 I_2/(I_1 + I_2)] + \alpha_1 \mu_2^2 + \alpha_2 \mu_1^2 + \frac{2}{3}(\mu_1 \mu_2)^2/kT}{[(\frac{3}{4}\alpha_1^2 I_1 + 2\alpha_1 \mu_1^2 + \frac{2}{3}\mu_1^4/kT)(\frac{3}{4}\alpha_2^2 I_2 + 2\alpha_2 \mu_2^2 + \frac{2}{3}\mu_2/kT)]^{1/2}}$$

where the subscripts 1 and 2 refer to the phases 1 and 2, respectively, V the molar volume of the interacting element, α the polarizability, I the ionization potential, and μ the permanent dipole moment.

Combination of Eqs. (38)–(40) gives the equation of Good and Girifalco [117–121]:

$$\gamma_{12} = \gamma_1 + \gamma_2 - 2\phi(\gamma_1 \gamma_2)^{1/2} \tag{41}$$

The utility of this approach is limited by the lack of information about the molecular constants for polymers. Another difficulty is that ϕ values must be accurately known. For instance, a 10% error in ϕ will result in a 50% error in the calculated γ_{12} since, for polymers, $\gamma_1 \simeq \gamma_2$.

However, Eq. (41) is a very useful formal relation linking interfacial tension with the surface tensions. Values of ϕ between polymers have been calculated from the interfacial and surface tension data [8, 20] and are found to be in the range of 0.8–1.0. Empirically, it has been shown [20] that

$$d\phi/dT = 0 \tag{42}$$

3. Theory of Fractional Polarity

Application of the concept of force additivity in a semicontinuum model has given simple and useful equations relating the interfacial tension to the surface tensions and the polarities of the two phases [15, 22, 64, 91]. Intermolecular energies are assumed to consist of additive nonpolar (dispersion) and polar components. A similar concept has been applied to solubility parameters [122, 123]. The surface tension of phase j is then given by

$$\gamma_j = \gamma_j^d + \gamma_j^p \tag{43}$$

where the subscripts d and p refer to nonpolar (dispersion) and polar components, respectively. Using certain approximations, two particularly useful equations can be obtained.

a. Harmonic-Mean Equation

The harmonic-mean equation is given by [22, 64, 91]

$$\gamma_{12} = \gamma_1 + \gamma_2 - \frac{4\gamma_1^d \gamma_2^d}{\gamma_1^d + \gamma_2^d} - \frac{4\gamma_1^p \gamma_2^p}{\gamma_1^p + \gamma_2^p} \tag{44}$$

which has been found to give good results for polymers. Equation (44) can be rewritten in the form of Eq. (41), and the interaction parameter ϕ can then be given by [64]

$$\phi = \frac{2x_1^d x_2^d}{g_1 x_1^d + g_2 x_2^d} + \frac{2x_1^p x_2^p}{g_1 x_1^p + g_2 x_2^p} \tag{45}$$

where x_j^p is the polarity of phase j, defined as

$$x_j^p = \gamma_j^p / \gamma_j = 1 - x_j^d \tag{46}$$

and

$$g_1 = (\gamma_1/\gamma_2)^{1/2} = 1/g_2 \tag{47}$$

b. Geometric-Mean Equation

The geometric-mean equation is given by [22, 64, 91]

$$\gamma_{12} = \gamma_1 + \gamma_2 - 2(\gamma_1^d \gamma_2^d)^{0.5} - 2(\gamma_1^p \gamma_2^p)^{0.5} \tag{48}$$

which can be rewritten in the form of Eq. (41), and the interaction parameter ϕ can then be given by [20]

$$\phi = (x_1^d x_2^d)^{0.5} + (x_1^p x_2^p)^{0.5} \tag{49}$$

If the polar term is neglected, Eq. (48) becomes the Fowkes equation [124]. It should, however, be stressed that the polar term plays a decisive role in determining the interfacial tension between polar polymers [64]. The geometric-mean equation is less accurate than the harmonic-mean equation for polymers, as will be seen later.

4. Mean-Field Theory of Helfand, Tagami, and Sapse [10–13]

Interfacial regions between two incompatible polymers are formed by interdiffusion of the segments of the two polymers. Assuming that the segments are in the average (mean) molecular environment and using the Gaussian random walk statistics to describe the interdiffusion process, Helfand and co-workers [10–13] obtained the interfacial tension as

$$\gamma_{12} = \tfrac{2}{3}kT\alpha^{1/2}\left[(\beta_1{}^3 - \beta_2{}^3)/(\beta_1{}^2 - \beta_2{}^2)\right] \tag{50}$$

and a characteristic interfacial thickness a_1 as

$$a_i = \left[(2/\alpha)(\beta_1{}^2 + \beta_2{}^2)\right]^{1/2} \tag{51}$$

where α is related to the interaction parameter χ of Flory and Huggins by

$$\alpha = \chi/V_r \tag{52}$$

where V_r is the molar volume of the reference segment, and the parameter β_j is defined as

$$\beta_j = \tfrac{1}{6}\rho_{0j}b_j{}^2 \tag{53}$$

where ρ_{0j} is the number density of pure phase j, and b_j is the length of a statistical unit defined such that

$$Z_j b_j{}^2 = \langle R^2 \rangle \tag{54}$$

where Z_j is the degree of polymerization and $\langle R^2 \rangle$ the mean square end-to-end distance. Figure 6 shows the interfacial density profile between

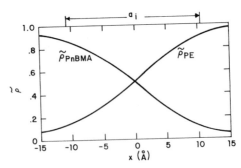

Fig. 6   Density profile across the interface and the interfacial thickness between polyethylene and poly(n-butyl methacrylate) at 140°C. The term $\tilde{\rho}_i$ is the relative density of phase i defined by ρ_i/ρ_{0i} where ρ_{0i} is the density of the pure phase i, a_i is the interfacial thickness defined by Eq. (70). (From Helfand and co-workers [10–13].)

polyethylene and poly(n-butyl methacrylate), as calculated by the mean-field theory. The reduced number density $\tilde{\rho}_j$ is defined as

$$\tilde{\rho}_j = \rho_j/\rho_{0j} \tag{55}$$

where ρ_j is the number density at a given location in the interfacial region. The coordinate x is the distance taken to be zero at $\tilde{\rho}_1 = \tilde{\rho}_2 = \frac{1}{2}$.

5. Lattice Theories of Helfand [13] and Roe [14]

Helfand [13] and Roe [14] have independently proposed lattice theories of the interfacial tensions between polymers. The theories consist in calculating the configurational entropy of the diffuse interfaces corresponding to the minimum free energy. Because of different assumptions used, Helfand and Roe obtained different results.

For infinite molecular weights, Helfand obtained [13]

$$\gamma_{12} = (kT/a)(m\chi)^{1/2} \tag{56}$$

and a characteristic interfacial thickness a_i as

$$a_i = 2(m/\chi)^{1/2} \tag{57}$$

where a is the cross-sectional area of a lattice cell, and m is a lattice constant defined such that if z is the number of nearest neighbors of a lattice cell, then the number of nearest neighbors in the same layer parallel to the interface is $z(1-2m)$ and in each of the adjacent layers is zm. Comparing Eqs. (56) and (57) with Eqs. (50) and (51) shows that Helfand's mean-field theory and his lattice theory are consistent in that both predict that γ_{12} is proportional to $\chi^{1/2}$, and a_i is proportional to $\chi^{-1/2}$.

On the other hand, for infinite molecular weights, Roe obtained [14]

$$\gamma_{12} = (2)^{-1/4}\tfrac{4}{3}(kT/a)m^{1/2}\chi^{3/4} \tag{58}$$

and a characteristic interfacial thickness a_i is given as

$$a_i = 2(2)^{3/4}m^{1/2}b\chi^{-1/4} \tag{59}$$

where b is the distance of separation between adjacent lattice layers. Eqs. (58) and (59) predict that γ_{12} is proportional to $\chi^{3/4}$ and a_i is proportional to $\chi^{-1/4}$. These are different from Helfand's results.

Estimations of the interfacial tensions by the lattice theories are difficult, since the values for the lattice parameters m and b are uncertain. Figure 7 shows the variation of the interfacial tension versus chain length (for $r = r_1 = r_2$) for three selected values of χ, as calculated by the general equation of Roe's theory. As the chain length r increases beyond the critical value $(2/\chi)$ for phase separation, the interfacial tension increases rapidly at first and then

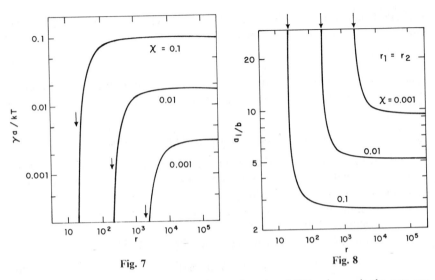

Fig. 7

Fig. 8

Fig. 7 Interfacial tension γ between polymers in units of kT/a where a is the area per lattice site, as a function of the degree of polymerization r. Arrows indicate the values of $r(= 2/\chi)$ at which incipient phase separation is expected to occur. (From Roe [14].)

Fig. 8 The effective interfacial thickness a_i, defined by Eq. (72) between polymers, in units of lattice spacing b, as functions of the degree of polymerization r. Arrows indicate the values of r for incipient phase separation. (From Roe [14].)

levels off to a plateau. Figure 8 shows the dependence of the interfacial thickness a_i on the degree of polymerization r, again showing the asymptotic nature of the variation.

C. Temperature Dependence

The interfacial tension varies more slowly with temperature than the surface tension. The temperature coefficient of the interfacial tension $\partial\gamma/\partial T$ is usually smaller than -0.03 dyn/cm-degree. Figure 9 shows the interfacial tension of some polymers as a function of temperature.

The temperature coefficient of interfacial tension can be obtained by differentiating Eq. (41) with respect to temperature,

$$\frac{d\gamma_{12}}{dT} = \frac{d\gamma_1}{dT} + \frac{d\gamma_2}{dT} - \phi\left[g_1\left(\frac{d\gamma_2}{dT}\right) + g_2\left(\frac{d\gamma_1}{dT}\right)\right] \qquad (60)$$

in which we have used Eq. (42). Thus, $d\gamma_{12}/dT$ can be calculated from the surface tensions of the individual phases and ϕ value which can be obtained if γ_{12} is known at a single temperature. Table VII compares the experimental and calculated values of $d\gamma_{12}/dT$ for some polymers. Good agreement can be seen.

Table VII

Comparison between Measured and Calculated $-(\partial\gamma_{12}/\partial T)$ values[a,b]

Polymer pairs	$-(\partial\gamma_{12}/\partial T)$	
	Measured	Eq. (74)
Polyethylene–poly(vinyl acetate)	0.027	0.025
Polyethylene–poly(methyl methacrylate)	0.018	0.022
Polyisobutylene–poly(vinyl acetate)	0.020	0.018
Poly(n-butyl methacrylate)–poly(vinyl acetate)	0.010	0.006

[a] The measured values are from Wu [8,20]. The term γ_{12} is the interfacial tension in dyn/cm per degree.
[b] From Wu [15].

D. Effect of Polarity

The interfacial tension arises mainly from the disparity between the polarities of the two phases. The extent of nonpolar (dispersion) interaction

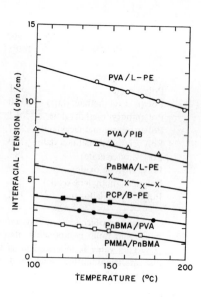

Fig. 9   Variation of interfacial tension with temperature. (After Wu [8, 15, 20, 22, 64].)

between two phases does not vary greatly from system to system, but that of polar interactions can vary greatly. Thus, the polar interactions largely determine the magnitude of the interfacial tension [64].

From Eq. (41), we can see that γ_{12} will have the minimum value when ϕ has the maximum value, and vice versa. As can be seen from Eqs. (45) and (49), ϕ has the maximum value when the polarities of the two phases are exactly equal, and the minimum value of zero when the polarities are exactly mismatched. Based on the harmonic-mean equation, the maximum value for ϕ is $2/(g_1 + g_2)$; based on the geometric-mean equation, the maximum value for ϕ is unity. Thus, the interfacial tension increases with increasing disparity between the polarities of the two phases.

The polarity of a polymer can be calculated from the interfacial tension between this polymer and a nonpolar polymer such as polyethylene, which can be regarded to be completely nonpolar. Thus, if the interfacial tension of a polar polymer against polyethylene is known, the polarity of the polar polymer can be calculated by Eq. (44) or Eq. (48). The polar interaction terms in these equations are zero, and the nonpolar (dispersion) component of the surface tension of the polar polymer can be obtained. The polar component is then obtained by difference.

Table VIII

Polarity Values for Some Polymers from Interfacial Tension Data[a,b]

Polymers	γ at 140°C	Eq. (52), harmonic-mean equation			Eq. (57), geometric-mean equation		
		γ^d	γ^p	x^p	γ^d	γ^p	x^p
Polychloroprene	33.2	29.6	3.6	0.11	29.5	3.7	0.11
Polystyrene	32.1	26.7	5.4	0.17	26.3	5.8	0.18
Poly(methyl methacrylate)	32.0	23.0	9.0	0.28	22.7	9.3	0.29
Poly(*n*-butyl methacrylate)	24.1	20.3	3.8	0.16	19.7	4.4	0.18
Poly(isobutyl methacrylate)	23.7	20.4	3.3	0.14	20.0	3.7	0.16
Poly(*tert*-butyl methacrylate)	23.3	20.5	2.8	0.12	19.2	4.1	0.18
Poly(vinyl acetate)	28.6	19.2	9.4	0.33	18.4	10.2	0.36
Poly(ethylene oxide)	33.8	24.2	9.6	0.28	24.5	9.3	0.28
Poly(tetramethylene oxide)	24.6	21.1	3.5	0.14	20.8	3.8	0.15
Poly(dimethylsiloxane)	14.1	13.5	0.6	0.04	12.1	2.0	0.14

[a] γ is the surface tension in dyn/cm at 140°C; γ^d and γ^p are its dispersion (nonpolar) and polar components, also at 140°C and x^p is the polarity, defined by γ^p/γ. This value is independent of temperature. The measured surface tension and interfacial tension values used are taken from Wu [8, 20, 22, 64].

[b] From Wu [15].

Table VIII lists the polarity values for some polymers as calculated from the interfacial tension data against polyethylene by both the harmonic-mean and the geometric-mean equations. Although the results are similar, the harmonic-mean equation is preferred since it can correctly predict the interfacial tension between polar polymers from the polarity values listed in the table, whereas the geometric-mean equation is not satisfactory, as will be shown later. The polarity values are found to be independent of temperature [15, 64]

$$dx^p/dT = 0 \qquad (61)$$

within the temperature range of 20–200°C investigated.

Organic polymers can be very polar. For instance, poly(vinyl acetate) is as much as 33% polar, poly(methyl methacrylate) 28%, polychloroprene 11%. It is because of the high polarity that poly(vinyl acetate) has an interfacial tension as high as 11.3 dyn/cm against polyethylene at 140°C. If poly(vinyl acetate) were nonpolar, the interfacial tension against polyethylene would have been nearly zero.

The polarity values obtained above are in good agreement with those estimated from solubility parameters [64]. In terms of the solubility parameter, the polarity may be defined by [64, 91]

$$\chi^p = E^p/E = (\delta_p/\delta)^2 \qquad (62)$$

where E is the cohesive energy, E^p the polar component of E, δ the solubility parameter, and δ_p the polar component of δ. The polarity values obtained by using Hansen's empirical solubility parameters [122, 123] agree well with those obtained from interfacial tension data [64].

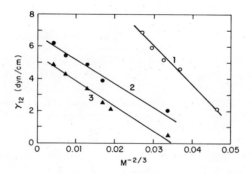

Fig. 10 Interfacial tensions (24°C) between *n*-alkanes and $C_{12.5}F_{27}$ (Curve 1), and between dimethylsiloxanes and $C_{12.5}F_{27}$ (Curve 2), or C_8F_{18} (Curve 3), plotted versus M^{-1} where M is the molecular weight of the alkanes or dimethylsiloxanes. (After LeGrand and Gaines [128].)

E. Molecular Weight Dependence

Applying Eq. (5) in Eq. (38) gives [128]

$$\gamma_{12} = k_0 - (k_1/M_1^{2/3}) - (k_2/M_2^{2/3}) \tag{63}$$

where k_1 and k_2 are constants independent of molecular weights, and k_0 is given by

$$k_0 = \gamma_{1\infty} + \gamma_{2\infty} - 2\phi[\gamma_{1\infty} - (k_1/M_1^{2/3})]^{1/2}[\gamma_{2\infty} - (k_2/M_2^{2/3})]^{1/2} \tag{64}$$

Thus, strictly k_0 is dependent on molecular weight. However, the property of Eq. (64) is such that k_0 varies little with molecular weight and is approximately a constant. Thus, if the molecular weight of one phase is kept constant, a plot of γ_{12} versus $M^{-2/3}$ will give a straight line. Such plots for the interfacial tensions between *n*-alkanes and a fluorocarbon ($C_{12.5}F_{27}$) and between dimethylsiloxanes and the same fluorocarbon are shown in Figure 10.

Equation (63) may be used to formulate a compatibility criterion. To be compatible, the interfacial tension must be zero or negative. Thus, a compatibility criterion is

$$\frac{k_1}{M_1^{2/3}} + \frac{k_2}{M_2^{2/3}} \geqslant k_0 \tag{65}$$

for compatibility. The compatibility limits for homologous series can thus be calculated from values of k_0. Some examples are given in Table IX.

Table IX

Miscibility Limit Estimated from Eq. (81)[a]

System	Estimated miscibility limit	k_0 dyn/cm
Perfluoroalkane–alkane	C_6F_{14}–$C_{6.5}H_{15}$	18.3
Perfluoroalkane–dimethylsiloxane	C_7F_{16}–Hexamethyldisiloxane	11.0
Perfluoroalkane–polyisobutylene	$C_{10.5}F_{23}$–isooctane	18.0
Polyisobutylene–dimethylsiloxane	($\overline{M}_n = 1250$)–Dodecamethyl-pentasiloxane	5.8

[a] After LeGrand and Gaines [128].

F. Effects of Additives

The interfacial tension between two polymers may be reduced with additives to improve the compatibility and adhesion between phases. This concept has been important in the technology of polymer blends [129] (see Volume 2, Chapter 12).

One important class of such additives is the block and graft copolymers with blocks or segments having the same chemical compositions as those of the two polymers to be blended [129]. An A–B block or graft copolymer will tend to accumulate at, orient, and bridge the interface between polymers A and B, and thus reduce the interfacial tension and improve the compatibility and adhesion. Block copolymers are usually more effective than graft copolymers, because the former can orient more readily than the latter. To be effective, the segmental molecular weight should conform to the specifications described in Volume 2, Chapter 12.

The block or graft copolymer may be added externally or formed in situ in the polymer blend. For instance, when polymer A containing carboxyl groups is blended with polymer B containing amino groups, block or graft copolymers of A–B may be formed in situ during the melt blending as a result of acid–base reaction. On the other hand, when a polyamide is melt blended with a polyester, block or graft copolymers of polyester–polyamide may be formed as a result of amine–ester or amide–ester interchange reaction. Numerous other chemical reactions may be utilized in the in situ formation of block or graft copolymers [129].

Figure 11 shows that only about 1–2% by weight of the block copolymer additive is sufficient to achieve the maximum effect on the reduction of interfacial tension [108]. Table X shows the effect of various block copolymers and homopolymers containing functional groups on the reduction of the interfacial tension between a poly(dimethylsiloxane) and a copolymer of polyoxyethylene–polyoxypropylene [108]. Block copolymers of poly(dimethylsiloxane)–polyoxyethylene and block copolymers of poly(dimethylsiloxane)–polyoxypropylene are found to reduce the interfacial tension of the system by 50–60%. Amino groups on poly(dimethylsiloxane) are less effective, causing only 18–28% reduction in interfacial tension. In contrast, hydroxyl groups on poly(dimethylsiloxane) are nearly totally effective [108].

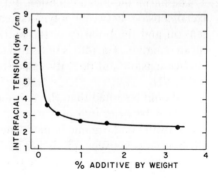

Fig. 11 Effect of addition of 60/40 poly-[(dimethylsiloxane)–*co*-polyoxyethylene] on interfacial tension between poly(dimethylsiloxane) (10,300 cP) and Ucon 75H copolyester fluid (11,000 cP) at 25°C. (From Patterson *et al.* [108].)

Table X

Effect of Additives on Interfacial Tension between DC-200-100,000 cS
Silicone and Ucon 75-H-14,000 Copolyether Fluid[a]

Additives at 2% by Weight	Interfacial tension (dyn/cm) at 25°C	Reduction % obtained with additive
None	8.3	—
60/40 Poly[(dimethylsiloxane)-*co*-polyoxyethylene]	2.3	72
25/75 Poly[(dimethylsiloxane)-*co*-polyoxyethylene]	3.0	64
25/75 Poly[(dimethylsiloxane)-*co*-polyoxypropylene]	4.1	51
Poly(dimethylsiloxane) with 10% carboxyl groups on alkyl side chains	3.1	63
Poly(dimethylsiloxane) with 20% carboxyl groups on alkyl side chains	3.6	57
Poly(dimethylsiloxane) with carboxyl end groups	4.2	49
Poly(dimethylsiloxane) with hydroxyl end groups	8.1	0
Poly(dimethylsiloxane) with 1% amino groups on alkyl side chains	6.0	28
Poly(dimethylsiloxane) with 6% amino groups on alkyl side chains	6.8	18

[a] From Patterson *et al.* [108].

G. Methods of Estimation

The interfacial tension between polymers can be predicted reasonably accurately from macroscopic properties of the two phases using the harmonic-mean equation. The mean-field theory can also give reasonable predictions.

1. Harmonic-Mean and Geometric-Mean Equations

The harmonic-mean equation, Eq. (44) can be used to predict the interfacial tension between polymers reasonably accurately from the surface tension and the polarity of the individual phases, whereas the geometric-mean equation, Eq. (48) is not satisfactory, as can be seen in Table XI. The surface tension and polarity values used in the calculation are taken from Table VII.

It should be noted that, the harmonic-mean equation is preferred for the interfaces between two polymers or between a polymer and a low-energy liquid (water and organic liquids), the geometric-mean equation is better for the interfaces between a polymer and a high-energy material (metals, metal oxides, inorganic glasses, and mercury) [15].

Table XI

Comparison between Measured and Calculated Interfacial
Tensions between Polymers at 140°C[a]

		Interfacial tension at 140°C	
		Calculated by	
		Harmonic-mean equation, Eq. (52)	Geometric-mean equation, Eq. (57)
Polymer pairs[b]	Measured		
PEO–PDMS	9.9	10.7	4.8
PTMO–PDMS	6.3	3.8	1.5
PVA–PDMS	7.4	8.5	3.6
PCP–PDMS	6.5	8.2	4.1
PnBMA–PDMS	3.8	3.7	1.4
PtBMA–PDMS	3.3	2.9	1.2
PCP–PnBMA	1.6	1.7	0.9
PCP–PS	0.5	0.5	0.4
PVA–PS	3.7	2.3	1.2
PVA–PnBMA	2.9	2.4	1.2
PMMA–PnBMA	1.9	2.2	1.0
PMMA–PtBMA	2.3	3.4	1.6
PMMA–PS	1.7	1.2	0.5
PEO–PTMO	3.9	3.1	1.4
PVA–PTMO	4.6	2.8	1.6

[a] After Wu [22, 64].
[b] See Table VI for abbreviations. The measured interfacial tension values are taken from Wu [8, 20, 22, 64].

Applications of the harmonic-mean equation to contact angles and surface energies of organic solids have been discussed elsewhere [22, 62, 63, 130].

2. Mean-Field and Lattice Theories [10–14]

The mean-field theory of Helfand and co-workers [10–13] can give reasonably good predictions of the interfacial tensions between polymers. Values of Flory and Huggins' interaction parameters between polymers are usually unknown but can be estimated from Hildebrand and Scott's relation [108]

$$\alpha = \chi/V_r = (1/kT)(\delta_1 - \delta_2)^2 \qquad (66)$$

in which the values of δ are assumed to be independent of temperature. Table XII lists the characteristic values of the parameters required for

Table XII

Parameters Used in the Mean-Field Theory Calculation of the
Interfacial Tension between Polymers[a]

Polymer	Abbreviation	δ (cal$^{1/2}$/cm$^{2/3}$)	Density (moles/cm^3)	$\langle R^2 \rangle^{1/2}/M^{1/2} \times 10^9$ (cm/gm$^{1/2}$)	b (Å)	$\beta \times 10^9$ (cm$^{-1/2}$)
Polystyrene	PS	9.0	0.0102	6.5	6.6	2.7
Poly(methyl methacrylate)	PMMA	9.1	0.0116	6.4	6.4	2.8
Poly(n-butyl methacrylate)	PnBMA	8.6	0.0073	5.1	6.1	2.1
Poly(vinyl acetate)	PVA	9.55	0.0139	7.1	6.5	3.2
Polyethylene	PE	7.8	0.0305	9.5	5.0	3.6
Polychloroprene	PCP	8.8	0.0140	7.5	7.1	3.4
Poly(tetramethylene oxide)	PTMO	8.6	0.0135	9.6	8.1	3.9
Poly(dimethylsiloxane)	PDMS	7.5	0.0132	6.7	5.8	2.7

[a] From Helfand and Sapse [11].

Table XIII

Comparison between the Measured and the Mean-Field
Theory Calculated Interfacial Tension Values
between Polymers[a]

	Interfacial Tension dyn/cm at 140°C		Interfacial thickness
Polymer pairs	Measured[b]	Eq. (50)	a_i (Å)
PE–PS	5.9	4.7	15
PE–PVA	11.3	7.4	11
PE–PMMA	9.7	5.1	14
PE–PnBMA	5.3	2.8	21
PS–PCP	0.5	0.7	88
PS–PVA	3.7	1.9	31
PS–PMMA	1.7	0.3	160
PDMS–PVA	7.4	7.8	8
PDMS–PCP	6.5	4.9	14
PDMS–PnBMA	3.8	3.3	13
PDMS–PTMO	6.3	4.4	17
PnBMA–PVA	2.9	3.1	16
PnBMA–PCP	1.6	0.7	82
PnBMA–PMMA	1.9	1.5	29

[a] From Helfand and Sapse [11].
[b] The measured values were taken from Wu [8, 15, 20, 22, 64].

predicting the interfacial tensions. Table XIII lists the predicted values of interfacial tension and interfacial thickness together with the experimental values of interfacial tension for some polymer pairs.

Applications of the lattice theories of Helfand [13] and Roe [14] are difficult, since the values of the lattice constants a, b, and m are uncertain; see Eqs. (56)–(59).

H. Tabulation of Interfacial Tension Data

Table XIV summarizes the available interfacial tension data.

Table XIV

Tabulation of Typical Values of Interfacial Tensions between Polymers

Polymers pairs[a]	Interfacial tension (dyn/cm)			$-(\partial \gamma / \partial T)$ dyn/cm-°C	Ref.[c]
	100°C	140°C	180°C		
Polyethylene versus others					
PE–PP	—	1.1	—	—	100
L-PE–PS	6.7	5.9	5.1	0.020	20 (100)
B-PE–PCP[b]	4.0	3.7	3.4	0.0075	22
L-PE–PVA[b]	12.4	11.3	10.2	0.027	8 (9, 101)
B-PE–PEVA, 25 wt % VA	1.6	1.4	1.2	0.005	9 (101)
L-PE–PMMA	10.4	9.7	9.0	0.018	20
L-PE–PnBMA[b]	5.9	5.3	4.7	0.015	20
B-PE–PiBMA	4.7	4.3	3.9	0.010	22
B-PE–PtBMA	5.2	4.8	4.4	0.009	22
B-PE–PEO	10.3	9.7	9.1	0.016	9
B-PE–PTMO	4.5	4.2	3.9	0.007	22 (9)
B-PE–PDMS	5.2	5.1	5.0	0.002	22 (9, 131)
Polypropylene versus others					
PP–PS	—	5.1	—	—	100
PP–PDMS	3.0	2.9	2.8	0.002	100
Polyisobutylene versus others					
PIB–PVA[b]	8.3	7.5	6.7	0.020	8
PIB–PDMS	4.4	4.2	4.0	0.006	131
Polystyrene versus others					
PS–PCP	0.6	0.5	0.4	0.0014	22
PS–PVA	3.9	3.7	3.5	0.0044	22
PS–PMMA	2.2	1.6	1.1	0.013	20
PS–PEVA, 38.7 wt % VA	—	5.6	—	—	101
PS–PDMS	6.1	6.1	6.1	0.000	131

(*continued*)

Table XIV (*continued*)

	Interfacial tension (dyn/cm)			$-(\partial\gamma/\partial T)$	Ref.[c]
Polymers pairs[a]	100°C	140°C	180°C	dyn/cm-°C	
Polychloroprene versus others					
PCP–P*n*BMA	1.8	1.6	1.4	0.0047	22
PCP–PDMS	6.7	6.5	6.3	0.0050	22
Poly(vinyl acetate) versus others					
PVA–P*n*BMA[b]	4.6	4.2	3.8	0.010	20
PVA–PDMS	7.7.	7.4	7.1	0.0081	20 (9, 131)
PVA–PTMO	4.9	4.6	4.3	0.0081	9
PVA–PEVA, 25 wt % VA	6.0	5.8	5.6	0.0043	9
Poly(methyl methacrylate) versus others					
PMMA–P*n*BMA[b]	2.4	2.0	1.5	0.012	20
PMMA–P*t*BMA	2.5	2.3	2.1	0.005	22
Poly(dimethylsiloxane) versus others					
PDMS/P*n*BMA	3.9	3.8	3.6	0.0037	22
PDMS/P*t*BMA	3.4	3.3	3.2	0.0025	22
PDMS/PEO	10.2	9.9	9.6	0.0078	9
PDMS/PTMO	6.3	6.3	6.2	0.0012	22 (9)
Polyethers versus others					
PEO–PEVA, 25 wt % VA	6.0	5.8	5.6	0.0045	9
PEO–PTMO	4.1	3.9	3.7	0.0051	9
PTMO–PEVA, 25 wt % VA	1.3	1.2	1.1	0.0023	9

[a] PCP = polychloroprene, PDMS = poly(dimethylsiloxane), PE = polyethylene, B-PE = branched polyethylene, L-PE = linear polyethylene, PEO = poly(ethylene oxide), PEVA = poly[ethylene-*co*-(vinyl acetate)], PIB = polyisobutylene, PMMA = poly(methyl methacrylate), P*n*BMA = poly(*n*-butyl methacrylate), PiBMA = poly(isobutyl methacrylate), P*t*BMA = poly(*tert*-butyl methacrylate), PP = polypropylene, PS = polystyrene, PTMO = poly(tetramethylene oxide) or polytetrahydrofuran, PVA = poly(vinyl acetate).

[b] See also Fig. 9.

[c] Data are taken from the first reference. The numbers in parentheses are additional references.

IV. ADHESION BETWEEN POLYMERS

A. Basic Concepts

Adhesion refers to bonding or joining of dissimilar bodies, while *autohesion* or *cohesion* refers to joining of identical bodies. If an adhesively bonded structure breaks under low stress, the structure is customarily said to have "poor adhesion." This usage can be misleading, since the fracture that occurred may not have been an adhesive (interfacial) failure, but rather may have been a cohesive or mixed failure. Careful investigations of the locus of fracture are important to determine the weak zones of the structure.

The free energy required to separate reversibly two phases from equilibrium separation to infinity at constant T and P is the work of adhesion, as defined in Eq. (38). The maximum force per unit area required in such a process is defined as the ideal adhesive strength, σ^a which is related to the work of adhesion W_a by [132]

$$\sigma^a = \left[\tfrac{16}{9}(3)^{1/2}\right](W_a/z_0) \qquad (67)$$

where z_0 is the equilibrium separation between the two phases.

For polymers, typically $W_a = 50$ ergs/cm^2, $z_0 = 5$ Å, so the ideal adhesive strength is calculated to be about 15,000 psi. This is at least one order of magnitude higher than the practical adhesive strength usually observed. This indicates that, if two phases are in perfect molecular contact, the van der Waals forces alone would be sufficient to give strong adhesive strength. However, perfect molecular contact is difficult to achieve in practice.

Several adhesion theories have been proposed. They have been a subject of much controversy, however, arising perhaps from overclaims of their merits [133]. Most theories deal with the formation of the adhesive bond. However, the fracture strength of an adhesive bond depends not only on the extent of bond formation but also on the mechanical response of the materials, as is discussed later.

Although each theory has been claimed to explain exclusively the phenomenon of adhesion, each theory, in fact, deals only with one certain aspect of the complex phenomenon. Thus, a composite theory should be more correct and useful. An adhesive bond is formed in the first stage by forming interfacial contact by wetting; in the second stage by interdiffusion and/or chemical bonding. The mechanical strength of the adhesive bond is determined by the extent of interfacial contact, the size of defects, and the extent of irreversible deformations occurring in the fracture zone, which is a function of the viscoelasticity of the bulk phases, interfacial chemical bonding, and interdiffusion.

B. Fracture Theory of Adhesion [134–135]

The discrepancy between the ideal adhesive strength and the practical fracture strength arises from the fact that the actual fracture process is not reversible, and that defects (microcracks) always exist in the interfacial zone and the bulk phases. The Griffith–Irwin theory [136] of cohesive fracture has been extended to the adhesive fracture [134, 135]. The fracture strength of an adhesive bond is determined by the size of the defect and the energy dissipated in the irreversible deformations occurring during the fracture process, that is, [135]

$$f = k(E\zeta/d)^{1/2} \qquad \text{for plain stress} \qquad (68)$$

where f is the fracture stress, k a constant $= (4/\pi)^{1/2}$, E the elastic modulus, d the crack (defect) length, and ζ the fracture energy, that is, the total work per unit area of fracture surface, which is stored as the surface free energy or dissipated in other irreversible processes such as plastic deformation, light emission, and electric discharge. Thus,

$$\zeta = W_a + \psi \qquad (69)$$

where W_a is the work of adhesion (Eq. 38) and ψ the total work for the irreversible processes. The term ψ is usually several orders of magnitude larger than W_a, about 10^5 ergs/cm^2 versus 10^2 ergs/cm^2 [137, 138] and so

$$\zeta \simeq \psi \qquad (70)$$

On the other hand, an adhesive bond may be reversibly separated at the

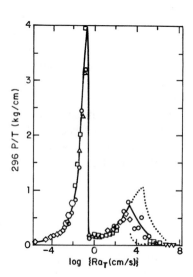

Fig. 12   Master relation for peel force P against rate of peel R, reduced to 23°C, for A polymer adhering to Mylar[®] Broken curves denote the extreme values when stick/slip peeling occurred. (From Gent and Petrich [140].)

interface when immersed in a liquid or vapor, and which can diffuse to and accumulate at the interface. In this case, ψ will be negligible and so [139]

$$\zeta \simeq W_a = \gamma_{1w} = \gamma_{2w} - \gamma_{12} \leqslant 0 \tag{71}$$

where the subscripts 1, 2, and w refer to phase 1, 2, and the liquid w, respectively.

The term ζ has been measured for some adhesive bonds [138]. In general, ζ is rate and temperature dependent. At slow rates, the fracture tends to be cohesive and change to adhesive fracture at higher rates [140]. The adhesive strength at various rates and temperature can be superimposed on a master curve by using the WLF shift factor [140]:

$$a_T = -17.4(T - T_g)/(52 + T - T_g) \tag{72}$$

where a_T is the time–temperature shift factor, T the temperature, and T_g the glass transition temperature. Figure 12 is such a plot.

Rivlin and Thomas' generalization [141] of the Griffith–Irwin theory has been utilized to formulate energy criteria of fracture for viscoelastic adhesives [142]. The fracture energy is given by [142]

$$\zeta = (\pi/\lambda^{1/2})(c/2)W_b \qquad \text{for simple tension} \tag{73}$$

$$= h_0 W_b \qquad \text{for pure shear} \tag{74}$$

$$= P/h_0 \qquad \text{for peeling} \tag{75}$$

where ζ is the fracture energy, λ the extension ratio at fracture, c the crack length, W_b the strain energy per unit volume at fracture, h_0 the unstrained thickness of the adhesive layer, and P the peel force per unit width. These criteria have been applied to the adhesion of a soft viscoelastic adhesive on a rigid substrate [142].

C. Weak Boundary Layer Theory [143]

The theory maintains that (a) a "properly" prepared adhesive bond can never fracture interfacially, and (b) fracture usually occurs cohesively in a weak boundary layer, which may be near the interface [143]. The first statement has been amply refuted, that is, true interfacial fracture can occur [135]. The second statement is doubtlessly true in some cases. Many examples have been given [143, 144]. However, systematic studies of the strength of weak boundary layer sandwiched between two strong phases show that the fracture toughness of the sandwich structure is essentially that of the strong phases until the thickness of the weak boundary layer exceeds 1 μm, at which point the fracture toughness falls by about 85% [145]. This tends to cast doubt on many previous interpretations based on the concept of weak boundary layer.

Several experimental techniques can be used fruitfully to locate the locus of fracture, including ESCA (electron spectroscopy for chemical analysis), SEM (scanning electron microscopy), TEM transmission electron microscopy), FT–IR (Fourier transform infrared spectroscopy, reflectance, or attenuated total reflectance), and contact angle measurements.

D.   Wetting–Contact Theory of Adhesion [64, 146–153]

The theory states that (a) van der Waals forces alone are sufficient to give strong adhesion provided the two phases are in molecular contact, and (b) the extent of interfacial contact and so the adhesive strength are affected by the energetics of wetting. These are certainly true but may not be the universal criteria since kinetics of wetting and weak boundary layers can preclude these principles.

The spreading coefficient has been suggested to be the driving force for wetting [64]:

$$\lambda_{12} = \gamma_2 - \gamma_1 - \gamma_{12} \tag{76}$$

where λ_{12} is the spreading coefficient for phase 1 on phase 2. If the size of unwetted interfacial void d can be related to λ_{12} by

$$d = d_0[1 - (\lambda_{12}/\gamma_2)]^n \tag{77}$$

where d_0 and n are positive constants, then from Eq. (68) we have

$$f = k[E\zeta/d_0]^{1/2}[1 - (\lambda_{12}/\gamma_2)]^{-n/2} \tag{78}$$

Thus, wettability is directly related to the adhesive strength. It should be noted that the mechanical properties of the systems E and ζ also play important roles.

A straight line is obtained when the adhesive strengths of a given phase bonded to different second phases are plotted against the surface tensions of the second phase for the system shown in Fig. 13. If E and ζ are similar and γ_{12} is relatively small, then the linearity suggests that the value of n is about 2. This is further supported by the fact that a plot of the adhesive strengths versus the interfacial tension gives a straight line for the system shown in Fig. 14.

In terms of the harmonic-mean equation, the spreading coefficient can be given by [64]

$$\lambda_{12} = 4\{[\gamma_1^d\gamma_2^d/(\gamma_1^d + \gamma_2^d)] + [\gamma_1^p\gamma_2^p/(\gamma_1^p + \gamma_2^p)] - (\gamma_1/2)\} \tag{79}$$

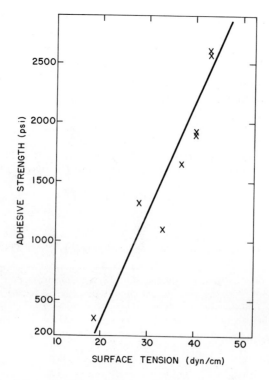

Fig. 13 Relation between surface tension and adhesive strength for various polymers using an epoxy adhesive. (From Levine *et al.* [153].)

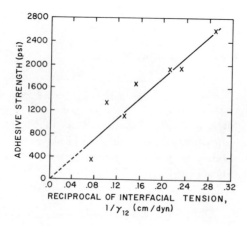

Fig. 14 Relation between interfacial tension and adhesive strength. (From Levine *et al.* [153].)

where $\gamma_j = \gamma_j^d + \gamma_j^p$. It has been shown that, for a given set of γ_1 and γ_2 values, λ_{12} will have a maximum value when $\gamma_1^p/\gamma_1 = \gamma_2^p/\gamma_2$. Thus, adhesion is favored by matching the polarities of the two phases, other factors being equal [64]. Table XV shows the correlation between λ_{12} and adhesion [64]. Similar correlations for release coatings have also been reported [130].

The rate of wetting of the interfacial voids may be given by an exponential decay function

$$d = d_\infty (1 - \alpha e^{-t/\tau})^{-2} \tag{80}$$

where d is the size of the interfacial void at time t, d_∞ that for infinite time, and α and τ are numerical constants. Then, from Eq. (68), we have

$$(f_\infty - f_0)/(f_\infty - f) = e^{t/\tau} \tag{81}$$

Table XV

Correlation between Spreading Coefficient and Adhesion of Some Polymer Pairs[a]

Polymer pairs[b]	Spreading coefficient[c] λ(ergs/cm² at 140°C)	Work of adhesion W_a(ergs/cm² at 140°C)	Adhesion[d]
PMMA–L–PE	−6.5	51.1	Poor
PMMA–PS	−1.6	62.4	Poor
PVA–PS	−0.2	57.0	Fair
PVA–PnBMA	+1.6	49.8	Fair
PMMA–PnBMA	+6.0	54.2	Good
PCP–PDMS	+12.0	40.8	Good

[a] After Wu [64].
[b] For the full names of the polymers, see Table VI.
[c] The spreading coefficient is the larger of the two values λ_{12} and λ_{21} at the bonding temperature of 140°C.
[d] The polymers pairs are bonded at 140°C. The adhesion was evaluated at the room temperature and rated qualitatively.

where f is the adhesive strength at time t, f_∞ that at infinite time, f_0 that at time zero, and τ the retardation time. Thus, the rate of adhesive strength development follows first-order kinetics. Plots of $\log[(f_\infty - f_0)/(f_\infty - f)]$ versus t have been shown to give straight lines [154, 155], as shown in Fig. 15, confirming the validity of Eq. (81), which has also been obtained by a different consideration [155].

E.   Diffusion Theory of Adhesion [156, 157]

The theory as proposed by Voyutskii [156] states that (a) diffusion of polymer chains across the interface is necessary for strong adhesive bond, and (b) molecular contact at the interface alone is not sufficient for strong adhesive bond [156, 158–161]. Controversy has centered on the question whether van der Waals forces alone are sufficient or interdiffusion is necessary for strong adhesion [158–161].

Despite the controversy, the concept of diffusion is nonetheless useful since interdiffusion of chain molecules can certainly increase the mechanical strength of the interfacial zone. It will be foolhardy to reject the practical utility of such a concept, especially in view of the fact that the wetting–contact theory of adhesion recognizes that complete wetting is often difficult to achieve.

Direct experimental evidence of polymer interdiffusion has been rather scanty. These include tracer studies [162] and electron microscopy [163, 164].

Compatibility is necessary for interdiffusion to occur. Yet, most polymer pairs are mutually incompatible [125]. However, statistical thermodynamic theories predict that various degrees of interdiffusion of the segments of polymers will occur in the interfacial layer as required to minimize the interfacial energy [10–14]. The thickness of the diffuse interface between polymers is given in Eqs. (51), (57), and (59) in terms of the χ parameter. As the compatibility increases (χ value decreases), the degrees of interpenetration increase (the interfacial thickness increases). According to Eq. (66), the χ

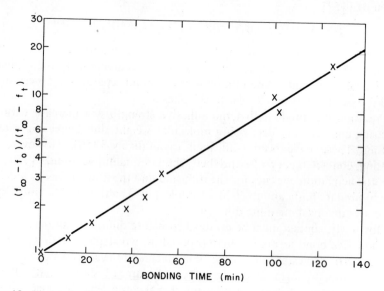

Fig. 15 First-order kinetic plot of the rate of development of adhesive strength. The term f_∞ is the adhesive strength at infinite time; f_0 that at zero time; f_t that at time t. (From Bright [154].)

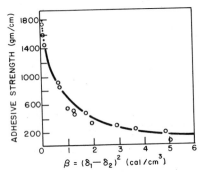

Fig. 16 Relation between adhesive strength and solubility parameter difference. The term δ is the solubility parameter. (From Voyutskii *et al.* [165].)

value is proportional to the square of the difference in the solubility parameter. The adhesive strengths between some polymers have been shown to decrease with increasing disparity in the solubility parameters of the two phases [165, 166], as shown in Fig. 16.

Similar relationships between the adhesive strength, the interfacial tension, and the solubility parameter are also predicted in the wetting–contact theory. As evident from Eqs. (50) and (66), matching of the solubility parameters will minimize the interfacial tension and so will increase the driving force for wetting and the adhesive strength. Thus, both the diffusion and the wetting–contact theories predict the same trend in this aspect.

The kinetics of the formation of adhesive bond by diffusion has been given by [157]

$$f = 5.55v[(2\rho_1/M_1)^{2/3}D_1^{1/2} + (2\rho_2/M_2)^{2/3}D_2^{1/2}]\,vt^{(1-\beta)1/2} \qquad (82)$$

where f is the adhesive strength at time t, v a frequency factor, ρ the density, M the molecular weight, D a constant characterizing the diffusion process of the system, v the rate of adhesive bond separation, β a constant, and the subscripts 1 and 2 refer to the phases.

Equation (82) predicts that the adhesive strength will increase with increasing contact time, decreasing molecular weight, and increasing rate of testing. These have been confirmed qualitatively [156]. However, the wetting–contact theory also predicts similar qualitative trends [159]. In fact, the activation energies for the diffusion and the flow processes are very much similar, both about 3000 cal/mole [159], although the molecular processes involved are quite different.

Obviously, contact must be established before diffusion can occur. Thus, the adhesive bond formation may proceed in two stages; the first stage involves wetting to establish interfacial contact, and the second stage involves interdiffusion to form diffuse interfaces. Figure 17 shows that adhesive strengths continue to increase after interfacial contact has been established, indicating that interdiffusion further strengthens the adhesive joint.

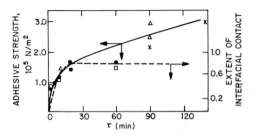

Fig. 17 Relation among adhesive strength, extent of interfacial contact, and bonding time for the autohesion of SKN-40 copolymer. (From Voyutskii [158].)

Equation (82) predicts that a plot of $\log f$ versus $\log t$ will give a straight line with a slope equal to $(1 - \beta)/2$. This has been shown to be true, and β has a value of about $\frac{1}{2}$. In fact, the rate of adhesive formation for many systems can be expressed by a power law equation:

$$f = at^b \tag{83}$$

where f is the adhesive strength at time t, a is a constant, and b is about $\frac{1}{5}$ to $\frac{1}{4}$.

Empirically, the adhesive strength has been found to increase with increasing applied pressure during bond formation. The diffusion process should, however, be independent of the pressure. A rationalization is that the applied pressure promotes the process of interfacial contact and thus provides larger interfacial areas for diffusion.

Although Eq. (82) predicts that the rate of adhesive bond formation will increase with decreasing molecular weight, the final strength of an adhesive bond should be greater with higher molecular weight, since the mechanical strength of a material increases with increasing molecular weight. Experimental results have, however, been erratic [156, 167].

F. Chemical Adhesion [168–170]

The bond energy of a chemical bond is typically about 80 kcal/mole, whereas that of van der Waals attraction is 2.5 kcal/mole [132]. Thus, interfacial chemical bonds could be utilized effectively to promote adhesion. The fact that chemical bonding can dramatically increase the adhesive strength has been shown experimentally [171]. The adhesive strength resulting from chemical bonding has been found to be about 35 times greater than that resulting from van der Waals attraction, coincident with the ratio of the bond dissociation energies.

Interfacial chemical bonding is well known for coupling agents based on silanes [172], titanates [173], and chrome complexes [174]. Reactive functional groups such as carboxyl [175, 176], amino [170], amide [176], hydroxyl [176], epoxide [176], and isocyanate have been found to promote adhesion to various substrates.

The adhesive strength tends to increase as the amount of functional groups increases, expressible by [177]

$$f = kC^n \tag{84}$$

where f is the adhesive strength, C the concentration of the functional group, and k and n are positive constants. The value of n has been found to be about 0.6 [177]. On the other hand, excessive amounts of functional groups may tend to lower the adhesive strength [178, 179].

Amino groups have been found to be very effective adhesion promoters [170]. Interfacial chemical bonding has been directly proved experimentally for the adhesion between a polymer containing amino groups and an aromatic polyester resin [170]. Amine–ester interchange reaction occurs at the interface between the amino group and the aromatic ester group, resulting in grafting of the polyester resin onto the amino polymer. Such interfacial reactions have been found to occur readily [170].

V. SUMMARY

Interfacial energy and structure between polymers are analyzed. The interfaces between polymers are diffuse, as the segments of polymer chains interpenetrate to minimize the free energy. Various factors affecting the formation and the fracture of adhesive bonds are analyzed. These subjects are to be discussed in greater detail elsewhere [91].

REFERENCES

1. S. Newman, *Polym. Plast. Technol. Eng.* **2**, 67 (1974).
2. H. VanOene, *J. Colloid Interface Sci.* **40**, 448 (1972).
3. L. P. McMaster, *Adv. Chem. Ser.* **142**, 43 (1975).
4. H. Keskkula, *Appl. Polym. Symp.* **15**, 51 (1970).
5. J. F. Kenney, *in* "Recent Advances in Polymer Blends, Grafts and Blocks" (L. H. Sperling, ed.), pp. 117–140. Plenum, New York, 1974.
6. H. Keskkula, S. G. Turley, and R. F. Boyer, *J. Appl. Polym. Sci.* **15**, 351 (1971).
7. M. Matsuo, C. Nozaki, and Y. Jyo, *Polym. Eng. Sci.* **9**(3), 197 (1969).
8. S. Wu, *J. Colloid Interface Sci.* **31**, 153 (1969).
9. R. J. Roe, *J. Colloid Interface Sci.* **31**, 228 (1969).
10. E. Helfand and Y. Tagami, *J. Polym. Sci. Part B* **9**, 741 (1971); *J. Chem. Phys.* **56**, 3592 (1972).

11. E. Helfand and A. M. Sapse, *J. Chem. Phys.* **62**, 1327 (1975).
12. E. Helfand, *Acc. Chem. Res.* **8**, 295 (1975).
13. E. Helfand, *J. Chem. Phys.* **63**, 2192 (1975).
14. R. J. Roe, *J. Chem. Phys.* **62**, 490 (1975).
15. S. Wu, *J. Macromol. Sci.* **C10**, 1, Marcel Dekker, Inc., New York (1974).
16. K. S. Pitzer and L. Brewer, "Thermodynamics," 2nd ed. McGraw-Hill, New York, 1961.
17. E. A. Guggenheim, *J. Chem. Phys.* **13**, 253 (1945).
18. L. Reidel, *Chem. Ing. Tech.* **27**, 209 (1955).
19. J. F. Padday, *in* "Surface and Colloid Science" (E. Matijevic, ed.), Vol. 1, pp. 39–99. Wiley (Interscience), New York, 1969.
20. S. Wu, *J. Phys. Chem.* **74**, 632 (1970).
21. R. J. Roe, *J. Phys. Chem.* **72**, 2013 (1968).
22. S. Wu, *J. Polym. Sci. Part C* **34**, 19 (1971).
23. R. H. Dettre and R. E. Johnson Jr., *J. Colloid Interface Sci.* **31**, 568 (1969).
24. D. B. MacLeod, *Trans. Faraday Soc.* **19**, 38 (1923).
25. A. Furguson and S. J. Kennedy, *Trans. Faraday Soc.* **32**, 1474 (1936).
26. F. J. Wright, *J. Appl. Chem.* **11**, 193 (1961).
27. S. Sugden, *J. Chem. Soc.* **125**, 32 (1924).
28. O. R. Quayle, *Chem. Rev.* **53**, 439 (1953).
29. R. Eötvös, *Ann. Phys.* **27**, 448 (1886).
30. J. R. Partington, "Advanced Treatise on Physical Chemistry," Vol. 2, pp. 140–141. Longmans, Green, New York, 1951.
31. T. G. Fox and P. J. Flory, *J. Appl. Phys.* **21**, 581 (1959).
32. T. G. Fox and P. J. Flory, *J. Polym. Sci.* **14**, 315 (1954).
33. T. G. Fox and S. Loshaek, *J. Polym. Sci.* **15**, 371 (1955).
34. K. Ueberreiter and G. Kanig, *Z. Naturforsch.* **A6**, 551 (1955).
35. R. Boyer, *Rubber Chem. Technol.* **36**, 1303 (1963).
36. J. R. Martin, J. F. Johnson, and A. R. Cooper, *J. Macromol. Sci.* **C8**, 57 (1972).
37. M. C. Phillips and A. C. Riddiford, *J. Colloid Interface Sci.* **22**, 149 (1966).
38. R. H. Dettre and R. E. Johnson Jr., *J. Phys. Chem.* **71**, 1529 (1967).
39. T. Hata, *Kobunshi (Jpn)* **17**, 594 (1968).
40. H. Edwards, *J. Appl. Polym. Sci.* **12**, 2213 (1968).
41. D. G. LeGrand and G. L. Gaines Jr., *J. Colloid Interface Sci.* **31**, 162 (1969).
42. D. G. LeGrand and G. L. Gaines Jr., *J. Colloid Interface Sci.* **42**, 181 (1973).
43. S. Wu (to be published).
44. G. W. Bender, G. D. LeGrand, and G. L. Gaines Jr., *Macromolecules* **2**, 681 (1969).
45. A. K. Rastogi and L. E. St. Pierre, *J. Colloid Interface Sci.*, **35**, 16 (1971).
46. P. H. Geil, "Polymer Single Crystals." Krieger Publ., Huntington, New York, 1973.
47. B. Wunderlich, "Macromolecular Physics," Vol. 1, Crystals Structure, Morphology, Defects. Academic Press, New York, 1973.
48. H. Schonhorn and F. W. Ryan, *J. Phys. Chem.*, **70**, 3811 (1966).
49. H. Schonhorn, *J. Polym. Sci. Part B* **5**, 919 (1967).
50. H. Schonhorn, *Macromolecules* **1**, 145 (1968).
51. H. Schonhorn and F. W. Ryan, *J. Polym. Sci.* **6**, 231 (1968).
52. H. Schonhorn and F. W. Ryan, *Adv. Chem. Ser* **87**, 140 (1968).
53. H. Schonhorn and F. W. Ryan, *J. Polym. Sci.* **7**, 105 (1969).
54. K. Hara and H. Schonhorn, *J. Adhes.* **2**, 100 (1970).
55. H. L. Frisch, H. Schonhorn, and T. K. Kwei, *J. Elastoplast.* **3**, 214 (1971).
56. D. R. Fitchmun and S. Newman, *J. Polym. Sci.* **7**, 301 (1969).
57. D. R. Fitchmun and S. Newman, *J. Polym. Sci. Part A-2* **8**, 1545 (1970).
58. D. G. Gray, *J. Polym. Sci. Part B* **12**, 509 (1974).

59. W. A. Zisman, *Adv. Chem. Ser:* **43**, 1 (1964); also *in* "Adhesion and Cohesion" (P. Weiss, ed.), pp. 176–208. Elsevier, Amsterdam, 1962.

60. E. G. Shafrin, *in* "Polymer Handbook" (J. Brandrup and E. H. Immergut, eds.), 2nd ed. Wiley (Interscience), 1975.

61. S. Wu, *J. Phys. Chem.* **72**, 3332 (1968).

62. S. Wu and K. J. Brzozowski, *J. Colloid Interface Sci.* **37**, 686 (1971).

63. A. F. El-Shimi and E. D. Goddard, *J. Colloid Interface Sci.* **48**, 242 (1974); **48**, 249 (1974).

64. S. Wu, *J. Adhes.* **5**, 39 (1973); also *in* "Recent Advances in Adhesion" (L. H. Lee, ed,), pp. 45–63. Gordon & Breach, New York, 1973.

65. D. K. Owens and R. C. Wendt, *J. Appl. Polym. Sci.* **13**, 1741 (1969).

66. D. H. Kaelble, *J. Adhes.* **2**, 66 (1970).

67. D. H. Kaelble and K. C. Uy, *J. Adhes.* **2**, 50 (1970).

68. M. Sherriff, *J. Adhes.* **7**, 257 (1976).

69. A. K. Rastogi and L. E. St. Pierre, *J. Colloid Interface Sci.* **31**, 168 (1969).

70. J. W. Belton and M. G. Evans, *Trans. Faraday Soc.* **41**, 1 (1945); see also R. S. Hansen and L. Sogor, *J. Colloid Interface Sci.* **40**, 424 (1972).

71. I. Prigogine and J. Marechal, *J. Colloid Sci.* **7**, 122 (1952).

72. G. L. Gaines Jr., *J. Phys. Chem.* **73**, 3143 (1969).

73. G. L. Gaines Jr., *J. Polym. Sci. Part A-2* **7**, 1379 (1969).

74. G. L. Gaines Jr., *J. Polym. Sci. Part A-2* **9**, 1333 (1971); **10**, 1529 (1972).

75. D. G. LeGrand and G. L. Gaines Jr., *J. Polym. Sci.* **C34**, 45 (1971).

76. T. C. Kendrick, B. M. Kingston, N. C. Lloyd, and M. J. Owen, *J. Colloid Interface Sci.* **24**, 135 (1967).

77. A. J. G. Allan, *J. Colloid Sci.* **14**, 206 (1959).

78. D. K. Owens, *J. Appl. Polym. Sci.* **8**, 1465 (1964); **14**, 185 (1970).

79. N. L. Jarvis, R. B. Fox, and W. A. Zisman, *Adv. Chem. Ser.* **43**, 317 (1964); R. C. Bowers, N. L. Jarvis, and W. A. Zisman, *I.E.C. Prod. Res. Develop.* **4**, 86 (1965).

80. D. G. LeGrand and G. L. Gaines Jr., *Polym. Prepr. Amer. Chem. Soc.* **11**(2), 442 (1970).

81. G. L. Gaines Jr. and G. W. Bender, *Macromolecules* **5**, 82 (1972).

82. R. G. Azrak, *J. Colloid Interface Sci.* **47**, 779 (1974).

83. Y. Tamai and S. Tanaka, *J. Appl. Polym. Sci.* **11**, 297 (1967).

84. R. E. Cuthrell, *J. Appl. Polym. Sci.* **11**, 1495 (1967); **12**, 1263 (1968).

85. E. Jenckel, E. Teege, and W. Hinrichs, *Kolloid Z.* **129**, 19 (1952).

86. R. J. Barriaut and L. F. Gronholz, *J. Polym. Sci.* **18**, 3933 (1955).

87. R. K. Elby, *J. Appl. Phys.* **35**, 2720 (1964).

88. J. H. Hildebrand and R. L. Scott, "Solubility of Nonelectrolytes." Van Nostrand-Reinhold, Princeton, New Jersey, 1950.

89. A. Beerbower, *J. Colloid Interface Sci.* **35**, 126 (1971).

90. P. Becher, *J. Colloid Interface Sci.* **38**, 291 (1972).

91. S. Wu (to be published).

92. P. A. Small, *J. Appl. Chem.* **3**, 71 (1953).

93. R. J. Roe, *J. Phys. Chem.* **69**, 2809 (1965).

94. I. Prigogine (with A. Bellemans and V. Mathot), "The Molecular Theory of Solutions," Chapter 16. North-Holland Publ., Amsterdam, and Wiley (Interscience), New York, 1957.

95. R. J. Roe, *Proc. Nat. Acad. Sci. U.S.* **56**, 819 (1966).

96. D. Patterson and A. K. Rastogi, *J. Phys. Chem.* **74**, 1067 (1970).

97. K. S. Siow and D. Patterson, *Macromolecules* **4**, 26 (1971).

98. R. H. Dettre and R. E. Johnson Jr., *J. Colloid Interface Sci.* **21**, 367 (1966).

99. H. Schonhorn and L. H. Sharpe, *J. Polym. Sci. Part B* **3**, 235 (1965).
100. Y. Oda and T. Hata, *Preprints Annu. Meeting High Polym. Soc. Jpn. 17th, May* p. 267 (1968).
101. Y. Oda and T. Hata, *Preprints, Symp. Adhes. and Adhesives, 6th, Osaka, Jpn., June* p. 69 (1968).
102. D. G. LeGrand and G. L. Gaines Jr., *J. Colloid Interface Sci.* **31**, 162 (1969).
103. G. W. Bender and G. L. Gaines Jr., *Macromolecules* **3**, 128 (1970).
104. W. Y. Lau and C. M. Burns, *Surface Sci.* **30**, 478 (1972).
105. W. Y. Lau and C. M. Burns, *J. Colloid Interface Sci.* **45**, 295 (1973).
106. W. Y. Lau and C. M. Burns, *J. Polym. Sci. Polym. Phys. Ed.* **12**, 431 (1974).
107. H. Schonhorn, F. W. Ryan, and L. H. Sharpe, *J. Polym. Sci. Part A-2* **4**, 538 (1966).
108. H. T. Patterson, K. H. Hu, and T. H. Grindstaff, *J. Polym. Sci. Part C* **34**, 31 (1971).
109. F. J. Hybart and T. R. White, *J. Appl. Polym. Sci.* **3**, 118 (1960).
110. N. Ogata, *Bull. Chem. Soc. Jpn.* **33**, 212 (1960).
111. H. W. Fox, P. W. Taylor, and W. A. Zisman, *Ind. Eng. Chem.* **39**, 1401 (1947).
112. T. Hata, *Hyomen (Surface Jpn.)* **6**, 659 (1968).
113. G. Antonow, *J. Chim. Phys.* **5**, 372 (1907).
114. G. Antonoff, *J. Phys. Chem.* **46**, 497 (1942).
115. G. Antonoff, M. Chanin, and M. Hecht, *J. Phys. Chem.* **46**, 492 (1942).
116. D. J. Donahue and F. B. Bartell, *J. Phys. Chem.* **56**, 480 (1952).
117. L. A. Girifalco and R. J. Good, *J. Phys. Chem.* **61**, 904 (1957).
118. R. J. Good, L. A. Girifalco, and G. Kraus, *J. Phys. Chem.* **62**, 1418 (1958).
119. R. J. Good and L. A. Girifalco, *J. Phys. Chem.* **64**, 561 (1960).
120. R. J. Good, *Adv. Chem. Ser.* **43**, 74 (1964).
121. R. J. Good and E. Elbing, *Ind. Eng. Chem.* **62**, 54 (1970).
122. C. M. Hansen, *Ind. Eng. Chem. Prod. Res. Develop.* **8**, 2 (1969).
123. C. M. Hansen and A. Beerbower, *in* "Kirk–Othmer Encyclopedia of Chemical Technology," 2nd ed., Suppl. Vol., pp. 889 ff. Wiley, New York, 1971.
124. F. M. Fowkes, *Adv. Chem. Ser.* **43**, 99 (1964).
125. S. Krause, *J. Macromol. Sci.* **C7**(2), 251 (1972).
126. P. J. Flory, *J. Chem. Phys.* **10**, 51 (1942).
127. P. J. Flory, "Principles of Polymer Chemistry," Chapter 12. Cornell Univ. Press, Ithaca, New York, 1953.
128. D. G. LeGrand and G. L. Gaines Jr., *J. Colloid Interface Sci.* **50**, 272 (1975).
129. N. G. Gaylord, *Adv. Chem. Ser.* **142**, 76 (1975).
130. D. J. Gordon and J. A. Colquhoun, *Adhesive Age* (June), p. 21 (1976).
131. Y. Kitazaki and T. Hata, *Preprints, Annu. Meeting High Polym. Soc. Jpn., 18th, May* p. 478 (1969).
132. R. J. Good, *in* "Treatise on Adhesion and Adhesives" (R. L. Patrick, ed.), Vol. I, pp. 9–68. Dekker, New York, 1967.
133. J. R. Huntsberger, *in* "Treatise on Adhesion and Adhesives" (R. L. Patrick, ed.), Vol. I, pp. 119–150. Dekker, New York, 1967.
134. M. L. Williams, *in* "Recent Advances in Adhesion" (L. H. Lee, ed.), pp. 381–422. Gordon & Breach, New York, 1973.
135. R. J. Good, *in* "Recent Advances in Adhesion" (L. H. Lee, ed.), pp. 357–380. Gordon & Breach, New York, 1973.
136. A. A. Griffith, *Trans. Roy. Soc. London Phil.* **A221**, 163 (1920); G. R. Irwin, *in* "Structural Mechanics" (J. N. Goodier and N. J. Hoff, eds.), pp. 557–594. Pergamon, Oxford, 1960.
137. E. H. Andrews, "Fracture in Polymers," Amer. Elsevier, New York, 1968.

138.   S. Mostovoy, E. J. Ripling, and C. F. Bersch, *J. Adhes.* **3**, 125, 145 (1971).
139.   D. H. Kaelble, *J. Appl. Polym. Sci.* **18**, 1869 (1974).
140.   A. N. Gent and R. P. Petrich, *Proc. Roy. Soc. London* **A310**, 433 (1969).
141.   A. N. Gent, *J. Polym. Sci. Part A-2* **9**, 283 (1971); R. S. Rivlin and A. G. Thomas, *J. Polym. Sci.* **10**, 291 (1953).
142.   A. N. Gent and A. J. Kinloch, *J. Polym. Sci. Part A-2* **9**, 659 (1971).
143.   J. J. Bikerman, "The Science of Adhesive Joints," 2nd ed. Academic Press, New York, 1968.
144.   H. Schonhorn and R. H. Hansen, *J. Appl. Polym. Sci.* **11**, 1461 (1967); **12**, 1231 (1968); H. Schonhorn and F. W. Ryan, **18**, 235 (1974); *J. Polym. Sci.* **7**, 105 (1969).
145.   R. E. Robertson, *J. Adhes.* **7**, 121 (1975).
146.   L. H. Sharpe and H. Schonhorn, *Adv. Chem. Ser.* **43**, 189 (1964).
147.   W. A. Zisman, *Adv. Chem. Ser.* **43**, 1 (1964); *Ind. Eng. Chem.* **55**, 18 (1963); *J. Paint Technol.* **44**(564), 42 (1972).
148.   Y. Kitazaki and T. Hata, *in* "Recent Advances in Adhesion" (L. H. Lee, ed.), pp. 65–76. Gordon & Breach, New York, 1973.
149.   C. A. Dahlquist, *in* "Aspects of Adhesion" (D. J. Alner, ed.), Vol. 5, pp. 183–201. CRC Press, Cleveland, Ohio, 1969.
150.   M. J. Barbarisi, *Nature* **215**, 383 (1967); J. E. McNutt, *Adhesives Age* **7** (10), 24 (1964).
151.   E. A. Boucher, *Nature* **215**, 1054 (1967).
152.   W. H. Smarook and S. Bonotto, *Polym. Eng. Sci.* **8**, 41 (1968).
153.   M. Levine, G. Ilkka, and P. Weiss, *Polym. Lett.* **2**, 915 (1964).
154.   W. M. Bright, *in* "Adhesion and Adhesives" (J. E. Rutzler Jr. and R. L. Savage, eds.), pp. 130–138. Wiley, New York, 1954.
155.   K. Kanamaru, *Kolloid Z. Z. Polym.* **192**, 51 (1963).
156.   S. S. Voyutskii, "Autohesion and Adhesion of High Polymers" (S. Kaganoff, translator). Wiley (Interscience), New York, 1963; S. S. Voyutskii and V. L. Vakula, *J. Appl. Polym. Sci.* **7**, 475 (1963).
157.   R. M. Vasenin, *Vysokomol. Soedin.* **3**, 679 (1961); Lavoisier Library translation, Tr. 10230, du Pont de Nemours & Co., Wilmington, Delaware; *Adhesives Age* **8**(5), 18 (1965); **8**(6), 51 (1965).
158.   S. S. Voyutskii, *J. Adhes.* **3**, 69 (1971).
159.   J. N. Anand, *J. Adhes.* **5**, 265 (1973).
160.   S. S. Voyutskii and B. V. Deryagin, *Kolloid. Zh.* **27**, 624 (1965); *Colloid J. USSR* **27**, 528 (1965).
161.   L. H. Sharpe and H. Schonhorn, *Kolloid. Zh.* **28**, 766 (1966); *Colloid J. USSR* **28**, 620 (1966).
162.   F. Bueche, W. M. Cashin, and P. Debye, *J. Chem. Phys.* **20**, 1956 (1952).
163.   A. N. Kamenskii, N. M. Fodiman, and S. S. Voyutskii, *Vysokomol. Soedin.* **7**, 696 (1965); **11**, 394 (1969); *Polym. Sci. USSR* **7**, 769 (1965); **11**, 442 (1969).
164.   J. Letz, *J. Polym. Sci. Part A-2* **7**, 1987 (1969).
165.   S. S. Voyutskii, S. M. Yagnyatinskaya, L. Ya. Kaplunova, and N. L. Garetovskaya, *Rubber Age* (February), p. 37 (1973).
166.   Y. Iyengar and D. E. Erickson, *J. Appl. Polym. Sci.* **11**, 2311 (1967).
167.   S. Gusman, *Off. Digest, Fed. Paint Var. Prod. Club* August, p. 884 (1962); S. W. Lasoski and G. Kraus, *J. Polym. Sci.* **18**, 359 (1955); G. Kraus and J. E. Mason, **6**, 625 (1951); **8**, 448 (1952); E. P. Dunigan, as quoted in P. Weiss, *J. Polym. Sci. Part C* **12**, 169 (1966).
168.   J. E. Rutzler Jr., *Adhesives Age* **2**(6), 39 (1959); **2**(6), 28 (1959).
169.   R. B. Dean, *Off. Digest* (June), p. 664 (1964).

170. S. Wu (unpublished results).
171. A. Ahagon and A. N. Gent, *J. Polym. Sci. Phys.* **13**, 1285 (1975).
172. E. P. Plueddemann, *in* "Interfaces in Polymer Matrix Composites" (E. P. Plueddemann, ed.), pp. 174–216. Academic Press, New York, 1974; P. E. Cassidy and B. J. Yager, *Rev. Polym. Technol.* **1**, 1 (1972).
173. S. J. Monte and P. F. Bruins, *Modern Plast.* (December), p. 68 (1974).
174. P. C. Yates and J. W. Trebilcock, *SPE Trans.* (October), p. 199 (1961).
175. W. H. Smarook and S. Bonotto, *Polym. Eng. Sci.* **8**, 41 (1968).
176. T. J. Mao and S. L. Reegen, *in* "Adhesion and Cohesion" (P. Weiss, ed.), pp. 209–217. Elsevier, Amsterdam, 1962.
177. C. H. Hofrichter Jr. and A. D. McLaren, *Ind. Eng. Chem.* **40**, 329 (1948).
178. A. D. McLaren and C. J. Seiler, *J. Polym. Sci.* **4**, 63 (1949).
179. H. P. Brown and J. F. Anderson, *in* "Handbook of Adhesives" (I. Skeist, ed.), pp. 255–267. Van Nostrand-Reinhold, Princeton, New Jersey, 1962.

Chapter 7

Rheology of Polymer Blends and Dispersions

H. Van Oene

Engineering and Research Staff
Ford Motor Company
Dearborn, Michigan

I. Introduction 296
II. Flow Behavior of Viscoelastic Fluids. 298
    A. Development of Constitutive Relationships 298
    B. Flow of a Second-Order Fluid through Tubes 301
    C. Secondary Flow Phenomena 303
III. Cocurrent Flow of Two Fluids 304
    A. Cocurrent Flow of Two Newtonian Fluids 304
    B. Cocurrent Flow of Two Viscoelastic Fluids 307
IV. Rheological Behavior of Suspensions of Rigid Axisymmetric Particles in Viscoelastic Fluids 310
    A. Suspensions of Rigid Spheroids in Newtonian Suspension Media . . 310
    B. Suspensions of Rigid Axisymmetric Particles in Viscoelastic Fluids . . 313
    C. Concentrated Dispersions of Rigid Particles 314
V. Rheological Behavior of Suspensions of Deformable Droplets 318
    A. Microrheology of Deformable Droplets 318
    B. Macrorheology of Suspensions of Deformable Particles 322
VI. Rheology of Filled Polymers 325
VII. Rheology of Polymer Blends 330
    A. Modes of Dispersion 330
    B. Viscosity of Polymer Blends 336
    C. Normal Stress Function of Polymer Blends 342
    D. Theoretical Considerations 344
VIII. Conclusion 347
    References 349

295

I. INTRODUCTION

Physical blends or "alloys" of dissimilar polymers are of considerable technological importance as blending provides a means for improving physical properties such as toughness and processability. Typical examples of polymer blends in current use are described by Jalbert and Smejkal [1]. Glass fiber-reinforced polymer composites are in widespread use, and increasing emphasis is placed on obtaining composites with higher ultimate strength by the use of graphite fibers. For optimum properties it is necessary to be able to control the fiber orientation and fiber distribution during processing.

For both polymer blends and reinforced composites the improvement in properties depends to a large extent on the degree of dispersion obtained and the precise orientation of fibers achieved. The flow behavior and morphology of these systems have not been investigated extensively; hence, the complex interplay between processing conditions, morphology, and improvement (or lack of it) in physical properties obtained remains a matter of empirical understanding. Rather than presenting a review of these empirical observations, an attempt will be made in this chapter to treat the flow behavior of these mixtures in terms of the flow behavior of disperse fluids.

From a rheological viewpoint one can consider polymer blends as dispersions of deformable, liquidlike particles, and reinforced composites as dispersions with axisymmetric rigid particles. In both instances the rheological behavior is governed by the state of dispersion and the shape and orientation of the dispersed phase, as well as particle–particle interaction. For polymer blends of an amorphous and semicrystalline polymer, the distinction between deformable or liquidlike and rigid or solidlike is not so clear since crystallization of the dispersed semicrystalline phase in flow will change the dispersed phase from a fluid to a solid. Hence, the rheological behavior of these blends will reflect features of either type dispersion.

The rheology of dispersions can be studied at two levels: the macrorheological level, which involves measurement of the rheological properties of the dispersion itself, such as the viscosity and normal stresses; and the microrheological level, which concentrates on the detailed motions of the individual particles themselves. In theory one should be able to predict the macrorheological properties from the observed microrheological behavior. Morphology, orientation, and distribution of the dispersed phase are essentially established by processes that take place at the microrheological level and reflect the hydrodynamic forces acting on a fluid element.

In order to describe the hydrodynamic forces, it is necessary to develop a relationship between the motion of the fluid and the forces required to maintain this motion. Such a relationship is expressed in terms of a constitutive equation. Constitutive equations therefore express a general relationship between those parameters that are characteristic of the motion, such as velocity gradient and acceleration, and the stresses generated by the motion. These general relationships are of great significance for the description of the flow behavior of polymer melts, because only in the light of the general constitutive equation can one account for the considerable normal stresses developed in the shear flow of polymer melts. Furthermore, the normal stresses affect in a fundamental way the forces acting on a fluid element. Hence, in disperse systems, where the motion and orientation of the dispersed particle are determined by the local forces acting on that particle, the normal stresses may have a decisive influence on microrheological phenomena.

Even though few systematic investigations have been carried out to elucidate the influence of normal stresses on the microrheology of dispersions, those observations that have been made, such as the rapid establishment of the equilibrium orientation of fibers and disks in a viscoelastic suspension medium by Gauthier, Goldsmith, and Mason [2] and the modes of dispersion of polymer blends [3], reveal striking differences between dispersions in viscoelastic media, dispersions in fluids that exhibit large normal stresses in shearing flows, and dispersions in Newtonian media, where normal stresses are essentially absent.

The main emphasis in this chapter is placed on the effects of normal stresses on the microrheology of dispersions, as the microrheological behavior determines the overall observed orientation of fibers and morphology of the composite or polymer blend.

In the first sections of this chapter the macrorheological behavior of general fluids, whose flow behavior is described in terms of a particular constitutive equation, is discussed. Explicit expressions for the stresses arising in capillary (Poiseuille) flow of such a fluid are given. As a first example of two-component flow, the side by side extrusion of two dissimilar polymers through slits is considered. Published studies reveal a great influence of nonequilibrium entrance conditions on the morphology of the bicomponent extrudate; the viscoelasticity of the two phases has a dominant influence on the stability of the interface.

Recent theoretical results for the flow behavior of dispersions are discussed next. The flow behavior of dispersions is found to be described by constitutive equations very similar to those employed for the flow of viscoelastic fluids. With the notable exception of dilute suspensions of rigid spheres, the flow behavior of suspensions of axisymmetric particles is described by three functions: the viscosity function and two, in general unequal, normal stress functions. The latter functions are found to depend on the axis ratio as well as the orientation of the particles. The microrheology

of dispersions of rigid and fluid droplets is briefly reviewed. It is found that
the viscoelasticity of the suspension medium for suspensions of rigid
asymmetric particles has a notable influence on the establishment of the
equilibrium orientation of the particle with respect to the flowfield. For
deformable particles, the finding is that a viscoelastic droplet appears less
deformable than an equivalent viscous droplet. At large concentrations of
suspended particles, particle–particle interaction may dominate the rheological behavior, resulting in very complex fluid characteristics. Migration
of particles away from the walls of the capillary or flow channel gives rise to
a so-called sheath–core configuration, the core consisting of densely packed
particles, the sheath being essentially the suspension medium itself. This
behavior is found for suspensions of both rigid and fluid particles.

Many of the effects found in dispersions of rigid particles in Newtonian
media are also observed when a polymer melt is used as a suspension
medium. For polymer–polymer blends, however, very little if any correspondence is found with the flow behavior of emulsions of Newtonian fluids.
The macrorheological behavior as well as the microrheology, as evidenced
by the modes of dispersion obtained, are both dominated by the viscoelastic
properties of the individual constituents. A detailed understanding of these
phenomena remains a theoretical as well as experimental challenge.

II. FLOW BEHAVIOR OF VISCOELASTIC FLUIDS

A. Development of Constitutive Relationships

In order to describe the flow behavior of a fluid it is necessary to employ
relationships that allow one to express, with complete generality, the influence of the imposed velocity gradients, accelerations, etc. on the state of
stress generated within the fluid by the motion. Such relationships are called
constitutive relationships. The constitutive equation is, therefore, a functional
relationship between a state of stress, described by a stress tensor, and the
velocity gradient and acceleration tensors.

It is outside the scope of this chapter to treat the derivation of constitutive
equations in detail. This has been reviewed in depth by Truesdell and Noll
[4]. More recent reviews are by Bird [5], Brenner [6], and Hinch and
Leal [7].

Even though the derivation of appropriate constitutive equations is based
on a very formal mathematical analysis, the power of such an analysis was
clearly illustrated in a paper by Reiner [8] on a mathematical theory of
dilatancy. Starting from the assumption that the stress T is given by a tensor

polynomial in the shearing, Reiner showed that for an isotropic fluid

$$\mathbf{T} = -p\mathbf{I} + a_0\mathbf{I} + a_1\mathbf{D} + a_2\mathbf{D}^2 \tag{1}$$

where p is the pressure that would correspond to equilibrium and where the coefficients a_i are scalar functions of the invariants of $\mathbf{D}$. For simple shear flow, $\mathbf{D}$ is given by

$$\mathbf{D} = G \begin{vmatrix} 0 & 1 & 0 \\ 1 & 0 & 0 \\ 0 & 0 & 0 \end{vmatrix} \tag{2}$$

Hence

$$\mathbf{T} = (-p + a_0)\mathbf{I} + a_1 G \begin{vmatrix} 0 & 1 & 0 \\ 1 & 0 & 0 \\ 0 & 0 & 0 \end{vmatrix} + a_2 G^2 \begin{vmatrix} 1 & 0 & 0 \\ 0 & 1 & 0 \\ 0 & 0 & 0 \end{vmatrix} \tag{3}$$

where the a_i are even functions of the magnitude of the shear rate G.

One obtains, therefore, the result that the flow behavior is characterized by three functions:

1. the function between the shear stress and magnitude of the shearing:

$$T_{12} = a_1 G \tag{4}$$

2. two deviatoric normal stress differences:

$$T_{11} - T_{33} = T_{22} - T_{33} = a_2 G^2 \tag{5}$$

The occurrence of the normal stress functions arises in the representation of the state of stress by a tensor polynomial. Their magnitude is second order in the shearing. Deviations from the linear relationship between shear stress and shearing is of third order. The magnitude of the second-order effect is, in theory, independent of the first-order effect.

Reiner's [8] paper has acted ever since as a catalyst for further development because the natural occurrence of the normal stress functions in the analysis of the general relationships between stresses and velocity gradients provides a completely new insight into many complex flow phenomena. Reiner's treatment was further generalized by Rivlin and Ericksen [9]. Assuming that the stress depends on the gradients of velocity and acceleration, they derived a constitutive equation of the form

$$\mathbf{T} = -p\mathbf{I} + \alpha_1\mathbf{A}_0 + \alpha_2\mathbf{A}_1 + \alpha_3\mathbf{A}_0^2 + \alpha_4\mathbf{A}_1^2 + \alpha_5(\mathbf{A}_0\mathbf{A}_1 + \mathbf{A}_1\mathbf{A}_0)$$
$$+ \alpha_6(\mathbf{A}_0^2\mathbf{A}_1 + \mathbf{A}_1\mathbf{A}_0^2) + \alpha_7(\mathbf{A}_0\mathbf{A}_1^2 + \mathbf{A}_1^2\mathbf{A}_0) \tag{6}$$
$$+ \alpha_8(\mathbf{A}_0^2\mathbf{A}_1^2 + \mathbf{A}_1^2\mathbf{A}_0^2)$$

where the α's are polynomials in the invariants of A_0 and A_1, the so-called Rivlin–Ericksen tensors:

$$A_{ij}^0 = \tfrac{1}{2}(v_{i,j} + v_{j,i})$$

$$A_{ij}^{(\alpha+1)} = DA_{ij}^{(\alpha)} + A_{ip}^{(\alpha)}v_{p,j} + A_{jp}^{(\alpha)}v_{p,i}$$

where v_i is the velocity and $v_{i,j} = \partial v_i/\partial x_j$ and where summation is implied over the repeated index.

In rectilinear laminar flow, with velocity gradient G and a flow field

$$v_1 = Gx_2, \qquad v_2 = v_3 = 0$$

the components of the stress tensor T assume the following form:

$$
\begin{aligned}
T_{11} &= -p' + G^2 f_1(G^2) \\
T_{22} &= -p' + G^2 f_2(G^2) \\
T_{33} &= -p' \\
T_{12} &= Gf_3(G); \qquad T_{23} = T_{31} = 0
\end{aligned}
\tag{7}
$$

and

$$f_1(G^2) = \tfrac{1}{4}\alpha_3$$

$$f_2(G^2) = (\alpha_2 + \tfrac{1}{4}\alpha_3) + (\alpha_4 + \tfrac{1}{2}\alpha_6)G^2 + \tfrac{1}{2}\alpha_8 G^4$$

$$f_3(G^2) = \tfrac{1}{2}(\alpha_1 + \alpha_5 G^2 + \alpha_7 G^4)$$

The α_i are polynomials in G^2 and p' is the isotropic hydrostatic pressure.

Following Markovitz [10], these equations can be rewritten in terms of an isotropic pressure p:

$$p = -\tfrac{1}{3}\text{Trace } T = -\tfrac{1}{3}(T_{11} + T_{22} + T_{33}) \tag{8}$$

and the deviatoric stress tensor t with components t_{ij} such that

$$t_{11} + t_{22} + t_{33} = 0 \tag{9}$$

The flow behavior then is characterized by three functions, the viscosity function

$$t_{12} = f_3(G) \cdot G \tag{10}$$

and two normal stress functions

$$t_{11} - t_{33} = [\tfrac{1}{4}\alpha_3]G^2 \tag{11}$$

$$t_{22} - t_{33} = [(\alpha_2 + \tfrac{1}{4}\alpha_3) + (\alpha_4 + \tfrac{1}{2}\alpha_6)G^2 + (\tfrac{1}{2}\alpha_8)G^4]G^2 \tag{12}$$

Other derivations of general constitutive equations have been considered by Oldroyd [11], Lodge [12], Giesekus [13], and Coleman and Noll [14].

For slow, steady flows these theories all reduce as a first approximation to a constitutive equation of the type

$$\mathbf{T} = -p\mathbf{I} + \alpha_1 \mathbf{A}_0 + \alpha_2 \mathbf{A}_1 + \alpha_3 \mathbf{A}_0^2 \tag{13}$$

The stresses are described by a viscosity function

$$\eta = t_{12}/G = \tfrac{1}{2}\alpha_1 \tag{14}$$

the normal stress functions

$$\sigma_1(G) = t_{11} - t_{33} = (\tfrac{1}{4}\alpha_3)\,G^2 \tag{15}$$

$$\sigma_2(G) = t_{22} - t_{33} = (\alpha_2 + \tfrac{1}{4}\alpha_3)\,G^2 \tag{16}$$

A fluid characterized by Eq. (13) is called an incompressible fluid of second grade or a second-order fluid, since the normal stress functions, which are of second order in the velocity gradient, are taken into account. Nonlinearity of the viscosity function (a third-order effect) can either be included or excluded, depending on whether one takes the function α_1 to be a constant or not. For many theoretical studies the approximation of general flow behavior given by the constitutive equation of a second order fluid provides a starting point for further analysis. In the limit $G \to 0$ the normal stress functions vanish, and the Navier–Stokes equations retain their validity. Furthermore, Coleman and Markovitz [16] obtained an important relationship between the coefficients appearing in the constitutive equation of a second-order fluid and quantities measured in a slowly varying periodic flow:

$$\lim_{G \to 0} (t_{11} - t_{22})/G^2 = +\eta_0^2 J_e \tag{17}$$

where η_0 is the zero shear viscosity and J_e is the steady state shear compliance.

The relationship between $(t_{11} - t_{22})$, the first normal stress difference, and the steady state shear compliance, a quantity that arises in the theory of viscoelasticity, shows that a viscoelastic fluid will exhibit normal stress effects in shearing flows. As will become readily apparent in subsequent sections, viscoelastic fluids are not the only class of fluids to exhibit normal stress effects. Most suspensions of axisymmetric rigid particles suspended in a Newtonian fluid exhibit normal stresses as do emulsions of Newtonian fluids suspended in a Newtonian fluid.

B. Flow of a Second-Order Fluid through Tubes

Explicit expressions for the components of stress tensor in Poiseuille flow of a second-order fluid have been derived by Langlois [17]. These are:

$$T\langle rr \rangle = -p + \prod z - \tfrac{1}{2}(\alpha_3 + 2\alpha_2)G^2$$

$$T\langle \theta\theta \rangle = -p + \prod z - \tfrac{3}{2}(\alpha_3 + 2\alpha_2)G^2$$

$$T\langle zz \rangle = -p + \prod z - \tfrac{1}{2}(\alpha_3 + 6\alpha_2)G^2 \qquad (18)$$

$$T\langle \theta z \rangle = T\langle r\theta \rangle = 0$$

$$T\langle rz \rangle = -\tfrac{1}{2}\prod r$$

where z is flow direction, r direction of the velocity gradient; θ neutral direction, η viscosity, $\prod$ pressure gradient in the flow direction, and G velocity gradient.

The normal stress functions are

$$T\langle zz \rangle - T\langle rr \rangle = -2\alpha_2 G^2$$
$$T\langle rr \rangle - T\langle \theta\theta \rangle = (\alpha_3 + 2\alpha_2)G^2 \qquad (19)$$

Separation of the isotropic part of the stress tensor and the deviatoric stresses yields

$$t\langle rr \rangle = +\tfrac{1}{3}(\alpha_3 + 4\alpha_2)G^2$$

$$t\langle \theta\theta \rangle = -\tfrac{2}{3}(\alpha_3 + \alpha_2)G^2$$

$$t\langle zz \rangle = +\tfrac{1}{3}(\alpha_3 - 2\alpha_2)G^2 \qquad (20)$$

$$p(r) = p - \prod z + \tfrac{1}{6}(5\alpha_3 + 14\alpha_2)G^2$$

It is seen that the isotropic pressure $p(r)$ is no longer a constant but depends on the radial position.

In order to obtain a feeling for the relative order of magnitude of the various terms, one may rewrite these expressions in terms of the normal stress differences, $(T\langle zz \rangle - T\langle rr \rangle)$ and $(T\langle rr \rangle - T\langle \theta\theta \rangle)$:

$$t_{rr} = -\tfrac{1}{3}[T\langle zz \rangle - T\langle rr \rangle][1 - E]$$

$$t_{\theta\theta} = -\tfrac{1}{3}[T\langle zz \rangle - T\langle rr \rangle][1 + 2E]$$

$$t_{zz} = +\tfrac{2}{3}[T\langle zz \rangle - T\langle rr \rangle][1 + \tfrac{1}{2}E] \qquad (21)$$

$$p(r) = p - \prod z - \tfrac{1}{3}[T\langle zz \rangle - T\langle rr \rangle][1 - \tfrac{5}{2}E]$$

$$E = (T\langle rr \rangle - T\langle \theta\theta \rangle)/(T\langle zz \rangle - T\langle rr \rangle)$$

Experimentally it is found that $(T\langle zz \rangle - T\langle rr \rangle)$ is positive; hence, a tension develops in the flow direction. Recent consensus is that $(T\langle rr \rangle - T\langle \theta\theta \rangle)$ is negative and smaller than $(T\langle zz \rangle - T\langle rr \rangle)$. The quantity E is therefore negative and smaller than unity. For polymer melts this ratio depends on shear stress and is of the order of -0.2 to -0.6 [18].

Last, the components of the stress tensor can be expressed as

$$T\langle rr \rangle = -p + \textstyle\prod z - \frac{1}{2}(T\langle rr \rangle - T\langle \theta\theta \rangle)$$

$$T\langle \theta\theta \rangle = -p + \textstyle\prod z - \frac{3}{2}(T\langle rr \rangle - T\langle \theta\theta \rangle) \tag{22}$$

$$T\langle zz \rangle = -p + \textstyle\prod z - \frac{1}{2}[(T\langle rr \rangle - T\langle \theta\theta \rangle) - 2(T\langle zz \rangle - T\langle rr \rangle)]$$

The variation of the isotropic pressure in a plane perpendicular to the flow gives rise to a pressure distribution on the walls, when flow through rectangular ducts and slits is considered [19]. For a slit with an aspect ratio of 6/1, such pressure differences have been observed by Han [20].

C. Secondary Flow Phenomena

Ericksen [21] and Green and Rivlin [22] have shown that for a second order fluid rectilinear flow through a tube of arbitrary, noncircular cross-section is in general only possible if the following condition is obeyed:

$$(T\langle rr \rangle - T\langle \theta\theta \rangle)/\eta(G)\,G^2 = \text{constant} \tag{23}$$

For a fluid with a second normal stress difference equal to zero, or for a Newtonian fluid this relationship holds. Experimentally it has been found, however, that most viscoelastic fluids have a finite (negative) second normal stress difference, which depends on the shear stress, hence rectilinear flow of such fluids through slits and rectangular ducts is dynamically not possible [23]. Illustrative solutions for the secondary flow in tubes of elliptical cross section are given by Green and Rivlin [22]. For noncircular pipes a steady motion is generally still mathematically possible. It consists of a simple rectilinear motion on which is superposed a vortexlike secondary motion in a plane perpendicular to the flow direction. The fluid elements follow spiral paths as they traverse the pipe. Such motion was observed by Giesekus [24] using solutions of polyisobutylene.

Secondary flow phenomena in viscoelastic fluids have also been demonstrated for other flow situations such as the case of flow around a rotating sphere [25] in which the fluid elasticity gives rise to a region of closed stream lines. The secondary flows arise essentially from the difference in pressure distribution even at low velocities, and hence, should be observed at all Reynolds numbers.

The onset of turbulence or flow instability at higher Reynolds numbers is strongly affected by the normal stress functions. For Couette flow, which is a shear flow generated between rotating cylinders, the onset of instability, the Taylor instability, is found to be determined by the sign of the second normal stress difference; when negative, fluid elasticity is stabilizing [26].

At the onset of the Taylor instability a reduction in viscous torque has been observed.

A decrease in viscous resistance has also been observed by Barnes and Walters [27] for the flow of a viscoelastic fluid through a *coiled* tube, which seemingly indicates that the presence of secondary flows may lead to a reduction in viscous resistance. A challenging discussion of the effect of the non-Newtonian properties on flow fields has been given by Pearson [29] and Astarita and Denn [28].

III. COCURRENT FLOW OF TWO FLUIDS

A. Cocurrent Flow of Two Newtonian Fluids

In cocurrent flow, two separate streams are made to pass side by side through a flow channel. The interface between the two fluids may be symmetric with the walls of the flow channel or asymmetric. Such stratified flows have been dealt with by numerous authors, Russell and Charles [30], Packham and Shail [31], Nubar [32], Jones [33], White and Lee [34], Everage [35], Han and Yu [36]. The Newtonian solution for cocurrent flow between parallel plates will be considered here in some detail in order to identify the salient characteristics of these flows. For convenience the treatment of White and Lee [34] is followed. The basic flow configuration is illustrated in Fig. 1.

For slow flows, the equations of motion for flow through an infinitely wide channel with parallel sides simplify to

$$0 = -(\partial p/\partial x) + \eta(\partial^2 u/\partial y^2) \qquad (24)$$

where p is the pressure, η the viscosity, and $u(y, x)$ the velocity. The boundary conditions are

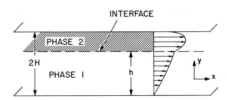

Fig. 1   Coordinate system for cocurrent flow through an infinitely wide channel of height $2H$.

at the walls:

in Phase 1:

$$u_1(0, x) = 0 \tag{25a}$$

in Phase 2:

$$u_2(2H, x) = 0 \tag{25b}$$

at the interface located at $y = h$:

i. $u_1(h, x) = u_2(h, x) = U$ (25c)

ii. continuity of shear stress:

$$\eta_1(\partial u_1/\partial y) = \eta_2(\partial u_2/\partial y) \tag{25d}$$

The velocity profiles in each phase are found to be given by

$$u_1 = U\left(\frac{y}{h}\right) + \frac{1}{2\eta_1}\left(-\frac{\partial p}{\partial x}\right)(hy - y^2) \tag{26}$$

$$u_2 = U\left(\frac{2H - y}{2H - h}\right) + \frac{1}{2\eta_2}\left(-\frac{\partial p}{\partial x}\right)\left[4H^2 - \frac{4H^2 - h^2}{2H - h}(2H - y) - y^2\right] \tag{27}$$

The interfacial velocity U may be determined by use of the boundary condition for the continuity of the shear stress at the interface:

$$U = \frac{h(2H - h)}{2\eta_2}\left[\frac{2H}{(\eta_2/\eta_1)(2H - h) + h}\right](-\partial p/\partial x) \tag{28}$$

The explicit expressions for the velocities in terms of the location of the interface are

$$u_1 = \frac{1}{\eta_1}\left(-\frac{\partial p}{\partial x}\right)\left\{2H\left[\frac{1 + \xi^2(R - 1)}{1 + \xi(R - 1)}\right]y - y^2\right\} \tag{29}$$

$$u_1 = \frac{1}{2\eta_2}\left(-\frac{\partial p}{\partial x}\right)\left\{\frac{4H^2\xi[1 - \xi][R - 1]}{1 + \xi[R - 1]} + 2H\left[\frac{1 + \xi^2(R - 1)}{1 + \xi(R - 1)}\right]y - y^2\right\} \tag{30}$$

$$U = \frac{2H^2}{\eta_1}\left(-\frac{\partial p}{\partial x}\right)\left\{\frac{\xi[1 - \xi]}{[1 + \xi(R - 1)]}\right\} \tag{31}$$

where $\xi = h/2H$ and $R = \eta_2/\eta_1$.

The volumetric flow rate per unit width of the channel of each component can now be obtained by integration of the velocities, which lead to

$$Q_1 = \frac{2}{3}\left(\frac{1}{\eta_1}\right)\left(-\frac{\partial p}{\partial x}\right)H^3\left\{\frac{3\xi^2 - 2\xi^3 + \xi^4[R - 1]}{1 + \xi[R - 1]}\right\} \tag{32}$$

$$Q_2 = \frac{2}{3}\left(\frac{1}{\eta_2}\right)\left(-\frac{\partial p}{\partial x}\right)$$

$$\times H^3\left\{\frac{3(1-\xi^2) - 2(1-\xi^3) + [R-1][4\xi - 9\xi^2 + 6\xi^3 - \xi^4]}{1 + \xi[R-1]}\right\}$$

(33)

Finally, the total volumetric flow rate/unit width $Q = Q_1 + Q_2$ is given by

$$Q = \frac{2}{3}\left(-\frac{\partial p}{\partial x}\right)H^3\left[\left(\frac{1}{\eta_2}\right)\left(\frac{[1-\xi^4]}{1+\xi[R-1]}\right) + \left(\frac{1}{\eta_1}\right)\left(\frac{[1-(1-\xi)^4 + \xi^4(R-1)]}{1+\xi[R-1]}\right)\right]$$

(34)

For $\xi = 1$ and $\xi = 0$, Eq. (34) reduces to the appropriate single phase flowrate of each component.

From these equations it is clear that, once Q_1 and Q_2 are specified (metered flow), the location of the interface is also determined. In general, the location of the interface is a function of the relative flow rates of each component and the respective viscosities. When the total volumetric flow rate/unit width is written as

$$Q = \tfrac{2}{3}(-\partial p/\partial x) H^3(1/\eta_{\text{mixt}})$$

the following mixture rule is obtained:

$$\frac{1}{\eta_{\text{mixt}}} = \left(\frac{1}{\eta_1}\right)\left[\frac{1-(1-\xi)^4 + \xi^4(R-1)}{1+\xi(R-1)}\right] + \frac{1}{\eta_2}\left[\frac{(1-\xi)^4}{1+\xi(R-1)}\right]$$

(35)

It should be noted that this relationship holds only for plane parallel Poiseuille flow.

The corresponding expressions for Poiseuille flow of stratified liquids with a symmetric interface, a so-called sheath–core configuration, through a capillary are [32]:

a. velocities:
    Sheath: $v_1 = k_1(1-\beta^2)R^2$ (36)

    Core: $v_2 = [k_2(\xi^2 - \beta^2) + k_1(1-\xi^2)]R^2$ (37)

where $k_1 = (1/4\eta_1)(-\partial p/\partial x)$, $k_2 = (1/4\eta_2)(-\partial p/\partial x)$, $\beta = r/R$; $\xi = r_1/R$, where r_1 is the location of the interface and R the radius of the capillary.

b. volumetric flow rate:
    $$Q_{\text{core}} = [k_2\xi^2 + 2k_1(1-\xi^2)]\xi^2(\pi/2)R^4$$ (38)

    $$Q_{\text{sheath}} = k_1(1-\xi^2)^2(\pi/2)R^4$$ (39)

c. total flow rate:

$$Q = [k_2 \zeta^4 + k_1(1 - \zeta^4)](\pi/2) R^4 \tag{40}$$

Hence

$$\frac{1}{\eta_{\text{mixt}}} = \left(\frac{\zeta^4}{\eta_2}\right) + \frac{(1 - \zeta^4)}{\eta_1} \tag{41}$$

These expressions may be used to calculate the volume fraction of each component and the location of the interface.

For plane Poiseuille flow, White and Lee [34] and Everage [35] have given a detailed analysis of the motion of the interface when initially at the inlet the interface is not at its equilibrium position. Both authors find that the shift of the interface to its equilibrium position takes place rapidly, that is, within a distance of the order of 2H, the height of the channel, and that the fluid of higher viscosity pushes into the fluid of lower viscosity. This can be readily understood as both authors consider an initial configuration with the interface at the center of the flow channel and equal flow rates for the individual components. The latter condition causes the pressure in the main of the more viscous phase to be larger than that in the less viscous phase. The interface distortion due to nonequilibrium inlet conditions is therefore dominated by the effect of the viscosities of the two phases.

B. Cocurrent Flow of Two Viscoelastic Fluids

Two-phase, annular, laminar flow of second-order fluids through a cylindrical tube has been treated by Slattery [37]. When the densities of the two fluids are equal, the boundary conditions reduce to those for the Newtonian case. The shape of the interface, however, is determined by the normal stress functions:

$$[t_{rr}(1) - t_{rr}(2)] = -\gamma/R_{12} \tag{42}$$

evaluated at the location of the interface whose radius of curvature is R_{12}, and γ is the surface tension.

Stability conditions for these flows have been obtained by Jones [33]. For conduits of arbitrary cross sections, the conditions are very restrictive for rectilinear flow, as might be expected on the basis of similar considerations for the flow of a single viscoelastic fluid. Four conditions have to be satisfied simultaneously, which are only satisfied if the two fluids have identical viscosities and second normal stress differences. In most cases swirling flows can be expected.

Recently, Han and Khan [38] made a numerical analysis of interfacial instability for cocurrent flow through a rectangular duct. Only two-dimensional disturbances could be treated. Instability in the 1–2 plane (the plane spanned by the flow direction and that of the velocity gradient) was found to depend only on the viscosity difference of the two fluids. Instability in the 2–3 plane, the plane normal to the flow direction, depends both on viscosity and viscoelasticity differences of the two fluids. Khan and Han [38] also considered the interface distortion for flow through square ducts and rectangular channels. A typical result is shown in Fig. 2 The displacement of the flat interface through the center of the duct can be regarded as similar in kind to that calculated by White and Lee [34] and Everage[35]. Differences in viscosity were found to be more decisive than differences in the elasticity parameters. The severe distortion (fingering) may be enhanced in square ducts by the fact that even in the Newtonian case the flow is no longer a plane Poiseuille flow. This is further reinforced by the observation of a similar effect in the corner region of the rectangular channel, where the flow is also not akin to plane Poiseuille flow.

In summary, cocurrent flow or bicomponent extrusion of two polymer melts is in general a nonstable flow, extremely sensitive to the precise inlet conditions. If initially the interface is not at the position required by the viscosities and relative flow rate of the two components, the interface will rapidly shift to this position. At the exit, relaxation of the velocities to the uniform takeoff velocity causes according to Everage [35], a reverse shift. An important consequence of the reversible interface motion is that the relative flow rate of the components cannot be determined from the observed extrudate morphology. As the maximum displacement takes place in the center region of the duct, that is, the region of highest velocities, the volumetric flow rate of the more viscous phase is underestimated.

Besides the reversible rapid interface motion, encapsulation of the one component by the other is observed (see Volume 2, Chapter 16). Observations

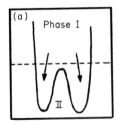

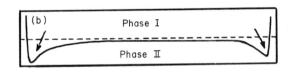

Fig. 2   Interface displacement and distortion for a cocurrent flow with an initially flat interface through a square duct and a rectangular channel. The viscosity and normal stress difference for Phase I is larger than those for Phase II. (From Khan and Han [38].)

by Southern and Ballman [39] and Everage [35] indicate that for their particular nylon–nylon system the encapsulation of the high-viscosity component by the low-viscosity component is complete after flow through a capillary with a length/diameter ratio of 120. The sheath–core configuration so obtained is the configuration consistent with minimum viscous energy dissipation [40].

Experimental studies of bicomponent extrusion of polymer melts through rectangular slits and capillaries have been reported by Yu and Han [36]. By measuring the wall pressures along a slit at the same position at either side of the interface, the continuity of pressure across the interface was verified. From the pressure measurements both the pressure gradient along the flow direction and the so-called "exit pressure" can be obtained, which enables one to obtain in a direct fashion both the magnitude of the shear stress and, for a slit, the first normal stress difference. Plots of pressure gradient as a function of flow rate are shown in Fig. 3 for extrusion through a slit and extrusion through a circular die. For a given flow rate, the pressure gradient reduction is much larger for circular dies than slits. Han [42] also explored the effect of initial flow geometry. When the two components were fed in a symmetric concentric configuration, little rearrangement of the interface took place after extrusion through a die with length/ diameter ratio equal to 18. When components were fed in an eccentric configuration considerable changes in morphology took place, as shown in

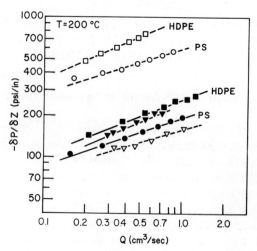

Fig. 3 Comparison of the pressure gradient as a function of flow rate of polystyrene and polyethylene and their cocurrent mixtures through a slit (filled symbols) and a capillary (open symbols). (∇) HDPE/PS capillary; ($\blacktriangledown$) HDPE/PS, slit. (Data from Yu and Han [36].)

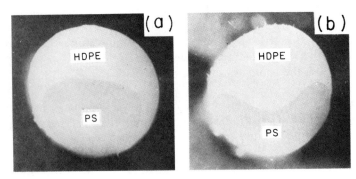

Fig. 4   Interface shape of the extrudate cross section of polystyrene (PS) and high-density polyethylene (HDPE) system coextruded through a circular die $(L/D = 11)$: (a) $T = 200°C$; (b) $T = 240°C$. (From Han [18, p. 280].)

Fig. 4. At 200°C, the viscosity of the polystyrene is larger than that of the polyethylene, at 240°C the viscosity of the polystyrene is lower. The difference in the prominent interface curvature in the center can therefore be explained by the differences in viscosity of the two phases. Regardless of the viscosity difference, however, the polyethylene phase tends to encapsulate the polystyrene phase.

IV.   RHEOLOGICAL BEHAVIOR OF SUSPENSIONS OF RIGID AXISYMMETRIC PARTICLES IN VISCOELASTIC FLUIDS

A.   Suspensions of Rigid Spheroids in Newtonian Suspension Media

The study of the flow behavior of suspensions of rigid spheroids has a long history. Reviews of the general rheology have been given by Rutgers [43], Thomas [44], and Jeffrey and Acrivos [45]. The microrheology, that is, the study of the individual motions of particles in various shear fields, has also been studied in great detail. A comprehensive summary of this work may be found in the article by Goldsmith and Mason [46]. In view of these excellent review articles it is unnecessary to deal with the prior work in detail. Instead attention will be focussed on those aspects that are needed to understand the difference observed between suspensions in viscoelastic and Newtonian suspension media.

When a rigid particle is immersed in a viscous fluid undergoing a simple shear flow, whose undisturbed velocity components are described by

$$u_1 = u_2 = 0 \quad \text{and} \quad u_3 = Gx_2 \tag{43}$$

where G is the velocity gradient, the particle in general starts to rotate. This rotation of a single spheroidal particle was expressed by Jeffery [47] in terms of the angular velocity of the axis of revolution of the particle as

$$d\theta_1/dt = [G(r_e^2 - 1)/4(r_e^2 + 1)] \sin 2\theta_1 \sin 2\phi_1 \tag{44}$$

$$d\phi_1/dt = [G/(r_e^2 + 1)](r_e^2 \cos^2 \phi_1 + \sin^2 \phi_1) \tag{45}$$

where θ_1 and ϕ_1 are Eulerian angles defined in Fig. 5 and r_e is the axial ratio of the spheroid. By convention $r_e > 1$ for rods, $r_e < 1$ for disks.
Integration of these expression yields

$$\tan \theta_1 = Cr_e/[r_e^2 \cos^2 \phi_1 + \sin^2 \phi_1]^{1/2} \tag{46}$$

$$\tan \phi_1 = r_e \tan [2\pi t/T + k] \tag{47}$$

where the integration constants C and k denote the orbit constant and the initial phase angle. T is the period of rotation of the spheroid about the vorticity axis (X_1) and is given by

$$T = (2\pi/G)(r_e + r_e^{-1}) \tag{48}$$

These relationships have been verified in detail for particles suspended in Newtonian suspension media [46]. It is found that in the absence of Brownian motion, the integration constants C and k are indeed constants of the motion. In a Newtonian suspension medium a single axisymmetric particle remains, therefore, in its initial orbit and rotates indefinitely never achieving a final equilibrium orientation. Okagawa *et al.* [48] have recently extended this work to a complete description of a collision free suspension of axisymmetrical particles with an arbitrary distribution of initial orientations.

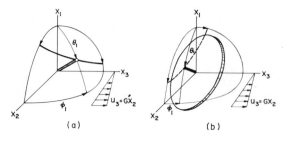

Fig. 5   Coordinates system to describe the orientation in shear of (a) rods and (b) disks.

Since the microrheology of suspended spheroids is well understood, it should be possible to derive a constitutive equation for the flow behavior of suspensions from first principles and to obtain equations where the unknown functions are explicitly expressed in terms of the properties of the particles themselves. Starting with the work of Kotaka [49] and Giesekus [50] this program has been essentially completed. A general dynamical rheological theory for axially symmetric particles has been formulated by Brenner [6]. The theory applies to any type of homogeneous shear flow, steady as well as unsteady, including extensional flows. The flow behavior for these suspensions is characterized by a viscosity function, already known from prior work, and two, unequal normal stress functions. Only for dilute suspensions of spherical particles do the normal stress functions vanish; hence, only such suspensions are Newtonian at finite shear.

Most other axisymmetric shapes require two normal stress functions for the specification of the flow behavior: $(T_{11} - T_{22})$ being positive and large and $(T_{22} - T_{33})$ being negative and small. An exceptional case is the "non-interacting" dumbbell, in which the second normal stress function vanishes for geometric reasons [6]. In these sections the indices of the normal stress functions, when explicitly stated, will follow the normal convention: 1—flow direction, 2—direction of velocity gradient, 3—neutral direction, in spite of the coordinate system shown in Fig. 5.

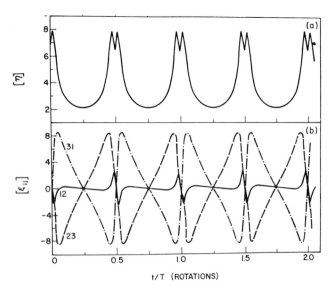

t/T (ROTATIONS)

Fig. 6   (a) Variations of the intrinsic viscosity $[\eta]$ for a system of prolate spheroids $(r_e = 10)$ and (b) variations of the three intrinsic normal stress differences ξ_{12}, ξ_{23}, and ξ_{31}. (From Okagawa *et al.* [48].)

Transient normal stresses and viscosity for a collisionless suspension of particles with initially random orientation in the absence of Brownian motion have been derived by Okagawa *et al.* [48]. A typical result for prolate spheroids with axial ratio equal to 10 is shown in Fig. 6. Both mean viscosity and mean normal stress differences fluctuate in time, the normal stress functions changing sign four times during a particle rotation. This fluctuation is caused by the strong dependence of the rate of rotation of each individual particle on the initial orientation of the particle. The normal stress functions vanish when the orientation distribution is symmetrical about the $X_1 X_3$ plane. The experimental investigation of Okagawa and Mason [51] confirmed the expected oscillations in the distribution function; these oscillations are damped because of the spread in the particle/axis ratio and particle–particle interactions. Therefore, after a few particle rotations equilibrium distributions were established in the plane described by the direction of flow and velocity gradient.

As a suspension at finite concentration can be expected to contain a distribution of orientations initially, the effects observed by Okagawa *et al.* [48] are important when the suspension medium is viscoelastic. The steady flow of the suspension medium may require a system of normal stresses incompatible with that arising from the orientation distribution of the axisymmetric particles; hence; unusual flow behavior may be anticipated for suspensions of axisymmetric particles in viscoelastic fluids.

B. Suspensions of Rigid Axisymmetric Particles in Viscoelastic Fluids

A number of striking differences between suspensions of disks and rods in Newtonian and non-Newtonian media have been documented by Gauthier *et al.* [52]. The first is a drift of the orbit constant C at low velocity gradients. Above a critical shear rate however, disks oriented themselves with their axes of revolution along the direction of the velocity gradient and ceased to rotate altogether. The alignment was maintained when the velocity gradient was increased further.

How quickly the equilibrium orientation is attained is illustrated by the data listed in Table I, in which the mean area, projected onto the coordinate planes, S_{ij} of the disks is taken as a measure of the degree of orientation. It is seen that the equilibrium orientation is almost achieved within 0.25 of a particle rotation.

Recently Leal [53] has shown that both orbit drift and the existence of a critical shear rate can be accounted for by the viscoelastic properties of the suspension medium. For slender bodies, suspended in a viscoelastic medium, considered to be a fluid of second grade, the Jeffery equations become

$$d(\tan \theta_1)/dt = G + KM_1(1 + 2E_1)G \sin^2 \theta_1 \tan \phi_1 (\sin^2\phi_1 - \cos^2\phi_1)$$

$$(49a)$$

Table I

Establishment of Orientation in Shear Flow for a
Suspension of Disks[a,b]

Suspension medium	Time t/T	Area		
		S_{32}	S_{31}	S_{21}
Viscoelastic	0	0.304	0.638	0.539
	0.25	0.034	0.981	0.132
	1.0	0.026	0.934	0.116
	35.0	0.013	0.996	0.049
Newtonian	360	0.215	0.861	0.221

[a] Data from Gauthier et al. [52].
[b] $r_e = 0.1340$.

$$dC/dt = -KM_1(1 + 2E_1) G \sin^2 \theta_1 \sin^2 \phi_1 C \qquad (49b)$$

where K is a parameter proportional to $(T_{11} - T_{33})/G^2$, M_1 a function that
depends on the shape of the particle only, and

$$2E_1 = -(T_{11} - T_{22})/(T_{11} - T_{33})$$

For a Newtonian fluid $K = 0$, and the equations reduce to Eqs. (44) and
(46) when $r_e \gg 1$. For most polymer solutions the factor $KM_1(1 + 2E)$ will be
positive; hence, for rods, these equations indicate that the orbit constant
should decay to an equilibrium value $C = 0$, as observed.
The value of the critical shear rate is given by

$$G_{\text{crit}}^2 = -(16/r_e)\left[\eta/\lim_{G \to 0} \frac{T\langle 22 \rangle - T\langle 33 \rangle}{G^2}\right] \qquad (50)$$

Even though the predictions of Leal's theory apply to slender rods, reasonable
agreement was obtained with both the orbit drift measured by Gauthier et al.
[52], and the critical shear rates observed for rods by Bartram et al. [54].

C.   Concentrated Dispersions of Rigid Particles

The rheological behavior of concentrated dispersions is extremely complicated due to effects of particle migration and particle–particle interactions.
The concentration range in which these effects become noticeable depends
on particle shape, suspension medium, type of particle interaction, and
temperature. Particle migration in Poiseuille flow is well established by the
studies of Karnis et al. [55].

Particle-free zones develop at the wall when suspensions of rigid spheres at volume concentrations ranging from 5 to 30% are subjected to steady or oscillating shear flow in rigid tubes. This effect influences markedly the values of the viscosity obtained by capillary flow [44] as the sheath–core structure lowers the viscosity markedly.

Particle–particle interactions have an all pervasive influence on the rheological behavior. At low shear stress the interactions may give rise to a yield value; they may lead to socalled shear thickening, an increase of viscosity with shear stress, or a more usual decrease of viscosity with shear stress. At higher concentrations, normal stress functions have been measured by Highgate and Whorlow [57, 58]. At concentrations above 40% by volume, discontinuities in the viscosity shear rate curves were observed by Hoffman [59], which are attributed to the influence of double layer repulsion, and London–van der Waals dispersion forces.

Haschisu and co-workers [60, 61] have found evidence for a "Kirkwood–Alder" transition, a kind of order–disorder transition induced by the long-range van der Waals forces, in suspensions of monodisperse lattices. Because of the complexity of the rheological behavior, no consensus exists as to how to study such highly "nonlinear" and unpredictable fluids.

Onogi, Matsumoto, and co-workers [62–64] have started a comprehensive program to characterize, at least, the rheological properties of concentrated suspensions, using a general nonlinear theory of viscoelasticity.

In viscoelastic theory all rheological properties can in principle be described in terms of a relaxation spectrum. The equilibrium shear modulus G is defined by

$$G^0 = \int_{-\infty}^{+\infty} H(\tau)\,d\ln\tau \tag{51}$$

where $H(\tau)$ is the distribution of relaxation times, the quantity $H(\tau)\,d\ln\tau$ is then that portion of the initial modulus, contributed by processes with relaxation times in the range $\ln\tau$ and $\ln\tau + d\ln\tau$. Time dependence can be described in terms of a stress relaxation modulus:

$$G(t) = \int_{-\infty}^{+\infty} H(\tau)e^{-t/\tau}\,d\ln\tau \tag{52}$$

The zero shear viscosity is given by

$$\eta_0 = \int_{-\infty}^{+\infty} \tau H(\tau)\,d\ln\tau \tag{53}$$

and the equilibrium compliance J_e^0 by

$$J_e^0 = \int_{-\infty}^{+\infty} \tau^2 H(\tau) d\ln\tau \bigg/ \left[\int_{-\infty}^{+\infty} \tau H(\tau) d\ln\tau \right]^2 \qquad (54)$$

Both the viscosity and equilibrium shear compliance are dominated by processes associated with long relaxation times. As shown by Coleman and Markovitz [16], the equilibrium shear compliance may be related to the first normal stress function. Similarly one may establish a relationship between quantities measured in an oscillating shear field of frequency ω and the slow steady flow parameters η_0 and J_e^0:

$$\eta_0 = \lim_{\omega \to 0} G''(\omega)/\omega \qquad (55)$$

$$J_e^0 = (1/\eta_0^2) \lim_{\omega \to 0} G'(\omega)/\omega \qquad (56)$$

where $G'(\omega)$ is the so-called shear storage modulus and $G''(\omega)$, the shear loss modulus; both are functions of frequency.

The relationship between steady flow parameters and viscoelastic moduli may be exploited to gain valuable insight as to the kinds of processes, their respective time scales and contribution to specific observed rheological behavior.

The study of Matsumoto et al. [64] on dispersions of cross-linked poly(styrene-co-divinylbenzene) particles dispersed in a 10% solution of polystyrene in diethylphthalate illustrates many of the aforementioned rheological complexities. Flow curves for these disperse systems are shown in Fig. 7. The suspension medium PS10 shows Newtonian behavior. The

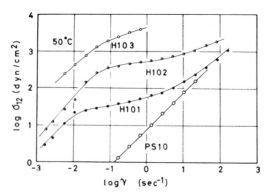

Fig. 7 Flow curves for the disperse systems H101, H102, and H103 containing, respectively, 10, 20, and 30 wt % styrene–divinylbenzene copolymer particles dispersed in a polystyrene solution PS10. (From Matsumoto et al. [64].)

dispersions are highly nonNewtonian; only at low shear rates does the viscosity become constant. At intermediate shear rates, the flow curve exhibits a plateau region. The authors note that the data obtained at high shear rates can be plotted according to the Casson equation [65]:

$$T_{12}^{1/2} = t_0^{1/2} + aG^{1/2} \tag{57}$$

to determine a yield value t_0. This "yield value" corresponds well with the value of the shear stress in the plateau region of the flow curve.

Results obtained in oscillating flow are shown in Fig. 8 as a function of frequency and strain amplitude. At low frequencies of oscillation the limiting behavior and amplitude independence required by the linear theory of viscoelasticity for G' and G'' is indeed achieved. Furthermore, the flow

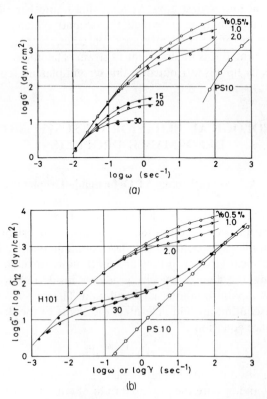

Fig. 8 (a) Frequency dependence of the storage modulus G' of suspension H101 as a function of strain amplitude γ_0 and the frequency dependence of the storage modulus of the suspending medium PS10 at 50°C. (b) Frequency dependence of the loss modulus G'' for the suspension H101 as a function of strain amplitude γ_0 at 50°C. Completely filled symbols designate the flow curve (shear stress versus shear rate) of H101; the curve marked PS10 represents the loss modulus of the suspending medium. (From Matsumoto et al. [64].)

curve merges with the G'' versus ω curve at frequencies smaller than 10^{-2} Hz. Unfortunately, no normal stress data were reported to verify the limiting behavior of the storage modulus.

The relaxation spectrum calculated from these data is shown in Fig. 9. Spectra of similar shape were obtained for dispersions at higher concentrations. The effect of concentration is primarily to extend the spectrum to longer times. The relaxation spectrum can be schematically represented by three regions (Fig. 10). A typical wedge-type spectrum due to the polymer molecules of the suspension medium is seen at short times and a (rubbery) plateau or box-type spectrum attributed to the interaction of entangled polymer chains and suspended particles is seen at intermediate times. The height in this region depends on polymer and particle concentrations and the magnitude of the imposed strains. At long times, another plateau region occurs. The relaxation time at which the second plateau disappears is dependent on temperature, concentration of particles, and the viscosity of the suspending medium, but is independent of strain. The apparent yield stress observed in steady flow is attributed to this second plateau region.

V. RHEOLOGICAL BEHAVIOR OF SUSPENSIONS OF DEFORMABLE DROPLETS

A. Microrheology of Deformable Droplets

When a suspension of fluid droplets is subjected to flow, the fluid stresses arising in the suspension medium tend to deform and orient the droplet. For a liquid droplet whose interface is not contaminated by impurities or surfactants, the shear stresses in flow are continuous through the interface thus establishing a system of velocity gradients inside the droplets, which in turn establish a pattern of internal circulation. The stresses normal to the interface are discontinuous and determine the shape of the droplet according to the Laplace equation

$$\Delta p = \gamma [b_{(1)}^{-1} + b_{(2)}^{-1}] \tag{58}$$

where Δp is the discontinuity in pressure across the interface; the interfacial tension and $b_{(1)}$ and $b_{(2)}$ are the principal radii of curvature of the interface.

The presence of surfactants may inhibit the internal circulation if the interfacial layer itself can support a shear stress, that is, can act as a solid membrane. Continuity of the shear stress across the interface is no longer required, and rheologically, the liquid droplet behaves as a rigid sphere [66]. As a consequence, the viscosity will *increase*.

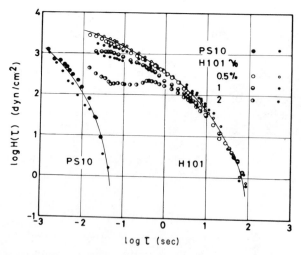

Fig. 9 Relaxation spectra for the suspension medium PS10 and the suspension H101 at various strain amplitudes at 50°C. The small circles are calculated from G'', large circles calculated from G'. (From Matsumoto *et al.* [64].)

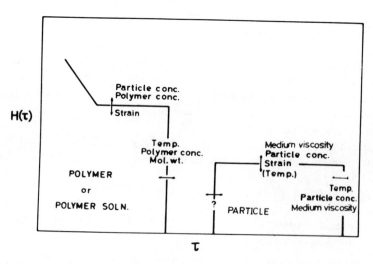

Fig. 10 Schematic diagram of the relaxation spectrum for a suspension of solid particles in a polymer melt or a concentrated polymer solution. (From Matsumoto *et al.* [64].)

By calculating the velocities inside and outside a fluid droplet and requiring the shear stress to be continuous, Taylor [67] showed that in Couette flow the pressure difference at the interface was given by

$$\Delta p = 4G\eta_0(19\lambda + 16)/(16\lambda + 16)\sin 2\phi_1, \qquad \lambda = \eta_1/\eta_0 \qquad (59)$$

where η_0 is viscosity of the suspension medium, η_1 the viscosity of the suspended fluid, and ϕ_1 orientation with respect to the direction of flow. The magnitude of Δp changes sign in each quadrant, the droplet being subjected to alternate tensile and compressive stresses.

This system of pressure differences can be accommodated if the cross section of the equatorial plane assumes an elliptical form, described by the polar equation

$$r = b(1 + D\sin 2\phi_1) \qquad (60)$$

where b is the radius of the undeformed droplet. For small deformations D, equal to the ratio of viscous forces to the surface tension forces, is given by

$$D = (Gb\eta_0/\gamma)(19\lambda + 16)/(16\lambda + 16) \qquad (61)$$

When the magnitude of p exceeds the force due to surface tension, that is, when

$$4G\eta_0(19\lambda + 16)/(16\lambda + 16) > 2\gamma/b \qquad (62)$$

the droplet will burst. In Couette flow this should occur for $D = 1/2$. Experiments by Rumscheidt and Mason [68] on a Newtonian system have confirmed the burst criterion for values ranging from 10^{-4} to 2. For larger values of λ, no bursting of droplets was observed in Couette flow. As the velocity gradient increased, the ellipsoidal droplet reached a limiting deformation and aligned itself with its long axis in the flow direction. Torza et al. [69] have confirmed this earlier observation and find that, on theoretical grounds, a droplet will not burst if $\lambda > 3$.

In different flow fields, the shape of the droplet will be different than that given by Eq. (60). The shape has been calculated by Hestroni and Haber [70] for Poiseuille flow.

By considering both the effect of viscosity differences and interfacial tension, Cox [71] was able to derive for small deformations both the deformation and orientation of the droplets. He obtained for the deformation D and orientation ϕ

$$D = 5(19\lambda + 16)/[4(\lambda + 1)\{(20k)^2 + (19\lambda)^2\}^{1/2}]$$

$$\phi = \tfrac{1}{4}\pi + \tfrac{1}{2}\tan^{-1}(19\lambda/20k), \qquad k = \gamma/\eta_0 Gb \qquad (63)$$

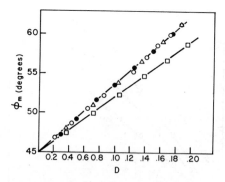

Fig. 11 The orientation angle ϕ_m of a viscoelastic droplet in shear as a function of deformation D calculated from Eq. (63): (●) experimental; (□) $\lambda = 2.84$, $\gamma = \gamma_0$; (△) $\lambda = 3.5$, $\gamma = \gamma_0$; (○) $\lambda = 3.5$, $\gamma = 1.66\gamma_0$. The parameters $\lambda = 2.84$ and $\gamma = \gamma_0$ are those used in Ref. [52].

Hence, when the interfacial tension is dominant, $\phi \to \pi/4$; when the viscosity is dominant, $\phi \to \pi/2$. The orientation in the flow field is therefore a sensitive measure of the relative importance of viscosity to interfacial tension. Analysis for viscoelastic fluids has not yet been carried out; hence, the effect of viscoelasticity can only be inferred from experimental observation.

Gauthier *et al.* [52] find that the deformation of viscoelastic droplets suspended in a Newtonian medium does not follow the Newtonian equations quantitatively. The droplets appear to be less deformable. When these data are analyzed in terms of Cox's theory, good agreement can be obtained, provided that the viscosity ratio and the magnitude of the interfacial tension are changed. As shown in Fig. 11, when the value of λ is changed from 2.84 to 3.5, good agreement with experiment is obtained for the dependence of orientation on deformation, which is independent of the value of the interfacial tension. This increase in viscosity ratio can be justified in view of the fact that the value of 2.84 arises from a measurement of the viscosity of the suspended fluid at a much higher shear rate than used in the deformation experiments. As Fig. 12 shows, agreement for the dependence of the orientation on the parameter Gb (the product of velocity gradient and particle radius) can now only be obtained by increasing the interfacial tension. Hence, it appears that the lack of deformability of the viscoelastic droplet suspended in a Newtonian fluid is due to an apparent increase in interfacial tension.

A further interesting observation by Bartram *et al.* [54] deals with the deformation of a viscous droplet in the absence of interfacial tension. In this case the theory of Cox [71] predicts that for $\lambda > 1$, the deformation and orientation oscillate without damping with a period $T = 2\pi/G$. This was confirmed for Newtonian droplets. A viscous viscoelastic droplet,

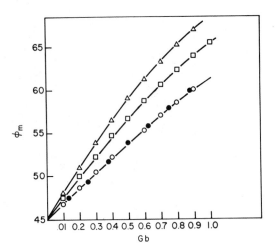

Fig. 12   The orientation angle ϕ_m of a viscoelastic droplet in shear as a function of the product of shear rate G and particle radius b calculated from Eq. (63); the symbols are the same as those in Fig. 11. The parameters $\lambda = 2.84$ and $\gamma = \gamma_0$ are those used in Ref. [52].

suspended in a less viscous viscoelastic medium demonstrated very different behavior. The initially spherical droplet became extended along the neutral (vorticity) axis. As this longitudinal deformation continued, the thread buckled, which resulted in the formation of a thin thread that finally broke up into thin cylindrical droplets. Similar behavior was also observed with viscoelastic droplets at finite interfacial tension. This is the first detailed observation that a viscous droplet is not deformed into an ellipsoid in the plane defined by the flow and velocity gradient direction, as all theory considered thus far predicts.

B.   Macrorheology of Suspensions of Deformable Particles

Constitutive equations for the flow of a dilute suspension of deformable droplets have been derived by Schowalter et al. [72]. The viscosity function is given by

$$\eta = \eta_0 \left[1 + \frac{5\lambda + 2}{2\lambda + 1} \cdot \phi \right] \tag{64}$$

where ϕ is the volume fraction. This result was first obtained by Taylor [73]. The normal stress functions are

$$T\langle 11 \rangle - T\langle 22 \rangle = \frac{1}{2} \left(\frac{b\phi}{\gamma} \right) \left[\frac{(19\lambda + 16)^2}{20(\lambda + 1)^2} \right] (\eta_0^2 G^2)$$

$$T\langle 22 \rangle - T\langle 33 \rangle = -\frac{1}{14}\left[\frac{58\lambda^2 + 122\lambda + 100}{(\lambda + 1)(19\lambda + 16)}\right]\left[T\langle 11 \rangle - T\langle 22 \rangle\right] \quad (65)$$

where b is the droplet radius and γ the interfacial tension. In the limit $\lambda \to \infty$, the viscosity function reduces to Einstein's result for rigid spheres; the normal stress functions, however, do not vanish. Constitutive equations for suspensions of solids, and viscoelastic spheres have been obtained by Roscoe [74].

Particle–particle interactions in Poiseuille flow have been studied by Gauthier *et al.* [75]. Both viscoelastic and Newtonian suspended droplets migrate toward the tube axis. The rates of migration increased with particle

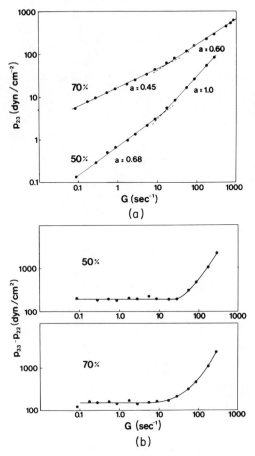

Fig. 13 Rheological functions of a 50 and 70% emulsion of amylacetate in aqueous glycerol: (a) flow curves; (b) normal stress difference. (From Vadas *et al.* [76].)

size and were larger than those measured for rigid particles of comparable size.

Recently Vada *et al.* [76] reported an extensive study of the flow behavior of concentrated emulsions of amylacetate in aqueous glycerol. The majority of the emulsion particles were in the size range of 1–3 μm; hence, deformation in shear can be neglected. The flow curves and the first normal stress difference, both measured in a cone-and-plate instrument, are shown in Fig. 13. The magnitude and shear dependence of the normal stress are highly unusual. Rather than decreasing with shear rate, the normal stress is constant and about an order of magnitude larger than the shear stress. Contrary to expectation, the normal stress is larger at a concentration of 50% by volume than at 70% by volume. On comparison with the results obtained by Onogi *et al.* [64] for suspensions of cross-linked polystyrene particles, one would perhaps expect that, if, in the present case, data could have been obtained at lower shear rates, the expected limiting behavior of viscosity and normal stress would have been obtained, and that the current low shear results would reflect a plateau region caused by particle–particle interaction.

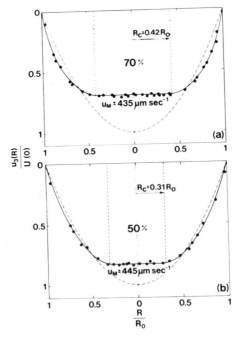

Fig. 14 Blunting of the velocity profile in shear flow of a 50 and 70% emulsion of amylacetate in aqueous glycerol: R_c, radius of the core region; u_m, maximum velocity. (From Vadas *et al.* [76].)

The velocity profiles in Poiseuille flow show (see Fig. 14) the expected blunting due to particle migration, resulting in a region of plug flow. Breakup of liquid droplets in flows other than shear flows has not been considered extensively. Mikami *et al.* [77] have recently considered the stability and breakup of an extending viscous liquid cylindrical thread suspended in an immiscible viscous fluid undergoing extensional flow. It is found that when λ is large, the viscous liquid thread will elongate to a thread of very small radius and remain stable for a long time before finally breaking up into droplets of very small size. Smaller extension rates give rise to thinner threads. These observations are important for polymer–polymer blends as the blend is inevitably subjected to an extensional flow in the entrance region to the capillary.

VI. RHEOLOGY OF FILLED POLYMERS

In most practical applications polymers are extended with rigid fillers in order to modify and control processing variables and physical properties of the final part. Fillers may be added to increase rigidity and reduce warpage of injection molded parts. Such fillers are essentially inert. Fillers may also be used to increase toughness, as in high-impact polystyrene or polypropylene, where dispersed deformable rubber particles are essential to realize the desired toughness. In the rubber industry the use of carbon black to improve processability is well established (see Volume 2, Chapter 19); depending on the type and structure of the carbon black, the interaction of the filler and polymer can be optimized.

Axisymmetric particles such as fibers and flakes, used to increase the modulus and heat distriction temperatures, are usually considered as reinforcing fillers. The control of orientation of these fillers during processing is crucial for controlling the final physical properties.

Last, highly filled polymer compositions (50% by volume) are gaining increasing importance as materials for their own sake, for example, glass fiber-reinforced, mineral-filled sheet molding compounds. The use of polymers as a vehicle for ceramic powders to obtain injection moldable compositions has made possible the fabrication of parts with complex shapes [78].

In view of the widespread application of filled-polymer compositions, it is surprising that the rheology of these systems has not been studied in great detail. This is in part due to the fact that the effects of rigid inert fillers are readily understood. For reinforcing fillers a greater amount of effort had to be expended in finding suitable surface treatments for these fillers in order to maximize physical properties. With increasing emphasis on light-weight, high-strength composites, control of orientation during processing will become increasingly important.

The behavior of an inert filler is illustrated by the measurements of Han [79] on calcium carbonate-filled polypropylene. The viscosities and first normal stress difference is shown in Fig. 15. As expected the viscosity increases considerably as the filler concentration increases; the normal stresses decrease, however. In data obtained at high shear, which is typical of processing conditions, the influence of a yield value is not apparent. Vinogradov and co-workers [80] studied the influence of carbon black on the rheology of polyisobutylene. The flow curves shown in Fig. 16 clearly show the development of a plateau at lower shear rates, indicative of the presence of a yield value. It was also established that the addition of carbon black reduced die swell, hence improving processability. The filler addition had only a minor influence on the onset of melt fracture, indicated by a dull, rough extrudate appearance. Similar results have been obtained by Lim [81] and White and Crowder [82].

Even though at high shear stresses the flow behavior is little influenced by a yield stress, it can manifest itself in other aspects, such as the filling of a mold in injections molding. In a study of the injection molding of highly filled phenolic compounds Beck [83] found that the mold filled by "jetting"; at the mold entrance the jet emerging from the gate traveled through the mold until it hit the opposite wall; mold filling occurred by the buckling and spreading of the solidlike jet.

The orientation of axisymmetric fillers such as fibers and flakes during processing has not been studied extensively, even though its importance is well recognized. The variation of the tensile strength of an injection molded sheet with respect to the direction of flow during mold filling is shown in Fig. 17, taken from a study by Hall and co-workers [84] on the injection molding of 40% glass fiber filled polycarbonate. It is seen that the ultimate tensile strength varies from 22,000 psi for a small tensile bar to 11,000 psi in the square plate perpendicular to the flow direction. Goettler [85–87] has studied the influence of flow through convergent channels on fiber orientation. By suitable modification of the mold gate, tensile strengths could be increased from 7 to 27×10^3 psi in matrices with a fiber loading of 40%.

The viscosity and tensile properties of mixed fiber–bead-reinforced composites have been reported by Roberts and Hill [88].

The work of Willermet et al. [78] on the formulation of injection moldable silicon carbide–phenolic compositions represents an interesting application in which the well-established tendency for plug flow in highly filled compounds is carefully exploited. The injection-molding technique makes possible the fabrication of complex shapes. In further processing steps, the phenolic resin first is carbonized at high temperature. When carbonizing

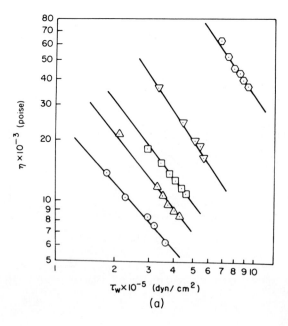

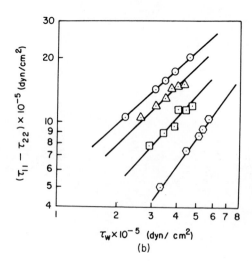

Fig. 15 Viscosity and normal stress difference for calcium carbonate filled polypropylene melts at various filler concentrations (wt % CaCO₃). (a) Viscosity: (○) 0.0: (△) 10; (□) 20; (▽) 40; (◊) 70. (b) Normal stress difference: (○) 0.0; (△) 10; (□) 20; (◊) 40. (From Han [18].)

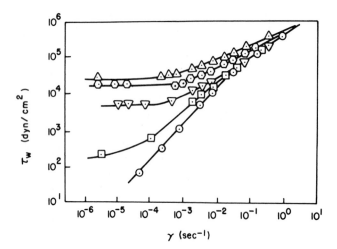

Fig. 16   Shear stress versus shear rate for carbon black-filled polyisobutylene at 60°C, for various carbon concentrations (vol %): (○) 0.0; (□) 2.5; (▽) 5.0; (◑) 9.0; (△) 13.0. (From Han [18, p. 182].)

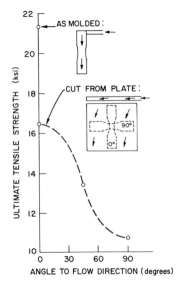

Fig. 17   Variation of the tensile strength of 40% glass fiber-filled polycarbonate with respect to the flow direction in the mold. (Data from Hall et al. [84].)

has been completed, the carbon is reacted further with silicon vapor to form in situ silicon carbide, which will encapsulate the silicon carbide already present. During all these operations it is crucial that the part retain its shape. Both moldability and dimensional change after siliciding was found to be dependent on the particle size distribution. Flow properties were studied using a spiral mold. In this technique, the length of the spiral, molded under conditions of constant injection pressure and mold temperature, is taken as a measure of moldability and flow.

As shown in Fig. 18, travel in the spiral mold depends strongly on the volume fraction of filler and filler size. At constant volume fraction of filler, travel can be further modified by use of bimodal particle size distributions as shown in Fig. 19. The influence of particle size distribution on flow is consistent with the earlier findings of Farris [89]. Overall shrinkage during further processing was found to vary between 1 and 2.4%, depending on particle size and particle size distribution. The small shrinkage is attributed to the formation of a densely packed solidlike plug during flow.

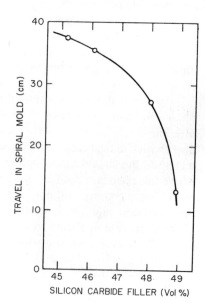

Fig. 18 Dependence of travel in a spiral mold of silicon carbide filled phenolic compositions as a function of volume fraction of silicon carbide. (From Willermert *et al.* [78].)

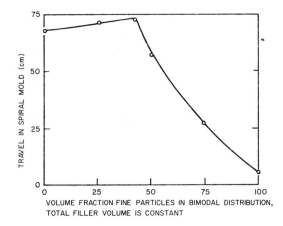

Fig. 19 Dependence of travel in a spiral mold of silicon carbide filled phenolic compositions as a function of particle size distribution at constant volume fraction of filler (volume fraction 0.47).

VII. RHEOLOGY OF POLYMER BLENDS

A. Modes of Dispersion

The mode of dispersion obtained when a finely divided mixture of two incompatible polymers are coextruded has been studied by van Oene [3]. Essentially two basic types of dispersion were found, a ribbon-type dispersion and the expected droplets.

Characteristic features of the ribbon-type (stratified) dispersions are illustrated in Fig. 20. In this figure, cross sections perpendicular to the flow direction are shown for extrudates of a mixture of 10% polyethylene (Marlex 6009) and 90% Ionomer (Surlyn A1558)—the ionomer phase is the phase of higher viscosity—at three different shear rates. At the lowest shear rate, the dispersed phase forms a lamellar morphology even though some fibers with elliptical cross section can be seen near the center; as the shear rate increases, the lamellar morphology becomes dominant. At the highest rate, where the extrudate itself is severely distorted by melt fracture, the lamellar morphology persists, outlining vortexlike swirls. Hence, with the exception of very low shear rates this type of morphology is independent of the magnitude of the shear rate or shear stress.

Similar stratification was observed in mixtures of 10% polyethylene–90% polystyrene (Styron 666) and 30% polyethylene–70% polystyrene. Again the mode of dispersion is stable in swirling type motions, as shown in Fig. 21, which is a section along the flow direction.

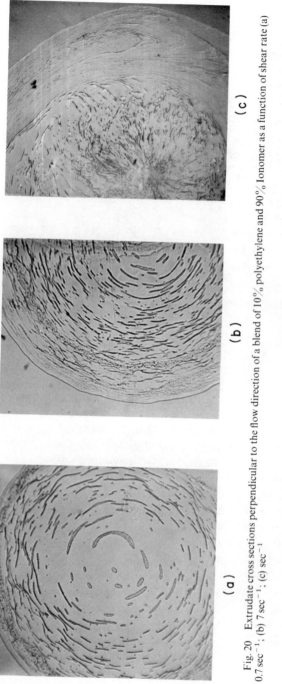

Fig. 20 Extrudate cross sections perpendicular to the flow direction of a blend of 10% polyethylene and 90% Ionomer as a function of shear rate (a) 0.7 sec⁻¹; (b) 7 sec⁻¹; (c) sec⁻¹

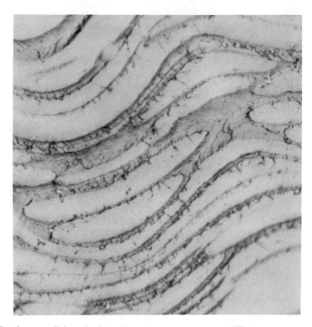

Fig. 21   Section parallel to the flow direction of the extrudate of a blend of 10% polyethylene and 90% polystyrene at high shear rates.

It should be stressed that this type of morphology is stable only in flow. When these extrudates are melted again, the ribbons break up rapidly to form small droplets. Hence, in practice, the morphology is only preserved by rapid quenching or crystallization of the dispersed phase during flow. Extrusion, followed by stretching, has produced ribbons as thin as 700–500 Å [90]. As the concentration of polyethylene is increased, the expected droplet type dispersions are found. Here it should be noted that the droplets are very large. Better dispersions could only be obtained by preextrusion through a 16-element Kenics mixer. An interesting effect was observed when comparing mixtures extruded at 170°C (here the viscosity ratio varies from 40 to 10, depending on the shear stress) and mixtures extruded at 225°C (where the viscosities of the two phases are approximately equal), as shown in Fig. 22. Extrusion at 170°C yielded droplets, whereas extrusion at 225°C yielded fibrils of polystyrene. Besides the droplet–fiber change, the mode of dispersion in these 70/30 mixtures was found to be independent of shear rate and temperature.

   The effect of viscosity and molecular weight on the mode of dispersion was explored using mixtures of poly(methyl methacrylate) (PMMA), and

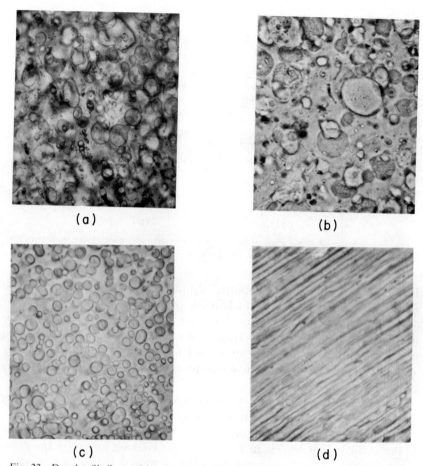

(a)   (b)

(c)   (d)

Fig. 22   Droplet–fibril transition in extruded blends of 30% polystyrene and 70% poly-ethylene at $T = 170$ and 225°C. (a) and (c) cut perpendicular to the flow direction; (b) and (d) cut parallel to the flow direction. In (d) the fibril direction is parallel to the flow direction. (From Van Oene [3].)

polystyrene by varying the molecular weight of the PMMA phase. When the molecular weight and viscosity of the PMMA is lower than that of polystyrene, stratification is observed, which gave the extrudates an expected luster and sheen. When the molecular weight of the PMMA is about equal to or greater than that of polystyrene, a dispersion of droplets is observed. Preblending by means of a static mixer again proved necessary to produce dispersions of finer texture. Representative electron micrographs are shown in Fig. 23. Stratification observed previously has finally disappeared; nevertheless, a difference in morphology still persists in that the mixture of 30/70

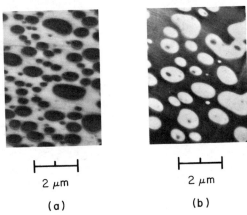

├──┤ 2 μm ├──┤ 2 μm

(a)   (b)

Fig. 23 Submicron dispersions of blends of polystyrene and poly(methyl methacrylate): (a) 30/70 PS/PMMA; (b) 30/70 PMMA/PS. (From Van Oene [3].)

PMMA/PS shows many "composite" droplets, while the mixture of 70/30 PMMA/PS shows essentially single-phase polystyrene droplets.

Hence, two distinct modes of dispersion are observed: stratification or droplet–fiber type dispersions. These modes of dispersion are independent of the magnitude of the shear stress and temperature. Low-shear mixing, such as that occurring in a static mixer, proved more successful in obtaining dispersions with a fine texture.

The absence of droplets in stratified mixtures and the surprising shear stability of this mode of dispersion is highly unusual, but may be related to the observation of Bartram *et al.* [54], who saw extension in the direction of the vorticity axis, which, in Poiseuille flow, is along an arc of the plane perpendicular to the flow direction. If this is indeed the general mode of deformation of droplets under these circumstances, no droplets would survive, and the substance would probably coalesce to form ribbons. Hayashida *et al.* [91] have observed similar stratified ("treelike") morphologies.

Han and Kim [92] made an extensive rheological study of blends of high-density polyethylene (HDPE) and polystyrene (PS). By making pressure measurements along a capillary ($L/D = 20$), the pressure drop and the exit pressure were obtained. This allows direct calculation of the shear stress and normal stress functions. The morphology of the HDPE–PS blends in ratios 25/75 and 50/50 was similar, both blends have polystyrene droplets dispersed in a continuous phase of HDPE. In the 75/25 blends no clear continuous phase was apparent; the two components seem interlocked. Typical results are shown in Fig. 24 for three shear stresses and three different temperatures.

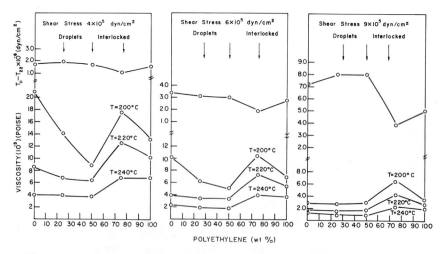

Fig. 24 Comparison of the composition and temperature dependence of the viscosity and normal stress of blends of polystyrene and polyethylene at various shear stresses. (Data from Han and Kim [92].)

The normal stress difference $(T_{11} - T_{22})$ was found to be independent of temperature when evaluated at constant stress. At a shear stress of 4×10^5 and 6×10^5 dyn/cm^2 the viscosity shows a sharp minimum and maximum, respectively, especially at $T = 200°C$. The normal stress difference has a shallow maximum and a pronounced minimum, especially at a shear stress of 9×10^5 dyn/cm^2. The normal stress difference tends to increase in those blends that have dispersed droplets and also tend to have a minimum viscosity. The blend with an "interlocked" morphology has a larger viscosity but a lower normal stress difference.

The composition dependence of the viscosity depends strongly on shear stress, as found by Ablazova and Vinogradov and co-workers [93]. Their results on mixtures of a copolymer of formaldehyde with 2% of 1,3-dioxalane (POM) and a mixed copolyamide (CPA), a copolymer of 44% caprolactam, 37% hexamethylene adipate, and 19% hexamethylene sebacate are shown in Fig. 25. It is clearly evident that the viscosity–composition relationship is extremely sensitive to the magnitude of the shear stress. In these blends the POM phase was found to form fibers. As the shear stress increased, the diameter of the fibers decreases to submicroscopic dimensions. A set of fine concentric fibrous tubes remained after removal of the CPA. The shear stresses at which the extrudate has a "telescopic" structure corresponds to the region of the sharp viscosity minimum.

These investigations deal with blends of incompatible polymers, with no attempt being made to stabilize a desired morphology. In this context one

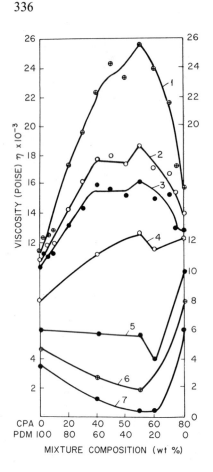

Fig. 25 Composition dependence of the viscosity of blends of CPA (copolyamide) and POM (polyoxymethylene) as a function of shear stress: (Curve 1) 1.27; (Curve 2) 3.93; (Curve 3) 5.44; (Curve 4) 6.30; (Curve 5) 12.59; (Curve 6) 19.25; (Curve 7) 31.62 $(10^{-5}$ (dyn/cm^2)). (From Vinogradov et al. [93].)

should note the high degree of order obtained by Folkes et al. [95] in extrusion and subsequent annealing of styrene–butadiene–styrene block copolymers. Here the dispersed polystyrene phase forms long cylinders, which form a hexagonal net in a matrix of polybutadiene. The perfection of the lattice formed by the polystyrene cylinders is such that these authors refer to the structures as "single crystals."

B. Viscosity of Polymer Blends

The rheological data for the flow behavior of polymer blends obtained thus far show a number of peculiarities that are difficult to understand. The occurrence of a viscosity maximum at a certain composition can probably be understood in view of differing morphologies; an absolute minimum is difficult to rationalize. Before any further analysis can proceed, however, one needs to clarify first how both temperature dependence and concentration dependence should be evaluated. Since the viscosity of a polymer melt exhibits a large shear dependence, the respective dependences can be evaluated either at constant stress or at constant shear rate. As Bestul and Belcher [96] have pointed out, one may derive a general relationship between the temperature dependence of the viscosity, evaluated at constant stress and the similar quantity, evaluated at constant rate, which is given by

$$(d\eta/dT)_G = (\partial\eta/\partial T)_{T_{12}}[1 + (\partial \ln \eta/\partial \ln G)_T] \tag{66}$$

Since $(d \ln \eta/d \ln G)_T$ is negative, at constant temperature the viscosity is a decreasing function of shear rate,

$$(d\eta/dT)_G < (\partial\eta/\partial T)_{T_{12}} \tag{67}$$

A similar relationship may be derived for the concentration dependence:

$$(\partial\eta/\partial C)_{T_{12}} = (\partial\eta/\partial C)_G[1 - G(\partial\eta/\partial T_{12})_C] \tag{68}$$

Again, since $(\partial\eta/\partial T_{12})_C$ is negative, the concentration dependence evaluated at constant stress will be larger than the concentration dependence evaluated at constant shear rate. Hence, if the concentration dependence is evaluated at constant shear stress, anomalies will be easily detected.

As already observed, when deriving expressions for the relative volumetric flow rate in the bicomponent flow Eq. (35), so-called mixture rules for the viscosity may be obtained. This is accomplished most readily when the pressure gradient can be eliminated. For Newtonian systems this is of no consequence; for non-Newtonian systems this would imply that the comparison is intended at constant shear stress. A further complication arises, however, from the shear dependence of the viscosity of the two fluids, which in general is not the same. It may be shown [97] that when for bicomponent flow of two power law fluids, the shear dependence of the viscosity can be represented as

$$T_{12} = KG^n \tag{69}$$

the mixture rule, derived for Poiseuille flow, becomes

$$\left(\frac{n_m}{3n_m + 1}\right)\frac{1}{\eta_{\text{mixt}}} = \frac{n_1}{3n_1 + 1}(\xi)^{(3n_1 + 1)/n_1}\frac{1}{\eta_1} + \frac{n_2}{3n_2 + 1}[1 - (\xi)^{(3n_2 + 1)/n_2}]\frac{1}{\eta_2} \tag{70}$$

where η_1 and η_2 and η_{mixt} are the viscosities evaluated at the same shear stress and n_1, n_2, and n_{mixt} are the respective power law exponents for the fluids. For certain combinations of the individual power law exponents and viscosities, the viscosity of the mixture of non-Newtonian fluids may be lower than that of Newtonian fluids of the same viscosity. Furthermore, since the boundary conditions at fluid–fluid interfaces require the continuity of the shear stress, both temperature dependence and concentration dependence will be evaluated at constant shear stress.

It is well established [98] that the shear-dependent viscosities of individual polymer melts obtained at different temperatures can, at constant shear stress, be superposed readily by vertical shifts only. The shift factor therefore reflects the temperature dependence of the viscosity, the shear dependence itself being independent of temperature in this representation. For polyethylene–polystyrene blends similar superposition was found possible, provided that each composition is considered separately. At shear stresses larger than 10^6 dyn/cm^2 the superposition was less satisfactory, probably because of the onset of melt fracture. As is evident from Fig. 26, superposition with respect to concentration is not possible due to the difference in shear dependence of the individual polymers.

The concentration dependence of the viscosity of polystyrene–polyethylene blends at $T = 170°C$ is shown for various shear stresses in Fig. 27. Blends containing 10 and 30% polyethylene are stratified; blends containing 10 and 30% polystyrene are dispersions of polystyrene droplets in polyethylene. (The morphology of the 30% PS–70% PE is shown in Fig. 22.) For the data

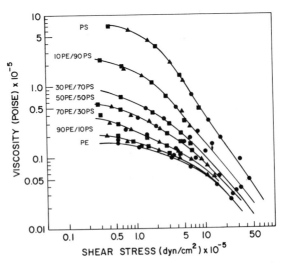

Fig. 26   Viscosity versus shear stress for polystyrene–polyethylene blends reduced to $T = 170°C$ (●) by a vertical shift only; (▲) data obtained at $T = 210°C$; (■) data obtained at $T = 230°C$.

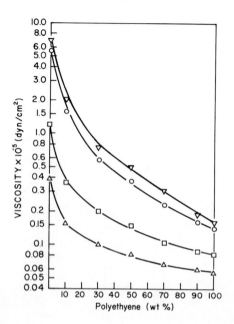

Fig. 27 Composition dependence of the viscosity of polystyrene–polyethylene blends at various shear stresses ($T = 170°C$): Shear stress (dyn/cm²) ($\triangledown$) 0.5×10^5; (◯) 1×10^5; (□) 5×10^5; (△) 10×10^5.

obtained at a shear stress of 10^5 and 10^6 dyn/cm² a detailed comparison is made in Fig. 28 with various theoretical predictions.

Hashin [99], using variational principles similar to those used successfully for the prediction of bounds for the modulus of elasticity of composite materials (see Chapter 9), obtained for the viscosity of Newtonian fluid mixtures the bounds:

$$\text{upper bound:} \quad \eta^*_{\text{mixt}} = \eta_2 + \frac{\phi_1}{1/(\eta_1 - \eta_2) + 2\phi_2/5\eta_2} \tag{71}$$

$$\text{lower bound:} \quad \eta^*_{\text{mixt}} = \eta_1 + \frac{\phi_2}{1/(\eta_2 - \eta_1) + 2\phi_1/5\eta_1} \tag{72}$$

For non-Newtonian fluids the coefficient $\frac{2}{5}$ in the denominator has to be replaced by $\frac{1}{2}$. Heitmiller *et al.* [100] derived an expression for the viscosity of a mixture with a morphology of n concentric layers of Fluid 1 in Fluid 2. When there are many layers, the viscosity expression reduces to

$$1/\eta_{\text{mixt}} = w/\eta_1 + (1 - w)/\eta_2 \tag{73}$$

where w is the weight fraction of Fluid 1.

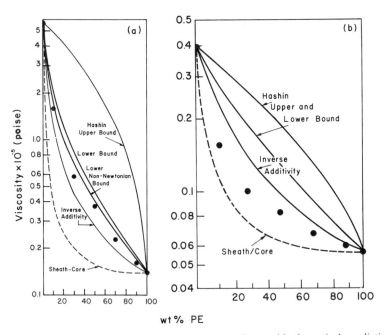

Fig. 28 Comparison of observed composition dependence with theoretical predictions, $T = 170°C$: (a) shear stress $= 1 \times 10^5$ dyn/cm^2; (b) shear stress $= 1 \times 10^6$ dyn/cm^2. Blends with 10 and 30% polyethylene have a ribbon-type morphology; blends with 70 and 90% polyethylene are dispersions of droplets of polystyrene in polyethylene.

As in dispersions, radial migration toward the tube axis may take place, leading to a sheath–core-type configuration, the lowest bound on the viscosity is given by assuming complete segregation, the fluid of lowest viscosity forming the outer annulus.

From Fig. 28 it is seen that the data obtained at a shear stress of 10^5 dyn/cm^2 straddle Hashin's lower bound for nonNewtonian fluids. For a shear stress of 10^6 dyn/cm^2 the experimental data are well below the lower bound and relationship predicted on the basis of inverse additivity. Hence, even when the viscosity of the two parent polymers differs substantially, and no viscosity minimum is observed, the viscosities of the blends are surprisingly low. As the shear stress is increased, the viscosities tend to fall below the lower bounds, even for dispersion of droplets. A value of the viscosity lower than that of either constituent is, however, not predicted, regardless of morphology.

A minimum in viscosity is observed more clearly, however, when the viscosities of the parent polymers are more nearly the same. A typical case is shown in Fig. 29 for data obtained on polyethylene–polystyrene mixtures

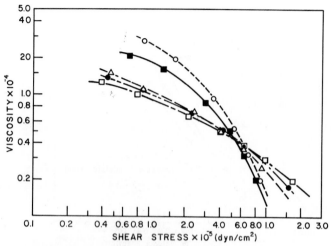

Fig. 29   Viscosity as a function of shear stress of polystyrene–polyethylene blends at $T =$ 210°C, showing a crossover at a shear stress of 6×10^5 dyn/cm^2: (O) polystyrene; (■) 90/10 PS/PE; (△) 50/50 PS/PE; (●) 30/70 PS/PE; (□) polyethylene.

at 210°C. At high shear stress the viscosity of polystyrene is lower than that of polyethylene (PE), at low shear stress the viscosity is higher. For shear stresses larger than 5×10^5 dyn/cm^2 the viscosity of the mixture of 10/90 PE/PS is clearly lower than that of polystyrene itself.

It should be remembered that in flow through a capillary the shear stress varies linearly over the cross section from zero in the center to maximum value at the wall. When the shear dependence of the individual fluids brings about a viscosity reversal, the value of the viscosity ratio becomes shear dependent and changes from larger than unity to smaller than unity. This need not affect the morphology in dispersions of droplets, but it should influence the value of the viscosity of the polymer–polymer blend. In Fig. 30 various predictions for the concentration dependence of dispersions of fluid spheres are compared as a function of the viscosity ratio. The predictions for the concentration dependence of dispersion of fluid spheres were derived by Brennan [101] using a cell model similar to that of Simha [102] for rigid spheres and that of Yaron and Gal-Or [103] for fluid spheres. In the case of fluid spheres, the particles are assumed to be constrained by surface tension and remain spherical. The viscosity is similar to that of rigid spheres. If the internal circulation within the droplets were inhibited, the results for rigid spheres would apply. Brennan [101] also considers the case of "infinitely deformable" droplets, which is applicable when interfacial tensions are negligible. In this case a viscosity decrease is predicted, but its magnitude

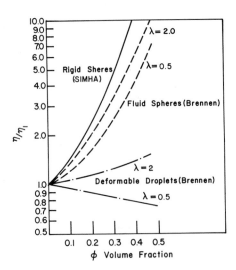

Fig. 30   Cell model prediction for the concentration dependence of dispersions of fluid spheres and deformable droplets.

is not such that the viscosity of the dispersion falls below that of the fluid with lowest viscosity.

The occurrence of absolute viscosity minima in polymer–polymer blends is therefore a feature that is not understood. Even in the case where no viscosity minima are observed, the viscosity of the polymer–polymer dispersions tends to be lower than predicted on variational grounds or by any special assumptions about particular morphologies.

C.   Normal Stress Function of Polymer Blends

In the work of Han [104] the first normal stress difference is obtained from the extrapolated value of the "exit" pressure:

$$T_{11} - T_{22} = p_{\text{exit}} + (T_{12})_w [\partial p_{\text{exit}}/\partial (T_{12})_w] \tag{74}$$

When the exit pressure is obtained from capillary data, the second normal stress function can also be obtained as

$$T_{22} - T_{33} = -(T_{12})_w [\partial p_{\text{exit}}/\partial (T_{12})_w] \tag{75}$$

Similar relations were also derived by Davis *et al.* [105]. As discussed in detail by Han [106], the exit pressure depends on the ratio of reservoir diameter D_r and the capillary diameter as well as the length–diameter ratio of the capillary itself. For polymer melts the exit pressure reaches a limiting value for $D_r/D = 12$ and $L/D = 15$–20. The exit pressure for polymer blends, however, shows a much stronger L/D dependence than observed for single component polymer melts. This is shown in Fig. 31 for blends of polystyrene and polyethylene, using data obtained by Han and Yu [107].

For $L/D = 4$ the blends have larger exit pressures than the pure components, while for $L/D = 20$, they are lower. The exit pressures of the 20/80 and 50/50 PE/PS blends decrease dramatically; the exit pressure of the 80/20 PE/PS blend decreases at first and then increases somewhat, a similar pattern being observed for polyethylene itself. It is, therefore, not certain from these data that a capillary with a $L/D = 20$ is sufficiently long to assure that the values of the exit pressure will be independent of any further increase in length and are characteristic of the shear flow in the capillary. The data do, however, indicate considerable pressure relaxation within the capillary. For the discussion of the data of Han and Kim [92] it will be assumed, however, that the data obtained with a capillary of $L/D = 20$ are representative of the shear flow itself. The most striking feature of the normal stress data obtained by Han and Kim [92], and shown in Fig. 24, is the absolute minimum observed for the 75/25 PE/PS blend, which has an interlocked morphology. As the shear stress increases, this minimum becomes more pronounced. Since the concentration dependence of the viscosity is opposite to that of the normal stress function, it is probably more pertinent

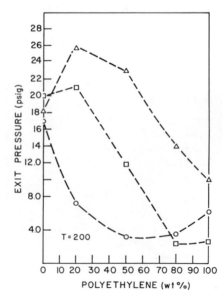

Fig. 31 Dependence of the exit pressure on L/D for polystyrene–polyethylene blends: ($\triangle$) $L/D = 4$; ($\square$) $L/D = 12$; ($\bigcirc$) $L/D = 20$. Shear stress is 8.3×10^5 dyn/cm^2. (Data from Han and Yu [107].)

to discuss the behavior of the normal stress functions in terms of the recoverable shear S_R given by the expression

$$S_R = \tfrac{1}{2}[(T_{11} - T_{22})/T_{12}] \tag{76}$$

rather than the normal stress function itself. The recoverable shear has been used extensively by Phillipoff and co-workers [108] for the treatment of viscoelastic effects in flow. The shear stress dependence of the recoverable shear for the polymer blends studied by Han and Kim [92] is shown in Fig. 32. The recoverable shear for the blend with an interlocked morphology is lower at all shear stresses than that of either one of the constituent polymers. The stress dependence resembles that of pure polyethylene itself. The blends, known to be a dispersion of polystyrene droplets suspended in polyethylene, have, at lower shear stresses, a recoverable shear comparable to that of polystyrene. As the shear stress increases, the recoverable shear exceeds that of either constituent. The stress dependence resembles that observed for polystyrene.

Since the magnitude of the recoverable shear is a measure of the stored free energy of deformation in flow [108], a possible inference is that dispersions of viscoelastic droplets have an additional mode for accumulation of free energy of deformation. This cannot be attributed directly to the disturbance of the shear flow by the particles themselves, as the viscosity of these dispersions tends to be significantly lower than that expected for dispersions.

D. Theoretical Considerations

The experimental observations of morphology and rheological behavior of polymer–polymer blends presents a surprising picture. Morphologically

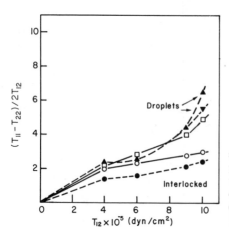

Fig. 32 Dependence of the recoverable shear on the magnitude of the shear stress for polystyrene–polyethylene blends and the individual polymers: ($\square$) polystyrene; ($\bigcirc$) polyethylene; ($\blacktriangle$) 25/75 PE/PS; ($\blacktriangledown$) 50/50 PE/PS; ($\bullet$) 75/25 PE/PS. (Data from Han and Kim [92].)

there are two distinct modes of dispersion: dispersions containing droplets and dispersions containing ribbons, which at higher concentrations yield an interwoven or interlocked web. Ultimately this type of dispersion also yields droplets, but in this case many droplets contain droplets within droplets. The modes of dispersion themselves are shear stable. Even when the flow has become turbulent, the ribbons tend to follow the swirling motion rather than break up in droplets.

Rheological behavior is characterized by viscosities, which at higher shear stresses are significantly below the lower bounds established by variational principles. Blends known to form dispersions of deformable droplets have been found to exhibit a viscosity lower than that of either one of its constituents. The first normal stress difference can be higher or lower than that of either one of its components.

None of these aspects is understood in detail. There is some indication, however, that the influence of viscoelasticity can manifest itself as an increase in interfacial tension, as noted when discussing the microrheology of a viscoelastic droplet in a Newtonian fluid.

Most theories of deformation in flow assume from the outset that a spherical droplet will be deformed into an ellipsoid. The shape of the deformed ellipsoid is such that, in a plane perpendicular to the flow direction, the cross section of the deformed ellipsoid remains circular. This is presently not the case. In the absence of a complete hydrodynamic theory, therefore, can one establish a criterion to decide what kind of dispersion will be formed? Such a criterion has been proposed by Van Oene [3], who analyzed the conditions that have to be satisfied for the formation of a droplet

In flow the shear stress acting at the fluid droplet interface is required to be continuous. This condition can be regarded as a definition of liquid-like behavior. In Poiseuille flow the shear stress is proportional to the pressure gradient. Hence, this boundary condition defines the local pressure within an additive to a constant. The shape of a droplet, however, is determined by the pressure difference acting across the interface. The magnitude of the pressure, p, is equal to:

$$p = -\tfrac{1}{3}\text{Trace } \mathbf{T} \tag{77}$$

At the interface of spherical droplet of radius a, the stress tensor has the following components:

1. the component normal to the interface T_n

$$T_n = -[p + 2\gamma/a] \tag{78}$$

2. the components in a plane tangential to the interface T_t

$$T_t = -[p + \gamma/a] \tag{79}$$

Hence, the magnitude of the isotropic hydrostatic pressure is continuous across the interface. For a plane interface, the curvature terms vanish.

In view of the expressions derived for the stress tensor in Poiseuille flow [17] for stratified cocurrent flow, the quantity

$$-\tfrac{1}{3}[T\langle zz\rangle - T\langle rr\rangle][1 - \tfrac{5}{2}E] \tag{80}$$

should be continuous. Since the normal stress functions are strongly dependent on material properties, continuity of the isotropic pressure across the interface is in general not satisfied.

Marucci [109] has shown that the quantity Trace **T** may also be given a thermodynamic interpretation as a free energy of deformation:

$$\Delta F_{\text{def}} = \tfrac{1}{2}\text{Trace } \mathbf{T} \tag{81}$$

This relationship is exact for a dilute solution of elastic dumbbells. A similar relationship was stated by Janeschitz-Kriegl [110]. Since for a solution of elastic dumbbells, the second normal stress difference vanishes (i.e., $E = 0$), the free energy of deformation in flow is equal to

$$\Delta F_{\text{def}} = \tfrac{1}{2}[T\langle zz\rangle - T\langle rr\rangle] \tag{82}$$

Hence the *recoverable* free energy of deformation is

$$\Delta F_{\text{rec}} = -\Delta F_{\text{def}} \tag{83}$$

Little is known about the "elastic free energy" of viscoelastic fluids. In a two-phase system, however, a possible mode of relaxation of the free energy of deformation is conversion into interfacial free energy. If contributions to the free energy of deformation by internal circulation is neglected, the formation of a droplet of a fluid α with a larger first-normal stress difference in a liquid β, with a lower first-normal stress difference should result in a free energy decrease of

$$\Delta F = \tfrac{1}{2}[(T\langle zz\rangle - T\langle rr\rangle)_\alpha - (T\langle zz\rangle - T\langle rr\rangle)_\beta] \tag{84}$$

If this decrease in free energy is recovered by formation of interfacial free energy, one obtains for the formation of n droplets of radius a

$$n4\pi a^2\gamma_{\alpha\beta} = n\tfrac{1}{2}[(T\langle zz\rangle - T\langle rr\rangle)_\alpha - (T\langle zz\rangle - T\langle rr\rangle)_\beta]\tfrac{4}{3}\pi a^3 \tag{85}$$

which formally can be regarded as the viscoelastic contribution to the interfacial tension $\gamma_{\alpha\beta}$. Taking the equilibrium interfacial tensions in the absence of flow $\gamma_{\alpha\beta}^0$ into account, the complete expression for the interfacial tension in flow is therefore [3]

$$\gamma_{\alpha\beta} = \gamma_{\alpha\beta}^0 + \tfrac{1}{6}a[(T\langle zz\rangle - T\langle rr\rangle)_\alpha - (T\langle zz\rangle - T\langle rr\rangle)_\beta] \tag{86}$$

The finding that, for submicron dispersions, differences in morphology still persists in the sense that a submicron dispersion derived from an originally stratified dispersion contains many composite droplets can also be accounted for in terms of Eq. (86) [3].

One can pursue further the possibility of conversion of free energy of deformation into interfacial free energy. It would imply that dispersions with a large interfacial area, such as those with an "interlocked" morphology, would exhibit a lower recoverable free energy of deformation after cessation of flow. In this context one may regard the recoverable shear S_R as the recoverable free energy of deformation, normalized with respect to shear stress, or the momentum flux. As shown in Fig. 32, the data of Han and Kim support such an interpretation.

Why dispersions of droplets should exhibit a larger recoverable shear is not accounted for by this argument. A possible clue may be that the viscosity of these dispersions is surprisingly low. As already indicated, once droplets are formed, the boundary conditions require internal circulation. The pattern of internal circulation in shear is that of a convective flow. The consequences of such a convective internal circulation can only be a matter for speculation. The closest analogy appears to be the onset of the so-called Taylor instability of Couette flow. From the analysis of Denn *et al.* [26], which was confirmed by experiment, the onset of the Taylor instability is accompanied by a torque reduction. A related observation is the experimental finding by Barnes and Walters [27] that at all Reynolds numbers a viscoelastic fluid flows through a coiled capillary tube *faster* than a Newtonian fluid of the same viscosity. Hence, it appears that it need not be necessary to generate a turbulent flow in order to observe "drag reduction;" it may be sufficient to generate a "secondary flow" situation.

It is therefore tempting to attribute the viscosity minima observed by Han and Kim [92] in dispersions of droplets as a consequence of internal circulation.

VIII. CONCLUSION

In this review of the rheology of polymer blends and polymer dispersions an attempt has been made to link the observed macrorheological behavior, expressed in terms of a shear-dependent viscosity and shear-dependent normal stress functions, and the microrheological processes, which establish the morphology of the resulting polymer blend. Most striking is the influence of domain size on morphology. When the size of the dispersed particle is in the submicron range, fluid particles will tend to remain spherical because of the influence of the interfacial tension; these dispersions will have many features in common with dispersions of rigid particles. At concentrations of practical interest (i.e., 10% by volume or higher) the macrorheological behavior tends to be dominated by particle–particle interaction, as evidenced by observed yield stresses in flow at small velocity gradients. Particle migration, enhanced when the suspending phase is a viscoelastic fluid [111], gives rise to a sheath–core morphology. Within the core region the velocity gradient is small; hence, the core is stabilized by the yield value. Velocity gradients tend to be large in the depleted wall layer.

As such, the sheath–core morphology, induced by flow leads to lower viscosities and hence a larger throughput for a given pressure gradient, when the dispersed particles are rigid; die swell is also reduced. Even though these factors contribute to ease of processing, an inevitable side effect will be that the surface layers will have a composition that differs markedly from that of the bulk of the material. Furthermore, the filling of a mold will be strongly affected as the jet of liquid entering the mold, being stabilized by the sheath–core morphology, will travel through the mold until it hits an obstacle, at which point it will buckle. With axisymmetric rigid fillers, the degree of orientation will in the core region be severely limited by particle–particle interaction; in the surface layers it will be more pronounced, leading to nonuniform physical properties of the composite.

When the domain size of the dispersed phase is larger than roughly 1 μm, dispersions of rigid particles still follow tractable rules, the particle size distribution becomes increasingly important for achieving maximum filler loadings at optimum flow properties. When both phases are fluid, as is the case in polymer blends, no general prediction of macrorheological behavior seems possible; viscosities are often lower than can be accounted for in terms of particle migration; normal stresses can either be significantly smaller or larger than those of the parent fluids. Two distinct morphologies are observed, with the dispersed phase dispersed either as ribbons or droplets. A particular type of morphology is independent of the magnitude of the shear stress. The droplets appear to be less deformable than an equivalent viscous Newtonian droplet. This resistance to breakdown in shear often makes intensive shear mixers ineffective [112]. In order to achieve submicron dispersions by melt blending it is necessary to employ flow fields other than shear. Converging flow, characterized by a superposition of a shear and an extensional flow, is more effective in reducing domain size, as the extensional flow will draw a droplet into a thin fiber (see Volume 2, Chapter 16), which on subsequent breakup will yield small droplets. Intermittent shear and extension, as occurs in some mixers and on a mill roll is in general more effective than a sustained shear flow.

It is difficult to formulate general blending rules. In general the morphology of a blend will depend on the detailed processing history. Ribbon-type morphologies are stable in flow; unless preserved by crystallization or quenching, for example, the ribbons will break up rapidly when flow ceases. Hence, molded composites may differ significantly in properties from extruded composites.

An area that has received some attention is the stabilization of a morphology by addition of an appropriate copolymer. Locke and Paul [113], Ide and Hasegawa [114], and Kowa [115] find that addition of a copolymer leads to more uniform dispersions of smaller domain size. Properties such as toughness generally improve upon addition of a copolymer. It is not clear, however, whether the improvement in physical properties so achieved is due to the smaller domain size or improved interfacial bonding (see Volume 2, Chapter 12). The extensive work of Kohler *et al.* [116] on blends of polystyrene and polyisoprene of differing molecular weight, with and without addition of copolymer, has shown that, if improved impact resistance is taken as a criterion, addition of a copolymer need not be necessary, provided the molecular weight of the components of the blend is carefully specified.

The great variety of morphologies observed in polymer–polymer blends has been attributed to the viscoelasticity of each component. In principle, therefore, these effects should also be present in blends of homopolymers of differing molecular weight and molecular weight distribution. Blends of homopolymers may therefore also show a great variability in ultimate properties and toughness.

REFERENCES

1. R. L. Jalbert and J. P. Smejkal, *in* "Modern Plastics Encyclopedia 1976–1977," p. 108. McGraw-Hill, New York, 1976.
2. F. Gauthier, H. L. Goldsmith, and S. G. Mason, *Kolloid Z.* **248**, 1000 (1971).
3. H. Van Oene, *J. Colloid Interface Sci.* **40**, 448 (1972).
4. C. Truesdell and W. Noll, The non-linear field theories of mechanics, *in* "Handbuch der Physik," Vol. III/3. Springer-Verlag, Berlin, 1965.
5. R. Byron Bird, Useful non-Newtonian models, *Ann. Rev. Fluid Mech.* **8**, 13 (1976).
6. H. Brenner, *Int. J. Multiphase Flow* **1**, 195 (1974).
7. E. J. Hinch and L. G. Leal, *J. Fluid Mech.* **71**, 481 (1975).
8. M. Reiner, *Am. J. Math.* **67**, 350 (1945).
9. R. S. Rivlin and J. L. Ericksen, *J. Rational Mech. Anal.* **4**, 323 (1955).
10. H. Markovitz, *Trans. Soc. Rheol.* **1**, 37 (1957).
11. J. G. Oldroyd, *Proc. Roy. Soc. London* **A200**, 523 (1950); **A245**, 278 (1958).
12. A. S. Lodge, *Trans. Faraday Soc.* **52**, 120 (1956).
13. H. Giesekus, *Rheol. Acta* **1**, 2 (1958).
14. B. D. Coleman and W. Noll, *Arch. Rational Mech. Anal.* **6**, 355 (1960).
15. See Truesdell and Noll [4, Sect. 123, p. 504].
16. B. D. Coleman and H. Markovitz, *J. Appl. Phys.* **35**, 1 (1964).
17. W. E. Langlois, *Trans. Soc. Rheol.* **8**, 33 (1964).

18. C. D. Han, "Rheology in Polymer Processing," Chapter 3, Academic Press, New York, 1976.
19. A. C. Pipkin and R. S. Rivlin, *ZAMP* **14**, 738 (1963).
20. C. D. Han, *AIChE J.* **17**, 1418 (1971).
21. J. L. Ericksen, *Q. Appl. Math.* **14**, 318 (1956).
22. A. E. Green and R. S. Rivlin, *Q. Appl. Math.* **14**, 299 (1956).
23. See Truesdell and Noll [4, Sect. 117, p. 471]; R. L. Fosdick and J. Serrin, *Proc. Roy. Soc. London* **A332**, 311 (1973).
24. H. Giesekus, *Rheol. Acta* **4**, 85 (1965).
25. H. Giesekus, *Rheol. Acta* **3**, 59 (1963).
26. M. M. Denn, Z. S. Sun, and B. D. Rushton, *Trans. Soc. Rheol.* **15**, 415 (1971).
27. H. A. Barnes and K. Walters, *Proc. Roy. Soc. London* **A314**, 85 (1969).
28. G. Astarita and M. M. Denn, *in* "Theoretical Rheology" (J. F. Hutton, J. R. A. Pearson, and K. Walters, eds., Chapter 20, p. 333. Halstead Press, Wiley, New York, 1975.
29. J. R. A. Pearson, *Ann. Rev. Fluid Mech.* **8**, 163 (1976).
30. T. W. F. Russel and M. E. Charles, *Can. J. Chem. Eng.* **37**, 18 (1959).
31. B. A. Packham and R. Shail, *Proc. Cambridge Phil. Soc.* **69**, 443 (1971).
32. Y. Nubar, *Ann. N.Y. Acad. Sci.* **136**, 33 (1966).
33. J. R. Jones, *Rheol. Acta* **14**, 397 (1975).
34. J. L. White and B. Lee, *Trans. Soc. Rheol.* **19**, 457 (1975).
35. A. E. Everage Jr., *Trans. Soc. Rheol.* **19**, 509 (1975).
36. T. C. Yu and C. D. Han, *J. Appl. Polym. Sci.* **17**, 1203 (1973).
37. J. C. Slattery, *AIChE J.* **10**, 817 (1964).
38. A. A. Khan and C. D. Han, *Trans. Soc. Rheol.* **20**, 595 (1976).
39. J. H. Southern and R. L. Ballman, *Appl. Polym. Symp.* **20**, 175 (1973).
40. A. E. Everage Jr., *Trans. Soc. Rheol.* **17**, 629 (1973).
41. See Han [18, Chapter 10].
42. C. D. Han, *J. Appl. Polym. Sci.* **17**, 1289 (1973).
43. R. Rutgers, *Rheol. Acta* **2**, 202, 305 (1962).
44. D. G. Thomas, *J. Colloid Sci.* **20**, 267 (1965).
45. D. J. Jeffrey and A. Acrivos, *AIChE J.* **22**, 417 (1976).
46. H. L. Goldsmith and S. G. Mason, The micro-rheology of dispersions, *in* "Rheology–Theory and Applications" (F. R. Eirich, ed.), Vol. 4. Academic Press, New York, 1967.
47. G. B. Jeffery, *Proc. Roy. Soc. London* **A102**, 161 (1922).
48. A. Okagawa, R. G. Cox, and S. G. Mason, *J. Colloid Interface Sci.* **45**, 303 (1973).
49. T. Kotaka, *J. Chem. Phys.* **30**, 1566 (1959).
50. H. Giesekus, *Rheol. Acta* **2**, 50 (1962).
51. A. Okagawa and S. G. Mason, *J. Colloid Interface Sci.* **45**, 330 (1973).
52. F. Gauthier, H. L. Goldsmith, and S. G. Mason, *Rheol. Acta* **10**, 344 (1971).
53. L. G. Leal, *J. Fluid Mech.* **69**, 305 (1975).
54. E. Bartram, H. L. Goldsmith, and S. G. Mason, *Rheol. Acta* **14**, 776 (1975).
55. A. Karnis, H. L. Goldsmith, and S. G. Mason, *J. Colloid Interface Sci.* **22**, 531 (1966).
56. A. Karnis, H. L. Goldsmith, and S. G. Mason, *Can. J. Chem. Eng.* **44**, 181 (1966).
57. D. J. Highgate and R. W. Whorlow, *in* "Polymer Systems, Deformation and Flow" (R. E. Wetton and R. H. Whorlow, eds.), p. 251. McMillan, New York, 1968.
58. D. J. Highgate and R. W. Whorlow, *Rheol. Acta* **9**, 569 (1970).
59. R. L. Hoffman, *Trans. Soc. Rheol.* **16**, 155 (1972); *J. Colloid Interface Sci.* **46**, 491 (1974). (1974).

60. S. Hachisu and Y. Kobayashi, *J. Colloid Interface Sci.* **46**, 470 (1974).
61. A. Kose and S. Hachisu, *J. Colloid Interface Sci.* **46**, 460 (1974).
62. S. Onogi, T. Masuda, and T. Matsumoto, *Trans. Soc. Rheol.* **14**, 275 (1970).
63. S. Onogi, T. Matsumoto, and Y. Warashina, *Trans. Soc. Rheol.* **17**, 175 (1973).
64. T. Matsumoto, C. Hitomi, and S. Onogi, *Trans. Soc. Rheol.* **19**, 541 (1975).
65. N. Casson, *in* "Rheology of Disperse Systems" (C. C. Mill ed.). Pergamon, Oxford, 1959.
66. See Goldsmith and Mason [46, p. 123].
67. G. I. Taylor, *Proc. Roy. Soc. London* **A146**, 501 (1934).
68. F. D. Rumscheidt and S. G. Mason, *J. Colloid Sci.* **16**, 238 (1961).
69. S. Torza, R. G. Cox, and S. G. Mason, *J. Colloid Interface Sci.* **38**, 395 (1972).
70. G. Hestroni and S. Haber, *Rheol. Acta* **9**, 488 (1970).
71. R. G. Cox, *J. Fluid Mech.* **37**, 601 (1969).
72. W. R. Schowalter, C. E. Chaffey, and H. Brenner, *J. Colloid Interface Sci.* **26**, 152 (1963).
73. G. I. Taylor, *Proc. Roy. Soc. London* **A138**, 41 (1932).
74. R. Roscoe, *J. Fluid Mech.* **28**, 273 (1967).
75. F. Gauthier, H. L. Goldsmith, and S. G. Mason, *Trans. Soc. Rheol.* **15**, 297 (1971).
76. E. B. Vadas, H. L. Goldsmith, and S. G. Mason, *Trans. Soc. Rheol.* **20**, 373 (1976).
77. T. Mikami, R. G. Cox, and S. G. Mason, *Int. J. Multiphase Flow* **2**, 113 (1975).
78. P. A. Willermet, R. A. Pett, and T. J. Whalen, *Amer. Ceram. Soc. Annu. Conf., April 24 1974*, Amer. Ceram. Soc., Chicago Illinois, 1974.
79. C. D. Han, *J. Appl. Polym. Sci.* **18**, 821 (1974); Han [18, p. 184].
80. G. V. Vinogradov, A. Ya. Malkin, E. P. Plotnikova, O. Yu. Sabsai, and N. E. Nikolayeva, *Int. J. Polym. Mater.* **2**, 1 (1972).
81. T. T. S. Lim, *Polym. Eng. Sci.* **11**, 240 (1971).
82. J. L. White and J. W. Crowder, *J. Appl. Polym. Sci.* **18**, 1013 (1974).
83. R. H. Beck Jr., *Annu. Tech. Conf., 29th* p. 392. Soc. Plast. Eng., Washington, D.C., 1971.
84. R. C. Hall, C. G. Cashulette, J. A. Valentine, and G. D. Chadderon, *Mater. Rev. 1975, Nat. SAMPE Tech. Conf., 7th* Vol 7. Soc. Advancement Mater. Process Eng., Azusa, California, 1975.
85. L. A. Goettler, *Mod. Plast.* **47**(4), 140 (1970).
86. L. A. Goettler, Flowfabrication of short fiber composites, *in* "Composite Materials in Engineering Design" (B. R. Noton, ed.). Amer. Soc. Metals, Tech. Div., Metals Park, Ohio, 1973.
87. L. A. Goettler, Molding of Oriented Short Fiber Composites II—Flow through Convergent Channels. Govt. Rep. AD 776591, Washington, D.C.
88. K. D. Roberts and C. T. Hill, Govt. Rep. AD/A 005989, Washington, D.C.
89. R. J. Farris, *Trans. Soc. Rheol.* **12**, 281 (1968).
90. Z. Mencik, H. K. Plummer, and H. Van Oene, *J. Polym. Sci. Part A-2* **00**, 507 (1972).
91. H. Hayashida, J. Takahashi, and M. Matsui, *Proc. Int. Congr. Rheol. 5th, Kyoto, Jpn., 1963* (S. Onogi, ed.), p. 525. Univ. of Tokyo Press, Tokyo, 1970.
92. C. D. Han and Y. W. Kim, *Trans. Soc. Rheol.* **19**, 245 (1975).
93. G. V. Vinogradov, B. V. Yarlykov, M. V. Tsebrenko, A. V. Yudin, and T. I. Ablazova, *Polymer* **16**, 609 (1975).
94. T. I. Ablazova, M. B. Tsebrenko, A. B. Yudin, G. V. Vinogradov, and B. V. Yarlykov, *J. Appl. Polym. Sci.* **19**, 1781 (1975).
95. M. J. Folkes, A. Keller, and F. R. Scalisi, *Kolloid Z.* **251**, 1 (1973).
96. A. B. Bestul and H. V. Belcher, *J. Appl. Phys.* **24**, 696 (1953).
97. H. Van Oene, Unpublished results.

98. R. S. Porter and J. F. Johnson, *J. Polym. Sci. Part C* No. 15, 65 (1966).
99. Z. Hasin, *in* "Second-order effects in Elasticity Plasticity and Fluid Dynamics" (M. Reiner and D. Abir, eds.). McMillan, New York, 1964.
100. R. F. Heitmiller, R. Z. Maar, and H. H. Zabusky, *J. Appl. Polym. Sci.* **8**, 873 (1964); H. L. Doppert and W. S. Overdiep, *in* "Multi Component Polymer Systems" (N. A. J. Platzer, ed.), Adv. Chem. Ser. No. 99. Amer. Chem. Soc. Washington, D. C., 1971.
101. C. Brennan, *Can. J. Chem. Eng.* **53**, 126 (1975).
102. R. Simha, *J. Appl. Phys.* **23**, 1020 (1952).
103. I. Yaron and B. Gal-Or, *Rheol. Acta* **11**, 241 (1972).
104. C. D. Han, *Trans. Soc. Rheol.* **18**, 163 (1974).
105. J. M. Davis, J. F. Hutton, and K. Walters, *J. Phys. D* **6**, 2259 (1973).
106. See Han [18], Chapter 5.
107. C. D. Han and T. C. Yu, *Polym. Eng. Sci.* **12**, 81 (1972).
108. W. Philippoff, *Trans. Soc. Rheol.* **5**, 163 (1961).
109. G. Marrucci, *Trans. Soc. Rheol.* **16**, 321 (1972).
110. H. Janeschitz-Kriegl, *Adv. Polym. Sci.* **6**, 170 (1969).
111. B. P. Ho and L. G. Leal, *J. Fluid Mech.* **76**, 783 (1976).
112. C. D. Han, Y. W. Kim, and S. J. Chen, *J. Appl. Polym. Sci.* **19**, 2831 (1975).
113. C. E. Locke and D. R. Paul, *J. Appl. Polym. Sci.* **17**, 2791 (1973).
114. F. Ide and A. Hagasawa, *J. Appl Polym. Sci.* **18**, 963 (1974).
115. S. Kowa, *J. Appl. Polym. Sci.* **19**, 1625 (1975).
116. J. Kohler, G. Riess, and A. Bandaret, *Eur. Polym. J.* **4**, 187 (1968).

Chapter 8

Mechanical Properties (Small Deformations) of Multiphase Polymer Blends

R. A. Dickie

Engineering and Research Staff
Ford Motor Company
Dearborn, Michigan

I.	Introduction	353
II.	Background	354
III.	Modulus–Composition Dependence	356
	A. Elastic Response	356
	B. Viscoelastic Response	367
IV.	Temperature Dependence of Composite Modulus	369
	A. Model Calculations	369
	B. Comparison with Experimental Data	374
V.	Time–Temperature Dependence	381
	A. Thermorheologically Simple Materials	381
	B. Thermorheologically Complex Materials	383
	References	388

I. INTRODUCTION

Small-deformation mechanical properties of amorphous, homogeneous polymers are largely determined by molecular relaxation processes. Changes in thermal and mechanical history, application of hydrostatic pressure, and addition of soluble diluents all affect mechanical response in ways that can be explained in terms of changes in the molecular environment of polymer chains. For composites or blends of incompatible polymers, the mechanical response reflects molecular relaxation processes characteristic of each constituent but is also profoundly influenced by blend composition and morphology (i.e., by blend structure on a scale generally much larger than molecular dimensions) and by new or modified relaxation processes characteristic of the blend (arising, for example, from intermolecular mixing or constraint of chain movement at the surface of a rigid filler). In principle, mathematical models describing the mechanical response in terms of properties of blend constituents allow identification of the features of response that can be accounted for on purely mechanical grounds, and hence, by elimination, also allow identification of those features that arise from changes in molecular relaxation processes. Exact solutions to the mechanical problems posed have been obtained in only a few cases, but many empirical and semiempirical formulae have been developed. In this chapter, after some fundamental background information has been presented, models for the composition and morphology dependence of elastic and isothermal viscoelastic response are discussed; application of these methods to isochronal viscoelastic response is reviewed; isothermal viscoelastic response is discussed with particular reference to time–temperature superposition.

II. BACKGROUND

For sufficiently small deformations, complete characterization of the (linear) elastic properties of a homogeneous, isotropic material requires two parameters. Those most commonly employed are the tensile or Young's modulus E, the lateral contraction or Poisson's ratio μ, the shear modulus G, and/or the bulk modulus K. Young's modulus and Poisson's ratio reflect material response to changes in shape (distortion) and volume; the shear modulus reflects response to changes in shape alone; and the bulk modulus reflects response to changes in volume alone. Experimentally, E and G are most amenable to direct determination. Poisson's ratio ranges from about 0.35 for glassy polymers to about 0.5 for elastomers. The elastic constants are related by

$$E = 2G(1+\mu) = 3K(1-2\mu) \tag{1a}$$

$$\mu = (3K - 2G)/(6K + 2G) \tag{1b}$$

Elastic properties can also be characterized in terms of compliances $D = 1/E$, $J = 1/G$, and $B = 1/K$. Further relations between the elastic constants are given in standard texts on elasticity [1].

For polymers, the response to mechanical excitation is generally time dependent (viscoelastic); time- (and frequency-) dependent response functions corresponding to the moduli and compliances of elastic solids are defined [2, 3]. For static deformation, the modulus is the ratio of observed time-dependent stress to imposed (constant) strain. In general terms,

$$M(t) = \sigma(t)/\varepsilon_0 \tag{2}$$

For dynamic deformation in which a sinusoidal strain of constant amplitude and frequency ω is imposed, the stress response and modulus are complex quantities. Again in general terms,

$$M^*(\omega) = \sigma^*(\omega)/\varepsilon(\omega) \tag{3}$$

$$M^*(\omega) = M'(\omega) + iM''(\omega) \tag{4}$$

where $i = \sqrt{-1}$. That is, stress response lags the imposed strain, and modulus can be decomposed into components M' and M'' which are respectively in phase and $\pi/2$ radians out of phase with the imposed strain. Component M' of M^* is often called the storage modulus, and component M'' is called the loss modulus. The tangent of the phase angle δ is given by

$$\tan \delta_M = M''/M' \tag{5}$$

Values of the loss tangent depend on the mode of deformation: the distinction is often ignored, but in general the loss tangent in shear ($\tan \delta_G = \tan \delta_J$) is larger than the loss tangent in tension ($\tan \delta_E = \tan \delta_D$) [4]. Poisson's ratio is also complex; $\tan \delta_\mu$ is usually much less than $\tan \delta_E$ [4, 5].

Experimentally, mechanical response is usually observed either at constant frequency (or time) over a broad range of temperature (isochronal measurement) or at constant temperature over a range of frequency (isothermal measurement). Resonance and free oscillation techniques (e.g., torsional pendulum experiments at a constant moment of inertia) are sometimes employed for measurement over a broad temperature range; such measurements involve changes in both frequency and temperature, which complicates the interpretation of results. Generally, isochronal measurements (preferably obtained using a forced-oscillation, fixed-frequency technique) over a broad temperature range allow most rapid characterization of the full range of viscoelastic behavior. Isothermal measurements normally cannot be made over a sufficiently broad range of time or frequency to encompass the full range of viscoelastic behavior. As a result, time–temperature superposition techniques have been used to broaden the apparent time (frequency) scale. Ferry [3] reviews experimental methods and time–temperature superposition. The first part of this chapter deals mostly with results of isochronal measurements; the final section deals with special problems posed by the application of time–temperature superposition to isothermal data on polymer blends.

Qualitatively, simple homogeneous polymers, whether homopolymers or random copolymers, display a single major transition in modulus at the glass transition temperature T_g. The loss tangent displays a maximum corresponding to this transition. The location of the transition on the temperature scale, the details of its shape, and the dependence of transition location and breadth on time scale depend on molecular relaxation processes characteristic of the individual polymer (see, e.g., Ferry [3] and Nielson [6]).

For a heterogeneous blend of two polymers, a broad range of response is possible, depending on blend morphology, on the degree of molecular mixing or interpenetration, and on the size of the phase separated regions— as well as on the molecular relaxation processes characteristic of blend constituents [6, 7]. Two transitions, corresponding to the glass transitions of the blend constituents, are observed. The basic theoretical problem is the deduction of the mechanical response of the blend (the magnitude of the modulus change at each transition, and the height and location of the loss tangent maxima) from properties of blend constituents and from assumptions about blend structure.

Heterogeneous blends vary widely in morphology. Classification schemes and detailed descriptions can be found in Nielsen [7], Manson and Sperling [8], Bever and Shen [9], and Sperling [10]. For the present discussion of isotropic blends, it is sufficient to distinguish:

1. simple dispersions of more or less spherical inclusions in a matrix;
2. compound dispersions in which a portion of the matrix polymer is occluded by the disperse phase polymer to form composite inclusions; and
3. random dispersions in which more than one distinguishable phase is continuous (interpenetrating or multiply continuous structures).

Region size, distribution of region size, and interaction (partial compatibility) at interfaces can also influence the observed mechanical response of each type of blend.

III. MODULUS–COMPOSITION DEPENDENCE

A. Elastic Response

There are three principal groups of models that have been applied to the problem of predicting modulus–composition dependence: mechanical coupling models, which provide an empirical representation of response in terms of constituent mechanical properties; "self-consistent" models, which yield approximate but single-valued expressions based on analysis of deformation and stress about a representative inclusion; and limits or bounds on the modulus, which specify for a given set of assumptions the possible range of modulus values.

1. Mechanical Coupling Models

Mechanical models like the Takayanagi Models I and II shown in Fig. 1 can be regarded as generalizations of the familiar spring and dashpot models sometimes used to discuss viscoelastic response [11]. In the models of Fig. 1, each block is assumed to have the mechanical properties of one of the blend constituents. Although they appear to be two-parameter models, the restriction imposed by

$$\lambda_\mathrm{I}\psi_\mathrm{I} = \lambda_\mathrm{II}\psi_\mathrm{II} = \phi_2 \tag{6}$$

means that each model involves just one adjustable parameter. It has often been claimed that one or the other of the two models allows better representation of mechanical response for a particular system; however, it is easy to show [12, 13] that the two models are exactly equivalent in the sense that exactly identical results can be obtained from each by proper choice of model parameters. There are several ways of writing expressions for the composite shear modulus G_c in terms of the model parameters; one convenient form introduces a single coupling parameter α:

$$\frac{G_c}{G_1} = \frac{\phi_1 G_1 + (\alpha + \phi_2) G_2}{(1 + \alpha\phi_2) G_1 + \alpha\phi_1 G_2} \tag{7}$$

where

$$\lambda_\mathrm{I} = \phi_2/\psi_\mathrm{I} = (\alpha + \phi_2)/(1 + \alpha) \tag{8}$$

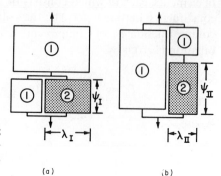

Fig. 1 Takayanagi models: (a) Model I; (b) Model II. As drawn, Models I and II correspond to $\phi_2 = 0.25$, $\alpha = 0.857$; values of $\lambda_\mathrm{I}, \psi_\mathrm{I}, \lambda_\mathrm{II}$, and ψ_II are given in Table I.

(a)   (b)

and

$$\lambda_{II} = \phi_2/\psi_{II} = \phi_2(1+\alpha)/(1+\alpha\phi_2) \tag{9}$$

for equivalent response [12]. These models have been applied extensively to the description of dynamic mechanical behavior of heterogeneous systems; in general, the major features of response can be reproduced satisfactorily. Takayanagi et al. [11], Kaplan and Tschoegl [13], Takayanagi et al. [14], Horino et al. [15], Okamoto and Takayanagi [16], Takayanagi [17], Matsuo et al. [18], Kraus et al. [19], and Marcincin et al. [20] are representative. For simple parallel coupling, $\alpha = 0$; for simple series coupling, $\alpha = \infty$. (The symbols λ and ϕ are usually used for the model parameters; ψ has here been substituted for ϕ to allow use of ϕ for volume fraction. Subscripts 1 and 2 denote the phases.)

A number of extensions and refinements to mechanical modeling techniques have been proposed and applied to experimental data. Kraus and Rollman [21] and Fujino et al. [22] have proposed generalized models based on more complicated coupling arrangements of elements having the mechanical properties of the blend components. For the Fujino et al. models, a distribution function in the model parameters is postulated and interpreted in terms of blend morphology. Charrier [23] has proposed a generalized form that allows for the effects of nonisometric filler particles. These models, like the Takayanagi models of Fig. 1, assume that the blend components are essentially noninteractive; that is, there are no new relaxation modes characteristic of the blend itself. Applied to dynamic data, these models usually predict two discrete transitions. For blends of polymers that do interact (are partially compatible), and for block copolymer systems in which the interfacial region between phases contributes significantly to the mechanical response, models have been proposed that introduce properties of copolymers or homogeneous blends into the calculation. Thus, Miyata and Hata [24] apply models comprising a simple parallel or series combination of elements, each of which consists of a homogeneous mixture of the blend constituents. Letting $\Phi_i(\phi_A')$ be the volume fraction of element i containing volume fraction ϕ_A' of Component A, Young's modulus of the parallel combination is given by

$$E_p = \sum_i \Phi_i(\phi_A')E(\phi_A') = \int_0^1 \Phi_i(\phi_A')E(\phi_A')\,d\phi_A' \tag{10a}$$

where

$$\int_0^1 \Phi(\phi_A')\,d\phi_A' = 1 \tag{10b}$$

and

$$\int_0^1 \Phi(\phi_A')\,\phi_A'\,d\phi_A' = \phi_A \tag{10c}$$

For the corresponding series model,

$$E_s = \left[\int_0^1 \frac{\Phi(\phi_A')\,d\phi_A'}{E(\phi_A')} \right]^{-1} \tag{11}$$

The analysis is applied [24] to dynamic data on blends of poly(vinyl acetate) and poly(methyl methacrylate) prepared in different ways; the function Φ is interpreted in terms of the degree of phase separation.

Models similar to Miyata and Hata's have been used by Kraus and Rollman [25] to elucidate the nature of the domain boundary in ABA block polymers and its influence on mechanical response. An orientation parameter γ is introduced and the mechanical response is calculated by combining the series and parallel models according to

$$E = (E_p)^\gamma (E_s)^{1-\gamma} \tag{12}$$

Takayanagi [26] has proposed a generalized model combining a graded series of random copolymer compositions in a partly series, partly parallel, model; this model predicts a single very broad transition in modulus qualitatively similar to results reported (e.g., Sperling *et al.* [27] and Dickie and Cheung [28]) on a variety of semicompatible systems.

Overall, equivalent mechanical models furnish a convenient framework for empirical curve fitting and systematic phenomenological description of blend behavior. They are not, however, morphologically or mechanically realistic models of blend structure or response. Hence, the detailed form of the models should not be interpreted too literally.

2. "Self-Consistent" Models

In the so-called self-consistent approximation, the mechanical response of an idealized representative composite structure of the type illustrated in Fig. 2 is compared with that of a homogeneous body having the (unknown) macroscopic elastic properties of the composite. The composite structure consists of a spherical dispersed phase particle embedded in a matrix shell which is in turn embedded in a continuous body having the elastic properties of the composite. (The averaging procedure implied was apparently first used for analysis of suspension mechanical properties by Bruggeman [29]; it is reminiscent of the self-consistent field methods of Hartree [30].) In Kerner's well-known derivation [31], the average tensile stress and strain in the model structure (subjected to tensile deformation) are equated to those in

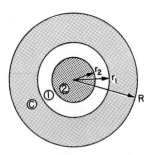

Fig. 2 Model composite inclusion for self-consistent scheme calculation of elastic properties. Calculation of shear modulus using the Kerner equation for the sphere as drawn yields the same prediction of response as the Takayanagi models of Fig. 1.

the corresponding homogeneous body. The integration involved requires a definition of the relative magnitude of r_1 and r_2; the value chosen is $(r_2/r_1)^3 = \phi_2$. The expression obtained for shear modulus is the same in form as the Takayanagi model result given in Eq. (7); the parameter α is no longer an empirical coupling parameter but is replaced by α_1, a function of the Poisson ratio of the continuous or matrix phase μ_1:

$$\alpha_1 = 2(4 - 5\mu_1)/(7 - 5\mu_1) \tag{13}$$

A similar analysis for bulk modulus K yields an equation similar in form to Eq. (7):

$$\frac{K_c}{K_1} = \frac{\phi_1 K_1 + (\beta_1 + \phi_2) K_2}{(1 + \beta_1 \phi_2) K_1 + \beta_1 \phi_1 K_2} \tag{14}$$

where $\beta_1 = (1 + \mu_1)/2(1 - 2\mu_1)$. Young's modulus E and Poisson's ratio μ for the composite can be calculated using Eqs. (1a) and (1b).

Results similar in form to Kerner's have appeared several times in the literature. The Halpin–Tsai [32–34] equation can be regarded as a generalization of Eq. (7) with the parameter $A(= 1/\alpha)$ interpreted as a measure of reinforcement and a function of inclusion geometry, packing, and loading conditions. By choosing appropriate values of A, moduli of fiber- and ribbon-reinforced composites have been estimated [32]. Uemura and Takayanagi [35] have obtained results identical to those of Kerner for shear modulus of particulate blends by a somewhat different method of analysis; similar results of Okano [36] are cited. Hashin [37] has given expressions equivalent to Eqs. (7), (13), and (14) as useful approximate forms. Kalfoglou and

Williams [38] combined Hashin's form of Eq. (14) for the bulk modulus with a simple mixture rule for Poisson's ratio to obtain an expression for shear modulus; the procedure was found to give a reasonable representation of data on epoxy–elastomer blends but is difficult to justify theoretically. As discussed subsequently, the very general Hashin–Shtrikman bounds [39] are similar in form to the Kerner equations.

Predictions of the Kerner analysis for the tensile modulus of two-component composites comprising materials of greatly different modulus (e.g., a rubbery and a glassy polymer) are schematically illustrated in Fig. 3. Simple, dispersed-sphere morphology is assumed. The modulus predictions are typically too high for rubber-modified plastics, and too low for hard-filler containing rubbers; that is, the dependence of modulus on concentration is always underestimated somewhat.

Better quantitative agreement with experimental data has been obtained (1) by empirical modification of the ratio r_2/r_1 and (2) by employing a different averaging assumption in comparing the special composite structure of Fig. 2 with the postulated homogeneous body of equivalent properties. For the empirical modification of the Kerner equation, it is usually assumed [40–45] that the concentration scale is somewhat compressed due to the limited volume fraction that can be occupied by a dispersed particulate

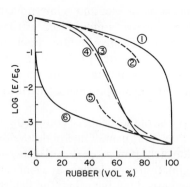

Fig. 3 Modulus–composition dependence for blends of a hard, glassy polymer and a soft rubber. Curves 1 and 6 are as predicted by the unmodified Kerner equation; Curves 2 and 5 are typical experimental data and predictions of modified Kerner and van der Poel equations. For Curves 1 and 2, the glassy polymer is the continuous phase; for Curves 5 and 6, the rubber is the continuous phase. Curve 3 shows modulus–composition dependence for a series of homogeneous copolymers corresponding to the heterogeneous blends in monomer composition. Curve 4 is the prediction of the Kerner packed-grain, Hill, and Budiansky models (Eq. 20). Curves 1 and 6 also correspond to the Hashin–Shtrikman bounds. E_G is Young's modulus of the hard (glassy) component.

phase. Analysis in terms of a mechanically effective volume fraction can be carried out by solving the Kerner composite shear modulus equation for volume fraction [45, 46]:

$$\phi_{2\text{eff}} = \frac{(G_c - G_1)(G_1 + \alpha_1 G_2)}{(G_2 - G_1)(G_1 + \alpha_1 G_c)} \tag{15}$$

The mechanically effective volume fraction $\phi_{2\text{eff}}$ then can be calculated from experimental values of modulus. In terms of the model of Fig. 2, Eq. (15) defines r_1 in terms of r_2 on an empirical, mechanical basis, rather than arbitrarily setting $r_2/r_1 = \phi_2^{1/3}$. Experimentally, $\phi_{2\text{eff}}$ can often be represented by

$$\phi_{2\text{eff}} = \phi_2 + \phi_2^2(1 - \phi_{2m})/\phi_{2m}^2 \tag{16}$$

a form originally proposed by Nielsen [40–42] in conjunction with a slightly different, but nearly equivalent [46], modified form of the Kerner equation. For hard fillers in rubbery polymers, the curve fitting parameter ϕ_{2m} is about 0.60–0.65; for rubbery particles in a glassy polymer matrix, ϕ_{2m} is about 0.80–0.83.

Ziegel and Romanov [47, 48] have proposed a similar treatment based on the Kerner equation; the form assumed for effective volume fraction is

$$\phi_{2\text{eff}} = B\phi_2 \tag{17}$$

B is termed an interaction parameter and is interpreted as a measure of the amount of (soft) polymer matrix immobilized at the (hard) filler surface. Experimental data on ethylene–vinyl acetate copolymer blends with high modulus filler polymers were reported [47] to be in agreement with Eq. (17), with B approximately independent of temperature and ranging from 1.6 to 2.3 for the systems studied. Data on nonpolymeric hard fillers have been similarly analyzed [44, 48].

The mechanical response of the idealized composite structure of Fig. 2 was analyzed by van der Poel [49] using a method developed by Fröhlich and Sack [50] for calculation of the flow properties of dispersions. In van der Poel's derivation, the approximation is made that, at a sufficiently large distance R from the center of the embedded inclusion, the deformations in the composite and the homogeneous body are the same except for terms of order higher than 3 in $1/R$. Results of van der Poel's calculations were compared with extensive data on filled urethane elastomers by Schwarzl and co-workers [51–53], and were found to be in excellent agreement with data for particles of sufficiently large size. Data on other filled systems have been compared with the van der Poel results and have been found to be in satisfactory agreement [53–55]. Hashin [56] has pointed out that the wrong elasticity solutions were employed in van der Poel's calculation. Smith

[57, 58] has corrected the van der Poel derivation and arrived at algebraic results much simpler in form. Numerically, the original and corrected van der Poel results are very similar. It has also been shown [45] that, for $G_2 \gg G_1$, the van der Poel–Smith result is virtually identical to the modified Kerner result with $\phi_{2m} = 0.63$; for $G_2 \ll G_1$, the van der Poel–Smith result is nearly identical to the modified Kerner result, with $\phi_{2m} = 0.8$. The modified Kerner equation can thus be regarded as a one-parameter empirical approximation to the van der Poel–Smith result for these limiting cases. For filler-to-matrix modulus ratio within an order of magnitude or so of unity, the equivalence breaks down. The modified Kerner equation is in fair agreement with experimental data in this range, but empirical modification of the van der Poel–Smith equation by introduction of an effective volume fraction of the form

$$\phi_{2\text{eff}} = \phi_2 + k(\phi_2/d)^{2/3} \tag{18}$$

where d is the dispersed particle size, and k is an empirical parameter of the order 1 for d in microns, allows accurate representation of a broader range of data [45].

The van der Poel–Smith prediction for shear modulus is obtained by solving the quadratic equation

$$AX^2 + BX + C = 0 \tag{19}$$

$$A = [4P(7 - 10\mu_1) - Sa^7][Q - (8 - 10\mu_1)(G_2/G_1 - 1)a^3]$$
$$\quad - 126P(G_2/G_1 - 1)a^3(1 - a^2)^2$$

$$B = 35(1 - \mu_1)P[Q - (8 - 10\mu_1)(G_2/G_1 - 1)a^3]$$
$$\quad - 15(1 - \mu_1)[4P(7 - 10\mu_1) - Sa^7](G_2/G_1 - 1)a^3$$

$$C = -525P(1 - \mu_1)^2(G_2/G_1 - 1)a^3$$

the quantities P, Q, and S are defined as

$$P = (7 + 5\mu_2)G_2/G_1 + 4(7 - 10\mu_2)$$

$$Q = (8 - 10\mu_1)G_2/G_1 + (7 - 5\mu_1)$$

$$S = 35(7 + 5\mu_2)G_2/G_1(1 - \mu_1) - P(7 + 5\mu_1)$$

and X is equal to $(G_c/G_1 - 1)$; $a^3 = \phi_2$.

For bulk modulus, the Kerner and van der Poel analyses yield the same result, Eq. (14). Excellent agreement with the bulk modulus result was obtained by Schwarzl et al. [51–53].

The results of Kerner and of van der Poel are appropriate for simple dispersions with a well-defined matrix phase and well-defined inclusions. For compound dispersions, calculation of the modulus can be accomplished in two stages [12]. The properties of the inclusions are estimated first; these results are then used in a second computation to obtain the composite modulus. For dispersions of two particulate species in a matrix, relative particle size appears to be an important parameter [44, 59, 60]. Again a two-stage calculation is performed; the modulus of a composite comprising the matrix and the smaller of the filler species is calculated and employed as the effective matrix modulus for a second-stage calculation of the composite modulus with the larger filler. This approach has been employed to describe the modulus–composition behavior of composites comprising a bimodal distribution of hard-filler particles in an elastomeric matrix [59, 60] and of a series of glass-bead-filled, rubber-modified thermoplastics [43].

For blends that do not have a well-defined matrix, in which both components can be regarded as continuous, but which are nevertheless isotropic and macroscopically homogeneous, a somewhat different approach is required. One method, proposed by Nielsen [7, 40], suggests that the modulus of such materials be estimated by applying the geometric rule of mixtures to values predicted by the empirically modified Kerner equation for composites with first one and then the other component as matrix. Another method is based on a little noticed section of Kerner's paper [31] dealing with so-called packed grain composites. Kerner's original derivation is couched in terms of an n-component composite comprising a matrix and $n-1$ dispersed particulate species. If $n = 2$, then the form given earlier (Eq. 7, with α as defined in Eq. 13) is obtained. If, however, $n = 3$ (i.e., there are two dispersed particulate species in a continuous matrix) and the volume fraction of the matrix tends toward zero, then the properties of the matrix appearing in the modulus equations can be replaced by the corresponding properties of the composite. A pair of simultaneous equations for G_c and K_c is obtained appropriate for a composite in which grains of Species 1 and 2 are packed together at random:

$$\phi_1(G_1 - G_c)(G_c + \alpha_c G_2) + \phi_2(G_2 - G_c)(G_c + \alpha_c G_1) = 0 \qquad (20a)$$

$$\phi_1(K_1 - K_c)(K_c + \beta_c K_2) + \phi_2(K_2 - K_c)(K_c + \beta_c K_1) = 0 \qquad (20b)$$

where

$$\alpha_c = 2(4 - 5\mu_c)/(7 - 5\mu_c) = 6(K_c + 2G_c)/(9K_c + 8G_c)$$

$$\beta_c = (1 + \mu_c)/2(1 - 2\mu_c) = 3K_c/4G_c$$

These results are algebraically identical to those obtained by Hill [61] and by Budiansky [62, 63]; qualitatively, the results are similar to those obtained using Nielsen's averaging procedure. Faucher [64] has applied Eq. (20) to the modulus–composition dependence of block copolymers. Calculations based on Eq. (20) are included in Fig. 3.

Highly anisotropic polymer blends can be produced by extrusion; in the case of certain block copolymers, highly ordered structures can be obtained [65–68]. The mechanical properties of extruded ABA block copolymers have been successfully analyzed by Arridge and Folkes [69] in terms of fiber reinforcement theories. A review of these theories is beyond the scope of the present chapter, but pertinent discussion and further references can be found in Ashton *et al.* [32] and Garg *et al.* [70]. Lamellar structures are appropriately analyzed in terms of simple series–parallel models of response [25, 32, 70].

3. Limits or Bounds on the Modulus

The self-consistent model approach relies on simplifying assumptions about morphology and physical behavior to obtain approximate but single-valued modulus predictions. The calculation of bounds or limits on the modulus represents an alternate approach based on the use of variational principles to bound the strain energy and thus to place limits on the composite moduli. Paul [71] and Hashin [37] have employed the principles of minimum potential energy and minimum complementary energy in this way, Paul to establish bounds for arbitrary phase geometry, and Hashin, to establish bounds for composites comprising spherical inclusions in a matrix (and, strictly, limited to a particular distribution of particle size). Hashin and Shtrikman [39] developed bounds on G and K for arbitrary phase geometry using additional variational principles [72, 73]; Hill [74] showed subsequently that the same bounds for K could be obtained for a two-component system without resort to the new variational principles. The results of these calculations are exact; however, the bounds coincide only in a few special cases [74], and a range of possible modulus values for a given composition is usually obtained. Often the bounds are too widely separated to provide a useful prediction of response, but potentially they provide a check on the basic assumptions that the blend or composite is isotropic, that the mechanical response of the blend constituents is the same in the blend as in bulk, and that there is perfect adhesion between phases.

The upper bound on composite shear modulus given by Paul [71] corresponds to a parallel combination of elements having the mechanical properties of the blend constituents; the lower bound corresponds to a series combination:

$$[(\phi_1/G_1) + (\phi_2/G_2)]^{-1} \leqslant G_c \leqslant \phi_1 G_1 + \phi_2 G_2 \tag{21}$$

Hashin [37] developed considerably more restrictive bounds for composites comprising spherical inclusions in a matrix; it is assumed that each dispersed phase particle can be regarded as enclosed in a shell of matrix phase material, that the composition of each composite sphere reflects the overall composition of the composite, and that the size distribution of the composite spheres is space filling (i.e., that the composite spheres may diminish to infinitesimal size). For the bulk modulus, the upper and lower bounds are found to coincide; the exact, single-valued prediction is identical to the Kerner (and van der Poel) result given in Eq. (14). For the shear modulus, the bounds do not coincide; explicit expressions for the bounds are not given. The bounds are stated, however, to bracket a simple approximate formula, which can easily be shown to reduce to the Kerner result given by Eqs. (7) and (13).

The most restrictive bounds thus far obtained without specification of blend morphology are those derived by Hashin and Shtrikman [39]. For a two-component composite, the bounds can be written in the form

$$K_l = K_1 \frac{\phi_1 K_1 + (\beta_1 + \phi_2) K_2}{(1 + \beta_1 \phi_2) K_1 + \beta_1 \phi_1 K_2}$$

$$\leqslant K_c \leqslant K_2 \frac{\phi_2 K_2 + (\beta_2 + \phi_1) K_1}{(1 + \beta_2 \phi_1) K_2 + \beta_2 \phi_2 K_1} = K_u \qquad (22a)$$

$$G_l = G_1 \frac{\phi_1 G_1 + (\alpha_1 + \phi_2) G_2}{(1 + \alpha_1 \phi_2) G_1 + \alpha_1 \phi_1 G_2}$$

$$\leqslant G_c \leqslant G_2 \frac{\phi_2 G_2 + (\alpha_2 + \phi_1) G_1}{(1 + \alpha_2 \phi_1) G_2 + \alpha_2 \phi_2 G_1} = G_u \qquad (22b)$$

where the subscripts 1 and 2 denote lower- and higher-phase moduli, respectively, and α and β are the previously defined functions of Poisson's ratio. The limits are identical in form to the results of Kerner's calculation, assuming for the lower bound that the lower-modulus material forms the matrix and for the upper bound that the higher-modulus material forms the matrix. Thus, referring to Fig. 3, the solid curves, previously identified as the Kerner equation predictions for the indicated phase morphology, can now be regarded as exact upper and lower limits on the modulus, independent of morphology. In this interpretation, it is only predicted that experimental values of the modulus should lie between the two solid curves. The discussion of bounds given here has been restricted to two-component systems; the results are readily generalized to multicomponent systems [37, 39, 70].

The widely differing mechanical response for hard- and soft-polymer continuous-phase, particulate blends predicted by the Kerner–van der Poel–Smith equations, and experimentally observed for simple dispersion composites, are readily encompassed by the Hashin–Shtrikman bounds. The response of homogeneous copolymers may (but need not) also fall within the

Hashin–Shtrikman bounds; depending on the temperature of measurement, the modulus composition dependence for homogeneous copolymers can strongly resemble the response predicted by Eq. (20), as shown in Fig. 3. Interpretation of modulus–composition data in terms of phase separation and morphology clearly must be undertaken cautiously. For cases in which complete phase separation is expected, analysis of modulus–composition data to determine the extent of compound inclusion formation has been suggested [12, 28, 75, 76]. Analysis of dynamic mechanical response over the complete range of viscoelastic behavior can provide additional information on phase separation and morphology, and is discussed in the next section of this chapter.

B. Viscoelastic Response

The equations of Section III.A have been written in terms of elastic moduli; the single-valued expressions obtained in mechanical model representation of elastic response and from self-consistent calculations can be extended to analysis of viscoelastic response by straightforward application of the correspondence principle, that is, by substitution of complex for elastic moduli [77–79]. The analysis of mechanical model response has in fact typically been given in terms of viscoelastic response; the correspondence principle can be used to recover the appropriate viscoelastic forms from the elastic expressions given here [e.g., Eq. (7)]. The correspondence principle has also been applied, for example, to the Kerner and van der Poel equations, and has been extended [80] to expressions for the moduli of fiber-reinforced materials. Conditions under which the correspondence principle is applicable have been reviewed by Schapery [79]. Explicit expressions for storage and loss moduli, and for loss tangent, have been given for the Kerner equation [12, 81], and, hence, effectively also for the Takayanagi models. If the Poisson ratio of the matrix is assumed to be real (i.e., $\mu_1'' \approx 0$), application of the correspondence principle to the Kerner equation yields

$$G_c' = G_1' A/C - G_1'' B/C \tag{23a}$$

$$G_c'' = G_1'' A/C + G_1' B/C \tag{23b}$$

where

$$A = \phi_1(1 + \alpha_1\phi_2)(G_1'^2 + G_1''^2) + \phi_1(\alpha_1 + \phi_2)\alpha_1(G_2'^2 + G_2''^2)$$
$$+ [\phi_1{}^2\alpha_1 + (\alpha_1 + \phi_2)(1 + \alpha_1\phi_2)](G_1'G_2' + G_1''G_2'')$$

$$B = (\alpha_1 + 1)^2\phi_2(G_2''G_1' - G_1''G_2')$$

$$C = (1 + \alpha_1\phi_2)^2(G_1'^2 + G_1''^2) + \phi_1{}^2\alpha_1{}^2(G_2'^2 + G_2''^2)$$
$$+ 2(1 + \alpha_1\phi_2)\phi_1\alpha_1(G_2'G_1' + G_2''G_1'')$$

also

$$(\tan \delta_G)_c = G_c''/G_c' = \frac{B/A + (\tan \delta_G)_1}{1 - (\tan \delta_G)_1 B/A} \qquad (24)$$

Limiting parallel and series models for viscoelastic response can be obtained from Eqs. (23) and (24) by setting $\alpha = 0$ or ∞, respectively.[†]

On the basis of a dimensional argument, Hashin [77] concluded that, for a composite comprising a viscoelastic matrix (Component 1) and an elastic filler (Component 2), the loss tangent of the composite should equal that of the matrix, provided that G_2/G_1 is either much greater or much less than unity. Similarly under these conditions $B/A \approx 0$ and $(\tan \delta_G)_c \approx (\tan \delta_G)_1$. If the phase moduli are of comparable magnitude, no simple restriction can be placed on $(\tan \delta_G)_c$, and discrepancies between composite and matrix loss tangent are not necessarily attributable to new or modified relaxation mechanisms.

Bounds on viscoelastic properties require special consideration. Explicit bounds for viscoelastic properties have been given by Christensen [78] and by Roscoe [82]. Christensen gives general bounds for bulk modulus; for shear modulus, the results are applicable only for suspensions of voids and rigid particles. The Christensen results resemble the Hashin composite sphere assemblage equations in form, but are of limited utility for application to polymer blends because of the restriction to limiting values of filler–matrix modulus ratio in the calculation of composite shear modulus. Roscoe gives lower bounds on the components of the complex dynamic shear modulus (G^*) and compliance (J^*) for arbitrary phase geometry and phase properties; these assume the form

$$P_c \geqslant 1/(\phi_1/P_1 + \phi_2/P_2) = P_l \qquad (25)$$

where property P is G', G'', J', or J''. Upper bounds are computed from the lowers bounds, viz., since

$$(G')^2 + (G'')^2 = 1/[(J')^2 + (J'')^2]$$

$$(G_c')^2 \leqslant 1/[(J_l')^2 + (J_l'')^2] - (G_l'')^2 \qquad (26)$$

[†] For the parallel model,

$$G_c' = \phi_1 G_1' + \phi_2 G_2', \qquad G_c'' = \phi_1 G_1'' + \phi_2 G_2''$$

For the series model, A, B, and C reduce to

$$A = \phi_1(G_2'^2 + G_2''^2)$$

$$B = \phi_2(G_1'^2 + G_1''^2)$$

$$C = \phi_1^2(G_2'^2 + G_2''^2) + \phi_2^2(G_1'^2 + G_1''^2) + 2\phi_1\phi_2(G_1'G_2' + G_1''G_2'')$$

and G_c' and G_c'' can be calculated from Eqs. (23a) and (23b).

In the elastic limit, these bounds degenerate to the simple parallel–series bounds given by Paul [71] (cf. Eq. 21).

Somewhat closer approximate bounds on G' can be obtained by application of the correspondence principle to the Hashin–Shtrikman bounds; algebraically, the results are identical in form to Eq. (22b) with substitution of G_1', G_2', and G_c' for their elastic counterparts. Loss tangent and loss modulus are not generally bounded by the elastic solutions [79]. Approximate bounds on loss modulus can be based on Hashin's results [83] for the viscosity coefficient by application of the correspondence principle and substitution of G'' for $\eta'\omega$, where η' is the real part of the complex dynamic viscosity and ω is radian frequency. The results are again identical to Eq. (22b) in form, with G_1'', G_2'', and G_c'' substituted for G_1, G_2, and G_c, respectively. For loss tangent one can write

$$G_l''/G_u' \leqslant \tan \delta_G \leqslant G_u''/G_l' \tag{27}$$

but these bounds are generally too widely spaced to be useful.

Bucknall and Hall [84] calculated bounds on G' and $\tan \delta_G$ for rubber modified plastics; since these results were based on Hashin's composite sphere assemblage results for elastic modulus, the bounds on G' are probably approximately correct, but those on $\tan \delta_G$ are probably not [79].

IV. TEMPERATURE DEPENDENCE OF COMPOSITE MODULUS

The prediction of isochronal modulus as a function of temperature has generally proceeded by an application of models for composition dependence point by point to isochronal data on blend constituents over the temperature range of interest. This procedure ignores differential thermal expansion effects, and assumes (just as in the initial composition dependence calculations) that constituent properties in the blend are the same as in bulk. Local hydrostatic pressure effects and ensuing relaxation phenomena are thus ignored in the calculation, even though they can be expected to affect the experimental results, for example, by introducing additional temperature dependence [42].

A. Model Calculations

Application and interpretation of models is most readily discussed in terms of sample calculations; Dickie [12], Bohn [75, 76] present representative results. Figures 4 and 5, adapted from Dickie [12], present results

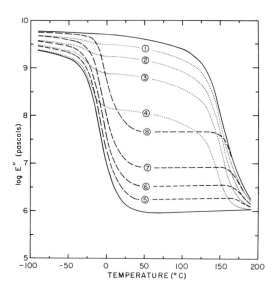

Fig. 4   Dependence of $\log E'$ at 110 Hz on temperature for poly(methyl methacrylate) (upper solid curve) and for poly(butyl acrylate) (lower solid curve). Intermediate curves are calculated predictions for blends using the parameters given in Table I. A detailed interpretation is discussed in text. (Adapted from Dickie [12].)

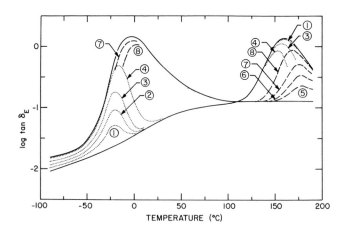

Fig. 5   Dependence of loss tangent at 110 Hz on temperature for poly(methyl methacrylate) and poly(butyl acrylate) (solid curves). Intermediate curves are calculated predictions for blends using the parameters given in Table I. Detailed interpretation is discussed in text. (Adapted from Dickie [12].)

of calculations based on dynamic Young's modulus data obtained on poly(methyl methacrylate) and on a poly(butyl acrylate) rubber. The calculations were performed using the complex form of the Kerner equation; since this form occurs so frequently in the analysis of modulus–composition dependence, several different interpretations may be placed on the curves obtained. Thus, in terms of the Takayanagi models, the curves correspond to the model parameters shown in Table I. Curve 1 is the response predicted for the models as drawn in Fig. 1. The Takayanagi model predictions are essentially arbitrary; even the relative series–parallel character of the response depends on the choice of model. The curves of Figs. 4 and 5 are also the predictions of expected behavior from the unmodified and modified Kerner equations for composites with simple dispersion morphology

Table I

Model Parameters for Figs. 4 and 5

Curve number (Figs. 4 and 5)	Kerner Eq. ϕ_2	Modified Kerner Eq. ϕ_2	$\phi_{2\text{eff}}$	Takayanagi Models I and II ϕ_2	λ_{I}	ψ_{I}	λ_{II}	ψ_{II}	Hashin–Shtrikman ϕ_1	ϕ_2
		Simple PBA[a] inclusions in PMMA[a] matrix							Upper bounds	
	$\mu_1 = 0.35$	$\mu_1 = 0.35,$ $\phi_{2m} = 0.83$			$\alpha = 0.857$				$\mu_1 = 0.35$	
1	0.25	0.236	0.25	0.25	0.596[b]	0.419[b]	0.382[b]	0.654[b]	0.75	0.25
2	0.50	0.450	0.50	0.50	0.731	0.684	0.649	0.769	0.50	0.50
3	0.75	0.647	0.75	0.75	0.865	0.867	0.848	0.885	0.25	0.75
4	0.95	0.794	0.95	0.95	0.973	0.976	0.972	0.977	0.05	0.95
		Simple PMMA[a] inclusions in PBA[a] matrix							Lower bounds	
	$\mu_1 = 0.50$	$\mu_1 = 0.50,$ $\phi_{2m} = 0.60$			$\alpha = 0.667$				$\mu_1 = 0.50$	
5	0.25	0.204	0.25	0.25	0.550	0.454	0.357	0.700	0.25	0.75
6	0.50	0.358	0.50	0.50	0.700	0.714	0.625	0.800	0.50	0.50
7	0.75	0.487	0.75	0.75	0.850	0.882	0.883	0.900	0.75	0.25
8	0.95	0.578	0.95	0.95	0.970	0.979	0.969	0.980	0.95	0.05

[a] PBA, poly(butyl acrylate) rubber; PMMA, poly(methyl methacrylate).
[b] These values of the Takayanagi model parameters were used to construct the models illustrated in Fig. 1.

assuming the values of volume fraction and effective volume fraction listed in Table I. Curves 1–4 in this interpretation are the response predictions for blends in which the poly(methyl methacrylate) is the matrix or continuous phase; Curves 5–8 are the response predictions for blends in which the poly(butyl acrylate) is the continuous phase.

The storage modulus curves of Fig. 4 can also be discussed as approximate bounds (at least in the regions of low loss) obtained by applying the correspondence principle to the Hashin–Shtrikman bounds. In this interpretation, ϕ_2 is the volume fraction of the softer component [here, the poly(butyl acrylate) rubber] and ϕ_1 is the volume fraction of the harder component [the poly(methyl methacrylate)]. No assumption of morphology or phase continuity is involved. Curves 1 and 7 are the bounds for blends comprising 0.25 volume fraction soft component; Curves 2 and 6 are the bounds for 0.5 volume fraction soft component; and Curves 3 and 5 are the bounds for 0.75 volume fraction soft component.

Examining the predictions more closely, and viewing the curves as approximate predictions of behavior for specific, simple phase morphology, the temperature dependence for blends is predicted to be similar to that of the matrix material in the region between the two principal transitions. The apparent glass transition temperature of dispersed phases (position of the loss tangent maximum) tends to be shifted somewhat; this does not reflect thermal expansion effects, which have been ignored in the theoretical development, nor (since phase mechanical properties have been assumed constant and equal to properties of the constituent materials in bulk) does it reflect a change in constituent modulus induced by mechanical deformation; it reflects rather a change in geometric, mechanical coupling that occurs as the ratio of phase moduli changes with temperature in the transition regions. The shifts are significant, especially for polymers of high T_g dispersed in polymers of low T_g: the high-temperature loss tangent maximum is shifted to higher temperatures than might be expected, and into a region that is difficult to access experimentally. Thus, the absence of a separate, high-temperature loss tangent maximum does not imply the absence of a separate, high glass transition temperature disperse phase in an elastomeric matrix. The position of the matrix loss peak, on the other hand, is not much affected by the presence of the dispersed phase except at volume loadings approaching ϕ_{2m}. The effects are less pronounced, but still significant, for blends with low T_g polymer dispersed in high T_g polymer matrix. For these materials, shifts in the location of loss maxima are not necessarily due to changes in T_g, or to interaction between blend constituents: only shifts that cannot be reasonably explained in terms of simple model calculations should be so interpreted [76, 85]. Changes in blend composition are reflected primarily in the height of the dispersed-phase loss maximum, and in the level of storage modulus between principal transitions; matrix loss maxima are barely affected by the presence of the dispersed phase.

Interpretation of the storage modulus curves of Fig. 4 as bounds does not appear to be too useful: the bounds are too widely separated over much of the temperature range to provide much information.

The discussion of dynamic response has thus far been limited to simple two-component dispersions. The presence of compound inclusions has been found experimentally to substantially increase the size of the low-temperature loss peak in rubber-modified, high-impact thermoplastics [86–88]. Some of the effects on mechanical properties attributed to particle size at constant rubber content [89] and to grafting [90] are likely due in part to changes in dispersed phase volume fraction resulting from changes in the amount of matrix phase material present as subinclusions in the dispersed phase. The effects of such changes in morphology have been explored by Bohn [75, 76] and by Dickie [12, 28, 43] in terms of mechanical model and self-consistent scheme predictions. The results of calculations published by Matonis and Small [91] on the effect of a soft boundary layer on stress distribution about an isolated hard inclusion are also of interest. Matonis and Small found that even a very thin, soft intermediate layer effectively decouples the hard filler from the matrix: the principal difference between simple and compound dispersions of the same composition can therefore be expected to be the apparent volume fraction of inclusions. Calculations of expected response for blends with compound inclusions, based on the Kerner equation, and presented in detail by Dickie [12], indicate that the modulus level in the temperature range between the principal transitions is essentially a function only of the volume fraction of inclusions, and is independent of the internal morphology of the inclusions. Comparison of the value of modulus in the intermediate temperature range with predictions of the modified Kerner or van der Poel equations should therefore provide an estimate of volume fraction of (compound) inclusions and, in combination with overall composition, an indication of blend morphology, provided that the properties of the blend constituents do not differ materially from bulk properties. The model calculations also indicate that the loss peak corresponding to the matrix polymer is little affected by phase morphology. The height of the dispersed phase loss peak is principally a function of inclusion volume fraction but can show significant dependence on blend morphology. The height of the dispersed phase loss peak is therefore likely to be less reliable than the level of modulus as a measure of inclusion volume fraction.

At some temperature intermediate to the two principal transitions, the loss tangents of the blend constituents are equal: defining this temperature as T_x,

$$[\tan \delta_G(T_x)]_1 = [\tan \delta_G(T_x)]_2 \equiv [\tan \delta_G]_x$$

At T_x, provided that μ_1'' and $\mu_2'' \approx 0$, the viscoelastic form of the Kerner equation predicts $(\tan \delta_G)_c = (\tan \delta_G)_x$ for simple and compound dispersion morphology [see Eq. 24; if $(\tan \delta_G)_1 = (\tan \delta_G)_2$, then $B = 0$ and the result given follows]. Calculated loss tangent curves based on the Kerner equation thus always pass through $(\tan \delta_G)_x$ at T_x (in Fig. 5 T_x is about 110°C).

For materials which do not have a well-defined matrix, calculation of response in terms of the viscoelastic form of Eq. (20) would seem to be appropriate, but does not appear to have been undertaken. The required data—measurements under comparable conditions of two complex moduli over the full range of viscoelastic behavior—also are apparently not available. The Hashin–Shtrikman bounds on the storage modulus for such blends are identical to those for simple dispersion composites of the same overall composition. It can be shown that at T_x Eq. (20) predicts $(\tan \delta_G)_c = (\tan \delta_G)_x$, provided that μ_1'', μ_2'', and $\mu_c'' \approx 0$.

B. Comparison with Experimental Data

Application of Hashin–Shtrikman bounds on storage modulus and of Kerner equation representations to dynamic data is illustrated in Figs. 6–9 on data (taken from Dickie and Cheung [28]) for which the model calculations of Figs. 4 and 5 are directly applicable. Figures 6 and 7 present data on a blend prepared by cocoagulation of two latexes. The relatively high modulus of the material suggests that the glassy polymer forms the continuous phase. The upper bound on storage modulus—which corresponds approximately to the unmodified Kerner equation prediction for a blend of this composition with simple rubbery inclusions in a glassy matrix— does not represent the storage modulus data well, but is consistent with it over most of the temperature range. Since the form of the experimental storage modulus–temperature curve is similar to that of the upper bound, and since $[\tan \delta_G(T_x)]_c = (\tan \delta_G)_x$, it is likely that representation in terms of the modified Kerner equation can be achieved with proper choice of blend morphology. In fact, the behavior of the blend can be represented over most of the temperature range by using the modified Kerner equation and assuming compound dispersion morphology, as shown in Figs. 6 and 7. The volume fraction, and hence composition, of the dispersed phase was estimated by comparison of storage modulus at 50°C with the modified Kerner equation prediction.

Data are presented in Fig. 8 on a simple dispersion of cross-linked poly(methyl methacrylate) particles in an uncross-linked poly(butyl acrylate) matrix. These data show almost no indication of a glassy polymer loss tangent maximum. This cannot be interpreted as implying an absence of

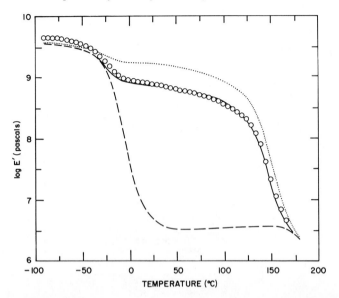

Fig. 6 Dependence of log E' at 110 Hz on temperature for a blend of equal parts by weight poly(methyl methacrylate) and poly(butyl acrylate). The broken curves are approximately the Hashin–Shtrikman upper and lower bounds, as well as predictions of the unmodified Kerner equation for simple phase morphology. The solid curve is the prediction of the modified Kerner equation assuming compound dispersion morphology (61 vol % compound inclusions comprising 14.4 vol % glassy polymer inclusions in rubber). (Adapted from Dickie and Cheung [28].)

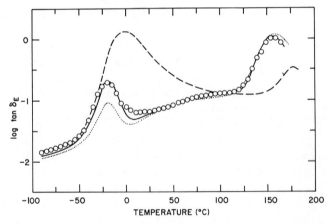

Fig. 7 Dependence of loss tangent at 110 Hz on temperature for the blend of Fig. 6. The broken curves are the predictions of the unmodified Kerner equation for simple dispersion morphology (dashed curve, rubber continuous phase; dotted curve, glassy polymer continuous phase). The solid curve is the prediction of the modified Kerner equation as described for Fig. 6. (Adapted from Dickie and Cheung [28].)

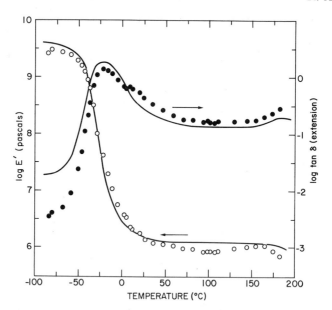

Fig. 8   Dependence of log E' and loss tangent at 110 Hz on temperature for a blend of 25 wt % poly(methyl methacrylate) in poly(butyl acrylate). Solid curves are predictions of the modified Kerner equation for simple dispersion morphology, rubber continuous phase. (From Dickie and Cheung [28].)

phase separation, however; calculations based on the modified Kerner equation indicate that the glassy polymer transition can reasonably be expected to be very weak and shifted substantially to the right on the temperature scale. Within experimental error, the storage modulus data are within the Hashin–Shtrikman bounds (not shown). The storage modulus and loss tangent can both be represented reasonably well by the modified Kerner equation assuming simple dispersion morphology.

Figure 9 reproduces data on a blend of the same composition as the blend of Figs. 6 and 7. In this case, however, the blend was prepared by a two-stage emulsion polymerization. The storage modulus displays a broad transition, well within the Hashin–Shtrikman bounds; the loss tangent substantially exceeds the loss tangent for both constituents over much of the intermediate temperature range and $[\tan \delta_G(T_x)]_c \neq (\tan \delta_G)_x$. The response cannot, therefore, be modeled in terms of simple-, compound-, or random-dispersion calculations based on Eq. (7) or Eq. (20) (assuming, as previously, that μ_1'', μ_2'', and $\mu_c'' \approx 0$). Similar mechanical response (character-ized by high damping over a broad temperature range) has been observed for so-called interpenetrating network systems; the properties of inter-

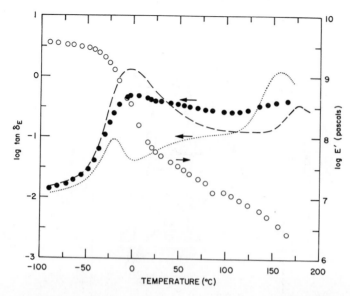

Fig. 9 Dependence of log E' and loss tangent at 110 Hz on temperature for a blend of equal parts by weight poly(methyl methacrylate) and poly(butyl acrylate). Curves are loss tangent predictions of modified Kerner equation for simple dispersion morphology (dashed curve, rubber continuous phase; dotted curve, glassy polymer continuous). (Adapted from Dickie and Cheung [28].)

penetrating networks are reviewed in Volume 2, Chapter 11 of this treatise and will not be discussed here.

Experimental results on physical blends in nonequilibrium mixing states suggest that broad damping characteristics result from extensive but incomplete mixing and are enhanced by random or multiply continuous phase geometry. Bauer *et al.* [92], for example, report broad-transition behavior in unannealed poly(ethyl acrylate)–poly(methyl methacrylate) mixtures; the behavior changes with annealing to typical two-phase, rubber inclusion–glassy matrix response. Miyata and Hata [24] report similar results (shown in Fig. 10) on a freeze dried poly(vinyl acetate)–poly(methyl methacrylate) blend. Kraus and Rollmann [21] report broad transition behavior for a blend of five styrene–butadiene copolymers ranging in styrene content from 10 to 50%. With complete mixing, one principal transition is observed at a temperature intermediate to those of the blend constituents [6, 24, 93]; the response then resembles that of a homogeneous, uniformly random copolymer [6, 21]. With more complete phase separation, discrete transitions characteristic of the phase composition are expected. Kollinsky and Markert's results [93] on multicomponent acrylic blends, for example, suggest

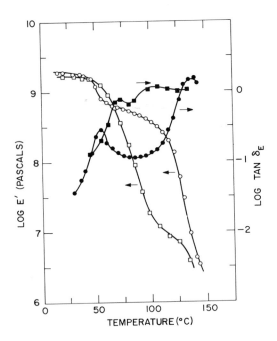

Fig. 10   Dependence of log E' and loss tangent at 25 Hz on temperature for a freeze-dried blend of equal parts by weight poly(vinyl acetate) and poly(methyl methacrylate). Measurements were performed on an unannealed specimen as temperature was gradually raised from room temperature to 160°C (squares) and then lowered to room temperature (circles). (Adapted from Miyata and Hata [24].)

the formation of matrix–inclusion morphology, with each phase being a blend of several components; the composition of the phase regions can be estimated from the position of loss tangent maxima on the temperature scale.

As previously discussed, Miyata and Hata [24], Kraus and Rollman [25], and Takayanagi [26] propose equivalent mechanical models that incorporate properties of intermediate composition copolymers to predict the response of partly compatible blends. Only by introducing a broad range of copolymer composition at substantial concentration can broad-transition behavior be predicted [8, 26]. An apparently unexplored approach to representation of broad transition behavior is use of the multicomponent form of Eq. (20):

$$\sum_{i=1}^{n} (K_i - K_c)\phi_i/(\beta_c K_i + K_c) = 0 \tag{28a}$$

$$\sum_{i=1}^{n} (G_i - G_c)\phi_i/(\alpha_c G_i + G_c) = 0 \tag{28b}$$

The summations are to be performed over the n species comprising the blend. Use of the properties of a graded series of copolymers in Eq. (28), as also used by Miyata and Hata [24] in Eqs. (10) and (11), leads to prediction of the broad, gradual transition in modulus [94]. A similar treatment based on the multicomponent form of Eq. (7) broadens the transitions somewhat, but the principal transition in modulus remains that of the assumed matrix material [94].

Both experimentally and theoretically, mechanical response is often remarkably insensitive to the presence of intermediate composition co-polymers. Matonis [95] analyzed the effect on stress distribution about an isolated inclusion of incorporating an "interphase" of intermediate proper-ties. The presence of the interphase was found to have very little effect on the form of the stress fields; the magnitude of the shear modulus was also little affected. Calculations based on a two-stage application of the Kerner equation indicate that presence of an interlayer of intermediate composition copolymer between inclusions and matrix has little effect on predicted response [94]. Kraus and Rollmann's results [25] on the effects of an inter-layer of graded composition on response of styrene–butadiene–styrene triblock copolymers suggest that only at interlayer volume fractions of at least 0.7 are transition locations substantially affected.

Intermediate transitions and new relaxation mechanisms of several types have been observed in heterogeneous triblock copolymers and in blends of homopolymers, random copolymers, and diblock copolymers with triblock copolymers. Modeling of the response of triblock systems (especially those in which the rubber phase is continuous) is complicated by the fact that the disperse glassy polymer regions act not only as discrete filler particles but also as cross links. Interpretation of mechanical response therefore requires consideration of network and network defect parameters in addition to phase structure and interface phenomena. For poly(styrene-*b*-1,4-butadiene-*b*-styrene) (SBS) intermediate losses have been attributed by Tschoegl and co-workers [96,97] to slippage of trapped entanglements and to interfacial effects. Kraus and Rollman [25] report no loss maxima ascribable to a mixed interlayer between phases; Shen and co-workers [98] report no relaxation peaks ascribable to mixed regions in blends of homo- and block co-polymers. Cohen and Tschoegl [97] indicate that location and intensity of secondary loss maxima in isochronal data may depend on the frequency of measure-ment. The blending of polybutadiene of low molecular weight with an SBS polymer leads to additional losses attributed to entanglement slippage [98, 99]. Added high molecular weight polybutadiene forms a separate phase; new losses observed in this case are attributed to crystallization and melting of the polybutadiene [98]. Addition of polystyrene to SBS does not lead to new intermediate transitions [100]. Addition of ethylene–vinyl acetate copolymer

(EVA) to SBS results in a strongly composition-dependent intermediate loss peak, suggesting that the EVA is present as discrete regions in an SBS matrix [101]; interspecies blending effects are not observed. Addition of diblock styrene–butadiene (SB) to SBS results in additional losses attributed to a combination of several entanglement slippage mechanisms; the SB–SBS blends are discussed as model network systems with controlled network defects [102–104]. Effects of blending homopolymers with block polymers are discussed in detail in Volume 2, Chapter 18. Examples of "compatibilizers" with interfacial activity are discussed in Volume 2, Chapter 12.

The discussion so far has centered on the behavior of blends of polymers with widely separated transitions. For blends of incompatible polymers with similar transition temperatures, Locke and Paul [85] suggest that loss peaks may appear to merge and be shifted from the location observed for blend constituents in bulk. Kraus and Rollman [21] report that a compositional difference of at least 20% is necessary to obtain two resolved loss maxima

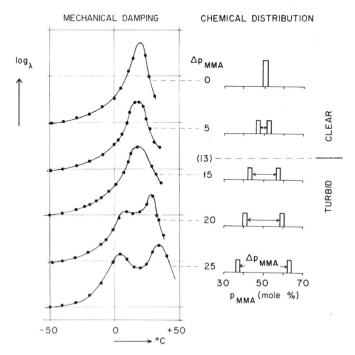

Fig.. 11 Dependence of mechanical damping on temperature for binary mixtures of methyl methacrylate–butyl acrylate copolymers. Each copolymer is chemically homogeneous; the mixtures are of constant composition. (From Kollinsky and Markert [93], *in* "Multicomponent Polymer Systems," Adv. Chem. Ser. No. 99, 1971. Copyright by the American Chemical Society.)

in blends of styrene–butadiene copolymers. Kollinsky and Markert [93] obtained similar results on butyl acrylate–methyl methacrylate copolymers; these results are illustrated in Fig. 11. Although the appearance of two loss maxima approximately coincides with development of turbidity for these copolymer blends, the absence of two distinct maxima is not a reliable indicator of phase separation when transition temperatures are similar.

V. TIME–TEMPERATURE DEPENDENCE

In the preceding discussion, the dependence of modulus on composition at constant time and temperature, and, subsequently, of viscoelastic response on temperature under isochronal conditions have been considered. The effects of changes in time scale on mechanical response, and the interrelationship of time and temperature dependence, are discussed in this section. Following a brief statement of the familiar superposition conditions for thermorheologically simple materials, the relationship of time–temperature dependence of blends to that of blend constituents is discussed. The principles and limitations of time–temperature superposition have been reviewed in detail by Ferry [3]. Analysis of time–temperature dependence for block copolymers and polymer blends has been dealt with extensively by Tschoegl and co-workers [13, 96, 97, 102–105].

The dependence of a modulus M or other mechanical response function on time and temperature is typically presented as a modulus at constant temperature as a function of a time variable, typically $\log t$ or $\log \omega$, or as a modulus at constant t (or ω) as a function of temperature. A third method of representation—a contour map tracing the loci of points of constant response as a function of time and temperature—has been used by Cohen and Tschoegl [97] to represent the complicated time and temperature dependence of block copolymer systems. Contour maps have also been used for example by Deutsch *et al.* [106] to represent response of poly(methyl methacrylate); Landel and Fedors [107] have discussed the representation of small-deformation moduli in terms of a three-dimensional response surface for thermorheologically simple materials.

A. Thermorheologically Simple Materials

For a thermorheologically simple material, time and temperature dependence are separable, and the response surface as a function of $\log t$ and T can be completely defined by a single isothermal section and a single constant response contour. Such a response surface—in this case constructed on the presumption that the material is thermorheologically simple—is illustrated in Fig. 12. All the response contours are of the same shape, and are displaced from one another by a simple translation along the ordinate, as a consequence of the assumption of thermorheological simplicity. All relaxation processes of a thermorheologically simple material have the same temperature dependence; that is, the ratio of relaxation time τ_i at temperature T to the corresponding relaxation time at a reference temperature T_r is constant for all relaxation times:

$$\tau_i(T)/\tau_i(T_r) = a_T \tag{29}$$

For a given reference temperature, $M = M(t, T)$ can be written as a function of q alone, where $q = t/a_T$ and $M = M(q)$. (It is assumed that any multiplicative shifts applicable to M have been taken into account separately.) From the usual rules of partial differentiation,

$$(\partial \ln M/\partial \ln t)_T = (\partial \ln M/\partial \ln q)_{T_r} \tag{30a}$$

$$(\partial \ln M/\partial T)_t = -(\partial \ln M/\partial \ln q)_{T_r}(\partial \ln a_T/\partial T)_t \tag{30b}$$

and

$$(\partial \ln t/\partial T)_M = (\partial \ln t/\partial T)_q = (\partial \ln a_T/\partial T)_t \tag{30c}$$

From Eq. (30a), $(\partial \ln M/\partial \ln t)_T$ is a function of q alone for a given reference temperature; since q is constant along contours of constant M,

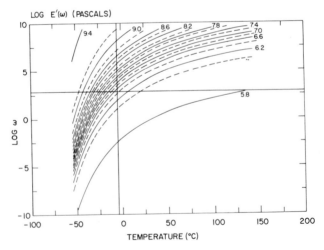

LOG E'(ω) (PASCALS)

Fig. 12 Dependence of log E' on frequency and temperature for poly(butyl acrylate) assuming thermorheological simplicity and employing the "universal" form of the Williams–Landel–Ferry equation (see Ferry [3]) with $T_g = -4°C$ in conjunction with isochronal dynamic data at 110 Hz. The response contours and isothermal sections are of constant shape.

$(\partial \ln M / \partial \ln t)_T$ is also constant. This is the condition allowing superposition of isothermal response curves by a (horizontal) translation along the time axis. From Eq. (30b), simple superposition of isochronal response curves by a translation along the temperature axis is not possible: the slope $(\partial \ln M / \partial T)_t$ is a function of T as well as M. From Eq. (30c), the slope $(\partial \ln t / \partial T)_M$ of response contours for a thermorheologically simple material is a function of T alone; the contours have the same shape, and can be superposed by a translation along the $\ln t$ axis. The contour shape is given by the temperature shift function: $\ln t = \ln q + \ln a_T$. Response contours of many different thermorheologically simple amorphous materials are of nearly the same shape above T_g since the shift function in this temperature range is approximately dependent only on T_g (cf. Ferry [3] and Landel and Fedors [107]).

B. Thermorheologically Complex Materials

For the general case $M = M(t, T)$ it is also possible to introduce the concept of a shift function a; the shift function in the general case is, however, a function of both time and temperature: $M = M(t, T) = M(q)$, where $q = t/a$, and $a = a(t, T)$. The following expressions corresponding to Eqs. (30a), (30b), and (30c) are obtained:

$$\left(\frac{\partial \ln M}{\partial \ln t} \right)_T = \left(\frac{\partial \ln M}{\partial \ln q} \right)_{T_r} \left[1 - \left(\frac{\partial \ln a(t, T)}{\partial \ln t} \right)_T \right] \tag{31a}$$

$$\left(\frac{\partial \ln M}{\partial T} \right)_t = - \left(\frac{\partial \ln M}{\partial \ln q} \right)_{T_r} \left(\frac{\partial \ln a(t, T)}{\partial T} \right)_t \tag{31b}$$

$$\left(\frac{\partial \ln t}{\partial T} \right)_M = \left(\frac{\partial \ln t}{\partial T} \right)_q$$

$$= \left(\frac{\partial \ln a(t, T)}{\partial T} \right)_t \left/ \left[1 - \left(\frac{\partial \ln a(t, T)}{\partial \ln t} \right)_T \right] \right. \tag{31c}$$

From Eq. (31a), $(\partial \ln M / \partial \ln t)_T$ is a function of time as well as of reduced time; from Eq. (31b), $(\partial \ln M / \partial T)_t$ is (as is also true of a thermorheologically simple material) a function of temperature as well as of reduced time; and from Eq. (31c), $(\partial \ln t / \partial T)_M$ is a function of time as well as temperature. Thus, no simple superposition by translation along one axis is possible in the general, thermorheologically complex case.

For a polymer blend, if a mechanical response function M_c is calculable from the corresponding mechanical response functions M_1, M_2 of the constituents of the blend and the composition of the blend—that is, if

$f_c(\phi_1, \phi_2, M_1, M_2)$—then, by applying the usual rules of partial differ-
n, it can be shown that the shape of contours of constant response
end of constant composition is defined by

$$
\left(t\right)_{M_c} = \frac{\left(\dfrac{\partial \ln M_c}{\partial \ln M_1}\right)_{M_2}\left(\dfrac{\partial \ln M_1}{\partial \ln t}\right)_T\left(\dfrac{\partial \ln t}{\partial T}\right)_{M_1} + \left(\dfrac{\partial \ln M_c}{\partial \ln M_2}\right)_{M_1}\left(\dfrac{\partial \ln M_2}{\partial \ln t}\right)_T\left(\dfrac{\partial \ln t}{\partial T}\right)_{M_2}}{\left(\dfrac{\partial \ln M_c}{\partial \ln M_1}\right)_{M_2}\left(\dfrac{\partial \ln M_1}{\partial \ln t}\right)_T + \left(\dfrac{\partial \ln M_c}{\partial \ln M_2}\right)_{M_1}\left(\dfrac{\partial \ln M_2}{\partial \ln t}\right)_T} \tag{32}
$$

he contour shape for the blend is a function of the contour shape
component, through $(\partial \ln t/\partial T)_{M_i}$; of the blending law, and hence
logy, through $(\partial M_c/\partial M_i)_{M_{j \neq i}}$; and of the steepness of the time de-
ce of the response function for each component, through $(\partial \ln M_i/$
Equation (32) also implies that, for a thermorheologically complex
l, response contours for different response functions do not necessarily
e same shape; that is, the shift functions are not necessarily identical
rent response functions. (Since blend moduli depend in general on
e dilational and the distortional properties of blend constituents, Eq.
not strictly correct; generalization is straightforward, but does not
lly alter qualitative conclusions concerning the dependence of blend
ctions on constituent properties.) Fesko and Tschoegl [105] give an
on analogous to Eq. (32) for $(\partial \ln a(t, T)/\partial T)_t$ based on a simple
e compliance model. Expressions similar to Eq. (32) can be written for
$(\partial \ln t)_T$ and $(\partial \ln M_c/\partial T)_t$.

blend comprising two constituents with well-separated transitions,
tour shape (or the shift function) tends to be dominated by one
on mechanism (principal transition) or the other [105]. In terms of
), this is a reflection of the fact that $(\partial \ln M_i/\partial \ln t)_T$ is generally
cept in the transition region. This means that the coefficients of
$l')_{M_i}$ in Eq. (32), and, in terms of empirical shifting, the coefficients of
tituent shift functions, can be approximated by step functions [105].
superposition may be approximately correct in regions dominated by
nponent.
simple dispersion composite, with Material 1 as the matrix, if either
l_1 or $M_1 \ll M_2$, then

$$
M_c \approx f(\phi_2)\, M_1 \tag{33}
$$

is a function of ϕ_2 but not of M_1 or M_2] and, from Eq. (32),

$$
(\partial \ln t/\partial T)_{M_c} \approx (\partial \ln t/\partial T)_{M_1} \tag{34}
$$

pe of the response contours for the blend is the same as that of the
e contours for the matrix for limiting values of modulus ratio.

For an elastic filler, $(\partial \ln M_2/\partial \ln t)_T = 0$, and the response contou
(shift factor) is the same as that of the matrix. A change in shift b
with addition of an elastic filler is thus an indication that matrix re
processes have been changed by introduction of the filler.

Empirical shifting has sometimes been applied to isothermal
thermorheologically complex polymer blends to obtain master
theoretical results such as the one given in Eq. (32) notwithstand
e.g., Smith and Dickie [108], Shen and Kaelble [109], Lim *et (*
and Bares and Pegoraro [111]). The validity of simple superpo
often difficult to evaluate since the experimental time scale a
commonly spans only a few decades, and changes in contour shape
be evident. Simple superposition in such cases can nevertheless be mi
The assumption of thermorheological simplicity imposes uniform r
contour shape on the experimental data; while the response su
generated may approximately represent material properties within t
of experimental measurements, response outside the range of meas
can be significantly misrepresented.

Data on blends can be superposed by comparison with cal
response curves based on models for composition dependence. Thus,
response surface for each component of a blend (defined, for exampl
presumption that the blend components are individually thermorheol
simple materials) a predicted blend response surface can be generate
an appropriate model for composition dependence. Tschoegl and co-
have explored two ways of comparing experimental data with ca
response. Fesko and Tschoegl [96, 105] generated shift factors point l
by comparing values of compliance calculated from constituent pr
at the time and temperature of each experimental measurement wit
culated isothermal master curve. The shift factors obtained were then
superpose experimental data. Failure of the calculated shift factors to
pose properly the experimental data over certain ranges of (reduced) t
interpreted as being due either to entanglement slippage or to p
characteristic of a mixed interlayer between phase regions. As expec
shape of the master curve—that is, the shape of an isothermal
through the response surface—was found to depend on the r
temperature chosen. Kaplan and Tschoegl [13] analyzed the data o
et al. [15] by shifting experimental blend response directly to a p
master curve, thus generating a temperature shift function. As expec
shape of the shift function obtained was substantially influenced
reference temperature chosen. In both cases, a fairly sharp transiti
behavior characteristic of one component to behavior characteristi
other was observed.

For thermorheologically complex materials, representation of mec response in terms of response contours can clearly reveal regions in which simple superposition is not valid, and regions in which new relaxation processes are active. Cohen and Tschoegl [97] have discussed data on block copolymers in this way. Their contour map for loss compliance of an SBS block copolymer is reproduced in Fig. 13. The extreme complexity of contour shape in this map results largely from entanglement contributions to mechanical response, as discussed in Cohen and Tschoegl [97]. The constancy of contour shape in the upper left portion of Fig. 13 reflects the approximate applicability of simple superposition in this region. A detailed interpretation of the features of this contour map has been given by Cohen and Tschoegl [97]. Isochronal sections through the response surface of Fig. 13, reproduced in Fig. 14, illustrate the change in apparent response that a change in time scale can produce; the location, magnitude, and breadth of the principal and intermediate transitions are significantly affected. The isothermal master curves obtained by Fesko and Tschoegl [96] are shown in Fig. 15; there is a dramatic change in the shape of the curve and in the extent of the intertransition plateau with changes in reference temperature.

Viscoelastic behavior is often discussed in terms of retardation or relaxation spectra; methods of calculation, and interrelations between functions can be found in Gross [2] and Ferry [3] (also see Chapter 7 and Volume 2, Chapter 21). To a rough approximation, the relaxation spectrum $H(\tau)$ is

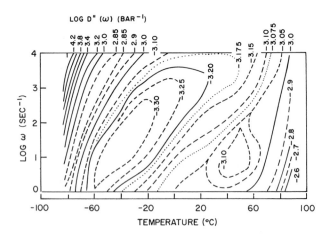

Fig. 13   Dependence of $\log D''$ (dynamic tensile loss compliance) on frequency and temperature for an SBS block copolymer. Solid contours are uniformly spaced at 0.2 log units; intermediate (dashed and dotted) contours are included in regions of relatively flat response. (Data of Fesko and Tschoegl [96]; figure adapted from Cohen and Tschoegl [97].)

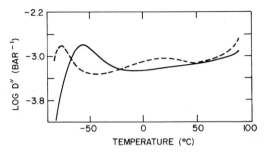

Fig. 14 Isochronal sections through response surface of Fig. 13 at $\log \omega = 1$ (dashed curve) and $\log \omega = 3$ (solid curve). (Adapted from Cohen and Tschoegl [97].)

given by

$$H(\tau) = dG(t)/(d \ln t)|_{t=\tau} \qquad (35)$$

Changes in $H(\tau)$ for filled systems have been attributed to changes in relaxation behavior, especially if the filler is nominally elastic (see, e.g., Lipatov *et al.* [112] and the references cited therein). If G_1 is the shear modulus of the continuous phase of a composite, and $G_1/G_2 \ll 1$ or $G_1/G_2 \gg 1$, then, from Eq. (33), $G_c = f(\phi_2) G_1$ and

$$H_c(\tau) = f(\phi_2) H_1(\tau) \qquad (36)$$

That is, the shape and location of the relaxation spectrum are unchanged unless new relaxation processes are active. If the constituent moduli are of comparable magnitude, however, the apparent relaxation spectrum for the blend $H_c(\tau)$ can display time dependence substantially different from that of $H_1(\tau)$ even if no additional relaxation mechanisms are involved. Thus, the application of Eq. (35) to the Kerner equation for G_c, with the assumption

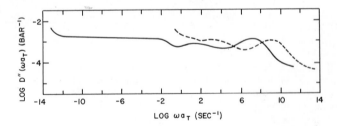

Fig. 15 Isothermal master curves constructed on the basis of the data presented in Fig. 12 and incorporating corrections for interlayer and entanglement effects. The reference temperature for the solid curve is 25°C; for the dashed curve it is 85°C. (Adapted from Fesko and Tschoegl [96].)

that Component 2 is elastic, yields

$$H_c(\tau) = H_1 \frac{\phi_1(1 + \alpha_1 \phi_2) G_1{}^2 + 2\phi_1{}^2 \alpha_1 G_1 G_2 + (\alpha_1 + \phi_2) \alpha_1 \phi_1 G_2{}^2}{[(1 + \alpha_1 \phi_2) G_1 + \alpha_1 \phi_1 G_2]^2}\bigg|_{t=\tau} \quad (37)$$

where H_1 is given by Eq. (35). Since the time-dependent relaxation modulus G_1 appears in the coefficient of H_1 in Eq. (37), the apparent time dependence of H_c will be different from that of H_1.

REFERENCES

1. A. E. H. Love, "A Treatise on the Mathematical Theory of Elasticity," 4th ed. Dover, New York, 1944.
2. B. Gross, "Mathematical Structure of the Theories of Viscoelasticity." Hermann, Paris, 1953.
3. J. D. Ferry, "Viscoelastic Properties of Polymers," 2nd ed. Wiley, New York, 1970.
4. P. S. Theocaris, *Kolloid Z.* **235**, 1182 (1969).
5. Z. Rigby, *Rheol. Acta* **5**, 28 (1966).
6. L. E. Nielsen, "Mechanical Properties of Polymers." Van Nostrand-Reinhold, Princeton, New Jersey, 1962.
7. L. E. Nielsen, "Mechanical Properties of Polymers and Composites." Dekker, New York, 1974.
8. J. A. Manson and L. H. Sperling, "Polymer Blends and Composites." Plenum, New York, 1976.
9. M. B. Bever and M. Shen, *Mater. Sci. Eng.* **15**, 145 (1974).
10. L. H. Sperling, *Fibre Sci. Technol.* **7**, 199 (1974); *Polym. Prepr.* **14**, 431 (1973); *Polym. Eng. Sci.* **16**, 87 (1976).
11. M. Takayanagi, H. Harima, and Y. Iwata, *J. Soc. Mater. Sci. Jpn.* **12**, 389 (1963); *Mem. Fac. Eng. Kyushu Univ.* **23**, 1 (1963).
12. R. A. Dickie, *J. Appl. Polym. Sci.* **17**, 45 (1973).
13. D. Kaplan and N. W. Tschoegl, *Polym. Eng. Sci.* **14**, 43 (1974).
14. M. Takayanagi, S. Uemura, and S. Minami, *J. Polym. Sci. Part C* **5**, 113 (1964).
15. T. Horino, Y. Ogawa, T. Soen, and H. Kawai, *J. Appl. Polym. Sci.* **9**, 2261 (1965).
16. T. Okamoto and M. Takayanagi, *J. Polym. Sci. Part C* **23**, 597 (1968).
17. M. Takayanagi, *Proc. Int. Cong. Rheol., 4th* (E. H. Lee, ed.), Part I, p. 161. Wiley (Interscience), New York, 1965.
18. M. Matsuo, T. K. Kwei, D. Klempner, and H. L. Frisch, *Polym. Eng. Sci.* **10**, 327 (1970).
19. G. Kraus, K. W. Rollman, and J. T. Gruver, *Macromolecules* **3**, 92 (1970).
20. K. Marcincin, A. Romanov, and V. Pollak, *J. Appl. Polym. Sci.* **16**, 2239 (1972).
21. G. Kraus and K. W. Rollman, in "Multicomponent Polymer Systems" (N. Platzer, ed.), Adv. Chem. Ser. No. 99, p. 189. Amer. Chem. Soc., Washington, D.C., 1971.
22. K. Fujino, Y. Ogawa, and H. Kawai, *J. Appl. Polym. Sci.* **8**, 2147 (1964).
23. J.-M. Charrier, *Polym. Eng. Sci.* **15**, 731 (1975).
24. S. Miyata and T. Hata, *Proc. Int. Cong. Rheol., 5th* (S. Onogi, ed.), Vol. 3, p. 71. Univ. Tokyo Press, Tokyo, 1970.
25. G. Kraus and K. W. Rollman, *J. Polym. Sci. Polym. Phys. Ed.* **14**, 1133 (1976).

26. M. Takayanagi, Private Communication (1972) cited in Bever and Shen [9].
27. L. H. Sperling, D. W. Taylor, M. L. Kirkpatrick, H. F. George, and D. R. Bardman, *J. Appl. Polym. Sci.* **14**, 73 (1970).
28. R. A. Dickie and M.-F. Cheung, *J. Appl. Polym. Sci.* **17**, 79 (1973).
29. D. A. G. Bruggeman, *Ann. Phys.* **29**, 160 (1937).
30. D. Hartree, *Rept. Progr. Phys.* **11**, 113 (1947); W. J. Moore, "Physical Chemistry," 4th ed., p. 509. Longmans Green, London, 1963.
31. E. H. Kerner, *Proc. Phys. Soc.* **69B**, 808 (1956).
32. J. E. Ashton, J. C. Halpin, and P. H. Petit, "Primer on Composite Materials: Analysis," Chapter 5. Technomic Publ., Stamford, Connecticut, 1969.
33. S. W. Tsai, Formulas for the Elastic Properties of Fiber-Reinforced Composites, AD834851. Nat. Tech. Informat. Serv., Springfield, Virginia, 1968.
34. J. Halpin and J. L. Kardos, *Polym. Eng. Sci.* **16**, 344 (1976).
35. S. Uemura and M. Takayanagi, *J. Appl. Polym. Sci.* **10**, 113 (1966).
36. K. Okano, *Rep. Progr. Polym. Phys. Jpn.* **3**, 69 (1960).
37. Z. Hashin, *J. Appl. Mech.* **29**, 143 (1962).
38. N. K. Kalfoglou and H. L. Williams, *J. Appl. Polym. Sci.* **17**, 1377 (1973).
39. Z. Hashin and S. Shtrikman, *J. Mech. Phys. Solids* **11**, 127 (1963).
40. L. E. Nielsen, *J. Appl. Polym. Sci. Appl. Polym. Symp.* **12**, 249 (1969).
41. L. E. Nielsen, *J. Appl. Phys.* **41**, 4626 (1970).
42. T. B. Lewis and L. E. Nielsen, *J. Appl. Polym. Sci.* **14**, 1449 (1970).
43. R. A. Dickie, M.-F. Cheung, and S. Newman, *J. Appl. Polym. Sci.* **17**, 65 (1973).
44. R. A. Dickie, *J. Appl. Polym. Sci.* **17**, 2509 (1973).
45. R. A. Dickie, *J. Polym. Sci. Polym. Phys. Ed.* **14**, 2073 (1976).
46. R. A. Dickie, in "Polymer Engineering Composites" (M. O. W. Richardson, ed.), p. 155, Appl. Sci. Publ., London, 1977.
47. K. D. Ziegel and A. Romanov, *J. Appl. Polym. Sci.* **17**, 1133 (1973).
48. K. D. Ziegel and A. Romanov, *J. Appl. Polym. Sci.* **17**, 1119 (1973).
49. C. van der Poel, *Rheol Acta* **1**, 198 (1958).
50. H. Fröhlich and R. Sack, *Proc. Roy. Soc. A* **185**, 415 (1946).
51. F. R. Schwarzl, H. W. Bree, C. G. Nederveen, G. A. Schwippert, L. C. E. Struick, and C. van der Wal, *Rheol. Acta* **5**, 270 (1966), reprinted in *Rubber Chem. Technol.* **42**, 557 (1969).
52. F. R. Schwarzl, H. W. Bree, and C. J. Nederveen, *Proc. Inter. Congr. Rheol. 4th* (E. H. Lee, ed.), Part 3, p. 241. Wiley (Interscience), New York, 1965.
53. F. R. Schwarzl, H. W. Bree, C. J. Nederveen, L. C. E. Struik, and C. van der Wal, in "Mechanics and Chemistry of Solid Propellants" (A. C. Eringen, H. Leibowitz, S. L. Koh, and J. M. Crowley, eds.), p. 503. Pergamon, Oxford, 1967; see also *TNO-Nieuws* **21**, 74 (1966).
54. S. R. Bodner and J. M. Lifshitz, in "Mechanics and Chemistry of Solid Propellants" (A. C. Eringen, H. Leibowitz, S. L. Koh, and J. M. Crowley, eds.), p. 663. Pergamon, Oxford, 1967.
55. J. C. Smith, *Polym. Eng. Sci.* **16**, 394 (1976).
56. Z. Hashin, in "Mechanics and Chemistry of Solid Propellants" (A. C. Eringen, H. Leibowitz, S. L. Koh, and J. M. Crowley, eds.), p. 201. Pergamon, Oxford, 1967.
57. J. C. Smith, *J. Res. Nat. Bur. Std. U.S.* **78A**, 355 (1974).
58. J. C. Smith, *J. Res. Nat. Bur. Std. U.S.* **79A**, 419 (1975).
59. F. R. Schwarzl, H. W. Bree, G. A. Schwippert, L. C. E. Struik, and C. W. van der Wal, *Proc. Int. Congr. Rheol., 5th* (S. Onogi, ed.), Vol. 3, p. 3. Univ. Tokyo Press, Tokyo, 1970.

60. L. C. E. Struik, H. W. Bree, and F. R. Schwarzl, Mechanical Properties of Filled Elastomers. VII. The Shear Modulus of Rubbers Containing Fillers with a Bimodal Size Distribution, AD662897. Nat. Tech. Informat. Serv., Springfield, Virginia, 1967.
61. R. Hill, *J. Mech. Phys. Solids* **13**, 213 (1965).
62. S. Budiansky, *J. Mech. Phys. Solids* **13**, 223 (1965).
63. B. Budiansky, *J. Composite Mater.* **4**, 286 (1970).
64. J. Faucher, *J. Polym. Sci. Polym. Phys. Ed.* **12**, 2153 (1974).
65. H. Hendus, K. Illers, and E. Ropte, *Kolloid Z.* **216**, 110 (1967).
66. G. E. Molau, *in* "Block Polymers" (S. L. Aggarwal, ed.), p. 79. Plenum, New York, 1970.
67. A. Keller, E. Pedemonte, and F. M. Willmouth, *Nature (London)* **225**, 538 (1970).
68. J. Dlugosz, A. Keller, and E. Pedemonte, *Kolloid Z.* **242**, 1125 (1970).
69. R. G. C. Arridge and M. J. Folkes, *J. Phys. D. Appl. Phys.* **5**, 344 (1972).
70. S. K. Garg, V. Svalbonas, and G. A. Gurtman, "Analysis of Structural Composite Materials," Dekker, New York, 1973.
71. B. Paul, *Trans. AIME* **218**, 36 (1960).
72. Z. Hashin and S. Shtrikman, *J. Franklin Inst.* **271**, 336 (1961).
73. Z. Hashin and S. Shtrikman, *J. Mech. Phys. Solids* **10**, 335 (1962).
74. R. Hill, *J. Mech. Phys. Solids* **11**, 357 (1963).
75. L. Bohn, *Angew. Makromol. Chem.* **29** ⩾ **30**, 25 (1973).
76. L. Bohn, *in* "Copolymers, Polyblends, and Composites" (N. Platzer, ed.), p. 66. Adv. Chem. Ser. No. 142. Amer. Chem. Soc., Washington, D.C., 1974.
77. Z. Hashin, *Int. J. Solids Structures* **6**, 539 (1970).
78. R. M. Christensen, *J. Mech. Phys. Solids* **17**, 23 (1969).
79. R. A. Schapery, *in* "Mechanics of Composite Materials" (G. P. Sendeckyj, ed.), p. 85. Academic Press, New York, 1974.
80. Z. Hashin, *Int. J. Solids Structures* **6**, 539 (1970).
81. M. G. Sharma and W. F. St. Lawrence, *Proc. Int. Congr. Rheol. 5th* (S. Onogi, ed.), Vol. 3, p. 47. Univ. Tokyo Press, Tokyo, 1970.
82. R. Roscoe, *J. Mech. Phys. Solids* **17**, 17 (1969).
83. Z. Hashin, *in* "Second Order Effects in Elasticity, Plasticity, and Fluid Dynamics" (M. Reiner and D. Abir, eds.), p. 434. Macmillan, New York, 1964.
84. C. B. Bucknall and M. M. Hall, *J. Mater. Sci.* **6**, 95 (1971).
85. C. E. Locke and D. R. Paul, *Polym. Eng. Sci.* **13**, 308 (1973).
86. H. Keskkula, *J. Appl. Polym. Sci. Appl. Polym. Symp.* **15**, 51 (1970).
87. H. Keskkula, S. G. Turley, and R. F. Boyer, *J. Appl. Polym. Sci.* **15**, 351 (1971).
88. G. Cigna, *J. Appl. Polym. Sci.* **14**, 11781 (1970).
89. M. Baer, *J. Appl. Polym. Sci.* **16**, 1109 (1972).
90. T. Ricco, A. Pavan, and F. Danusso, *Polymer* **16**, 685 (1975).
91. V. A. Matonis and N. C. Small, *Polym. Eng. Sci.* **9**, 90 (1969).
92. P. Bauer, J. Hennig, and G. Schreyer, *Angew. Makromol. Chem.* **11**, 145 (1970).
93. F. Kollinsky and G. Markert, *in* "Multicomponent Polymer Systems" (N. Platzer, ed.), Adv. Chem. Ser. No. 99, p. 175. Amer. Chem. Soc., Washington, D.C., 1971.
94. R. A. Dickie, unpublished calculations.
95. V. A. Matonis, *Polym. Eng. Sci.* **9**, 100 (1969).
96. D. G. Fesko and N. W. Tschoegl, *Int. J. Polym. Mater.* **3**, 51 (1974).
97. R. E. Cohen and N. W. Tschoegl, *Trans. Soc. Rheol.* **20**, 153 (1976).
98. L. Toy, M. Niinomi, and M. Shen, *J. Macromol. Sci.–Phys.* **B11**, 281 (1975).
99. G. Choi, A. Kaya, and M. Shen, *Polym. Eng. Sci.* **13**, 231 (1973).
100. C. W. Childers, G. Kraus, J. T. Gruver, and E. Clark, *in* "Colloidal and Morphological

Behavior of Block and Graft Copolymers" (G. Molau, ed.), p. 193. Plenum, New York, 1971.

101. U. Mehra, L. Toy, K. Biliyar, and M. Shen, in "Copolymers, Polyblends, and Composites" (N. Platzer, ed.), Adv. Chem. Ser. No. 142, p. 399. Amer. Chem. Soc., Washington, D.C., 1971.

102. R. E. Cohen and N. W. Tschoegl, Int. J. Polym. Mater. 2, 49 (1972).

103. R. E. Cohen and N. W. Tschoegl, Int. J. Polym. Mater. 2, 205 (1973).

104. R. E. Cohen and N. W. Tschoegl, Int. J. Polym. Mater. 3, 3 (1974).

105. D. G. Fesko and N. W. Tschoegl, J. Polym. Sci. Part C No. 35, 41 (1971).

106. K. Deutsch, E. A. W. Hoff, and W. Reddish, J. Polym. Sci. 13, 565 (1954).

107. R. F. Landel and R. F. Fedors, in "Fracture Processes in Polymeric Solids" (B. Rosen, ed.), p. 361. Wiley (Interscience), New York, 1964.

108. T. L. Smith and R. A. Dickie, J. Polym. Sci. Part C No. 26, 163 (1963).

109. M. Shen and D. H. Kaelble, Polym. Lett. 8, 149 (1970).

110. C. K. Lim, R. E. Cohen, and N. W. Tschoegl, in "Multicomponent Polymer Systems" (N. Platzer, ed.), Adv. Chem. Ser. No. 99, p. 397. Amer. Chem. Soc., Washington, D.C., 1971.

111. J. Bares and M. Pegoraro, J. Polym. Sci. Part A-2 9, 1287 (1971).

112. Yu. S. Lipatov, V. F. Babich, and V. F. Rosovizky, J. Appl. Polym. Sci. 20, 1787 (1976).

Chapter 9

Optical Behavior of Polymer Blends

R. S. Stein

Polymer Research Institute
and
Department of Chemistry
University of Massachusetts
Amherst, Massachusetts

I.	Introduction	393
II.	Interaction of Radiation with Polymers	394
	A. General Comments	394
	B. Polarization	395
	C. Refraction	396
	D. Absorption	396
	E. Raman Polarization and Fluorescence	397
	F. Light, x-Ray, and Neutron Scattering	398
III.	The Transparency and Scattering of Blends	401
	A. The Relationship between Transparency and Light Scattering	401
	B. The Model Approach for Rayleigh–Gans–Debye Scattering	402
	C. The Model Approach Using Mie Theory	409
	D. The Statistical Description of Scattering	413
	E. Scattering from Blends with Crystalline Components	428
	F. Scattering Arising from Anisotropy	431
IV.	Other Optical Properties of Blends	433
	A. Birefringence	433
	B. Infrared Dichroism	439
	C. Raman Polarization	442
V.	Conclusions	442
	References	442

I. INTRODUCTION

The study of the optical properties of polymer blends is of interest for two reasons. A blend, being composed of two or more components, will scatter

393

light. This may result in its being cloudy or turbid in appearance, a property that affects the marketability of a polymer, particularly for such applications as containers and films. Also, the scattering, as well as other optical properties such as birefringence and dichroism, convey information about the structure of the blend that is desirable for the understanding of its mechanical, electrical, thermal, and other physical properties.

II. INTERACTION OF RADIATION WITH POLYMERS

A. General Comments

Light and x rays consist of electromagnetic waves in which an electrical and a magnetic field, perpendicular to each other and to the direction of propagation of the wave, vary periodically with position and time (Fig. 1). The direction of polarization of the radiation is taken as that of the electrical field which, for plane polarized light, is confined to a plane. For unpolarized light, the plane of polarization varies randomly about the incident direction. Polarized radiation may be produced from unpolarized radiation by transmission through a dichroic filter or by reflection at a particular (Brewster's) angle. For purposes of considering the interaction of the wave with matter, it is usually sufficient to consider the effect of the electrical field.

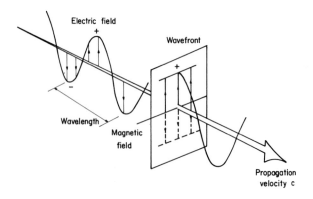

Fig. 1   The variation of the electric and magnetic field for plane polarized light. (From Kakudo and N. Kasai [1a].

B. Polarization

The electrical field of the light wave exerts a force on the electrons of the matter; displacing them from their equilibrium positions and inducing a dipole moment. **M**. For moderate field strengths, the dipole moment is a linear function of the electrical field **E** and may be given by

$$\mathbf{M} = |\alpha| \mathbf{E} \tag{1}$$

The dipole moment and field strength vectors are not necessarily collinear so the proportionality factor, the polarizability $|\alpha|$, is generally a tensor. For isotropic bodies, it is a scaler. It is dependent upon the amount and kind of matter. To a first approximation, the polarizability of an atom is equal to the cube of the atomic radius. Thus, larger atoms are more polarizable. The polarizability of a diatomic molecule is anisotropic, generally with a principal component b_1 lying along the molecular axis greater than that perpendicular to the axis. Such values may be obtained, for example, from light-scattering polarization or Kerr effect measurements on gases.

As a first approximation, the scaler polarizability of a collection of atoms may be taken as the sum of the atomic polarizabilities. This is not quite true since dipoles exert fields on each other. In fact, it is this internal field effect which for the most part leads to the anisotropy of molecules [1b].

It is a good approximation to represent the principal polarizability of a molecule as a sum of contributions from its constituent bonds [2] as

$$\alpha_1 = \sum_i [(b_1 - b_2)_i \cos^2 \theta_i + (b_2)_i] \tag{2}$$

where $(b_1)_i$ and $(b_2)_i$ are the principal polarizabilities of the ith bond along and perpendicular to its axis and θ_i is the angle between the axis of the ith

Table I

Principal Polarizabilities of
Some Chemical Bonds[a]

Bond	$b_1(Å^3)$	$b_2(Å^3)$
C—C	18.8	0.2
C=C	28.6	10.6
C—H	7.9	5.8
C—Cl	36.7	20.8
C—C (aromatic)	22.5	4.5
C—N	15.9	8.9
Phenyl	123.0	63.8

[a] According to Denbigh [2].

and the principal axis of the molecule. Table 1 gives some typical
of bond principal polarizability differences obtained from measure-
on small gaseous molecules. Departures from such additivity arising
ernal field effects have been discussed [3,4].

C. Refraction

croscopic property related to polarizability is refraction. An electro-
wave in vacuum travels with a velocity $c = 3.0 \times 10^{10}$ cm/sec.
raveling through a medium the velocity is reduced to $v = c/n$, where
index of refraction. This leads to the bending of a beam of light on
obliquely through a boundary between phases as described by Snell's

dex of refraction is related to the polarizability by the Lorenz–
equation

$$(n^2 - 1)/(n^2 + 2) = \tfrac{4}{3}\pi N\alpha \tag{3}$$

is the number of molecules per cubic centimeter of polarizability α.
sotropic media the difference between the refractive index in two
l directions, the birefringence Δ may be approximately related to
rizability difference by differentiating Eq. (3) to give for perfectly
molecules

$$\Delta = n_1 - n_2 = \tfrac{2}{9}\pi[(n^2 + 1)^2/n]\, N(\alpha_1 - \alpha_2) \tag{4}$$

is the average refractive index $[(n_1 + n_2 + n_3)/3]$. Thus, the bire-
e is a measure of the molecular orientation and may be obtained
combination of Eqs. (2) and (4) to give [5]

$$\Delta = (2\pi/9)[(n^2 + 2)^2/n] \sum_i [(\alpha_1 - \alpha_2)_i\, f_i] \tag{5}$$

is the Hermans orientation function [6] defined by

$$f_i = [3\overline{\cos^2 \theta_i} - 1]/2 \tag{6}$$

D. Absorption

ules absorb radiation by virtue of being excited to a higher electronic
cular energy level through absorption of a photon of frequency v

$$h_p v = E_2 - E_1 \tag{7}$$

is Planck's constant and E_1 and E_2 are the energies of the lower and higher levels. From a classical viewpoint, the absorption may sidered to occur when the frequency of the radiation equals a re frequency of electronic or molecular motion. Electronic transitions absorption of x rays (inner electrons), ultraviolet, or visible light, molecular vibrational or rotational transitions lead to absorption of far infrared radiation. The absorption may be characterized by a tr. moment, a vector in the direction of the polarization of the radiatior is absorbed. This is dependent upon the symmetry of the electr molecular motion associated with the absorption. For a vibrational of a diatomic molecule, for example, the transition moment lies al molecular axis. A polyatomic molecule may absorb at several freq each corresponding to processes in a particular part of a molec having its characteristic transition moment.

The extinction coefficient ε of a thickness l of material is defined

$$\varepsilon = (1/l)\ln(I_0/I_t)$$

where I_t and I_0 are the transmitted and incident intensities. This related to the absorbence a_i along the associated transition moment

$$\varepsilon_i = a_i\overline{\cos^2\theta_i}$$

where θ_i is the angle between the transition moment axis and the d of polarization of the radiation.

For an oriented material the extinction coefficient depends u direction of polarization of the radiation. For two principal direc absorption, the dichroism is defined as

$$D = \varepsilon_1/\varepsilon_2$$

This is related to the orientation function for the transition moment d by [7]

$$(D-1)/(D+2) = f_M$$

The dichroism depends upon the particular absorption band stud characterizes the orientation of a particular part of the molecule.

E. Raman Polarization and Fluorescence

Absorbed radiation may be reemitted at a different frequenc energy may be lost or gained from a molecular or electronic ener change. The change in energy is

$$h_P(v_a - v_e) = \Delta E$$

where v_a and v_e are the absorbed and emitted frequencies. If the absorbed radiation is polarized in the direction of the unit vector **P** and the emitted radiation is viewed with an analyzer transmitting radiation polarized along the unit vector **A**, the intensity of the emitted light is

$$I = \overline{K(\mathbf{M}_a \cdot \mathbf{P})^2 (\mathbf{M}_e \cdot \mathbf{A})^2} \tag{13}$$

where $\mathbf{M}_a$ and $\mathbf{M}_e$ are unit vectors along the transition moment directions for absorption and emission. If these are parallel and **P** and **A** are parallel to each other, this equation becomes [8]

$$I_{\|} = \overline{K \cos^4 \theta_M} \tag{14}$$

The term θ_M is the angle between the common transition moment direction and the polarization direction. Thus, the state of polarization of the emitted light characterizes a different average of the orientation distribution.

For fluorescence, the transition is between electronic energy levels, whereas Raman spectra involves transitions between molecular rotational and vibrational energy levels so that the displacement of a Raman line from the exciting frequency corresponds to the frequency of a molecular transition. An excellent discussion of the relationship of the polarization of Raman spectra to molecular or orientation has recently been published [9a].

Fluorescence studies are restricted to those molecules having fluorescent chromophores in their structures. This is not generally the case for polymers, although in some cases, special polymers have been synthesized for studies that have fluorescent labels such as stilbene or anthracene residues. Alternatively, fluorescent dyes have been dissolved in polymers as orientation labels [8]. The latter technique suffers from the uncertain orientation of the dye with respect to the polymer. While experimentally more difficult, the Raman technique is more generally applicable since most polymers have Raman active transitions.

F.   Light, x-Ray, and Neutron Scattering

When incident electromagnetic radiation impinges upon matter, an oscillating electrical field is induced. For an electrical field described by the equation

$$E = E_0 e^{i\omega t} \tag{15}$$

where ω is the angular frequency of the wave given by $2\pi v$, where v is the frequency in hertz; the magnitude of the scattered electric field at a distance r from the scattering object is

$$E_s = (\ddot{m} \sin \psi)/rc^2 \tag{16}$$

where c is the velocity of light in vacuum and ψ is the angle between the dipole $\mathbf{m}$ and the vector $\mathbf{r}$ extending from the dipole to the observer.

For the case of Rayleigh scattering in which the frequency of the light is small as compared with the natural frequency of the electrons, Eqs. (15) and (16) lead to

$$E_s = (\alpha E_0 \, \omega^2/rc^2) \sin \psi \, \exp[i(\omega t - \phi)] \tag{17}$$

The polarizability of the scatterer is α, and ϕ is a phase factor (discussed later) accounting for the phase lag (experienced by the wave as a consequence of the distance that it travels to the observer).

For x-ray scattering, the frequency of the radiation is higher than the natural frequency of the electrons, in which case one obtains for this case of Thomson scattering

$$E_s = (-q^2 E_0/m_0 \, rc^2) \sin \psi \, \exp[i(\omega t - \phi)] \tag{18}$$

where q is the electronic charge and m_0 its mass. It is noted that for Rayleigh scattering E_s depends upon frequency, while for Thomson scattering it is independent. In both cases, the scattering may be represented in the form

$$(E_s)_j = K_j \exp[i(\omega t - \phi)] \tag{19}$$

where the subscript j refers to the scattering from the jth object.

For a collection of scattering objects the total scattered amplitude must be obtained by summing the amplitudes of the contributors and is given by

$$E_s = e^{i\omega t} \sum_j K_j e^{i\phi j} = e^{i\omega t} F \tag{20}$$

F is referred to as the form factor and is a property of the structure of the scattering system.

The intensity of the scattered wave is given by

$$I_s = (c/4\pi) E_s E_s^* = (c/4\pi) F F^* \tag{21}$$

(where the asterisk represents the complex conjugate).

The scattering power may be conveniently expressed in terms of the Rayleigh ratio defined by

$$\mathscr{R} = I_s r^2/I_0 V_s \tag{22}$$

where I_0 is the incident intensity and V_s is the volume of the scattering system.

The problem of describing the scattering from a system is one of calculating the form factor in terms of its geometry. There have been two approaches to this problem. The first is the model approach in which the summation of Eq. (20) is specifically evaluated based upon a model of the scattering object. This approach is of greatest value for a system consisting of a dilute, randomly dispersed arrangement of discrete scattering particles of definite shape. Interparticle interference effects can be included if the distribution of interparticle distance can be specified.

For a more ill-defined scattering system in which the scattering regions are diffuse and may even be fluctuations from the uniform properties of a medium, a statistical approach is preferable. In such a case, the system is defined in terms of the mean-squared fluctuation in scattering power and by means of correlation functions, which describe the probability that fluctuation occurring in volume elements separated by a given distance r will be correlated. These fluctuations for polymer blends are primarily fluctuations in composition. However, when components may exhibit local orientation, fluctuations in the magnitude and principal axis direction of anisotropy may occur. The relative contributions of composition and anisotropy fluctuations may be determined from an analysis of the polarization of the scattered radiation. Since anisotropy of outer electrons of atoms does occur by valence interactions, such anisotropy scattering is seen for visible light but not with x rays, which primarily involve inner electrons.

Neutron scattering involves interactions of a neutron beam with atomic nuclei. The neutrons are associated with a wavelength described by the deBroglie equation

$$\lambda = h_p/M_n v \tag{23}$$

where M_n is the mass of the neutron. Hence, the wavelength depends upon the neutron velocity v, which is affected by the temperature of the neutron source and is usually selected with a suitable rotating channel monochromter. A principal advantage of neutron scattering is that the scattering power may be changed by isotopic substitution, usually the replacement of a hydrogen atom by a deuterium atom, which has negligible effects on the chemical structure. Thus, the scattering is selectively dependent upon the location of the labelled groups.

Elastic scattering phenomena lead to scattered radiation of the same frequency as the incident radiation. However, because of the motion of the scattering atoms frequency shifts may occur because of the Doppler effect. This effect may be seen when using single-frequency laser light and leads to broadening of the frequency spectrum of the scattered light in a manner that characterizes the spectrum of motion of the scattering atoms or molecules. When such motions are quantized, as in a solid, discrete frequency shifts occur, leading to Brillouin spectra. The frequency shifts of the Brillouin lines may be regarded as the result of the interaction between a photon and a phonon and are related to the elastic constants of the solid.

III. THE TRANSPARENCY AND SCATTERING OF BLENDS

A. The Relationship between Transparency and Light Scattering

A light wave impinging upon a solid may be transmitted, absorbed, or scattered. These phenomena may be described by

$$I_t = I_0 \exp[-(\varepsilon + \tau)l] \tag{24}$$

where I_t and I_0 are the transmitted and incident intensities of a sample of thickness l, ε is the absorption coefficient (previously discussed, which represents the conversion of the energy of the incident radiation into molecular or electronic motion), τ is the turbidity representing the loss of incident resulting from scattering. This is related to the Rayleigh ratio for scattering at angles θ and μ (Fig. 2) by the equation

$$\tau = \int_{\theta=0}^{\pi} \int_{\mu=0}^{2\pi} \mathcal{R}(\theta,\mu) \sin d\mu \, d\theta \tag{25}$$

In real situations, the instrument (or eye) measuring the transmission has finite resolving power and measures not only the transmitted light but also the light scattered at very small angles. Thus, Eq. (25) gives for the effective turbidity (for cylindrically symmetrical scattering)

$$\tau_{\text{eff}} = 2\pi \int_{\theta_m}^{\pi} \mathcal{R}(\theta) \sin \theta \, d\theta \tag{26}$$

where θ_m represents the maximum angular deviation from the incident beam for the light seen by the transmission measuring device. Thus, attention must be given to the optics of the spectrophotometer used for turbidity

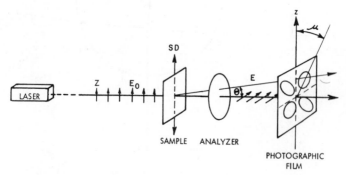

Fig. 2   The scattering angles θ and μ. (From Samuels [9b]. Reprinted with permission from Hercules, Incorporated.)

measurements and the value of θ_m must be specified for a quoted value of turbidity to be meaningful.

Thus, the description of the clarity of a blend is related to the definition of its scattering power. As discussed previously, this problem may be treated in two ways. If the blend may be considered as a dilute dispersion of particles or molecules of one type in a matrix of a second type, then the model approach is best. However, at higher concentrations or where the dispersed regions cannot be described in terms of an object of definite shape, the statistical approach is preferred.

B. The Model Approach for Rayleigh–Gans–Debye Scattering

If a particle of a given refractive index is dispersed in a medium of different refractive index, the light beam will be refracted in crossing the interface. For large particles of definite shape, this effect may be calculated using geometric optics (Fig. 3). However, when the particle has dimensions comparable with the wavelength of light, one must approach the problem by solving the electromagnetic equations subject to the boundary conditions of the particle shape. This has been done in Mie theory for particles of simple shape, as discussed later (Section III.C) [10, 11]. However, if the particle is sufficiently small and the refractive index difference between it and its surroundings is sufficiently small, the effect may be neglected. This approximation is referred

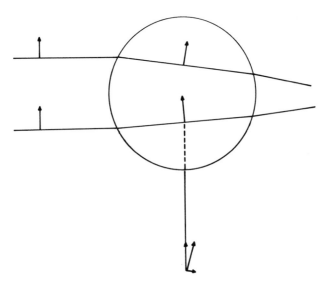

Fig. 3 The refraction of a light beam in passing through a spherical particle. (From Zimm *et al.* [9c].)

to as Rayleigh–Gans–Debye (RGD) scattering. Deviations from this approximation are dependent upon the refractive index ratio $m_r = n_1/n_0$ (where n_1 and n_0 are the refractive index of the particle and the medium, and for spheres of a radius a, a reduced size $\alpha_r = 2\pi a/\lambda$. Graphs of percentage error resulting from this approximation have been published [10–12]. For example, the turbidity will be in error by 5% at $m_r = 1.30$ for $a_r/\lambda = 0.1$ [11].

Using the RGD approximation for a collection of scattering elements, Eqs. (17) and (20) lead to

$$E_s = -(E_0\, \omega^2/rc^2) \sin \psi \, \exp(i\omega t) \sum_j \alpha_j \exp(-{}^i\phi_j) \qquad (27)$$

where α_j is the polarizability of the jth particle and ϕ_j is the phase angle for scattering from this particle. The term ϕ_j may be related to the location of the particle by

$$\phi_j = (2\pi/\lambda)(\mathbf{r}_j \cdot \mathbf{s}) + \phi_0 \qquad (28)$$

where $\mathbf{r}_j$ is a vector from an arbitrary origin to the jth particle and $\mathbf{s}$ is the scattering vector defined by

$$\mathbf{s} = \mathbf{s}_0 - \mathbf{s}_1 \qquad (29)$$

where $\mathbf{s}_0$ and $\mathbf{s}_1$ are unit vectors along the incident and the scattered rays. The term ϕ_0 is a fixed phase shift dependent upon the location of the origin.

For a continuous distribution of scattering volume elements with a center of symmetry, Eqs. (21), (22), (27), and (28) lead to

$$\mathscr{R} = (\omega^4/c^4 V_s) \sin^2 \psi \left[\int \alpha(\mathbf{r}) \exp\left[(2\pi i/\lambda)(\mathbf{r} \cdot \mathbf{s})\right] d\mathbf{r} \right]^2 \qquad (30)$$

The $\alpha(\mathbf{r})$ is the polarizability of a volume element at position $\mathbf{r}$. The integration is carried out over the volume of the scattering system. For spherically symmetrical scatters, integration of Eq. (30) over the angle gives

$$\mathscr{R} = (16\pi^2/V_s)(\omega^4/c^4) \sin^2 \psi \left[\int_0^\infty \alpha(r) \left[\sin (hr)/hr\right] r^2\, dr \right]^2 \qquad (31)$$

where $h = (4\pi/\lambda) \sin (\theta/2)$.

For isotropic spheres of polarizability α_s and radius R immersed in a medium of polarizability α_0, this gives

$$R = N_s(\omega^4/c^4)(\sin^2 \psi)\, B^2(\alpha_s - \alpha_0)^2 \left[(3/U^3)(\sin U - U \cos U\right]^2 \qquad (32)$$

and $B = \frac{4}{3}\pi R_s^{\,3}$ is the volume of the sphere, and $U = hR$; N_s is the number of spheres per unit volume. A plot of the variation of relative scattered intensity with U is shown in Fig. 4. The maxima and minima are associated with various orders of interference.

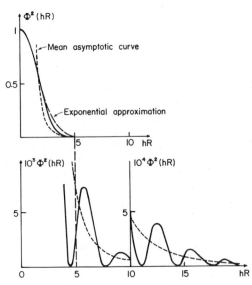

Fig. 4   The variation of relative scattered intensity with $U = hR$ for an isotropic sphere. The curve is drawn with different scale for the various range of U. (From Guinier and Fournet [13], Fig. 6.)

Various features of this equation may be seen which are common to many scattering phenomena:

1. The scattered intensity varies as ω^4. Thus, higher frequency radiation is scattered much more than lower.

2. The scattering varies as the square of the volume of the sphere. Larger spheres are much more efficient scatterers.

3. The scattering depends on the square of the polarizability difference (or refractive index difference) between the sphere and its surroundings. The scattering vanishes when refractive indices are matched.

4. The angular dependence of scattering depends on the variable U, which depends on the product of (R_s/λ) with $\sin(\theta/2)$. Thus, the scattering from larger spheres decreases more rapidly with scattering angle than for smaller.

For a collection of randomly located spheres, the scattering from different spheres will not be related in phase so that intensities are additive and

$$\mathscr{R}(\theta) = \int N(R)\,\mathscr{R}_1(R,\theta)\,dR \tag{33}$$

where $N(R)\,dR$ is the number of spheres with radii between R and $R + dR$ and $\mathscr{R}_1(R,\theta)$ is the Rayleigh ratio for scattering from single spheres of radius

R. Since $\mathscr{R}_1(R)$ depends upon B^2 (or $R_s{}^6$), the scattering is heavily weighted in terms of the larger spheres in the distribution. It is possible to introduce an empirical distribution function for $N(R)$ and fit its parameters to the experimental data, or, if the distribution is sufficiently narrow, it is possible to invert the $\mathscr{R}(\theta)$ data to obtain $N(R)$. An effect of a distribution of sizes is to smear out the scattering maxima and minima and lead to a monotonic decrease in scattered intensity with increasing scattering angle. In fact, the maxima and minima are usually only experimentally observed when the distribution in sizes is very narrow. If the spheres are arranged in space in some definite manner (as is the case, for example, with the spherical domains of block copolymers), then the scattering is coherent and amplitudes must be added. This leads to the result that

$$E_s = \sum E(\mathbf{r}_j)\exp[ik(\mathbf{r}_j \cdot \mathbf{s})] \tag{34}$$

where $E(\mathbf{r}_j)$ is the scattered amplitude from a single sphere with its center at position $\mathbf{r}_j$ and $k = 2\pi/\lambda$. For regularly arranged spheres, this leads to interference maxima, which evolves to diffraction for a lattice arrangement. The summation may be replaced by an integral over the distribution function of sphere centers. This has been evaluated, for example, for the case of a hard-sphere fluid and leads to the development of an interference maximum with increasing volume fraction of spheres [13] as shown in Fig. 5. While for a

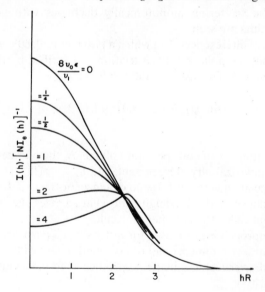

Fig. 5 The variation of scattered intensity with angle for collections of noninteracting hard spheres of various volume fractions. v_0/v_1 is the volume fraction of spheres and ε is a constant close to unity. (From Guinier and Fournet [13], Fig. 12.)

single sphere the intensity is a maximum at zero angle, the intensity at this angle decreases with increasing concentration of spheres, leading to the building up of a maximum at an angle corresponding to the interference from pairs of spheres at their distance of closest approach. It has been shown that, if growing spheres do not exclude volume from each other but can merge together, this interference maximum is not seen [14].

These equations for sphere scattering are not only restricted to scattering by light but also apply to x-ray, electron, and neutron scattering, with different constant terms. For example, for x-ray scattering, the intensity depends upon the square of the electron density rather than the polarizability difference.

Equation (31) is generally applicable to the case in which the polarizability or the scattering power varies with radius. This may occur, for example, if a boundary layer occurs at the surface of a phase leading to a gradient of polarizability. A particular application of interest is that of a gaussian polarizability variation of the form

$$\alpha(r) = \alpha_0 \exp(-r^2/a^2) \tag{35}$$

which leads to the scattering variation

$$\mathscr{R} = (\omega^4/c^4)(\sin^2\psi)\alpha_0\,\pi^3 a^6 \exp(-ksa^2/2) \tag{36}$$

In this case the scattering monotonically decreases with angle and no maxima or minima are seen.

Another case of interest occurs in which a polymer molecule is molecularly dispersed so that the polarizability variation is described by the molecular distribution function. This leads to the result

$$\mathscr{R} = N_c(\omega^4/c^4)(\sin^2\psi)[N_s(\alpha_s - \alpha_0)]^2 \{(2/U_c^2)[\exp(-U_c) - 1 + U_c]\} \tag{37}$$

where N_c is the number of coils per unit volume, N_s the number of segments per coil, α_s the polarizability of a segment, and $U_c = 4\pi(R_c/\lambda)\sin(\theta/2)$, where R_c is the root-mean-squared end-to-end distance for the coil. In this case the angular dependence of the Rayleigh ratio characterizes the mean-squared end-to-end distance for the polymer molecule.

The RGD approach may be used for other shaped particles than spheres. For example, for a two-dimensional rod of length L and width W oriented at an angle Ω to the vertical in a plane perpendicular to the incident beam, the scattered intensity is

$$\mathscr{R} = N_R(\omega^4/c^4)(\sin^2\psi)(\alpha_R - \alpha)^2 L^2 W^2 \left[\frac{\sin(aL/2)}{(aL/2)}\right]^2 \left[\frac{\sin(bW/2)}{(bW/2)}\right]^2 \tag{38}$$

where N_R is the number of rods per unit area, α_R and α_0 are the polarizabilities per unit area of the rods and the surroundings, and

$$a = (2\pi L/\lambda)\sin\theta\,\sin(\Omega + \mu) \tag{39}$$

$$b = (2\pi W/\lambda)\sin\theta\,\cos(\Omega + \mu) \tag{40}$$

This result differs from that of the sphere in that, while for the sphere the scattered intensity is cylindrically symmetrical about the incident beam, the scattering from an oriented rod depends upon the azimuthal angle μ. The intensity falls off more rapidly with θ in the direction of the length of the rod, giving to a scattering pattern extended in a direction perpendicular to the direction of extension of the rod. Of course, for randomly oriented rods, one must average the intensity over Ω leading to a pattern which is independent of μ.

In this treatment of RGD scattering from isotropic particles, the direction of polarization of the scattered light is the same as that of the incident light, and so there is no depolarized scattering. This is not so for anisotropic particles. The scattering from anisotropic spheres [15] and anisotropic rods [16] has been described by Stein and Rhodes. This leads to polarization-dependent scattering, which depends upon μ even for a sphere.

When a polymer containing spherical domains is stretched, the spheres deform and become ellipsoidal. The scattering from such an isotropic ellipsoid with its principal axis in the vertical direction can be described by Eq. (32), provided that U is replaced by U' defined by [17].

$$U' = U[1 + (v^2 - 1)\cos^2(\theta/2)\cos^2\mu]^{1/2} \tag{41}$$

where v is the axial ratio. An affine deformation with constant volume is assumed. This equation predicts a μ-dependent scattering pattern with a principal extension perpendicular to the stretching direction.

An application of these principles to the structure and deformation of rubber toughened plastics containing polystyrene, polyisoprene, and a block copolymer of the two has been given by Inoue et al. [18]. Circular patterns are seen for the unstretched polymer, which become ellipsoidal upon stretching.

The case of deformation of polymers containing polystyrene domains in a rubbery matrix has also been considered by Inoue et al. [19]. In this case, it is not thought that the spherical polystyrene inclusions deform upon stretching the rubber. However, it is postulated that an ellipsoidal density-depleted region forms about the polystyrene particles, giving rise to enhanced scattering.

The model approach for the study of the scattering mixtures of polystyrene, poly(methyl methacrylate), and a block copolymer of the two has been

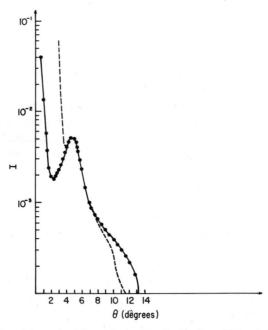

Fig. 6 Experimental angular light-scattering distribution for a 45% copolymer sample and a theoretical intensity distribution for polydisperse spheres (a gaussian size distribution centered on $R_0 = 3.75$ μm, $a = 0.8$): (–––) theoretical curve; (——) experimental curve. (From Duplessix et al. [20], Fig. 3.)

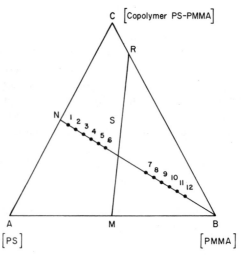

Fig. 7 Ternary diagram representing the composition of a blend of polystyrene block copolymer with poly(methyl methacrylate). (From Duplessix et al. [20], Fig. 1.)

described by Duplessix *et al.* [20]. A typical variation of scattered intensity with angle is shown in Fig. 6. The data has been interpreted in terms of RGD scattering from an assembly of spheres with sizes described by a gaussian distribution. The position of the scattering maximum was shown to vary in a regular way with composition as shown in Figs. 7 and 8.

In the above applications one is dealing with blends of polymers having appreciably different refractive indices. Consequently, the application of RGD theory may not be a good approximation, and results should be interpreted with caution.

C. The Model Approach Using Mie Theory

The results of scattering calculations from spheres in the Mie region cannot be obtained in closed form and consist of infinite sums of terms involving spherical Bessel functions and associated Legandre polynomials. Their properties are well reviewed elsewhere [10, 11] and will not be repeated here. We are mainly concerned with their consequences relevant to

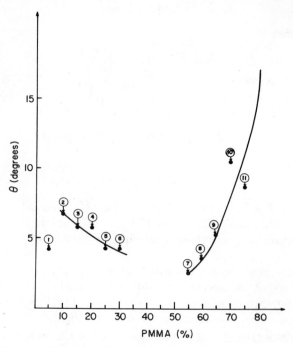

Fig. 8   Variation of θ_{max} with composition along BN on the ternary diagram. The abscissa is the weight content of PMMA homopolymer. (From Duplessix *et al.* [20], Fig. 4.)

the scattering by polymer blends. General discussion of this subject here have been published by Conaghan and Rosen [21, 22]. They consider single scattering from spheres of radius R, described in terms of the scattering coefficients K, that is, related to the turbidity by

$$\tau = N\pi R^2 K \tag{42}$$

where N is the number of particles per unit volume. The specific turbidity is defined as

$$\tau/\phi_1 = 3K/2d_p \tag{43}$$

where ϕ_1 is the volume fraction of particles and d_p the particle diameter. Generally, K, which is a function of the refractive index ratio m_r and the reduced size α_r, is obtained as a solution of Mie's equations. Kerker presents useful three-dimensional plots of K as a function of m_r and α_r. While values of K have not been extensively tabulated in the region close to $m_r = 1$, approximate expressions for K may be readily obtained for this region.

In the Rayleigh–Gans–Debye region where $d_p < \lambda/10$, one may write the approximate equation

$$K = \tfrac{8}{3}\pi\alpha_r^4 \left|(m_r^2 - 1)/(m_r^2 + 2)\right|^4 \tag{44}$$

which becomes

$$K = (32/27)\left|m_r - 1\right|^2 \alpha_r^4 \tag{45}$$

when m_r is close to one.

For values of m_r in the range of 0.9–1.1, Eq. (44) is a good approximation [10] obtained the approximate equation for m_r close to one

$$K = 2 - (4/\rho)\sin\rho + (4/\rho^2)(1 - \cos\rho) \tag{46}$$

where

$$\rho = 2\alpha_r \left|m_r - 1\right| \tag{47}$$

For values of m_r in the range of 0.9–1.1, Eq. (44) is a good approximation for $\alpha_r < 0.4$, while Eq. (46) may be used for $\alpha_r > 5.0$. For intermediate values of α_r the correction factors to these equations published by Moore [23] may be employed.

Using these procedures Conaghan and Rosen [21] have published calculated curves of the specific turbidity versus particle diameter for $\lambda = 0.5893$ μm for various values of m_r, as shown in Fig. 9, while a cross plot as a function of m_r for various α_r is given in Fig. 10. The approximate validity of the Rayleigh expression is indicated in Fig. 9. It is noted that K passes through a maximum when the particle diameter is of the order of the wavelength and goes to zero as m_r approaches one. These authors

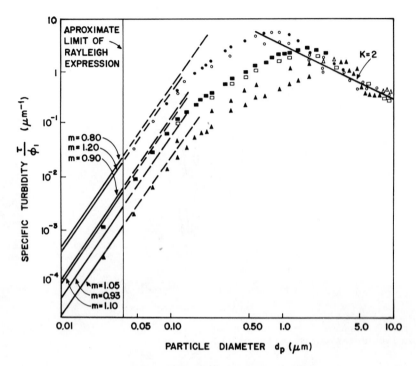

Fig. 9 Variation of specific turbidity with particle diameter for various values of m_r ($\lambda = 0.5093$ μm, $n_0 = 1.60$): ($\triangle$) 0.93; ($\square$) 0.90; ($\bigcirc$) 0.80; ($\blacktriangle$) 1.05; ($\blacksquare$) 1.10; ($\bullet$) 1.20. (From Conaghan and Rosen [21], Fig. 2.)

present a useful table of the maximum particle diameter permissible for a given level of transparency for various values of ϕ_1 and m_r, demonstrating that reasonable transparency requires a quite close match of refractive index. They also discuss the possibility that m_r may be temperature dependent due to the different coefficients of refractive index with temperature for the two phases.

A case of interest is that of coated particles, often encountered in practice when a bonding agent is used to increase the adhesion between the particle and the matrix. Such situations may be considered in terms of the idealized cases treated by Kerker [11] of Mie scattering from spheres or cylinders coated with a shell having a different refractive index from the particle or the matrix. Many papers have been published dealing with the determination of particle size and its distribution by comparing experimental measurements of the angular dependence of scattered intensity with Mie theory calculations. For example, Robillard and Patitsas [24] report studies on a Dow latex, and Levit and Rowell [25] report time dependent studies on sulfur sols.

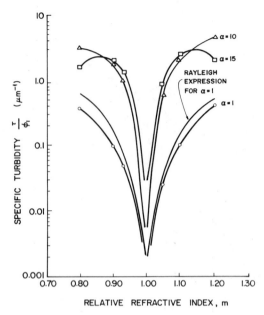

Fig. 10   Variation of specific turbidity with m_r for various values of α_r ($\lambda = 0.5093$ μm, $n_0 = 1.60$). (From Conaghan and Rosen [21], Fig. 3.)

Usually a form for the particle size distribution is assumed in terms of parameters that are evaluated so as to minimize deviations between experimental and calculated intensities. Such procedures are applicable to the analysis of scattering of dilute dispersions of spheres of one polymer in another.

It should be recognized that the above theories are single-scattering theories. That is, it is assumed that the incident ray is only scattered once in reaching the observer. This is a good approximation for films of low turbidity and thickness. However, for thicker and more highly scattering samples, the effects of multiple scattering must be considered, theories for which have been described [26]. The effect of multiple scattering is to smear out the intensity distribution, leading to decreased dependence of intensity on θ and μ. While it is desirable to minimize this effect by studying samples that are as thin as possible, it cannot always be eliminated so that theoretical correction for the effect is desirable.

Scattering due to surface roughness may also be appreciable. This may be reduced for measurement by placing the sample between flat glass plates using a fluid between the sample and the glass of intermediate refractive index.

D. The Statistical Description of Scattering

A perfectly homogeneous medium does not scatter because destructive interference leads to cancellation of the wave scattered from a volume element by that from another. Real media scatter because of fluctuations from ideality expressed in terms of η_j for the jth element defined by

$$\alpha_j = \alpha + \eta_j \tag{48}$$

(For x-ray scattering, this is defined as $\rho_j = \bar{\rho} + \eta_j$.) If Eq. (48) is substituted into Eq. (27) and combined with Eqs. (21) and (22), one obtains for the Rayleigh ratio

$$\mathcal{R} = \frac{(\omega^4/c^4)\sin^2\psi}{V_s}\left\{\sum_j(\bar{\alpha}+\eta_j)\exp(i\phi_j)\sum_j(\bar{\alpha}+\eta_l)\exp(-i\phi_l)\right\}$$

$$= \frac{(\omega^4/c^4)\sin^2\psi}{V_s}$$

$$\times\left\{\alpha^2\sum_j\sum_l\exp(i\phi_{jl})+\bar{\alpha}\sum_j\sum_l(\eta_j+\eta_l)\exp(i\phi_{jl})+\sum_j\sum_l\eta_j\eta_l\exp(i\phi_{jl})\right\}$$

$$\tag{49}$$

where $\phi_{jl} = \phi_j - \phi_l$. The first term represents the scattering from a homogeneous medium and is zero. The second is also zero because η_j or η_l are not correlated with the phase difference ϕ_{jl}. Thus, only the last term remains.

On the average, $\eta_j\eta_l$ depends only upon the vector separation of the volume elements $\mathbf{r}_{jl}$. Thus, it is possible to define a correlation function $\gamma(\mathbf{r}_{jl})$ by

$$\langle\eta_j\eta_l\rangle_{\mathbf{r}_{jl}} = \gamma(\mathbf{r}_{jl})\overline{\eta^2} \tag{50}$$

where $\langle\ \rangle_{\mathbf{r}}$ designates an average over all pairs of volume elements separated by distance $\mathbf{r}$. Obviously $\gamma(\mathbf{r}_{jl})$ must be one when $\mathbf{r}_{jl} = 0$ since then $\langle\eta_j\eta_l\rangle_0 = \overline{\eta^2}$ by definition. When volume elements are very far apart, their fluctuations will not be correlated, and $\gamma(\infty) = 0$. Thus, $\gamma(\mathbf{r}_{jl})$ will be a function which decreases from one to zero in a manner dependent upon structure of the system. Substitution of Eq. (50) into (49) and introducing the specific expression for ϕ_{jl}

$$\mathcal{R} = \frac{(\omega^4/c^4)\sin^2\psi}{V_s}\overline{\eta^2}\sum_j\sum_l\gamma(\mathbf{r}_{jl})\exp[ik(\mathbf{r}_{jl}\cdot\mathbf{s})] \tag{51}$$

Upon replacing the sums by integrals one obtains

$$\mathcal{R} = \frac{(\omega^4/c^4)\sin^2\psi}{V_s}\overline{\eta^2}\int\int\gamma(r_{jl})\,\exp[ik(\mathbf{r}_{jl}\cdot\mathbf{s})]\,dr_j\,dt_l$$

$$= \frac{(\omega^4/c^4)\sin^2\psi}{V_s}\overline{\eta^2}\int\gamma(r_{jl})\exp[ik(\mathbf{r}_{jl}\cdot\mathbf{s})]\,dr_{jl}\int dr_j$$

$$= (\omega^4/c^4)(\sin^2\psi)\overline{\eta^2}\int\gamma(r_{jl})\,\exp[ik(\mathbf{r}_{jl}\cdot\mathbf{s})]\,dr_{jl} \qquad (52)$$

The integral over dr_j and dr_l was replaced by one over dr_j and dr_{jl} in the second step. Since the integral depends only upon r_{jl}, the integral over dr_j gives the volume of the scattering system V_s.

For spherically symmetrical systems, integration over the angular coordinates of r_{jl} leads to

$$\mathcal{R} = 4\pi(\omega^4/c^4)(\sin^2\psi)\overline{\eta^2}\int_0^\infty\gamma(r)[\sin(hr)/hr]\,r^2\,dr \qquad (53)$$

where r is the magnitude of r_{jl}, and the correlation function is independent of angle in this case. This is the famous equation of Debye and Bueche [27] that has proved so useful for describing the scattering from heterogeneous media. The scattering is defined in terms of two parameters, $\overline{\eta^2}$, which describes the mean squared deviation of the scattering power from the average, and $\gamma(r)$, which describes the spatial extent of the fluctuations. The parameter $\gamma(r)$ may be determined from Fourier inversion of the scattered intensity by

$$\gamma(r) = (A/r)\int_0^\infty\mathcal{R}(h)\sin(hr)\,dh \qquad (54)$$

A may be determined by normalization of $\gamma(0) = 1$. While in principal it is necessary to know $\mathcal{R}(h)$ for all values of h to infinity, this, of course, is not possible in practice leading to a termination error. However, intensity usually decreases quite rapidly with increasing h so that the error is not large. Fourier inversion indicates for many systems a rather smooth decrease in $\gamma(r)$ with r, which can be often fitted by an exponential equation

$$\gamma(r) = \exp(-r/a_c) \qquad (55)$$

where a_c is called a correlation distance. In fact, Debye et al. [28] have shown that such a function is theoretically expected for a randomly arranged two-phase system with sharp boundaries. Upon substituting Eq. (55) into (53), integration gives

$$\mathscr{R} = \frac{8\pi(\omega^4/c^4)(\sin^2\psi)\overline{\eta^2}a_c^3}{(1+h^2a_c^2)^2} \tag{56}$$

Thus, a plot of $\mathscr{R}^{-1/2}$ versus h^2 should give a straight line, the ratio of slope to intercept of which equals a_c^2 (for this purpose it is not necessary to calibrate $\mathscr{R}$ in absolute units). The results of integration for six other empirical correlation functions has been described [29].

The correlation function contains much information about the system. Debye *et al.* [20] show that the ratio of surface area to volume of a two-phase system with sharp phase boundaries is given by

$$S/V = -4\phi_1\phi_2(\partial\gamma(r)/\partial r)_{r=0} \tag{57}$$

where ϕ_1 and ϕ_2 are the volume fractions of the two phases. For a random system with an exponential correlation function, this becomes

$$S/V = 4\phi_1\phi_2/a_c \tag{58}$$

Good agreement has been achieved between specific surface areas of porous inorganic materials determined in this way and those obtained from gas adsorption.

In general, a correlation distance can be defined in terms of the correlation function by

$$a_c = \int_0^\infty \gamma(r)\,dr \tag{59}$$

This definition holds, regardless of whether or not an exponential correlation function is applicable.

Useful parameters defined by Porod and Kratky are l_1 and l_2, the average transversal lengths through the phases, which are illustrated in Fig. 11. These are the average lengths through the phases of random lines drawn through the system. For a random system with sharp boundaries, those are given by [30]

$$\bar{l}_1 = a_c/\phi_2 \tag{60}$$

$$\bar{l}_2 = a_c/\phi_1 \tag{61}$$

An alternative approach to describing the size of a component for a dilute dispersion of a component may be obtained by expanding the $\sin(hr)$ term of Eq. (53) in a power series, giving

$$\int \gamma(r)\,[\sin(hr)/hr]\,r^2\,dr$$

$$= \int \gamma(r)r^2\,dr\left[1 - \frac{h^2}{3!}\frac{\int r^2(r)r^2\,dr}{\int\gamma(r)r^2\,dr} + \cdots\right] \tag{62}$$

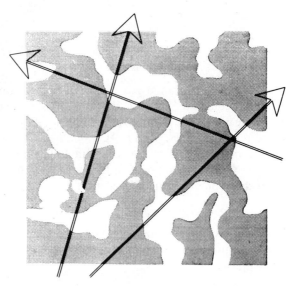

Fig. 11   The average transversal lengths through the phases of a random two-phase system. (From Kratky [30], Fig. 19.)

Now for a dilute dispersion of particles, it may be shown that the mean-squared radius of gyration is defined by

$$\overline{R_g^2} = \sum_{i=1}^{N} \overline{x_i^2}/N \qquad (63)$$

(where x_i is the distance of the ith volume element from the center of gravity of the distribution), and may be related to the correlation function by

$$R_g^2 = \frac{1}{2}\left[\int r^2 \gamma(r) r^2\, dr \bigg/ \int \gamma(r) r^2\, dr\right] \qquad (64)$$

Thus, combining Eqs. (53), (62), and (64) gives

$$\mathcal{R} = C\left[1 - (\overline{R_g^2}/3)h^2 + \cdots\right] \qquad (65)$$

Consequently, the ratio of the initial slope to the intercept of a plot of $\mathcal{R}$ versus h^2 gives the mean squared radius of gyration.

If one neglects higher powered terms, Eq. (65) is equivalent to the expansion of

$$\mathcal{R} = C\exp(-h^2\overline{R_g^2}/3) \qquad (66)$$

Equation (64) is known as the Guinier equation (13) according to which the radius of gyration may be obtained from the slope of a plot of $\ln \mathcal{R}$ versus h^2.

For an exponential correlation function, it is evident that

$$\overline{R_g^2} = 6a_c^2 \tag{67}$$

For dilute dispersions of particles of specified shape, it is possible to obtain specific equations for the correlation function. Thus, for spheres of radius R_s it may be shown that

$$\gamma(r) = 1 - \tfrac{3}{4}(r/R_s) + \tfrac{1}{16}(r/R_s)^3 \tag{68}$$

Substitution of Eq. (68) into (53) leads to Eq. (32) for Rayleigh ratio for isotropic spheres.

An application of this approach to the study of the scattering from a dilute suspension of glass spheres ($n = 1.515$) in a styrene–butadiene copolymer ($n = 1.534$) has been given by Caulfield et al. [31] in which comparisons are given between particle diameters determined from Guinier plots, Debye–Bueche plots, and microscopy. Also, comparisons of surface areas obtained from the correlation function with those measured by BET gas

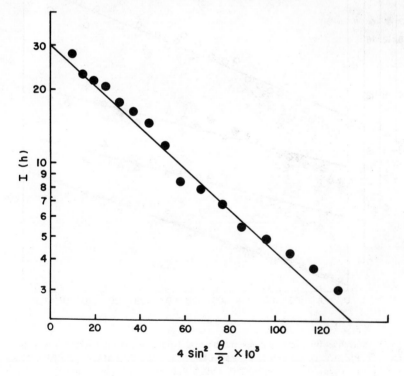

Fig. 12   A Debye–Bueche plot for a polystyrene latex E in glycerine ($d = 1.31\ \mu$m). (From Caulfield et al. [31], Fig. 9.)

adsorption isotherms are given. Differences are discussed in terms of the different averages given by the various methods. Data is also given for polystyrene lattices suspended in glycerine. A typical Debye–Bueche plot for one of these is given in Fig. 12, where the particle diameter of 1.31 μm determined from its slope may be compared with 1.17 μm as determined by electron microscopy.

The correlation function technique was used for the study of blends of poly(methyl methacrylate) (PMMA) and polystyrene (PS) by Yuen and Kinsinger [29]. A typical Debye–Bueche plot obtained by them is shown in Fig. 13. They fitted their data to a combination correlation function of the sort

$$\gamma(r) = f \exp(-r/a_1) + (1-f)\exp(-r^2/a_2{}^2) \qquad (69)$$

For a blend containing 0.005% PS, the data was fitted with $a_1 = 3070$ Å,

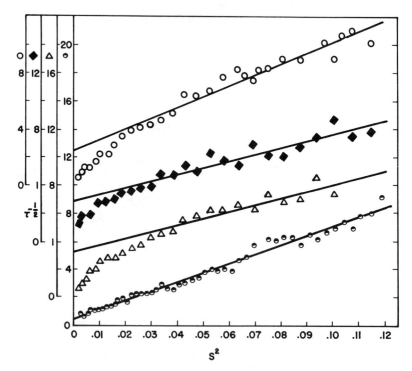

Fig. 13 A Debye–Bueche plot for some blends of poly(methyl methacrylate) with polystyrene: (●) PMMA; (△) 0.0001% PS/PMMA; (♦) 0.001% PS/PMMA; (○) 0.005% PS/ PMMA. (From Yuen and Kinsinger [29], Fig. 2. Reprinted with permission from *Macromolecules*. Copyright by the American Chemical Society.)

$a_2 = 1.69$ μm, and $f = 0.742$. A correlation length of $a_c = 12{,}300$ Å was calculated. They find an increase in the correlation distances with increasing concentration of polystyrene, suggesting an increased incompatibility.

Correlation functions obtained from light scattering were used by Visconti and Marchessault [32] for the study of structure in elastomer reinforced epoxy resins. Typical results of the variation of intensity with angle for various concentrations of a carboxyl-terminated butadiene–acrylonitrile copolymer rubber in an epoxy resin are shown in Fig. 14. A novel feature is the existence of a scattering maximum at higher rubber concentrations. A Debye–Bueche analysis of the data led to a correlation function of a gaussian form characteristic of a correlation distance of 1–2 μm. The maximum was associated with higher-order interference of the sphere-scattering function. Its position was not strongly dependent upon rubber concentration, indicating that it was not a consequence of interparticle interference.

Moritani *et al.* [33] applied correlation function analysis to blends of a styrene–isoprene diblock copolymer with polyisoprene and polystyrene, also

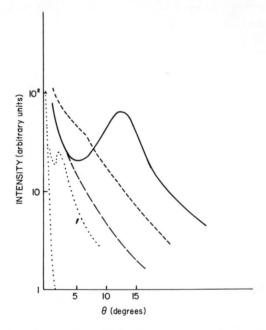

Fig. 14 The variation of scattered intensity with scattering angle for a series of elastomer-reinforced epoxy resins with different rubber concentrations: percentage CTBN. (———) 28%; (— —) 28% extrapolated; (– –) 20%; (...) 14%; (...) 0%. (From Visconti and Marchessault [32], Fig. 3. Reprinted with permission from *Macromolecules*. Copyright by the American Chemical Society.)

using the same type of combination correlation function. For a typical sample of a 50% 70/30 styrene/isoprene block copolymer with an equal weight ratio of the homopolymers it was found that $a_1 = 1870$ A, $a_2 = 1.6$ mμ, and $f = 0.983$. Generally, a_1 is associated with particle sizes and a_2 with interparticle correlations. Specific surface areas were calculated using Eq. (57), giving, for example, 4.42 μm^{-1} for the above blend. A heterogeneity length of $a_c = 4000$ A was calculated as well as a volume of the heterogeneity defined by

$$\bar{v}_c = \int \gamma(r) 4\pi r^2 \, dr \tag{70}$$

of 3.7×10^{11} Å^3. Based on a spherical model, an effective radius of the heterogeneity, R_{scatt} was calculated from $R_{scatt} = 3/(S/V)$ and compared with values determined from electron microscopy, as shown in Fig. 15. A good correlation is seen. These authors use the quantities a_c and $\bar{v}_c$ as measures of the heterogeneity of the dispersion which became greater as the dispersion becomes less uniform.

The correlation function approach has also been used by Blundell *et al.* [34] for examining an interstitially polymerized poly(methyl methacrylate)–polyurethane system, employing both small-angle x-ray scattering (SAXS)

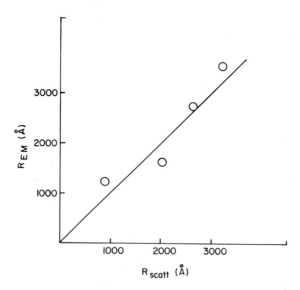

Fig. 15   A comparison of the radius of domains in blends of a 70/30 styrene/isoprene block copolymer with an equal weight ratio mixture of the homopolymers as determined by electron microscopy and light scattering. (From Moritani *et al.* [33], Fig. 8. Reprinted with permission from *Macromolecules.* Copyright by the American Chemical Society.)

and light scattering. Typical Debye–Bueche plots for their x-ray data are given in Fig. 16. The correlation distances were determined from the slopes and the persistance lengths, using Eqs. (60) and (61). Light scattering was determined from the turbidity measured by using a spectrometer, which was related to the integrated Rayleigh ratio using Eq. (26). For a two-phase model, this gives (assuming an exponential correlation function)

$$\tau = (2\pi\phi_1\phi_2/\lambda)[(n_1 - n_2)^2/\overline{n^2}]\,B(y) \tag{71}$$

where

$$B(y) = [(y^2 + 2)/y][(y^2 + 2)/(y^2 + 1) - (2/y^2)\ln(y^2 + 1)] \tag{72}$$

and

$$y = (4\pi/\lambda)\bar{l}_1\phi_2 \tag{73}$$

Complete incompatability is assumed so that ϕ_1 and ϕ_2 correspond to the

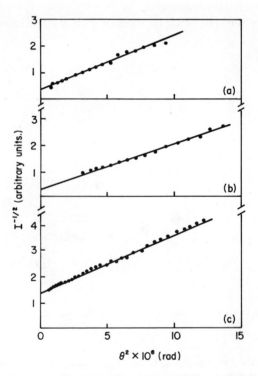

Fig. 16 Typical Debye–Bueche SAXS plots for three different PMMA compositions of interstitial polymers with polyurethane: (a) 80%; (b) 60%; (c) 30% PMMA. (Reproduced from Blundell *et al.* [34], Fig. 1, by permission of the publishers, IPC Business Press Ltd. ©.)

added volume fractions of the two components and n_1 and n_2 are their macroscopic refractive indexes. A typical comparison gives values for the persistance length of the disperse phase, and 85 wt % PMMA composite, of $\bar{l}_1$ of 200 nm from turbidity, 210 nm from SAXS, and 280 nm from electron microscopy. A plot of $\bar{l}_1$ versus wt % PMMA as measured by SAXS and by electron microscopy is given in Fig. 17.

The statistical analysis was applied to the SAXS from blends of poly(vinyl chloride) (PVC) with polycaprolactone (PCL) (see Volume 2, Chapter 22) by Khambatta et al. [35]. These blends are amorphous for compositions with more than about 60 wt % PVC and were analyzed from Debye–Bueche plots from the SAXS data. Values of the correlation distances and transversal lengths through the two phases are shown in Fig. 18. It is evident that the transversal length through the dispersed PCL decreases with PCL content and reaches a value of about 30 A for the blend containing about 10% PCL. Since this is of the order of the dimensions of a single PCL molecule, this observation is taken to indicate that there is molecular dispersion and that the components are compatible.

For such systems, the interpretation of $\overline{\eta^2}$ is straightforward. It may be given by

$$\overline{\eta^2} = \phi_1 \eta_1^2 + \phi_2 \eta_2^2 \tag{74}$$

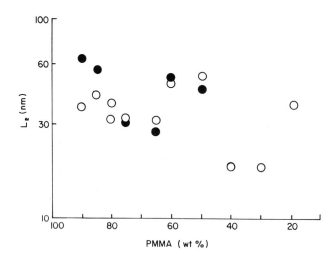

Fig. 17   Variation of the persistence length of the disperse phase, $\bar{l}_2$ with wt % PMMA in an interstitial composite with a polyurethane as measured by (O) SAXS and (●) electron microscopy. (From Blundell et al. [34], Fig. 3, by permission of the publishers, IPC Business Press Ltd. ©.)

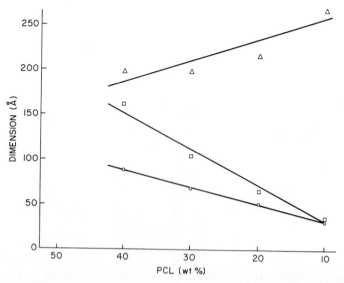

Fig. 18   The variation of correlation distance (l_c) and transversal lengths through the two phases ($\bar{l}_{PVC}$ and $\bar{l}_{PCL}$) with composition for poly(vinyl chloride)–polycaprolactone blends: ($\triangle$) $\bar{l}_{PVC}$; ($\square$) $\bar{l}_{PCL}$; ($\bigcirc$) l_c. (From Khambatta *et al.* [35], Fig. 15.)

where η_1 is the value of η within Phase 1 given by

$$\eta_1 = n_1 - \bar{n} \tag{75}$$

where n_1 is the refractive index (or electron density) of Phase 1 and $\bar{n}$ is the average refractive index of the medium given by

$$\bar{n} = \phi_1 n_1 + \phi_2 n_2 \tag{76}$$

Thus, from Eqs. (73) and (74)

$$\eta_1 = \phi_2(n_1 - n_2) \tag{77}$$

and

$$\eta_2 = \phi_1(n_2 - n_1) \tag{78}$$

so from Eq. (72)

$$\overline{\eta^2} = (n_1 - n_2)^2 \phi_1 \phi_2 \tag{79}$$

Thus, $\overline{\eta^2}$ is zero for the pure components and passes through a maximum when $\phi_1 = \phi_2 = \tfrac{1}{2}$.

The equation must be modified when there is a diffuse boundary between phases. Under these conditions Eq. (79) is modified to give [36]

$$\overline{\eta^2} = (n_1 - n_2)^2(\phi_1\phi_2 - ES/6V) \tag{80}$$

or in terms of the formulation of Khambatta *et al.* [35]

$$\overline{\eta^2} = (n_1 - n_2)^2(\phi_1\phi_2 - \phi_3/6) \tag{81}$$

where E is the thickness of the transition layer, ϕ_3 its volume fraction, and S/V the specific surface area of the phase boundary. Thus, a diffuse transition layer leads to a lowering of η^2. It is evident that for spherical domains having a transition layer that is thin compared with their radius R, $\phi_3 = 3E\phi_2/R$ so that

$$\overline{\eta^2}/\phi^2 = (n_1 - n_2)[(1 - 3E/R) - \phi_2] \tag{82}$$

Thus, for constant E/R, a plot of $\overline{\eta^2}/\phi_2$ versus ϕ_2 should give a straight line with an intercept over slope ratio of $[1 - 3E/R]$.

The variation of $\overline{\eta^2}$ with composition as determined by SAXS for PVC–PCL blends is plotted in Fig. 19. The maximum expected for intermediate compositions is evident. It is noted that, while Eq. (70) indicates that $\overline{\eta^2}$ should be zero for the pure components, finite values are found. For the PCL, this

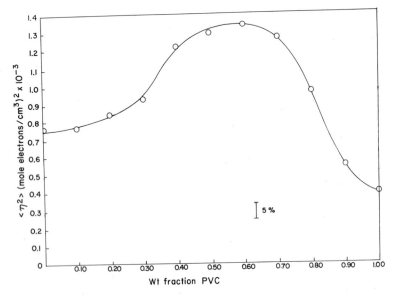

Fig. 19 The variation of $\overline{\eta^2}$ with composition for PVC–PCL blends as determined by SAXS. (From Khambatta *et al.* [35], Fig. 15.)

is primarily a result of the electron density fluctuations arising because of its crystalline structure. For the PVC this is associated with gel or paracrystalline structure within the polymer.

Despite this excess scattering for the pure components the experimentally measured values of η^2 for the amorphous blends are less than that predicted using Eq. (77). For example, for a blend containing 70 wt % PVC, $\overline{\eta^2_{exp}} = 1.28 \times 10^{-3}$ (moles electrons/cm^3)^2, while $\overline{\eta^2_{theor}} = 3.18 \times 10^{-3}$. This suggests that the two-phase model with sharp boundaries is not appropriate. The theoretical and experimental values may be rationalized either by assuming that the phases are not pure components or else that there exists a diffuse boundary between the phases. Using the latter hypothesis, a plot according to Eq. (82) gives a straight line with an intercept–slope characteristic of an interface thickness that is comparable with the radius of the domain itself. This is consistent with the hypothesis that molecules of the blend are dispersed at the molecular level such that considerable molecular interpenetration occurs.

A similar comparison between the experimental and calculated values of η^2 as determined from light scattering has been made by Yuen and Kinsinger [29] for the PS–PMMA system. They find that at low concentrations of PS where partial compatability occurs, $\overline{\eta^2_{exp}} < \overline{\eta^2_{theor}}$. For example, at 0.005% PS, $(\overline{\eta^2_{exp}})^{1/2} = 7.41 \times 10^{-4}$ but at higher concentrations of PS where the separated components are essentially pure, the values are approximately equal (both 23.4×10^{-4} at 0.05% PS). On the basis of these observations, they conclude that the critical concentration of this blend for compatibility is 0.008 wt % PS.

The effect of diffuse interfaces may be studied by examining the behavior of the correlation function at small r or of the scattered intensity at large θ. Debye *et al.* [28] show that the thickness of the interphase may be given by

$$E = \frac{4}{3} \frac{(\partial \gamma / \partial r)_{r=0}}{(\partial^2 \gamma / \partial r^2)_{r=0}} \tag{83}$$

Vonk has shown in this way that E has values of 0.5–1.0 nm between the amorphous and crystalline regions of typical crystalline polymers (PE, PET, nylon 6) and 1–2 nm between domains of a methyl methacrylate–isoprene triblock copolymer. For a phase-separated polyurethane, Bonart and Müller [37] conclude in this way that the interphase is typically the order of 2 nm between the hard and soft segment domains.

The analysis of domain boundary thickness for highly ordered samples of a styrene–isoprene diblock copolymer was attempted by Hashimoto *et al.* [38]. These samples possessed lamellar structure and were highly oriented so that the lamellar planes were aligned parallel to the machine direction and with the lamellar normals parallel to the film sample surface. The SAXS from such samples exhibited many maxima and were analyzed in terms of a scattering theory based upon the one-dimensional paracrystalline model of Blundell [39]. In addition to providing information about the mean lattice spacing, the paracrystalline disorder, and the volume fractions of the phases, qualitative information was obtained about the grain size of the ordered domains, and the thickness is presented based upon analysis of the tail of the SAXS intensity distribution at large scattering angles. While the technique was applied to block copolymers, the same methods are applicable to the study of polymer blends.

The analysis is made in terms of a one-dimensional, electron density fluctuation along the z axis $\eta(z)$. This may be represented in terms of the convolution product

$$\eta(z) = g(z) * h(z) = \int_{-\infty}^{\infty} g(u)h(z-u)\,du \tag{84}$$

where $g(z)$ is the electron density fluctuation for an ideal two-phase system with sharp boundaries and $h(z)$ is a smoothing function related to the actual electron density distribution in the boundary region.

The structure factor $F(s)$ (proportional to the scattering amplitude) is obtained from the Fourier transform of $\eta(z)$ given by

$$F(s) = \mathscr{F}[\eta(z)] = \int_{-\infty}^{\infty} \eta(z)\exp(2\pi i s z)\,dz \tag{85}$$

It follows that

$$F(s) = \mathscr{F}[g(z)]\,\mathscr{F}[h(z)] \tag{86}$$

and the scattering intensity is given by

$$I_s = I_e |F(s)|^2 = I_g(s)I_h(s) \tag{87}$$

where I_e is the Thomson scattering intensity from an electron and

$$I_g(s) = \left| \mathscr{F}[g(z)] \right|^2 \tag{88}$$

and

$$I_h(s) = |\mathscr{F}[h(z)]|^2 \tag{89}$$

$I_g(s)$ is the scattering intensity from an ideal two-phase system given (for isotropic systems) by Porod's law [40] and by Debye et al. [28] as

$$I_g(s) = G_1 s^{-4} \tag{90}$$

where

$$C_1 = I_e(\rho_1 - \rho_2)^2 (s/2\pi)^3 \tag{91}$$

Alternatively, for an oriented lamellar system, Porod's law is modified to

$$I_g(s) = C_1 s^{-2} \tag{92}$$

For a smoothing function $h(z)$ given by a trapezoidal or a gaussian function, $I_h(s)$ is given respectively as

$$I_h(s) = 1 - (\pi st)^{2/3} + 0(\pi st)^4 \tag{93}$$

or

$$I_h(s) = \exp(-4\pi^2\sigma^2 s^2) = 1 - 4\pi^2\sigma^2 s^2 + 0(\pi\sigma s)^4 \tag{94}$$

where t is the width of the trapezoid and σ is the standard deviation of the gaussian function. Consequently a plot of $I(s)s^2$ for an isotropic system or $I(s)$ for an oriented system versus s^{-2} should yield a linear plot at small s, the intercept of which determines t or σ, which is a measure of the width of the domain boundary. Such a plot for annealed and as-cast films of the block copolymer is shown in Fig. 20, which corresponds to values of t in the range of 1.9–2.2 nm.

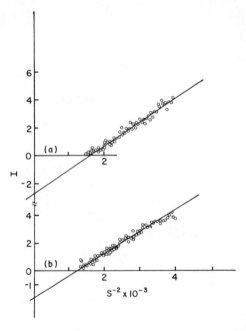

Fig. 20 The variation of the scattered intensity with s^{-2} for the SAXS from an oriented lamellar sample of a styrene–isoprene oriented diblock copolymer: (a) as cast; (b) annealed. (From Hashimoto *et al.* [41], Fig. 7. Reprinted with permission from *Macromolecules*. Copyright by the American Chemical Society.)

E. Scattering from Blends with Crystalline Components

The formation of crystals from one or both components of a blend leads to additional scattering arising from the density and anisotropy differences between the crystalline and amorphous regions. When the crystallizable component is present in large concentration, its morphology often dominates that of the blend. For example, in the PVC–PCL blend containing more than about 50% PCL, the scattering is dominated by the spherulitic morphology of the PCL. The scattering pattern obtained for the sample held between crossed polaroids (H_v pattern) exhibits the typical four-leaf clover pattern characteristic of spherulites, as shown in Fig. 2. These patterns may be described on the basis of the simplified model of the scattering from anisotropic spheres for which the H_v intensity is given by [15, 42]

$$I_{H_v} = K\{B_s(n_r - n_t)(\sin \mu)(\cos \mu)(3/U^3)[4\sin U - U\cos U - 3Si(U)]\}^2$$

(95)

where

$$U = 4\pi(R_s/\lambda)\sin(\theta/2)$$

(96)

B_s is the volume of the spherulite $[\frac{4}{3}\pi R_s^3]$ and R_s is the spherulite radius. The radial and tangential refractive indices of the spherulite, are designated n_r and n_t. The scattering angles θ and μ are defined in Fig. 2. This equation predicts an intensity maximum at $\mu = 45°$ and at an angle θ_{max} given by

$$U_{max} = 4.1 = 4\pi(R_s/\lambda)\sin(\theta_{max}/2)$$

(97)

which may be used to determine R_s from experimental measurements of θ_{max}.

Quantitative analysis of the patterns such as that in Fig. 21 shows that in the range of 0–50% PCL, the blends consist of volume filling spherulites with no major variation of spherulitic size with composition. However, the intensity of scattering decreases with increasing PVC content, primarily because of a decrease in the degree of crystallinity of the spherulites, resulting in a decrease in the $(n_r - n_t)$ term since, to a good approximation

$$(n_r - n_t) = \phi_{c,s}f_{c,s}\Delta_c^0$$

(98)

where $\phi_{c,s}$ in the degree of crystallinity of the spherulite, $f_{c,s}$ in the orientation degree of the crystals with respect to the spherulite radius and Δ_c^0 is the intrinsic birefringence of the crystals.

While light scattering permits the characterization of the structure of the polymer, the shorter wavelength of x rays permits their use for probing the internal arrangement of crystals within the spherulite. The SAXS from PCL exhibits a maximum characteristic of the repeat period of the alternating

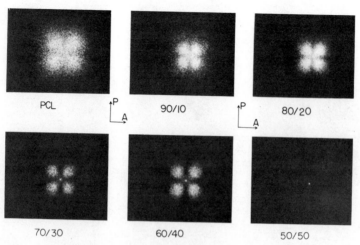

Fig. 21 H_v scattering patterns for PCL–PVC blends of differing composition. (From Khambatta *et al.* [35], Fig. 16.)

layers of crystalline and amorphous PCL. With increasing PVC content this repeat period increases, as shown in Fig. 22, by just the right amount to accommodate the PVC in the amorphous region in between the crystalline PCL lamella. Also shown in this figure is the thickness of these lamella, which increases slightly with increasing PVC content, and the linear crystallinity, the ratio of the lamellar thickness to the repeat period, which decreases substantially.

The crystalline blend may be regarded as a three-phase system consisting of crystalline PCL and possibly two amorphous phases. For such a three-phase system, Eq. (77) may be generalized to give

$$\overline{\eta^2} = (n_1 - n_2)^2 \phi_1 \phi_2 + (n_1 n_3)^2 \phi_1 \phi_3 + (n_2 - n_3)^2 \phi_2 \phi_3 \qquad (99)$$

A comparison of the experimental value of $\overline{\eta^2}$ calculated from the invariant with the value calculated from Eq. (97), assuming incompatibility of the PVC and PCL in the amorphous phase, reveals too large a calculated value. On the other hand, too small a value is found on assuming complete compatibility of the amorphous component so that the amorphous phase would be considered as a single phase with an intermediate electron density between that of PCL and PVC. Consequently, experimental and theoretical values may be rationalized by either assuming two mixed-composition amorphous phases, or else assuming the existence of a suitable diffuse boundary between the amorphous phases.

The blend of partly crystalline poly(vinylidene fluoride) (PVF$_2$) and amorphous poly(methyl methacrylate) (PMMA) has been studied by Nishi

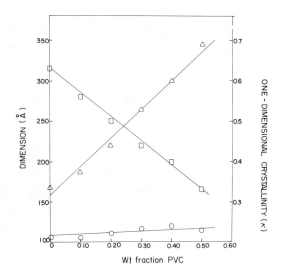

Fig. 22   The variation of the repeat period, the crystal thickness and the linear crystallinity k with composition of a PCL–PVC blend: (△) long period; (○) crystalline site; (□) crystallinity. (From Khambatta *et al.* [35], Fig. 6.)

and Wang [43]. As with the PVC–PCL system, crystallinity is found when the PVF_2 composition is greater than 50 wt %. The crystallites form volume-filling spherulites [44], suggesting that, as with the PVC–PCL blend, the PMMA is included within the spherulites, most likely in the interlamellar regions. The melting point of the crystals is depressed by the PMMA, suggesting compatibility within the amorphous phase.

Similar behavior has been reported by Andrews [45] for the blend of *cis*- and *trans*-1,4-polyisoprene from which the trans component crystallizes. The shift of T_g of the amorphous phase with composition may be explained upon application of the law of mixtures to the amorphous phase, assuming compatibility within this phase.

Wenig *et al.* [46] have studied the Noryl type blends of poly(2,6-dimethyl-1,4-phenylene oxide) (PPO) with isotactic and atactic polystyrene (PS) [46–48]. In this case, the iPS may be thermally crystallized down to 50% concentration leading to spherulites of iPS containing the PPO in inter-lamellar regions. Beyond about 30% PPO, the spherulites are no longer volume filling. The interlamellar spacing increases with increasing PPO content in a manner that is in quantitative agreement with the inclusion of the amorphous PPO between the iPS lamellae. By treatment with 2-butanone vapor, the PPO phase may also be induced to crystallize.

Crystalline blends of poly(ethylene terephthalate) (PET) and poly-(butylene terephthalate) (PBT) have recently been studied [49]. For the

PET- or PBT-rich blends the structure is volume filled with the spherulites of the component present in excess containing an interlamellar amorphous phase of the other component. At intermediate compositions, a non-spherulitic mixed crystalline structure is formed.

F. Scattering Arising from Anisotropy

The scattering from spherulitic polymers referred to in the last section is a particular case in which scattering arises primarily from fluctuations in anisotropy rather than density in that it occurs because of the difference between the radial and tangential refractive indices of the spherulite. This is a particular example of a general phenomena in which scattering may arise from variations in the anisotropy magnitude or optic axis orientations of optically anisotropic region, as well as from variations in the density or average refractive index of the medium. Such fluctuations may occur because of correlated orientation of anisotropic molecules or their aggregates, from fluctuations of form birefringence, or from variations in strain.

When dealing with scattering from anisotropic media, the tensor nature of the polarizability in Eq. (1) must be considered. This leads to a tensor polarizability in Eq. (30). For polarized light scattering at a not too large forward angle, it may be shown that the effective polarizability to be inserted in Eq. (30) is [15]

$$|\alpha(\mathbf{r})| = (\mathbf{m} \cdot \mathbf{O})/|\mathbf{E}| \qquad (100)$$

where $\mathbf{m}$ is the induced dipole moment given by

$$\mathbf{m}(\mathbf{r}) = \delta(\mathbf{r})\{[\mathbf{a}(\mathbf{r}) \cdot \mathbf{E}]\mathbf{a}(\mathbf{r}) + \alpha_2(\mathbf{r})\mathbf{E}\} \qquad (101)$$

where $\delta(\mathbf{r})$ is the anisotropy at $\mathbf{r}$ defined by

$$\delta(\mathbf{r}) = \alpha_1(\mathbf{r}) - \alpha_2(\mathbf{r}) \qquad (102)$$

where $\alpha_1(\mathbf{r})$ and $\alpha_2(\mathbf{r})$ are the principal polarizabilities parallel to and perpendicular to the optic axis in the direction of the unit vector $\mathbf{a}(\mathbf{r})$. Uniaxial symmetry of polarizability is assumed. The term $\mathbf{O}$ is a unit vector in the polarization direction admitted by an analyzer in the scattered ray.

To obtain Eq. (95), it is assumed that the principal axis is along the radius of the spherulite $[\mathbf{a}(\mathbf{r}) = \mathbf{r}/r]$ or perpendicular to it (with random rotation about the radius) [15].

The scattering arising from stresses within a polymer may be obtained by assuming that there is a local anisotropy related to the local stress $\sigma(r)$

$$\delta(r) = C\sigma(r) \qquad (103)$$

where C is the stress optical coefficient. This approach has been used by Goldstein [50] for calculating the scattering arising from the stresses occurring about voids resulting from radiation damage in glasses. A discussion of such scattering arising from voids or inclusions in polymers has been offered by Wilkes and Stein [51]. A specific application and test of these theories has been made by Picot et $al.$ [52] and Ong and Stein [53] for the scattering arising from the stress pattern surrounding glass spheres in a rubber, following the swelling or stretching of the rubber. Such scattering serves as a model for similar scattering that may result from stress surrounding phases in phase-separated systems.

The statistical theory of anisotropy scattering was first described by Goldstein and Michalek [54] and then presented in a more simple form by Stein and Wilson [55]. This leads to an equation analogous to Eq. (53) for H_v and V_v polarization

$$\mathcal{R}_{H_v} = 4\pi\left(\frac{\omega^4}{c^4}\right)(\sin^2\psi)(1/15)\overline{\delta^2}\int_0^\infty f(r)\frac{\sin(hr)}{hr}r^2\,dr \qquad (104)$$

and

$$\mathcal{R}_{V_v} = 4\pi\left(\frac{\omega^4}{c^4}\right)\sin^2\psi\left\{\overline{\eta^2}\int_0^\infty \gamma(r)\frac{\sin(hr)}{hr}r^2\,dr + (4/45)\overline{\delta^2}\int_0^\infty f(r)\frac{\sin(hr)}{hr}r^2\,dr\right\} \qquad (105)$$

$\overline{\delta^2}$ is the mean squared anisotropy and $f(r)$ is a correlation function for orientation defined by

$$f(r) = [3\langle \mathbf{a}_1 \cdot \mathbf{a}_2\rangle_r^2 - 1]/2 \qquad (106)$$

where $\mathbf{a}_1$ and $\mathbf{a}_2$ are the optic axis unit vectors for pairs of volume elements separated by distance r. Like $\gamma(r)$, $f(r)$ is unity at $r=0$ and decreases toward zero as r increases in a manner dependent upon the size and shape of correlated regions. For many systems it may also be represented by an empirical exponential function.

For an isotropic system where $\overline{\delta^2} = 0$, $\mathcal{R}_{H_v} = 0$ and $\mathcal{R}_{V_v}$ reduces to that given by the Debye–Bueche equation, Eq. (53). It is evident that $\mathcal{R}_{H_v}$ is dependent upon anisotropy fluctuations and its angular dependence is characterized by $f(r)$. It may be determined from a Fourier transform of $\mathcal{R}_{H_v}$.

It is seen from Eqs. (104) and (105) that

$$\mathcal{R}_{V_v} - \tfrac{4}{3}\mathcal{R}_{H_v} = 4\pi\left(\frac{\omega^4}{c^4}\right)(\sin^2\psi)\overline{\eta^2}\int_0^\infty \gamma(r)\frac{\sin(hr)}{hr}r^2\,dr \qquad (107)$$

Thus, $\gamma(r)$ may be obtained from a Fourier transform of $(\mathcal{R}_{V_v} - \tfrac{4}{3}\mathcal{R}_{H_v})$.

It should be pointed out that the Stein–Wilson theory predicts scattering that is circularly symmetrical about the incident beam (independent of azimuthal angle). This is a consequence of an assumption implicit in its derivation of "random orientation correlations." That is, it is assumed that the probability of optic axes being parallel is only dependent upon their separation and not upon the angle between **a** and **r**, leading to spherically symmetrical correlated regions. A more general treatment such as that of Goldstein and Michelek [54] or Erhardt *et al.* [56] allows for more general correlations and lead to azimuthally dependent scattering.

This statistical theory can account for turbidity of samples arising from anisotropic regions resulting from (1) correlated assemblies of crystals, (2) statistically distributed strains or, (3) form birefringence arising from correlated regions of anisotropically shaped domains. Scattering from the latter cause was observed by Wilkes and Stein [51] for domains in styrene–butadiene triblock copolymers. The individual polystyrene cylinders or lamella in these polymers have dimensions of the order of a few tens of nanometers in size, too small to scatter light, and are individually isotropic. However, appreciable H_v scattering characteristic of regions of several hundred nanometers is observed. It has been observed by Lewis and Price [57] that these domains occur in grains of the order of several hundred nanometers in size within which the domains are locally parallel, although they are randomly arranged with respect to each other so that the sample is macroscopically isotropic. This parallel ordering of the domains leads to local form birefringence [58] with an optic axis direction corresponding to the direction of local orientation of the domains. This may lead to scattering of the sort described by the Stein–Wilson theory in which the orientation correlation function corresponds to the correlation of domain orientation within the grains [59]. For this reason, such copolymers scatter light even though the domain size is considerably smaller than its wavelength. It is likely that similar scattering may also arise with blends in which there may be local correlated orientation of domains.

IV. OTHER OPTICAL PROPERTIES OF BLENDS

A. Birefringence

The birefringence of a multicomponent system may be expressed as a sum of the contributions from the components or

$$\Delta = \sum_i \phi_i f_i \Delta_i^0 + \Delta_F \tag{108}$$

where ϕ_i is the volume fraction of component i, f_i is its orientation function, and Δ_i^0 is the intrinsic birefringence of the component. The form birefringence Δ_F is negligible if the components are compatible but may be important if there is phase separation. It is dependent upon the modification of the field of the light wave in traversing the phase boundary and depends upon the square of the refractive index difference between the phases, the shape of the phase and its orientation. For example, for a dilute dispersion of volume fraction ϕ_2 of a long, cylindrical, perfectly oriented phase of refractive index n_2 in a medium of refractive index n, Weiner [60] gives for the form birefringence

$$\Delta_F = [\phi_1\phi_2(n_1{}^2 - n_2{}^2)]/\{2n_a[(\phi_1 + 1)n_2{}^2 + \phi_2 n_1{}^2]\} \tag{109}$$

where $n_a = \phi_1 n_1{}^2 + \phi_2 n_2{}^2$. The validity of this equation has recently been demonstrated by Folkes and Keller [58] for an extruded styrene–butadiene block copolymer, which may be represented as a rubber matrix filled with oriented polystyrene cylinders. A consequence of this theory is that the form birefringence for crystalline polymers was first demonstrated by Hermans [61] and then applied by Bettelheim and Stein [62] by swelling the amorphous regions with solvents of refractive index and obtaining a parabolic curve of total birefringence versus solvent refractive index, with a minimum occurring when $\Delta_F = 0$. In this way, it was shown that with the relatively small refractive index difference between the crystalline and amorphous regions of crystalline polymers, Δ_F is relatively small (5–10%). However, for the larger refractive index differences that may occur between components of a block copolymer or blend, this contribution may be appreciable.

For a rubber homopolymer, the theory of Kuhn and Grün [63] and Treloar [64] demonstrate that the birefringence is proportional to the stress (Eq. 103), where the stress–optical coefficient is given by

$$C = (2/45)[(n^2 + 2)^2/nkT](b_1 - b_2)_s \tag{110}$$

where $(b_1 - b_2)_s$ is the difference between the principal polarizabilities of the statistical segment. More recent theories such as that of Flory [65] relate this to bond polarizabilities, molecular geometry, and bond rotation potentials. While the equation was originally derived for cross-linked networks, Read has shown [66] by extending normal mode theory that the theory may also be applied to uncross-linked amorphous rubbers.

Use of this result in Eq. (108) for a blend gives

$$\Delta = \sum_i \phi_i C_i \sigma_i + \Delta_F \tag{111}$$

where C_i and σ_i are the stress–optical coefficients and stress is on component i. One may represent σ_i by

$$\sigma_i = \kappa_i \sigma \tag{112}$$

where κ_i is a stress-concentrating factor for the ith component, which depends upon homogeneity and geometry so that [67-69]

$$\Delta = \left[\sum \phi_i C_i \kappa_i\right]\sigma + \Delta_F \tag{113}$$

Neglecting Δ_F, this equation may be written in the form

$$\Delta = \bar{C}\sigma \tag{114}$$

where $\bar{C}$ is the average strain–optical coefficient of the blend defined by

$$\bar{C} = \sum \phi_i C_i \kappa_i \tag{115}$$

For the special case of a two-component system, Livingston and Brown [68] define the parameter r as the ratio of the stresses on the two phases as

$$r = \sigma_1/\sigma_2 \tag{116}$$

If one assumes that stress on the phases added in proportion to their volume fractions as

$$\sigma = \phi_1 \sigma_1 + \phi_2 \sigma_2 \tag{117}$$

then combination of Eqs. (116) and (117) with (115) for a two-phase system gives

$$r = (\bar{C} - C_r)/(C_1 - \bar{C})(\phi_2/\phi_1) \tag{118}$$

Thus, a measurement of $\bar{C}$ for different compositions serves to characterize r.

This equation was applied to an incompatible blend of two styrene–butadiene rubbers (SBR) containing 16 and 60 wt % styrene. Plots of the birefringence versus the stress for various compositions are given in Fig. 23. The values of $\bar{C}$ are obtained from slopes of these lines, and are plotted as a function of composition in Fig. 24. The resulting curve may be fitted to Eq. (118) with a constant value of $r = 0.621$. This demonstrates that for an incompatible blend, the two components bear unequal fractions of the load.

On the other hand, the variation of the stress–optical coefficient with composition for a compatible blend of SBRs containing 16 and 23.5% styrene is shown in Fig. 25. In this case, the curves are fitted with $r = 1.0$, indicating that in the compatible blend the components bear equal fractions of the stress. In fact in this case

$$\sigma = \sigma_1 = \sigma_2 \tag{119}$$

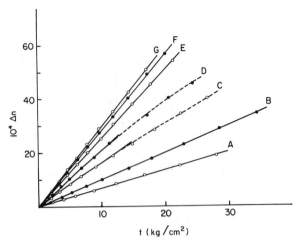

Fig. 23   The variation of the birefringence (Δn) with stress (t) for blends of 16 and 50% styrene–butadiene rubber containing the following weight percentages of the 16% rubber: (a) 0; (b) 10; (c) 30; (d) 50; (e) 70; (f) 90; (g) 100. (From Livingston and Brown [68], Fig. 6.)

so $\kappa_1 = \kappa_2 = 1$. Thus, Eq. (115) gives

$$\bar{C} = \phi_1 C_1 + \phi_2 C_2 \qquad (120)$$

and so the average stress optical coefficient varies linearly with composition. This was also shown to be the case by Fukuda *et al.* [70] for a network

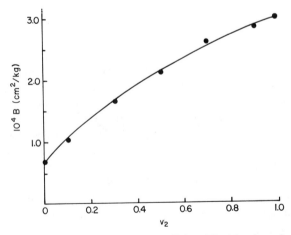

Fig. 24   The variation of the stress–optical coefficient (B) with volume fraction (v_2) of 16% SBR in mixtures with 50% SBR. (From Livingston and Brown [68], Fig. 7.)

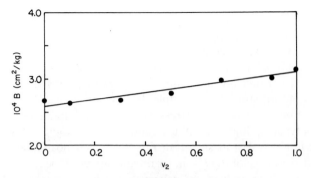

Fig. 25 The variation of the stress–optical coefficient (B) with volume fraction (v_2) of 16% SBR in mixtures with 23.5% SBR. (From Livingston and Brown [68], Fig. 11.)

consisting of covulcanized *cis*- and *trans*-1,4-polybutadiene chains. However, if one attempts to apply this result to copolymers, deviation from linearity with composition may be expected. The theory of this dependence was developed by Shindo and Stein [71] who extended the Kuhn–Grün derivation to polymer chains consisting of N_1 and N_2 segments of two types having segment lengths L_1 and L_2 and obtained

$$\bar{C} = (N_1 C_1 L_1{}^2 + N_2 C_2 L_2{}^2)/(N_1 L_1{}^2 + N_2 L_2{}^2) \tag{121}$$

giving a nonlinear variation of $\bar{C}$ with N_1, except for the special case of $L_1 = L_2$. This is a consequence of the fact that stiffer portions of the chain (represented by larger L's) orient to a greater degree than the more flexible

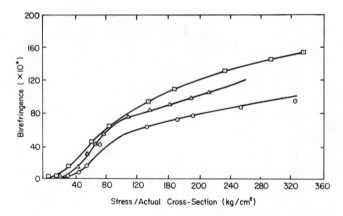

Fig. 26 Variation of the birefringence with stress for SBS copolymers cast from different solvents: 36 mole % styrene deposited from (○) MEK; (△) toluene; (□) heptane. (From Henderson *et al.* [67], Fig. 4.)

parts resulting in a greater weighting of the stress–optical coefficient of the stiffer component. The application of this equation was demonstrated for partly dehydrohalogenated poly(vinyl chloride) [72].

The values of κ_1 and κ_2 for an incompatible blend or a phase-separated block copolymer depend upon morphology, which may depend upon the conditions of formation. For example, Henderson *et al.* [67] show that the variation of birefringence with stress for a styrene–butadiene–styrene (SBS) block copolymer is highly nonlinear and depends upon the solvent from which the film was cast as shown in Fig. 26. This is postulated as resulting from the dependence of the κ's on strain and solvent type.

Wilkes and Stein studied the relaxation of birefringence and stress for such copolymers [69] and found, as shown in Fig. 27, that the stress–optical coefficient was constant with line for SBS films cast from toluene but, as shown in Fig. 28, vary greatly with time for films cast from methyl ethyl ketone (MEK). The explanation is that in

$$\bar{C} = \phi_S C_S \kappa_S + \phi_B C_B \kappa_B \qquad (122)$$

(where S and B designate the polystyrene and polybutadiene phases), the butadiene is the continuous phase for the toluene cast films so that $\kappa_S \approx 0$ and $\kappa_B \approx 1$ and $\bar{C} = \phi_B C_B$. However, for the MEK cast films, both phases are believed continuous so that both κ's are finite and vary with time during the relaxation.

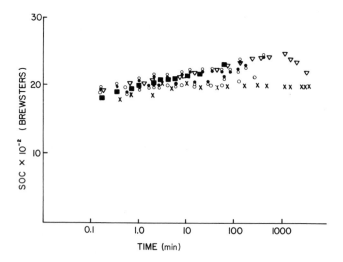

Fig. 27   The variation with time at different temperatures of the stress–optical coefficient of SBS copolymer (Kraton 101) films cast from toluene: (×) 24°C; (○) 35°C; (◐) 41°C; (▽) 51°C; (●) 63°C; (■) 74°C. (From Wilkes and Stein [69], Fig. 6.)

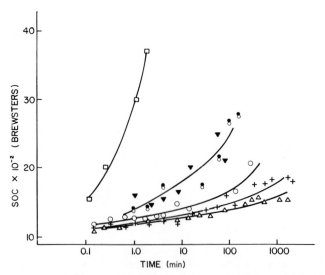

Fig. 28   The variation with time at different temperatures of the stress–optical coefficient of SBS copolymer (Kraton 101) films cast from methyl ethyl ketone: (△) 31°C; (+) 48°C; (○) 59°C; (●) 71°C; (▼) 75°C; (□) 85°C. (From Wilkes and Stein [69], Fig. 7.)

B.   Infrared Dichroism

The interpretation of birefringence is often ambiguous because of its being a composite property reflecting the combined orientation of both phases, as shown in Eq. (108). On the other hand, infrared dichroism has the potential of selectively measuring the orientation of a particular phase. By choosing the wavelength of the infrared radiation, one may selectively examine an absorption bond arising from a particular phase. Thus, for a measurement of the dichroism of Phase i designated as D_i, one may apply Eq. (11) to give

$$(D_i - 1)/(D_i + 2) = f_i \tag{123}$$

where f_i is the orientation function for the transition moment of a bond arising from the ith phase.

A principal application of this approach has been to crystalline polymers [73, 74], where studies of dichroism have been made for absorption bands arising from crystalline and amorphous phases, which have been used to characterize the orientation of those phases. A number of studies have been made, using this technique, for copolymers. For example, Uemura *et al.* [75] studied Surlyn-type ionomers, which are copolymers of ethylene with methacrylic acid and its salts. It was possible to compare the dichroism of bands such as that at 720 cm^{-1} arising from the ethylene residues and that

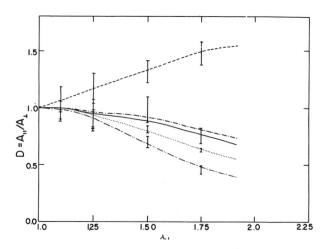

Fig. 29 The variation of the dichroism with elongation for the various absorption bands of the copolymer acid at 40°C: (——) 1700 cm^{-1}; (---) 1470 cm^{-1}; (–·–) 1263 cm^{-1}; (———) 935 cm^{-1}; (—··—) 720 cm^{-1}. (From Uemura et al. [75]. Fig. 2. Reprinted with permission from Macromolecules. Copyright by the American Chemical Society.)

at 1700 cm^{-1} arising from the carbonyl group of the methacrylic acid residue. For example, Fig. 29 is a comparison of the change in dichroism with elongation for different absorption bands for the copolymer acid at 40°C. In this way, it has been possible to distinguish the orientation of the two components of the copolymer and study their dependence on temperature and degree of ionization.

The technique has been applied extensively to segmented polyurethanes by Cooper and co-workers [76] in which orientation of the hard and soft segments characterized by absorption bands arising from NH and CH groups have been compared. These results have been correlated with the results of birefringence studies on these same polymers [77, 78].

There has been limited application of this technique to the study of blends. For example, Onogi and co-workers [79] studied the deformation mechanisms of blends of polypropylene with ethylene–propylene rubber (EPR) using the 720-cm^{-1} band from the ethylene of the EPR and the 998-cm^{-1} band arising from polypropylene (PP) crystals. Typical results are shown in Fig. 30 for the variation of the orientation of the c axis of the PP crystals with strain for different blend compositions. For the higher PP contents, the orientation function greatly increases with strain in a region corresponding to the yield point of the polymer, approaching and levelling off at a value close to unity. With increasing EPR content, the increase in PP orientation with strain becomes less, showing no significant increase with strain for PP

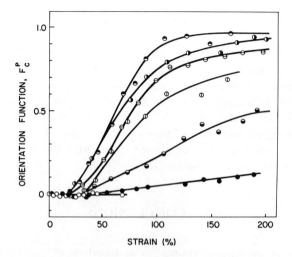

Fig. 30   The variation of the orientation function of the c axis of polypropylene crystals with strain for different composition of PP–EPR blends: (◐) 10/0; (◑) 9/1; (⊖) 8/2; (⊕) 7/3; (◓) 6/4; (●) 5/5; (◎) 4/6; (○) 3/7. (From Onogi *et al.* [79], Fig. 8.)

volume fractions less than 0.4. The EPR orientation function is almost zero for all strains and compositions. This data is interpreted in terms of there being a continuous phase of PP for its volume fractions greater than ~0.5. Below this concentration the PP crystals are believed to be embedded in a continuous phase of the soft matrix of the EPR and consequently undergo little orientation.

Recent studies have been reported by Cooper [80] on studying the relative orientation of the components of PCL–PVC blends by examining the infrared dichroism of bands arising from the respective components. The similar orientation of the components confirm that the amorphous phases of the two components are compatible.

Studies have been made using infrared dichroism of the relative orientation of PVC and plasticizers of dioctyl phthalate (DOP) and diethyl phthalate (DEP) by Gotoh and co-workers [81]. It was found that with less than about 15% added DOP, the DOP oriented upon stretching along with the PVC suggesting binding between the two, but above 15% little orientation of the DOP occurred suggesting that his excess DOP was dispersed freely in the PVC medium.

Folkes and Keller [58] have attempted to study the orientation of the polystyrene and polybutadiene in stretched triblock copolymers by using infrared dichroism and have found negligible orientation, indicating, as previously discussed, that the observed birefringence for these systems arose from the form birefringence contribution rather than from molecular orientation.

C. Raman Polarization

In principle, the Raman polarization technique can be used to selectively study the orientation of components of a blend in a similar manner to infrared dichroism. However, little work has been done using this technique, largly because of the large amount of Rayleigh scattering arising because of the heterogeneity of the blends. The Raman technique has the possible advantage over the infrared technique in that samples of small size and appreciable thickness may be employed rather than the thin films usually required for infrared dichroism studies.

V. CONCLUSIONS

The study of the optical properties of blends is of interest for several reasons. A knowledge of the factors leading to turbidity of a blend leads to an understanding of factors leading to their scattering and hence to the design of blends of superior appearance. Furthermore, the analysis of the scattering of both light and x rays provides information about the morphology of the blends, leading to measures of the sizes and shape of the domains, as well as the difference in refractive index or electron density between them and their surroundings. The study of other optical properties such as birefringence and infrared dichroism provides information about the deformation mechanisms of the blends that are dependent upon their morphology.

REFERENCES

1a. M. Kakudo and N. Kasai, "X-Ray Difraction by Polymers," Fig. 1. American Elsevier, New York, 1972.
1b. L. Silberstein, *Phil. Mag.* **33**, 92, 215, 521 (1927).
2.   K. G. Denbigh, *Trans. Faraday Soc.* **36**, 936 (1940).
3.   R. L. Rowell and R. S. Stein, *J. Chem. Phys.* **47**, 2985 (1967).
4.   S. D. Hong, C. Chang, and R. S. Stein, *J. Polym. Sci. Polym. Phys. Ed.* **13**, 1447 (1975).
5.   R. S. Stein and F. H. Norris, *J. Polym. Sci.* **21**, 381 (1956).
6.   P. H. Hermans and P. Platzek, *Kolloid Z.* **88**, 68 (1939).
7.   R. D. B. Fraser, *J. Chem. Phys.* **21**, 1511 (1953).
8.   Y. Nishijima, Y. Onogi, and T. Asai, *Rep. Progr. Polym. Phys. Jpn.* **8**, 131 (1965).
9a. D. I. Bower, *J. Polym. Sci. Polym. Phys. Ed.* **10**, 2135 (1972).
9b. R. J. Samuels, *Hercules Chem.* No. 56, p. 14, Fig. 4 (1968).
9c. B. H. Zimm, R. S. Stein, and P. Doty, *Polym. Bull.* **1**, 90, Fig. 10 (1945).
10. H. C. Van de Hulst, "Light Scattering by Small Particles." Wiley, New York, 1957.
11. M. Kerker, "The Scattering of Light and Other Electromagnetic Radiation." Academic Press, New York, 1969.

12. W. Heller, *J. Chem. Phys.* 1609 (1965).
13. A. Guinier and G. Fournet, "Small Angle Scattering of X-Rays," Chapter 2. Wiley, New York, 1955.
14. D. T. Sturgill, "Advances in Nucleation and Crystallization in Glasses" (L. L. Hench and S. W. Freiman, eds.), Amer. Ceram. Soc. Spec. Publ. No. 5, p. 33, 1972.
15. R. S. Stein and M. B. Rhodes, *J. Appl. Phys.* **31**, 1873 (1960).
16. R. S. Stein and M. B. Rhodes, *J. Polym. Sci. Part A-2* **7**, 1539 (1969).
17. L. C. Roess and C. G. Schull, *J. Appl. Phys.* **18**, 308 (1947).
18. T. Inoue, H. Ishihara, H. Kawai, Y. Ito, and K. Kato, *Proc. Int. Conf. Mech. Behavior Mater., 1971* Vol. III, p. 419. Soc. Mater. Sci., Japan, 1972.
19. T. Inoue, M. Moritani, T. Hashimoto, and H. Kawai, *Macromolecules* **4**, 500 (1971).
20. R. Duplessix, C. Picot, and H. Benoit, *J. Polym. Sci. Polym. Lett. Ed.* **9**, 321 (1971).
21. B. F. Conaghan and S. L. Rosen, *Polym. Eng. Sci.* **12**, 134 (1972).
22. B. F. Conaghan and S. L. Rosen, *SPE Tech. Papers* **22**, 225 (1974).
23. D. M. Moore, D. Bryant, and P. Latimer, *J. Opt. Soc. Amer.* **58**, 281 (1968).
24. F. Robillard and A. J. Patitsas, *Can. J. Phys.* **52**, 1571 (1974).
25. A. B. Levit and R. L. Rowell, *J. Colloid Interface Sci.* **50**, 162 (1975).
26. D. H. Woodward, *J. Opt. Soc. Amer.* **54**, 1325 (1964).
27. P. Debye and A. M. Bueche, *J. Appl. Phys.* **20**, 518 (1949).
28. P. Debye, H. R. Anderson, and H. Brumberger, *J. Appl. Phys.* **28**, 679 (1957).
29. H. K. Yuen and J. B. Kinsinger, *Macromolecules*, **7**, 329 (1974).
30. O. Kratky, *Pure Appl. Chem.* **12**, 483 (1966).
31. D. Caulfield, Y. F. Yao, and R. Ullman, *in* "X-Ray and Electron Methods of Analysis," Chapter VII. Plenum, New York, 1968.
32. S. Visconti and R. H. Marchessault, *Macromolecules* **7**, 913 (1974).
33. M. Moritani, T. Inoue, M. Motegi, and H. Kawai, *Macromolecules* **3**, 433 (1970).
34. D. J. Blundell, G. W. Longman, G. D. Wignall, and M. J. Bowden, *Polymer* **15**, 33 (1974).
35. F. H. Khambatta, F. Warner, T. Russell, and R. S. Stein, *J. Polym. Sci. Polym. Phys. Ed.* **14**, 1391 (1976).
36. C. G. Vonk, *J. Appl. Crystallogr.* **6**, 81 (1973).
37. R. Bonart and E. H. Müller, *J. Macromol. Sci. Phys.* **B10**, 177, 345 (1974).
38. T. Hashimoto, K. Nagatoshi, A. Todo, H. Hasegawa, and H. Kawai, *Macromolecules* **7**, 364 (1974).
39. D. J. Blundell, *Acta. Crystallogr. Ser. A* **26**, 472, 476 (1970).
40. G. Porod, *Kolloid Z. Z Polym.* **124**(2), 83 (1951); **125**(1), 51 (1952); **125**(2), 108 (1952).
41. T. Hashimoto, A. Todo, H. Itoi, and H. Kawai, *Macromolecules* (in press).
42. R. S. Stein, *in* "Structure and Properties of Polymer Films" (R. W. Lenz and R. S. Stein, eds.). Plenum, New York, 1973.
43. T. Nishi and T. T. Wang, *Macromolecules* **8**, 909 (1975).
44. T. T. Wang, private communication (1976).
45. E. H. Andrews, private communication (1976).
46. W. Wenig, R. Hammel, W. J. MacKnight, and F. Karasz, *Macromolecules* **9**, 253 (1976).
47. R. Hammel, W. J. MacKnight, and F. E. Karasz, *J. Appl. Phys.* **46**, 4199 (1975).
48. W. Wenig, F. E. Karasz, and W. J. MacKnight, *J. Appl. Phys.* **46**, 4194 (1975).
49. A. Escala, E. Balizer, and R. S. Stein, unpublished work.
50. M. Goldstein, *J. Appl. Phys.* **33**, 3377 (1962).
51. G. L. Wilkes and R. S. Stein, *J. Polym. Sci. Part A-2* **7**, 1695 (1969).
52. C. Picot, M. Fukuda, C. Chou, and R. S. Stein, *J. Macromol. Sci. (Phys.)* **B6**, 263 (1972).

53. C. S. M. Ong and R. S. Stein, *J. Polym. Sci. Polym. Phys. Ed.* **12**, 1899 (1974).
54. M. Goldstein and E. R. Michalek, *J. Appl. Phys.* **26**, 1450 (1955).
55. R. S. Stein and P. R. Wilson, *J. Appl. Phys.* **33**, 1914 (1962).
56. P. Erhardt, S. Clough, G. Adams, and R. S. Stein, *J. Appl. Phys.* **37**, 3980 (1966).
57. P. R. Lewis and C. Price, *Polymer* **12**, 258 (1971).
58. M. J. Folkes and A. Keller, *Polymer* **12**, 222 (1971).
59. R. S. Stein, *J. Polym. Sci. Part B* **9**, 747 (1971).
60. O. Wiener, *Abh. Kgl. Sachs. Ges. Wiss. Math. Phys. Kl.* **32**, 509 (1912).
61. P. H. Hermans, "Contributions to the Physics of Cellulose Fibers." Elsevier, Amsterdam, 1946.
62. F. A. Bettelheim and R. S. Stein, *J. Polym. Sci.* **27**, 567 (1958).
63. W. Kuhn and F. Grün, *Kolloid Z.* **101**, 248 (1942).
64. L. R. G. Treloar "The Physics of Rubber Elasticity," 2nd ed., Chapter X. Oxford Univ. Press, London and New York, 1958.
65. P. J. Flory, "The Statistical Mechanics of Chain Molecules." Wiley (Interscience), New York, 1969.
66. B. E. Read, *Polymer* **3**, 143 (1962).
67. J. F. Henderson, K. H. Grundy, and E. Fischer, *J. Polym. Sci. Part C* No. 16, 3121 (1968).
68. D. I. Livingston and J. E. Brown, Jr., *Proc. Int. Congr. Rheol., 5th* (S. Onogi, ed.), Vol. 4, p. 25. Univ. Tokyo Press, Tokyo, 1970.
69. G. L. Wilkes and R. S. Stein, *J. Polym. Sci. Part A-2* **7**, 1525 (1969).
70. M. Fukuda, G. L. Wilkes, and R. S. Stein, *J. Polym. Sci. Part A-2* **8**, 1917 (1971).
71. Y. Shindo and R. S. Stein, *J. Polym. Sci. Part A-2* **7**, 2115 (1969).
72. Y. Shindo, B. E. Read, and R. S. Stein, *Makromol. Chem.* **118**, 272 (1968).
73. B. E. Read and R. S. Stein, *Macromolecules* **1**, 116 (1968).
74. Y. Fukui, T. Asada, and S. Onogi, *Polym. J.* **3**, 100 (1972).
75. Y. Uemura, R. S. Stein, and W. J. MacKnight, *Macromolecules* **4**, 452 (1971).
76. G. M. Estes, R. W. Seymour, and S. L. Cooper, *Macromolecules* **4**, 490 (1971).
77. G. M. Estes, R. W. Seymour, D. S. Huh, and S. L. Cooper, *Polym. Sci. Eng.* **9**, 383 (1969).
78. R. W. Seymour, G. M. Estes, D. S. Huh, and S. L. Cooper, *J. Polym. Sci.* **10**, 1521 (1972).
79. S. Onogi, T. Asada and A. Tanaka, *J. Polym. Sci. Part A-2* **7**, 171 (1969).
80. S. L. Cooper, Private Communication.
81. J. Umemura, T. Takenaka, and S. Hagashi, *Bull. Inst. Chem. Res. Kyoto Univ.* **46**, 228 (1968).

Chapter 10

Transport Phenomena in Polymer Blends

H. B. Hopfenberg

Department of Chemical Engineering
North Carolina State University
Raleigh, North Carolina

D. R. Paul

Department of Chemical Engineering
University of Texas
Austin, Texas

I. Introduction 445
    A. Classification of Polymer Blends 447
    B. General Principles of Mass Transport in Polymeric Materials . . . 448
II. Permeation through Multiphase (Heterogeneous) Polymeric Systems . . 454
    A. Useful Analyses 454
    B. Experimentally Determined Permeability and Barrier Properties of Polymer Blends and Related Composites 460
III. Sorption of Vapors and Liquids in Polymer Blends 468
    A. Limiting Transport Mechanisms 468
    B. Thermodynamic Characterization of Polymer Blends by Vapor Sorption Techniques 474
IV. Membranes Prepared from Polymer Blends 479
    A. Uncharged Membranes 479
    B. Charged Membranes Including Polyelectrolyte Complexes 483
V. Summary 485
    References 487

I. INTRODUCTION

Virtually all polymeric materials are, in a certain sense, polymer blends. Except for scrupulously fractionated polymers or "living polymers" prepared by the anionic polymerization process [1], thermoplastic polymers are a blend of widely distributed molecular weight components, and, frequently, components that are branched to a lesser or a greater extent. There is, moreover, a superposition between the distributions of species of varying molecular weight and molecules with varying degrees of branching, producing a rather complex blend of molecular species.

Semicrystalline homopolymers are most simply described, morphologically, as a dispersion of a well-defined and ordered polymer within a continuous matrix of amorphous polymer. Typically, the amorphous regions of a semicrystalline polymer are less dense, and, in many cases, the crystalline material does not dissolve penetrating species. Many semicrystalline polymers are therefore blends of crystalline filler, which are nonsorbing and therefore nonconducting, dispersed within a matrix of sorbing and permeable polymer.

In a less subtle way, laminated film composites (see Volume 2, Chapter 15) or coated plastics are valid examples of polymer blends. Between the extremes of distributed homopolymers and macroscopic laminates lies the broad class of materials conventionally referred to as polymer blends, involving the comixing of at least two chemically distinguishable polymer molecules. The most narrowly defined use of the term polymer blends would involve a mixture of noncovalently bonded, chemically distinguishable polymers.

Typically, however, block-and-graft copolymers share a common morphological feature with noncovalently bonded blends. Specifically, domains rich in one homopolymer species are distributed through a matrix largely composed of the second polymeric constituent. Recognition of the domain structure is useful for developing models that correlate observed properties with the structure of the polymeric phase or phases.

In this chapter, the phenomena of low molecular weight penetrant transport, in and through polymer blends, are discussed in terms of the miscibility of the blend system. Currently accepted models for unsteady-state and steady-state permeation of polymeric materials are presented. A discussion of unsteady-state absorption and desorption behavior observed in a variety of polymer blends complements the treatment of permeation behavior. Throughout, relationships between the polymer structure and observed transport properties, common to blends, block, and graft copolymer systems, are emphasized. The emerging principles regarding criteria for formation of miscible blend systems are related, in turn, to the use of blending as a useful means for alteration of transport behavior in polymeric barriers and permselective membranes.

The history of humanity and its very evolution have been intimately related to the successful blending of materials. The epoch in prehistory known as the bronze age is a consequence of the development of an alloy composed primarily of copper and tin. Biological evolution is a consequence of the formation of spectacularly functional blends of polymeric materials comprising semipermeable membranes and barriers within living species; these composite membranes are responsible for the coordination and maintenance of life processes. For example, the currently accepted structure of animal cell membranes, involving a lipid bilayer, is an elegant, albeit prehistoric, development of polymeric blends. Clearly, novel polymeric blends for ever-expanding industry is inevitable in light of the relative importance that synthetic organic materials have achieved in the emerging science and technology of materials development.

A. Classification of Polymer Blends

A polymer blend is often considered to be compatible if the resulting mixture meets the use criteria of the intended application. For example, a blend might be rather casually considered to be compatible if the resulting mixture is transparent. Transparency, however, could be developed in a two-phase system in which the refractive indices of the phases are matched or if one of the phases is small compared with the wavelength of the transmitted light.

Even more casually, a blend might be considered compatible if the mechanical properties of the resulting composite meet rather arbitrarily imposed performance criteria. In contrast, the term miscibility, as applied to polymer blends, is as restrictive as the requirements for miscibility imposed upon mixtures of low molecular weight species.

Most useful blends are, in fact, immiscible; however, a rather large number of polymer pairs form a miscible single phase characterized by monotonic variation in most properties as a function of composition. Typically, the glass transition temperature of the blend varies regularly with polymer composition for miscible mixtures (see Chapter 5).

Alternatively, a domain, rich in one polymer species, is frequently dispersed within a matrix, rich in the second polymeric component. For the case of a plastic material dispersed in an elastomeric phase, reinforcement of the elastomer is a useful consequence of the two phase nature of the immiscible composite. On the other hand, rubber dispersed within a continuous plastic phase affords improved impact resistance to the resulting composite. High-impact polystyrene (HIPS) is a classic example of this effect (see Volume 2, Chapters 12–14).

The term domain implies that the dispersed phase is relatively small, presumably in the colloidal size range. Alternatively, a polymer blend might be composed of macroscopic phases. The simplest case of macroscopic blends might be a laminate. The laminate could be formed either in series or in parallel, and the steady-state transport properties, in particular, will be a sensitive and easily predicted function of the detailed arrangement of the polymeric phases. A more conventional composite may be formed from one polymeric component, of noncolloidal dimensions, dispersed within a second polymeric phase. In general, there should be no restriction regarding the geometry or relative dimensions of the phases. Interpenetrating polymer networks (IPNs) (see Volume 2, Chapter 11) have been prepared in which each of the cross-linked phases is colloidal; however, both phases are a continuous reticulum trapping the corresponding second phase [2].

The entropy per molecule of a randomly coiled, high molecular weight polymer is exceedingly high, owing to the vast number of conformations available to the coiling chain, but a small entropy increase results from the mixing of two high molecular weight polymers. For any isothermal process, the free energy change associated with that process is given by

$$\Delta G = \Delta H - T \Delta S \tag{1}$$

For the case of a rather small, albeit positive, change in entropy of mixing, a zero or negative enthalpy of mixing would be required to produce a significant free-energy driving force for the mixing. This criterion is not, in fact, excessively restrictive. For example, the polystyrene–poly(phenylene oxide) mixture is generally regarded as miscible, and the density of the blends are typically higher than the densities predicted by simple volume additivity, as shown in Fig. 1 [3]. This densification consequent to blending is consistent with strong polymer–polymer interactions, which would be similarly manifested by an exothermic mixing and, in turn, a significant free-energy driving force for mixing.

B. General Principles of Mass Transport in Polymeric Materials

Sir Thomas Graham [4], in 1866, described accurately the qualitative features governing the transport of gases through rubbery polymer membranes. Graham recognized that the overall transport processes involved the solution of gas at one surface followed by diffusion through the membrane and subsequent evaporation of the diffusing species from the downstream membrane surface. The nonporous nature of rubber and its similarity to a liquid was recognized. The transport rate of the penetrating gas was correlated directly with the solubility of the gas in the rubber. In addition, the use of rubber membranes for the separation of gas mixtures was cited and demonstrated quantitatively.

In 1879 Wroblewski [5] extended the qualitative discussions of Graham, providing a quantitative formulation for the steady-state gas flux in terms of

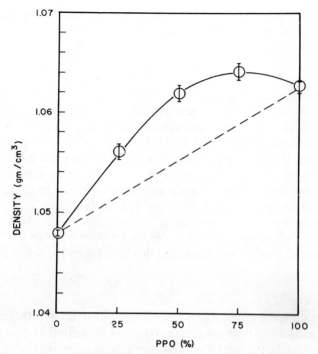

Fig. 1   Density of solution cast blends of polystyrene and poly(2,6-dimethyl-1,4-phenylene oxide) as a function of blend composition. (From Jacques and Hopfenberg [3].)

the kinetic and equilibrium parameters describing the gas–rubber system. Wroblewski speculated that Fick's diffusion law for gases and liquids might be successfully applied to the transport processes controlling penetration of rubbers by fixed gases. Starting with the relationship

$$J = -D \, \partial C / \partial x \tag{2}$$

where J is the gas flux, D the diffusion coefficient, C the penetrant concentration in the rubber, and x the position coordinate, one can substitute the Henry's law relationship

$$C = k_D p \tag{3}$$

relating the concentration of gas in the membrane with the imposed pressure p in the contiguous gas phase. The resulting expression relates the flux to the diffusion coefficient and the Henry's law solubility coefficient k_D:

$$J = D k_D \, \Delta p / l \tag{4}$$

where Δp is the pressure drop across the membrane of thickness l.

A permeability coefficient is conveniently defined as the flux per unit pressure difference. Combining this definition with Eq. (2) results in the concise equation

$$\bar{P} = Dk_D \tag{5}$$

where $\bar{P}$ is the defined permeability coefficient.

Unsteady-state analyses were provided by Daynes [6] and later by Barrer [7]. Their combined analyses indicated that, if a previously degassed membrane were contacted with a step increase in concentration of a penetrating gas, the downstream emission of penetrating gas molecules would initially be quite slow but would increase until a steady state was achieved in which the flux of penetrating gas molecules into and out of the membrane was constant, independent of time. Extrapolation of this steady-state relationship between the amount of penetrated gas versus time to the time axis results in an intercept that is a characteristic "time lag" for the system. The diffusion coefficient is related to this time lag by the following relationship:

$$\Theta = l^2/6D \tag{6}$$

and, therefore, the single time lag experiment of Barrer and Daynes provides a direct measure of both the permeability and the diffusivity, and in addition by using Eq. (5), the solubility coefficient can be calculated.

Most of the published research prior to 1950 was concerned primarily with diffusion in rubbery polymers. A rather consistent transport mechanism involving the solution of penetrant in the polymer followed by diffusion through the polymer characterizes virtually all of the rubbery polymer-penetrant systems. Small deviations from this model are encountered for rubbery systems containing crystalline regions that are susceptible to reordering by an interacting or plasticizing vapor or liquid penetrant [8]. In general, vapors and liquids do not obey Henry's law even in rubbery polymers. Furthermore, the diffusion coefficients are often highly dependent upon the concentration of the penetrant in the polymer.

Whereas Fick's diffusion laws and conventional equilibrium relationships describe transport in rubbery polymers, the transport of penetrants in polymeric organic glasses is not characterized by a simple, unifying mathematical or conceptual model [9]. Experimental observations with glassy polymers are frequently considered to be "anomalous" since they do not fit the pattern expected from the simple Fickian diffusion model that describes transport in rubbers. Hopfenberg and Stannett [10] have reviewed the various glassy state transport anomalies, which include: (a) dual sorption modes even for inert gases; (b) time-dependent boundary conditions for vapor transport; (c) diffusion coefficients characterized by an apparent time dependence;

(d) polymer relaxations that provide the rate determining transport step; (e) polymer fracture or microfracture (crazing) accompanying polymer relaxation; and (f) a significant change in the transport mechanism as the glass transition temperature is traversed.

For certain pairs of gases and glassy polymers the sorption isotherm is described by the following relationship:

$$C = k_D p + C_H' bp/(1 + bp) \tag{7}$$

where C_H' is a hole saturation constant and b is a hole-affinity constant. Vieth *et al.* [11] have presented an extensive review of systems obeying this simple linear combination of a Henry's Law isotherm and a Langmuir isotherm. This isotherm suggests, therefore, that two distinct populations or "modes" of sorbed gas molecules exist within the confines of a glassy polymer. Equation (7) is, therefore, frequently referred to as a dual-mode isotherm. One population is merely dissolved in the polymer and is described by the first term of Eq. (7). This term is equal to C_D, the concentration of dissolved species. The second term is considered to be equal to C_H, the concentration of species adsorbed in holes or selective sites according to the Langmuir adsorption isotherm. The second population apparently vanishes when the temperature of sorption is raised to T_g [12, 13]. For permeation through glassy membranes, obeying idealized dual-mode sorption, Paul [14] developed the following expression for the protraction of the time lag, due to hole or site filling:

$$\Theta = (l^2/6D)[1 + Kf(bp_2)] \tag{8}$$

where $K = C_H' b/k_D$ and $f(bp_2)$ is a tabulated function that ranges between a value of 1.0 at $p_2 \to 0$ and a value of 0 at $p_2 \to \infty$. The time lag is therefore more significantly protracted at very low pressures corresponding to conditions under which a relatively large fraction of the absorbed gas species occupy immobilizing sites. At high pressures, the immobilizing sites are saturated and the time lag is unaffected by the dispersed sorption sites since the ratio of dissolved to immobilized penetrant species can increase without limit consequent to complete saturation of the Langmuirian sites.

Paul and Koros [15a,b] have introduced an important refinement to the dual-mode sorption model which relaxes the requirement that the Langmuirian species are totally immobilized. This description replaces Eq. (2) with

$$J = -D\, \partial C_D/\partial x - FD\, \partial C_H/\partial x \tag{9}$$

where D is the diffusion coefficient of penetrant obeying Henry's law and F is a fraction that characterizes the degree of partial mobility of the gas sorbed by the Langmuir mode.

Combining the transport equations with the equilibrium pressure dependencies of C_D and C_H permits integration of the resulting differential equation for steady-state permeation conditions, yielding a general result for the pressure dependence of the steady-state permeability coefficient:

$$\bar{P} = Dk_D[1 + FK/(1 + bp_2)] \qquad (10)$$

For total immobilization ($F = 0$) there is no pressure dependence of $\bar{P}$, as predicted by Eq. (5); however, as the value of F increases toward 1.0, the permeability coefficient tends to increase with decreasing upstream pressure since the solubility component of the permeability is nonlinear in pressure. Pressure dependent permeabilities have, in fact, been determined for CO_2 in poly(ethylene terephthalate) by Koros et al. [13] that fit the pressure-dependent model embodied in Eq. (10). Similarly, as the site sorbed molecules gain mobility the protraction of the time lag vanishes as $F \rightarrow 1.0$. The general expression for the asymptotic value of the time lag (for very small values of p_2) is given by

$$\Theta = (l^2/6D)(1 + K)/(1 + FK) \qquad (11)$$

For total immobilization, $F = 0$, and Eq. (11) simplifies to the asymptotic low-pressure solution, given by the low-pressure limit of Eq. (8). Conversely, as $F \rightarrow 1$, the time lag expression simplifies into an identity with Eq. (6) in which no protraction is associated with the presence of a second population of freely mobile species. Equations (10) and (11) are therefore general expressions for the permeability coefficient and diffusion coefficient for all cases, considered to date, describing gas transport in rubbery and glassy polymers.

Fickian diffusion kinetics frequently describes gas, vapor, and liquid absorption in polymeric materials. For the case of sorption into or desorption out of a slab, the integral sorption kinetics are given by

$$\frac{M_t}{M_\infty} = 1 - \frac{8}{\pi^2} \sum_{n=0}^{\infty} \frac{1}{(2n+1)^2} \exp\left[\frac{-(2n+1)^2 \pi^2 Dt}{l^2}\right] \qquad (12)$$

where M_t is the amount of sorption in time t and M_∞ the amount of penetrant sorbed at equilibrium. The half-time, corresponding to $M_t/M_\infty = 0.5$ is approximated by

$$t_{1/2} = 0.049l^2/D \qquad (13)$$

and, therefore, diffusion coefficients can be calculated simply by Eq. (13). The fractional sorption is typically linear in $t^{1/2}$ for at least half of the total sorption (or desorption) for systems obeying Fickian transport kinetics [16].

This behavior is typical for penetration of rubbers and semicrystalline polymers above T_g; however, vapors and liquids that cause significant volumetric swelling of organic glasses typically follow a second limiting transport mechanism, completely independent from diffusional sorption kinetics. The second limiting transport mechanism, designated as Case II transport by Alfrey *et al.* [17], is characterized by the following distinctive features, resulting in a sorption rate that is ideally constant over the entire course of sorption. For penetration of a slab obeying idealized Case II sorption kinetics:

1. A sharp boundary separates the outer swollen regions from an essentially unpenetrated central core.
2. The boundary separating swollen and unpenetrated polymer advances at a constant rate toward the film midplane.
3. The concentration of penetrant in the swollen outer regions is essentially constant and equal to the equilibrium concentration of penetrant in the polymer.
4. Diffusion to the advancing boundary is rapid compared with rate-determining osmotically induced, stress–relaxation processes occurring at the boundary between swollen and unswollen polymer.

Hopfenberg [18] has derived a general expression that describes Case II transport in spheres, cylinders, and slabs:

$$1 - M_t/M_\infty = (1 - k_0 t/C_0 a)^n \qquad (14)$$

where k_0 is the relaxation rate constant, C_0 is the equilibrium concentration of penetrant in the polymer, a is the radius of the sphere or cylinder (or the half thickness of the slab, and n is an integer equal to 3 for a sphere, 2 for a cylinder, and 1 for a slab.

Hopfenberg *et al.* [19] have coined the term Super Case II transport as a third limiting case of transport for vapors or liquids in polymeric glasses. Whereas Fickian kinetics result in a linear relationship between M_t, and $\sqrt{t}$, and Case II transport predicts a linear relationship between M_t and t, Super Case II transport describes experiments in which M_t is proportional to t^n, where $n > 1.0$. The acceleration of sorption at long times, characteristic of Super Case II sorption in films, is apparently a dimensional anomaly resulting from the overlap of the Fickian tails that precede the advancing Case II boundaries. If the Case II boundaries advance sufficiently slowly and the film is suitably thin, then overlap of the Fickian tails precedes equilibration and the absorption accelerates since the long-term* relaxations, occurring at the advancing boundary, are accelerated by the increased concentration of plasticizing penetrant ahead of the relaxing front.

Recent studies in the general area of transport in polymeric materials have emphasized the role of long-term polymer relaxations on the change in apparent transport coefficients in polymeric glasses. The pioneering work by Berens describing vinyl chloride monomer transport in poly(vinyl chloride)(PVC)[20, 21] is typical of the more recent studies. The studies were motivated by the early successes in using penetrating molecules as probes to characterize the nature of the glassy state and the time dependence of glassy state properties. In addition, the study of vinyl chloride monomer transport in poly(viny chloride) was required to design suitable engineering systems for the successful removal of potentially toxic vinyl chloride monomer from the parent homopolymer. Dual sorption modes were used to explain Berens' sorption data describing the interactions between vinyl chloride monomer and PVC.

II. PERMEATION THROUGH MULTIPHASE (HETEROGENEOUS) POLYMERIC SYSTEMS

A.   Useful Analyses

Polymeric composites can be constructed in a variety of geometrical forms. One limiting case for composite geometry involves the dispersion of a completely nonsorbing and, therefore, nonconducting phase within the confines of a conducting polymeric matrix. The dispersed phase may be present over a wide range of volume fractions, particle sizes, and particle shapes. Moreover, the dispersed phase can be arranged randomly or in a more regular fashion. Each of these parameters has an effect on the overall permeability of the composite, and the simpler cases have been treated analytically. These analyses are primarily related to steady-state behavior since meaningful analysis of unsteady-state behavior is exceedingly difficult.

An impermeable filler imposes a more tortuous path upon the steady-state permeation of a penetrant across a polymer film. In addition, the filler reduces the effective cross-sectional area of the membrane. A tortuosity factor τ is therefore conveniently defined as the ratio of the effective path length required for permeation divided by the actual composite film thickness. The area reduction is typically proportional to the volume fraction of filler.

Maxwell [22] derived an approximate limiting relationship for the tortuosity affecting the electrical conductivity of a composite comprised of a continuous conductor phase, filled with nonconducting, spherical filler particles. His expression is

$$\tau \simeq 1 + \phi_d/2 \tag{15}$$

where ϕ_d is the volume fraction of filler.

Nielsen [23] presented a general approximation for estimating the effective tortuosity of systems containing asymmetric filler particles with length L and width W:

$$\tau \simeq 1 + (L/2W)\phi_d \tag{16}$$

where the platelet length is taken to be parallel to the composite surface. A commonly used approximation for the ratio of the permeability of the conducting, continuous phase as a function of the tortuosity and the volume fraction of matrix polymer ϕ_m has been suggested earlier by Michaels and Bixler [24] and Barrer *et al.* [25]. Their relationship is

$$\bar{P}_c/\bar{P}_m \simeq \phi_m/\tau \tag{17}$$

where $\bar{P}_c$ is the permeability of the composite and $\bar{P}_m$ is the permeability of the matrix polymer. Combining Eqs. (16) and (17) results in

$$\frac{\bar{P}_c}{\bar{P}_m} \simeq \frac{\phi_m}{1 + (L/2W)\phi_d} \tag{18}$$

Similarly, combining Eq. (15) with Eq. (17) provides a relationship for the effect of spherical nonconducting filler particles on the composite permeability:

$$\frac{\bar{P}_c}{\bar{P}_m} = \frac{\phi_m}{1 + \phi_d/2} \tag{19}$$

Equation (18) provides for the maximum filler effect since it assumes that the filler particles are oriented with their flat surface parallel to the film surface. Conversely, if the plates are oriented perpendicular to the surface, and L is still taken to be the large dimension of the plate, then the tortuosity would be given by

$$\tau \simeq 1 + (W/2L)\phi_m \tag{20}$$

rather than by Eq. (16), and the tortuosity is markedly reduced. Although more comprehensive treatments of inert filler effects have been performed by Barrer [26], the important implications are revealed by Eqs. (16)–(20). Specifically, the asymmetry and orientation of the filler particle has a powerful effect upon the permeability of the filled composite. Thin platelets with a large but finite L/W ratio can reduce the permeability by an order of magnitude for $\phi_d = 0.1$ if the platelets are oriented so that their flat surfaces are parallel to the film surfaces. If the plates are oriented at any other angle, the tortuosity factor will be correspondingly reduced and, in turn, the permeability ratio $\bar{P}_c/\bar{P}_m$ will be increased toward unity.

Alternatively, the dispersed filler phase might be highly sorbing although nonconducting. Paul and Kemp have prepared composites of this type by loading activated molecular sieves into rubbery silicones [27]. They demonstrated that this immobilizing sorption, quite analogous to the dual-mode sorption that frequently describes gas equilibration with glassy polymers, results in a marked protraction of the time lag, consistent with the predictions of Eq. (8). The steady-state permeability is affected modestly, which is consistent with the effects of increased tortuosity and decreased effective membrane area predicted by the Maxwell equation.

A polymeric phase might be dispersed within the confines of a continuous second polymeric component. In this case both phases would be conducting, although, in general, the phases would be characterized by different transport and equilibrium parameters. This case would also include polymer blends involving domains rich in one of the polymeric species confined in a continuous matrix rich in the second polymeric component. Steady-state transport in composites, in which both the dispersed and continuous phases are conducting, follows the analysis first presented by Maxwell [22]. One of the more useful analyses based upon the early work of Maxwell is presented by Robeson et al. [28] to aid interpretation of oxygen permeability data obtained on (AB)$_n$-type block copolymers of polysiloxanes (PSX) and polysulfones (PSF). They treat not only the case of dispersed domains within a polymeric matrix but also compare and contrast the effects of laminates arranged in series and parallel, and encompass the effect of interpenetrating reticula of continuous phases that might be developed consequent to the formation of an interpenetrating polymeric network.

Robeson's summary of limiting models starts with the simple series and parallel laminates as the extremes of two phase structure. For laminates composed of two phases designated by the subscripts 1 and 2, the series model result is (see also Volume 2, Chapter 15)

$$\bar{P}_c = \bar{P}_1 \bar{P}_2 / (\phi_1 \bar{P}_2 + \phi_2 \bar{P}_1) \tag{21}$$

and the parallel model result is

$$\bar{P}_c = \bar{P}_1 \phi_1 + \bar{P}_2 \phi_2 \tag{22}$$

where $\bar{P}_c$ is the composite permeability, P_1 and P_2 the permeabilities of the respective phases, and ϕ_1 and ϕ_2 their volume fractions.

To describe the effect of a conducting, spherical filler on the overall composite permeability, Robeson reports Maxwell's result as

$$\bar{P}_c = \bar{P}_m \left[\frac{\bar{P}_d + 2\bar{P}_m - 2\phi_d(\bar{P}_m - \bar{P}_d)}{\bar{P}_d + 2\bar{P}_m + \phi_d(\bar{P}_m - \bar{P}_d)} \right] \tag{23}$$

where the subscripts m and d refer to the continuous matrix phase and the dispersed phase, respectively.

Substituting $\bar{P}_d = 0$ into Eq. (23), corresponding to the case of non-conducting spherical filler, as expected, results in the limiting relationship presented earlier in this discussion as Eq. (19).

Robeson extends Maxwell's analysis by assuming, at intermediate concentrations, both phases contribute continuous and discontinuous characteristics. Addition of Maxwell's equations for both phases, weighted to their fractional contribution (x_a and x_b) to the continuous phase, result in

$$\bar{P}_c = x_a \bar{P}_1 \left[\frac{\bar{P}_2 + 2\bar{P}_1 - 2\phi_2(\bar{P}_1 - \bar{P}_2)}{\bar{P}_2 + 2\bar{P}_1 + \phi_2(\bar{P}_1 - \bar{P}_2)} \right]$$
$$+ x_b \bar{P}_2 \left[\frac{\bar{P}_1 + 2\bar{P}_2 - 2\phi_1(\bar{P}_2 - \bar{P}_1)}{\bar{P}_1 + 2\bar{P}_2 + \phi_1(\bar{P}_2 - \bar{P}_1)} \right] \tag{24}$$

where $x_a + x_b = 1$, and Eq. (24) applies to two cases:

Case A: Component 1 continuous ($\bar{P}_m = \bar{P}_1$; $\bar{P}_d = \bar{P}_2$);
Case B: Component 2 continuous ($\bar{P}_m = \bar{P}_2$; $\bar{P}_d = \bar{P}_1$).

Phase inversion is assumed to occur when the fractional contributions

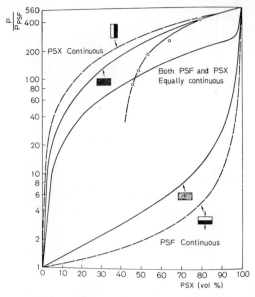

Fig. 2 Normalized oxygen permeability plotted semilogarithmically versus composition for PSF–PSX composites. Limiting models for series and parallel laminates as well as graphical representations of Maxwell's equation describing the normalized permeability of dispersed-continuous PSF–PSX composites are included in the graphical presentation. (From Robeson *et al.* [28].)

to the continuous phase are equal, that is, $x_a = x_b = 0.5$. For block copolymers as well as polymer blends, phase inversion is not a sharp transition
at a specific volume fraction. The transition is apparent over a broad range
of volume fractions as demonstrated by the comparison of the experimental
permeability data with the graphical representations of the limiting models
corresponding to Eqs. (21)–(24) for the block copolymer system PSX–PSF
presented in Fig. 2.

A replotting of the limiting models for parallel, series, and dispersed models
on the linear coordinates of Figs. 3 and 4 indicates clearly that the
continuous phase properties dominate the observed transport behavior of
two-phase, continuous–discontinuous blends. Specifically, if the relatively
low permeability PSF is continuous, then the composite permeability remains
low up to the volume concentration (74%) corresponding to closest packing
of spheres. When the PSX is continuous, the composite permeability,

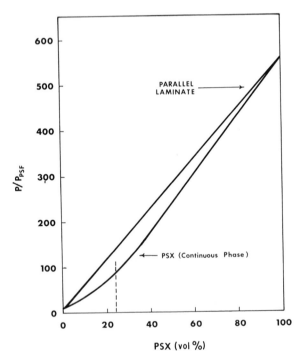

Fig. 3 Normalized oxygen permeability plotted as a linear function of blend composition
for idealized parallel laminate and also for a dispersed-continuous model with the relatively
high-permeability PSX phase continuous. The dispersed model is only valid above 24% PSX
(vertical dashed line) corresponding to closest packing of dispersed PSF.

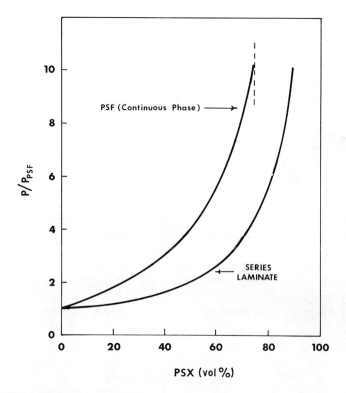

Fig. 4   Normalized oxygen permeability plotted as a linear function of blend composition for idealized series laminate and also for a dispersed-continuous model with the relatively low-permeability PSF phase continuous. Dispersed model only valid below 74% PSX (vertical dashed line) corresponding to closest packing of dispersed PSX.

although nonlinear in PSX volume percent, varies dramatically and closely approximates the predicted results for the parallel model, which consists of a linear relationship between permeability and blend composition.

The O_2 permeability data of Fig. 2 for the two-phase PSX–PSF system suggest that phase inversion occurs at 0.53 volume fraction PSF since the data intersect the S-shaped curve corresponding to $x_a = x_b = 0.5$ (phase inversion) at this point. These results are consistent with an independent analysis of phase inversion effects on elastic modulus behavior. At high volume fractions of PSX, the permeability data are consistent with the pure PSX phase as continuous; however, as the PSF composition increases, the permeability behavior reflects a varying distribution of the PSX and PSF between the continuous and dispersed phases. An alternate and even more complex model has been presented by Higuchi and Higuchi [29] for analyzing transport through heterogeneous barriers.

For a completely miscible polymer pair, the logarithm of the composite permeability frequently correlates linearly with the volume fraction of either of the components comprising the polymer blend [30, 31]. This linear, semilogarithmic behavior is consistent with the empirical model represented by the following additivity law:

$$\ln \bar{P}_c = \phi_1 \ln \bar{P}_1 + \phi_2 \ln \bar{P}_2 \tag{25}$$

Although frequently cited as a criterion for miscibility, Eq. (25) has no a priori theoretical basis.

B. Experimentally Determined Permeability and Barrier Properties of Polymer Blends and Related Composites

There is a growing body of literature on experimental determinations of the permeation behavior of polymer blends and related copolymer systems. In most cases the data are analyzed in terms of the various limiting models describing two-phase structure of polymer blends and block copolymer systems.

The most detailed analysis of permeation data appears to be that of Robeson et al. [28], described in Section II.A. Although Robeson considered oxygen and nitrogen permeability of block copolymers, their analysis stands as a model for a description of the nature of the continuous and dispersed phases in a blend.

Ranby and co-workers have reported the most extensive experimental study of permeability behavior in polymer blend systems [30, 32–38]. Ranby focused his early work on mechanical blends of poly(vinyl chloride) (PVC) with ethylene–vinyl acetate copolymers (EVA) and with acrylonitrile–butadiene copolymers (NBR), respectively.[†] The PVC–EVA blends were prepared from EVA copolymers containing 45% vinyl acetate and 65% vinyl acetate, respectively. The effect of varying vinyl acetate content on the miscibility of the resulting blends and, in turn, on the transport behavior was the primary focus of the study. Typical results from the experiments are presented in Fig. 5 in which the permeability coefficients of oxygen and nitrogen are plotted as a function of the EVA content of the PVC–EVA blends. The data of Fig. 5 show the effects of varying milling conditions, as well as the effects of vinyl acetate content in the EVA copolymer. The skewed, S-shaped nature of the curves for the blends involving the 45%

[†] See systems **d40** and **d38**, respectively, in Chapter 2 for a summary of the compatibility behavior of these systems. [—EDITORS]

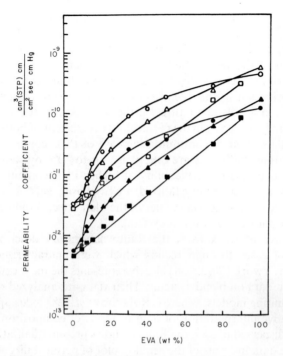

Fig. 5 Permeability coefficients versus blend composition of PVC–EVA blends at 25°C: (○, ●) PVC–EVA 45, milled at 160°C; (△, ▲) PVC–EVA 65, milled at 160°C; (□, ■) PVC–EVA 65, milled at 185°C. The open symbols denote oxygen; closed symbols denote nitrogen. (From Rånby [30].)

vinyl acetate–EVA have been interpreted to suggest a two-phase system with a "phase inversion" in the region 5–7 wt % EVA. The linear relationship between the log of the permeability coefficient and the blend composition for the 65% vinyl acetate–EVA blend, milled at 185°C, was interpreted as formation of a miscible blend. Identical blends milled at 160°C exhibit a curve on the semilog plot that is concave toward the composition axis, which may be the result of poor mixing at this low temperature.

Similar experiments were performed with blends of PVC and NBR with varying acrylonitrile contents. Results with the NBR system closely parallelled those obtained with the EVA blends. Specifically, as the nitrile content of the NBR phase increased, the plots relating the logarithm of the permeability coefficient with the blend composition tended toward linearity. The permeability coefficients and diffusion coefficients were measured in both systems for oxygen, nitrogen, carbon dioxide, and helium. Rånby extended his studies with Shur [33, 34] to include blends of PVC and chlorinated polyethylene [35] and PVC and acrylonitrile–butadiene styrene terpolymer (ABS). Similar experiments were performed on blends of PVC and poly(ε-caprolactone) (PCL) [37]. Although linear behavior was observed between the logarithm of the permeability of both oxygen and nitrogen and the composition of PCL-containing blend, the permeability coefficient for both oxygen and nitrogen in this blend system reached a maximum at 30% by weight of PCL in the blend mixture. Further increases in PCL concentration resulted in a linear decrease in the permeability coefficient of both oxygen and nitrogen as a function of continuing increase of PCL content. Rånby and Shur suggest that the blends are one phase up to 30% by weight of PCL; however, the decrease in permeability at higher PCL contents reflects the formation of a separate, crystallized PCL phase. This hypothesis was supported by x-ray diffraction studies of the blends (see Volume 2, Chapter 22 for further information on this system).

The most recent work from the Rånby laboratory describes selective permeation of gases through blends which were characterized in earlier studies. Rånby's work [30, 32–38] clearly represents the most extensive body of permeation data on blend systems. Their data are analyzed qualitatively in terms of limiting models. Whereas Robeson et al. [28] focus upon detailed and quantitative interpretation of data in terms of compositon of the continuous and dispersed phases, the Rånby studies present qualitative interpretations of the data in terms of the semilog plot of permeability versus blend content. Suggestions of incompatibility, consequent to observation of an S-shaped semilog permeability–blend composition plot are typically supported by companion dynamic mechanical and/or x-ray diffraction and density studies on these blends.

Ito [39] has determined the permeability and solubility coefficients for blends of polyethylene–polypropylene, polyethylene–polyisobutylene, and blends of high-density (HDPE) and low-density (LDPE) polyethylene. Ito reports that, whereas the permeability coefficient for water, CO_2, and nitrogen all decrease monotonically as the HDPE content increased in the HDPE–LDPE blend, a distinct maximum is observed in a plot of the permeability coefficient for both CO_2 and nitrogen as a function of polypropylene content in the polyethylene–polypropylene blends. These results are presumably a consequence of the interference of the second polymer with the crystallization processes that otherwise would proceed in the original homopolymer specimen. Intermediate compositions result in lower levels of crystallinity and, therefore, higher overall permeabilities.

The gas transmission behavior of natural rubber, blended with a variety of synthetic elastomers, has been reported by Barbier [40]. The data of Barbier have not been subjected to scrutinizing analysis in his published work. If, for example, his data describing the permeability to nitrogen of blends of natural rubber and Paracril-26 (a butadiene–acrylonitrile copolymer containing 26% by weight of acrylonitrile) are analyzed in terms of the Maxwell model, assuming two-phase structure, the experimental data are quite consistent with the natural rubber forming a dispersed phase over virtually all of the available composition range of the blend.

Pieski has reported the effect of blending high density and low density polyethylene on the permeability to water vapor. The results are presented in the rectilinear plot of Fig. 6. The permeability of these blends to water vapor is substantially lower than the values corresponding to the linear tie line between the permeabilities of the parent homopolymers. A replotting of this data in Fig. 7, however, indicates that a linear relationship between the logarithm of the permeability and blend composition is obeyed. Empirically this suggests blend miscibility; however, fundamentally there is no a priori reason to expect a linear relationship for either a rectilinear plot or a semilog plot. Pieski [41] speculates that such blends may be of special interest for producing polyethylene films with reduced moisture vapor transmission. He also reports on the water vapor permeability of blends of polyethylene–styrene–isobutylene copolymers. In this case, intermediate blend compositions of the rubber-modified polyethylene actually exhibit lower permeability than either of the parent homopolymers.

Continental Can Company, in a British letter patent, discloses the properties of polyethylene blended with nylon 6 [42]. In principle, the nylon should impart oil resistance and the polyethylene component could impart moisture resistance to the resulting blend. Mesrobian and Ammondson [43] report on the permeability of polyethylene–nylon blends to *n*-heptane, methyl salicylate, and methyl alcohol. The results were not sufficiently extensive to provide insight into the mechanisms controlling the variation of permeability with blend composition, however, permeabilities varied by as much as a factor of 100 over the complete range of compositions for the polyethylene–nylon 6 blends. Kohan [44] points out that, in general, hydrocarbon polymers that are immiscible with the nylons frequently yield compositions that have improved permeability and are processible into film, filaments, and bottles (see Volume 2, Chapter 21).

The properties of polyolefin–nylon composites, intended for applications in bottles and molded products, are reviewed by Komatsu and Ikegami [45]. Polyethylene, containing approximately 1 wt % of vinyl acetate, was laminated to poly(vinylidene chloride) and, subsequently, blow molded to provide two-layer bottles. The bottles had low oxygen permeability, and good adhesion between the respective polymeric layers was obtained.

Barnabeo *et al.* [46] have studied the gas permeability of nitrile containing block copolymers and compared the behavior of these apparently two-phase structures with random copolymers of the same monomers over

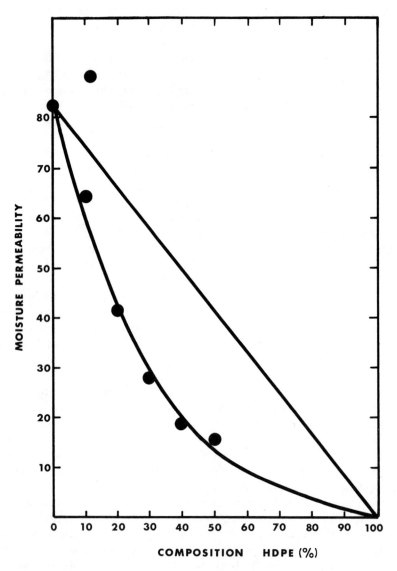

Fig. 6 Permeability of blends of low-density (0.9137) polyethylene (LDPE) with high-density (0.9757) polyethylene (HDPE) plotted rectilinearly versus blend composition. The linear tie line between permeabilities of LDPE and HDPE is also included. (From Pieski [41].)

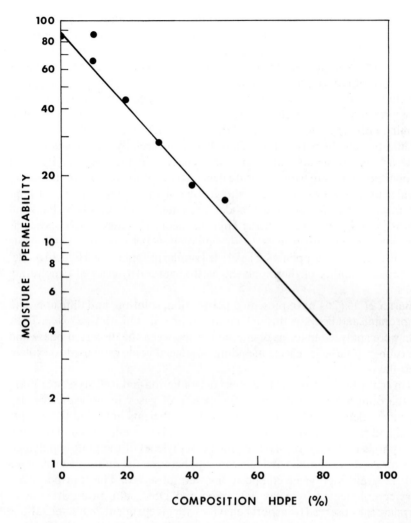

Fig. 7   Semilogarithmic presentation of permeability versus the blend composition data of Fig. 6.

the entire range of blend and copolymer composition. They use the analyses presented earlier in Section II.A to interpret the permeability data of the block and random copolymers. They conclude that, since a plot of the logarithm of the oxygen permeability versus volume fraction of the nitrile monomer is represented extremely well by an S-shaped curve, that the block copolymer is in fact a two-phase structure that is continuous in polystyrene at low nitrile compositions and conversely is continuous in nitrile polymer at high nitrile compositions. In contrast, the semilogarithmic plot of permeability versus volume fraction of the random nitrile copolymer is nearly linear and does not have an inflection point (46).

Peterson [47] has characterized the oxygen permeability of heterogeneous films prepared from mixtures of various polymer lattices dispersed in film-forming polymers. He applies the Higuchi–Higuchi model [29] as a means for interpreting the permeability data. Peterson carefully analyzes unsteady-state as well as steady-state transport behavior in the systems. Peterson recognized that there was a pressure dependence for the oxygen permeability coefficients in glassy polystyrene and glassy poly(methyl methacrylate), consistent with the observation and models presented more recently by Paul and Koros [15]. Deviations from the Higuchi theory in heterogeneous systems were interpreted with respect to the interaction between the components constituting the major phases. Ziegel [48] similarly applies the Higuchi model to the permeability of thermoplastic elastomers for a series of penetrating gases.

Barrer et al. [25] have presented permeation, solution, and diffusion data for propane and benzene through natural rubber filled with zinc oxide. Their data were analyzed in terms of interactions between the dispersed phase and the rubber. Interfacial effects including interfacial voids were used to explain their data.

Toi et al. [49] have studied the effect of irradiation-induced styrene grafting on the diffusion and solubility of a variety of gases in polyethylene. Independent determinations of the solubility coefficients indicated that simple linear additivity was observed between the solubility coefficients and the total amorphous fraction (polystyrene plus polyethylene) of the graft copolymer. In contrast, more complicated relationships between the diffusion coefficients and the graft copolymer compositions were observed. The transport data were complimented by x-ray diffraction and DSC data, in an attempt to correlate the observed transport behavior with independent characterizations of the resulting morphology of the graft copolymers. Their study confirms the notion that the detailed nature of the individual phase morphology affects both properties of two-phase block graft, and polymer blend systems in a rational manner. The authors demonstrate that, by extracting homopolymer from the grafted copolymer, the solubility coefficients for argon are not only increased, but a more linear relationship between the solubility coefficient and the polystyrene volume fraction is observed consequent to extraction.

Odani et al. [50, 51] present permeation, diffusion, and solubility data for gases in styrene–butadiene–styrene block copolymers. They employ simple limiting series and parallel models for the interpretation of their data. They implicitly consider the block copolymer system to be represented morphologically as a two-phase system. In this regard, they employ analyses used by Michaels and Bixler [24] related to the sorption and diffusion of gases in crystalline polyethylene. Specifically they employ a tortuosity factor and a chain immobilization factor to interpret the diffusion data and they correlate the solubility coefficients with the Leonard–Jones force constants of the respective gases according to the earlier suggestions of Michaels and co-workers.

Stallings *et al.* [52] have studied the transport of fixed gases in solution cast blends of poly(2,6-dimethyl-1,4-phenylene oxide) (PPO) and polystyrene (PS). The PPO–PS blend system is generally conceded to be miscible (see Chapters 2 and 5) and the solution cast blends are transparent in all proportions; each of the respective blend compositions are characterized by a single T_g. The glass transition temperature of the blends varies monotonically with blend composition. The permeability coefficients reported by Stallings *et al.* [52] vary monotonically with blend composition for argon and krypton; however, more complex behavior was reported for neon in the various blends.

A discernable minimum was observed in the plots of the time lag diffusivity versus blend composition; the relative magnitude of the minimum became more pronounced as the penetrant diameter decreased for the rare gas penetrants studied in their work. Stallings *et al.* [52] recognized that transport in this glassy blend system is, most likely, confounded by the dual-mode sorption anomalies. They suggest that the complex relationship between diffusion coefficient and blend composition may be a consequence of the effect of blend composition on the contribution of site sorption in these glassy blends. The Stallings work, although compromised by a lack of high-pressure sorption and permeability data, stands as the only work to date on gas transport in a glassy blend system which is apparently miscible in all proportions.

In summary, unsteady- and steady-state gas permeation behavior has been studied in a rather wide variety of polymer blend systems. In general, the systems are complex, and in many cases, one of the polymers is complicated by the development of crystallinity. In most cases, limiting transport models, involving the polyphasic nature of the blends, are recognized although few studies have systematically and quantitatively examined the detailed morphology of the blends by suitable analysis of transport data in conjunction with ancillary experiments. Transport in miscible, glassy blends awaits systematic study over a wide range of pressures so that dual mode sorption contributions can be incorporated into the analysis.

III. SORPTION OF VAPORS AND LIQUIDS IN POLYMER BLENDS

A. Limiting Transport Mechanisms

Cates and White were among the first to report sorption behavior in polymer blends. Their three-part series, which appeared in 1955 [53–55], presented water sorption data in blends of polyacrylonitrile (PAN) and cellulose, silk, and cellulose acetate respectively. In this manner, three individual two-component blend fibers were prepared. The sorption of water in the PAN–cellulose acetate and PAN–cellulose varied linearly with blend composition. Conversely, the blend of PAN–silk resulted in more complicated sorption behavior. The suggested two-phase structure of the PAN–silk composite was substantiated by the observation that PAN-rich blends absorbed very little anthraquinone based dye whereas the pure silk rather rapidly absorbed the identical dye. Homopolymer PAN does not absorb this dye. Apparently, in the PAN-rich compositions the dye is not able to penetrate the presumably continuous PAN matrix and, consequently, is not absorbed into the dispersed silk phase.

Cabasso et al. [56] studied the sorption of benzene–cyclohexane mixtures in polymer blend membranes composed of poly(phosphonates) and acetyl cellulose. The authors state that this blend represents a miscible pair. The blends selectively absorb benzene from benzene–cyclohexane mixtures.

Although the authors recognize that limiting Case II transport could in fact contribute to the observed sorption behavior, their sorption results are presented in the form of fractional sorption as a function of the square root of time. These plots are most useful for systems obeying Fickian, diffusion controlled transport, even if the diffusion coefficient is concentration dependent. Cabasso et al. [56] indicate that there is a dramatic concentration dependence of the diffusion coefficient, with the diffusion coefficient of benzene increasing with the concentration of benzene in the blend membrane. The fractional sorption of benzene from benzene–cyclohexane mixture into poly(phosphonate)–acetyl cellulose blends is presented for a series of blend membranes in Fig. 8, where the square root of time is selected as the abscissa. The authors suggest that several qualitative features of the sorption process are consistent with limiting Case II sorption; however, they do not present a plot of fractional sorption versus linear time. The authors state that their data result in an exponent of time greater than one consistent with Super Case II transport [19]. Moreover, replotting of the data presented in Fig. 8 as fractional sorption versus linear time in Fig. 9 suggests that Case II transport is indeed observed for all of the blends and that Fickian absorption predominates in the acetyl cellulose homopolymer. The authors

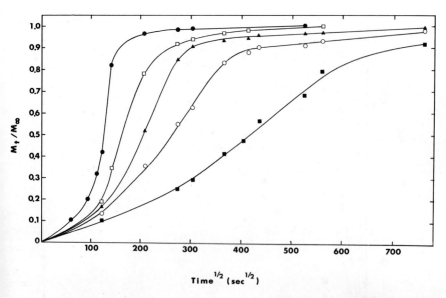

Fig. 8   Fractional sorption versus $\sqrt{t}$ for various poly(phosphonate) (P)–cellulose acetate (A) blends determined at 30°C from 1/1 (w/w) benzene/cyclohexane mixture: (●) P/A, 50; (□) P/A, 40; (▲) P/A, 30; (○) P/A, 20; (■) cellulose acetate. (The numbers denote wt % P.) (From Cabasso *et al.* [56].)

emphasize that there is a monotonic variation in sorption rate as a function of blend composition.

The sorption kinetics and equilibria of some normal alkanes, especially normal hexane, in solution cast blend films of atactic polystyrene (PS) and poly(2,6-dimethyl-1,4-phenylene oxide) (PPO) have been studied extensively by Hopfenberg and co-workers [3, 19, 57–59]. The effects of temperature, penetrant activity, blend composition, and thermal history on the sorption kinetics and equilibria of *n*-hexane in the PS–PPO blends have been examined systematically. This blend system has been reported to be compatible over the entire range of blend composition [59].

Jacques *et al.* [57] observed that the equilibrium solubility of *n*-hexane in the PS–PPO blends [60] increased monotonically with increasing poly-(phenylene oxide) content at a vapor activity (*n*-hexane partial pressure divided by the *n*-hexane vapor pressure) of 0.775. In marked contrast, the equilibrium sorption goes through a dramatic minimum with blend composition at higher penetrant activities [3]. This contrasting behavior is presented in Fig. 10. The explanation of this seeming anomaly was provided by comparing the sorption isotherms for the various blend compositions and by recognizing the effect of *n*-hexane content in the blends on the resulting

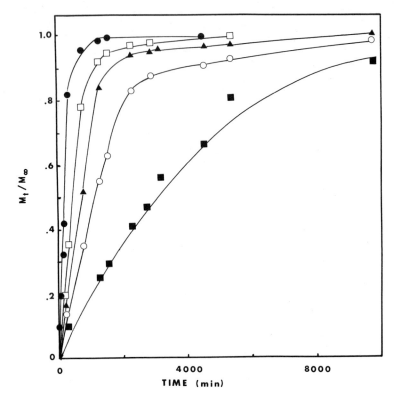

Fig. 9   Data of Fig. 8 replotted on linear (Case II) plot of fractional sorption versus time for various P–A blends.

depression of the T_g. The resulting explanation is completely general and does not depend on any behavior peculiar to this particular blend.

Sorption isotherms for n-hexane in the parent homopolymers and in a 50/50 blend of the two parent polymers are presented in Fig. 11. The isotherms for PS and the 50/50 blend intersect at high penetrant activities. As postulated earlier by Kambour [61], the equilibrium solubility of n-hexane in the glasses is essentially a linear function of penetrant activity until a penetrant concentration sufficient to lower the T_g to the experimental temperature is traversed; beyond this activity, there is a relatively dramatic increase of penetrant solubility with further increase of penetrant activity. This change in behavior between the low activity and high activity regimes in the sorption isotherms is not observed for n-hexane sorption in the pure poly(phenylene oxide) because a penetrant concentration corresponding to the effective T_g is apparently not reached at this temperature over the entire

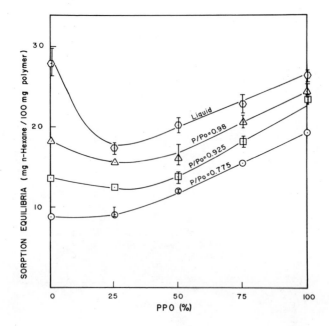

Fig. 10   Equilibrium sorption of *n*-hexane as a function of blend composition in PS–PPO blends for various vapor activities at 40°C. (From Jacques and Hopfenberg [3].)

range of possible activities. The effective T_g's were measured on *n*-hexane-equilibrated samples by differential thermal analysis and these experimentally determined values correspond quite well with the estimates of the reduced T_g's provided by analysis of the shape of the respective isotherms [3].

Over the range of temperatures and activities studied, Case II transport of sorbing penetrant described virtually all the observed sorption kinetics. In addition, the relationship between sorption rate and blend composition was qualitatively identical to the relationship between sorption equilibria and blend composition. This qualitative similarity between sorption behavior and kinetic behavior is a consequence of the observed linear relationship between the logarithm of the sorption rate and the equilibrium penetrant concentration. This relationship was observed to be independent of either blend composition or penetrant activity, since the various solubilities were determined in two different blend samples at activities sufficient to generate the various equilibrium solubilities. The observed behavior, which relates sorption rate uniquely with the equilibrium *n*-hexane concentration, is reasonable since the osmotic swelling stresses generated by this sorption increase with

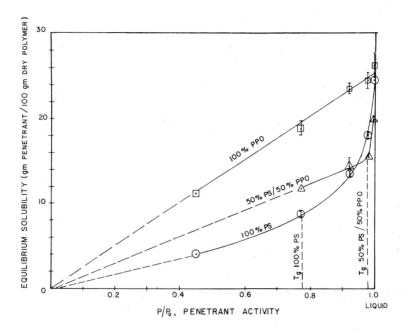

Fig. 11   Sorption isotherms at 40°C for PS, PPO and a 50/50 PS/PPO blend in *n*-hexane. (From Jacques and Hopfenberg [3].)

equilibrium penetrant concentration. Therefore, the stress-biased, relaxation-controlled transport is, in turn, a strong function of the volumetric swelling of the polymer. For instance, a twofold increase in the equilibrium penetrant solubility results in a ten-thousandfold increase in the sorption rate.

The observed results are, therefore, related to the somewhat serendipitous intersection of the sorption isotherms and the dramatic and monotonic coupling of sorption rates with equilibrium solubility. Specifically, at low activities the equilibrium sorption varies monotonically with blend composition. In the activity range where the isotherms cross, a minimum is observed for both the equilibrium sorption and the kinetics when each are plotted versus blend composition. Moreover, Jacques and Hopfenberg [58] show in Fig. 12 that the front velocity describing penetration of *n*-hexane in the blend sample varies monotonically with blend composition when all swollen samples are rubbery ($T = 85°C$, $P/P° = 1.0$, *n*-heptane) or when all samples are glassy ($T = 30°C$, $P/P° = 0.775$, *n*-hexane). However, minima are apparent in these same plots at those activity–temperature combinations which render the PS-rich compositions rubbery and the PPO-rich samples glassy [58]. Possibly even more striking is the observation that, whereas at an activity of 0.775 the PPO-rich samples sorbed up to 10,000 times faster than the PS homopolymer, under conditions corresponding to complete rubbery behavior for the equilibrated blend samples (liquid *n*-heptane, 85°C), the PS actually sorbed more rapidly than the PPO.

Hopfenberg *et al.* [59] later studied the effect of thermal annealing on the kinetics and sorption equilibria in this identical system. The annealing was performed 20°C above the respective T_g's of the various polyblends. Sorption rates were significantly affected in all compositions except for the pure polystyrene. This behavior was explained in terms of reduction of free volume

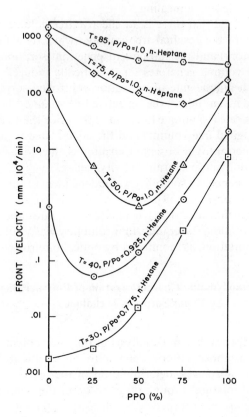

Fig. 12 Front velocity for Case II sorption at various activities plotted versus blend composition for PS–PPO blends. (From Jacques and Hopfenberg [58].)

consequent to the arbitrary annealing cycles. Apparently, in the case of pure polystyrene, densification was sufficiently rapid during the original drying process so that very little additional density change was achieved in subsequent thermal annealing. In all other compositions examined [58], annealing produced progressively more densification as the PPO content increased Case II sorption rates of *n*-hexane in these blends similarly decreased progressively with the PPO content consequent to annealing [59]. Specifically, the sorption rate was decreased by two orders of magnitude in pure poly(phenylene oxide) after 24 hr of annealing, whereas the effect of annealing progressively vanished as the polystyrene increased. The reduction in free volume retarded the sorption rate to the extent that Super Case II transport [19] was observed in all samples after annealing, except, of course, Case II kinetics persisted for the pure polystyrene since no densification resulted from the imposed annealing.

These results suggest that there is an effective free volume resulting from the superposition of the residual free volume (which is a function of the prior thermal history) and the subsequent penetrant sorption. Presumably, this effective free volume generates an osmotically induced stress, which, in turn, leads to a unique rate of relaxation-controlled sorption that only depends upon the magnitude of the "effective free volume." In principle, the observed sorption rate is a unique function of the prior thermal history (providing a given original free volume) and the added "free volume" provided by the imbibed penetrant. The observed sorption is, therefore, independent of polymer composition and penetrant activity, per se.

The equilibrium penetrant solubility was independent of annealing over the entire range of composition [59]. These results are consistent with earlier findings [62], which indicated that, whereas sorption rate varied dramatically with residual orientation, the equilibrium solubility was virtually unaffected by changes in orientation, as monitored by optical birefringence.

B. Thermodynamic Characterization of Polymer Blends by Vapor Sorption Techniques

In principle, a thermodynamic analysis of polymer–polymer interactions can be used to establish criteria related to blend miscibility. For low molecular weight substances, determination of the activity of each component in a liquid mixture through observation of the appropriate vapor–liquid or solid–liquid equilibria provides a measure of the solute–solute interaction. Unfortunately, the nonvolatility of polymers rules out vapor–liquid techniques for studying polymer–polymer mixtures; however, the determination of the melting point depression (solid–liquid equilibrium) has recently been used [63–65] to obtain useful information regarding polymer–polymer interactions (see Chapters 4 and 5). This technique is compromised somewhat, however, by kinetic and morphological complications and is not applicable when neither of the polymeric components is crystalline.

For low-viscosity liquids, calorimetric measurements can be used to obtain the heat of mixing directly. This approach is not directly applicable to mixtures of two high molecular weight polymers for a variety of obvious reasons; however, it has been used in a few instances for mixtures of oligomeric components with relatively low viscosities [66]. Calorimetry, although somewhat difficult and tedious, has been used to infer the heat of mixing of two polymers indirectly [67–71]. In this approach, the heat of dilution of the blend by solvent is compared with the heat of diluting each pure polymer by analyzing a thermodynamic cycle analogous to Hess' law to arrive at the desired heat of polymer–polymer mixing. This technique is compromised by an accumulation of errors since the calculation involves differences of difficult to measure values of low inherent accuracy.

An alternate approach, involving the equilibrium sorption of a vapor by the solid polymer blend (similar in concept to analysis of the thermodynamics of liquid solutions of polymer pairs), has been reported recently [72–74] (see also Chapter 4). Analysis of equilibrium sorption of a vapor by the blend can provide information regarding polymer–polymer interactions. The amount of vapor sorbed is related to its interaction with the blend. By comparing this interaction with the interaction between the vapor and each of the pure polymers, it is possible to obtain information about the interaction between the polymer components of the blend.

This analysis is most easily visualized by the use of a thermodynamic model for characterizing the sorption equilibrium and the blend interactions. The Flory–Huggins theory [75] is convenient for this purpose. A simple version of this analysis, given here, will point out the advantages, limitations, and some considerations of experimental design of this potentially useful technique for studying blend thermodynamics.

In terms of the Flory–Huggins theory, the chemical potential of each component μ_i of a mixture of Polymers 2 and 3 may be described by the following equations:

$$\mu_2 - \mu_2{}^\circ = RT[\ln(1 - \phi_{30}) + (1 - \tilde{V}_2/\tilde{V}_3)\phi_{30} + \chi_{23}\phi_{30}^2] \qquad (26)$$

$$\mu_3 - \mu_3{}^\circ = RT[\ln \phi_{30} + (1 - \tilde{V}_3/\tilde{V}_2)(1 - \phi_{30}) + \chi_{32}(1 - \phi_{30})^2] \qquad (27)$$

where $\mu_i{}^\circ$ is the chemical potential of pure polymer i, ϕ_{i0} the volume fraction of polymer i in the penetrant-free, binary polymer–polymer blend, and the χ_{ij} the appropriate interaction parameters, which are related by [75]

$$\chi_{32} = (\tilde{V}_3/\tilde{V}_2)\chi_{23} \qquad (28)$$

with $\tilde{V}_i$ being the molar volume of polymer i. When both polymers have very large molecular weights (infinite in the limit), χ_{23} and χ_{32} must be negative for the two polymers to be miscible in all proportions and thus to form a homogeneous phase (see Chapter 2).

If vapor species 1 is sorbed into either pure polymer i to an equilibrium volume fraction ϕ_1, the chemical potential μ_1 and activity a_1 of the vapor will be given by

$$\mu_1 - \mu_1^\circ = RT \ln a_1 = RT \left[\ln \phi_1 + (1 - \phi_1) + \chi_{1i}(1 - \phi_1)^2 \right] \qquad (29)$$

according to Flory–Huggins theory, provided the molecular weights, or molar volumes, of both polymers are very large. Here, χ_{1i} is the appropriate interaction parameter for solvent vapor with polymer i. It is necessary for this analysis to select a system in which χ_{1i} does not depend on ϕ_1. The activity of vapor sorbed into a homogeneous blend of Polymers 2 and 3 (again assuming molecular weights are sufficiently large) is given by

$$\ln a_1 = \ln \phi_1 + (1 - \phi_1) + (\chi_{12}\phi_2 + \chi_{13}\phi_3)(1 - \phi_1) - \chi'_{23}\phi_2\phi_3 \qquad (30)$$

where the interaction parameter between 2 and 3 used here is related to those in Eqs. (1) and (2) by [64]

$$\chi'_{23} = (\tilde{V}_1/\tilde{V}_2)\chi_{23} = (\tilde{V}_1/\tilde{V}_3)\chi_{32} \qquad (31)$$

In Eq. (30), ϕ_2 and ϕ_3 represent volume fractions of each polymer in the ternary system and are related to the solvent-free composition of the blend by

$$\phi_2 = \phi_{20}(1 - \phi_1) \qquad (32)$$

$$\phi_3 = \phi_{30}(1 - \phi_1) \qquad (33)$$

It is easy to show that Eq. (30) can be recast into the form of Eq. (29) if the χ_{1i} is replaced by

$$\chi_{1b} = \chi_{12}\phi_{20} + \chi_{13}\phi_{30} - \chi'_{23}\phi_{20}\phi_{30} \qquad (34)$$

where χ_{1b} is the effective interaction parameter of the solvent with the homogeneous blend treated as a single component. In the case where $\chi'_{23} = 0$, the effective interaction parameter for the solvent with the blend is the composition weighted average of its interaction with the individual polymers. If $\chi'_{23} < 0$, the effective χ_{1b} is larger than this value. The amount of vapor sorption ϕ_1 gets smaller as the value of the interaction parameter gets larger. Thus, it is evident, that as the interaction between Components 2 and 3 becomes more favorable (i.e., χ'_{23} becomes more negative), the amount of vapor sorption is decreased. As Components 2 and 3 become more attracted to each other, their blend should become relatively less attractive to solvent compared to pure Polymers 2 and 3. Figure 10 for the miscible PPO–PS system is an example of this. The opposite case, $\chi'_{23} > 0$, would increase the vapor sorption beyond the neutral case, $\chi'_{23} = 0$, but this situation is unstable since such blends, in the limit of very large molecular weights, will separate into two polymer phases at equilibrium.

To pursue the question of phase separation, it will be informative to considering the limiting case of a physical composite whose phases are pure Components 2 and 3. It is assumed that the sorption of vapor by each phase is not affected by the other phase as might occur by mechanical constraints in some cases. The average equilibrium vapor sorption by the composite at equilibrium is given by

$$\phi_1 = \frac{[\phi_{12}/(1-\phi_{12})]\,\phi_{20} + [\phi_{13}/(1-\phi_{13})]\,\phi_{30}}{1/(1-\phi_{12}) + 1/(1-\phi_{13})} \tag{35}$$

where ϕ_{12} and ϕ_{13} are the vapor sorptions in pure Polymers 2 and 3 at this activity given in accordance with Eq. (29). If both ϕ_{12} and ϕ_{13} are quite small compared to one or if $\phi_{12} = \phi_{13}$, Eq. (35) reduces to a simpler form,

$$\phi_1 = \phi_{12}\phi_{20} + \phi_{13}\phi_{30} \tag{36}$$

It is not mathematically possible to recast the equations for this situation into the form of Eq. (29). This means there is no single effective value of χ_{1c} for the composite. If Eq. (29) is used to calculate an effective χ_{1c} point by point along the experimental sorption isotherm, the results will be a χ_{1c} function dependent on ϕ_1 even though χ_{12} and χ_{13} are not. This ϕ_1 dependence indicates that the blend is not one phase. Consequently, in designing such experiments, it is important to select a solvent and activity ranges such that χ_{12} and χ_{13} are independent of ϕ_1, or to use another model that yields constant and physically meaningful parameters. Otherwise, the analysis may give misleading results. Likewise, the analysis becomes confused if χ'_{23} depends on ϕ_1. It is not essential, however, that χ_{23} be independent of ϕ_{30}.

This example may be continued with the simplifying restriction that ϕ_1 is very small, which is attainable if a_1 is sufficiently low. In this case [75], the sorption isotherm, ϕ_1 versus a_1, is linear, and Eq. (29) reduces to

$$a_1 = \phi_1 \exp(1 + \chi_{1i}) \tag{37}$$

Equation (37) applies appropriately to Polymers 2 and 3 and combined with Eq. (36) yields

$$\phi_1 = \frac{a_1}{\exp(1 + \chi_{12})}\phi_{20} + \frac{a_1}{\exp(1 + \chi_{13})}\phi_{30} \tag{38}$$

or

$$a_1 = \phi_1 \left[\frac{\phi_{20}}{\exp(1+\chi_{12})} + \frac{\phi_{30}}{\exp(1+\chi_{13})} \right]^{-1} \tag{39}$$

In this limit, it is possible to define an effective χ_{1c} by Eq. (37) so that Eq. (39) is satisfied. This is done by eliminating a_1 between Eqs. (37) and (39) to obtain

$$\exp(\chi_{1c}) \equiv \left[\frac{\phi_{20}}{\exp \chi_{12}} + \frac{\phi_{30}}{\exp \chi_{13}} \right]^{-1} \tag{40}$$

or after some rearranging

$$\chi_{1c} = \chi_{12} - \ln\{1 + [\exp(\chi_{12} - \chi_{13}) - 1]\phi_{30}\}. \tag{41}$$

If the difference $(\chi_{12} - \chi_{13})$ is not very large, a simpler form can be obtained by expanding the exponential and logarithm into power series, which (after truncating third and higher terms in this difference) gives

$$\chi_{1c} = \chi_{12}\phi_{20} + \chi_{13}\phi_{30} - \tfrac{1}{2}(\chi_{12} - \chi_{13})^2\phi_{20}\phi_{30} \tag{42}$$

By comparing this with Eq. (34) it is seen that the "apparent" polymer–polymer interaction parameter is

$$(\chi'_{23})_c = \tfrac{1}{2}(\chi_{12} - \chi_{13})^2 \tag{43}$$

This value can never be negative. For high molecular weight polymers, a single-phase blend is not stable if χ'_{23} is positive. Thus, it is apparent that this analysis would result in a positive χ'_{23}, provided $\chi_{12} \neq \chi_{13}$, for a two-phase composite and would thus be a signal that this is the case. However, in the case where $\chi_{12} = \chi_{13}$, vapor sorption would indicate $\chi'_{23} = 0$ for a two-phase blend even though the true parameter might be quite positive. Since it is possible, in this case, to have a one phase blend when the true χ_{23} is very near zero, there is the opportunity to be misled. Thus, for a system of unknown miscibility it would be quite important to select a solvent such that $\chi_{12} \neq \chi_{13}$ to avoid this possibility. However, if the blend is known to be a stable single phase, this does not apply. It might be convenient to select a solvent such that $\chi_{12} = \chi_{13}$ to make the differences in blend sorption versus pure polymer sorption caused by negative values of χ'_{23} more readily apparent.

If the polymer molecular weights are not very large, additional terms must be incorporated into the above equations [75]. These have not been included here for simplicity. In this case, accurate values are needed for the appropriate molecular weight average of each component.

In many cases the blend may consist of a miscible amorphous phase plus a crystalline phase of one component. Usually, vapors do not penetrate the crystal, and if the amounts sorbed are small, the degree of crystallinity will not be changed. Within these limits, this technique can be applied to crystalline blends; however, it must be recognized that the apparent compositions based on the sample as a whole, that is, ϕ_1, ϕ_{20}, and ϕ_{30}, are not the ones of interest. Rather, the analysis should be based on the compositions in the amorphous phase where sorption occurs. If only Component 2 crystallizes and if the volume fraction of its crystallinity based on the solvent free blend is ϕ_c, then the proper compositions to use in Eqs. (26)–(42) are

$$(\phi_1)_a = \phi_1/(1 - \phi_c) \tag{44}$$

$$(\phi_{20})_a = (\phi_{20} - \phi_c)/(1 - \phi_c) \tag{45}$$

$$(\phi_{30})_a = \phi_{30}/(1 - \phi_c) \tag{46}$$

IV. MEMBRANES PREPARED FROM POLYMER BLENDS

A. Uncharged Membranes

Membranes that are useful as selective barriers for the rectification of multicomponent, gaseous or liquid feeds must provide high product throughputs with sufficient selectivity to permit, in many cases, the completion of the desired separation in a single stage. In addition to these stringent requirements, the membranes must be chemically stable at the temperature and pH of the feed solution. Although membranes are usually supported, they must exhibit sufficient mechanical stability to permit normal handling during installation, use, and maintenance. Frequently, selective membranes, in addition, must be resistant to biological attack.

Although polymers are, in many cases, sufficiently permselective to provide for a single stage rectification of a binary or multicomponent feed, polymeric films that are thicker than 0.001 inch do no permit commercially interesting product throughputs.

In 1960, Loeb and Sourirajan [76a] successfully resolved this seeming dichotomy by synthesizing cellulose acetate membranes that were at once highly permeable and sufficiently permselective to permit one-stage desalination of brackish water. In substance, their invention embodied the complex casting of a cellulose acetate membrane, which comprised a thin and dense cellulose acetate top layer, overlying a much thicker porous support. The dense top layer, usually less than 1 μm in thickness, provided for selective transport of water with respect to rejected sodium chloride.

The porous sublayer, typically in excess of 100 μm in thickness, did not contribute to the selective transport, but did provide mechanical stability to the resulting composite to permit handling. The notion of asymmetry, in which one layer of the composite membrane provides mechanical stability and a top layer provides selective transport properties, is frequently the key to the spectacular performance achieved in biological membranes. Moreover, the more recent development of membranes in the form of fine hollow fibers (see Volume 2, Chapter 16) involves the controlled synthesis of fiber walls, which are in fact asymmetric, composed of a nonselective porous sublayer underlaying a rate determining and selective top layer that may be of different polymeric compositions, as illustrated in Fig. 13a.

The synthesis of asymmetric selective membranes represents the successful reduction to practice of the development of superimposed properties within a polymeric composite. Flux, selectivity, mechanical stability, and requisite thermal and pH stability are all provided by judicious selection of the polymer and techniques for controlling the polymer morphology.

The emerging technology of controlled delivery [76b] of biologically active chemicals from polymeric delivery devices also includes construction of laminated composites (see Fig. 13b). These laminated composites are, in their simplest embodiment a central core (containing the biologically active solute to be delivered to the selected environment) surrounded by rate-controlling polymeric membranes with significantly lower permeability to the solute than the polymeric matrix comprising the continuous phase of the central core. Constant rate delivery is provided from these laminated structures when the diffusional resistance of the core material is extremely low compared with the rate controlling transport resistance afforded by the exterior membranes.

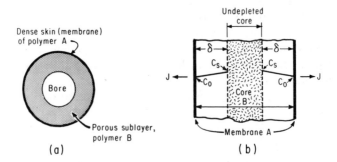

Fig. 13 Examples of composite membrane devices: (a) cross-sectional view of composite hollow fiber; (b) Higuchi-type matrix device (solute in excess of saturation suspended in a polymer matrix) with a rate controlling membrane at the surface.

In many cases, although the inherent permeability of the polymeric core matrix is exceedingly high compared with the permeability of the rate-determining barrier layer, the relative thickness of the core material provides for significant diffusional resistances as the biologically active component is depleted from the central core. Typically the solute is initially present in the core as dispersed particles within a matrix that is saturated by the solute. Usually, as solute is eluted from the device, a sharp boundary is established that divides the matrix into two regions (see Fig. 13b). A central region in the core matrix will still contain suspended particles and a uniform concentration of the dissolved and saturated solute. However, solute is no longer suspended in the region adjacent to the external membrane. In the pseudosteady state, a linear concentration profile of dissolved solute will be established in this region as a result of the diffusional resistance to solute transport within the core. A mathematical model that adequately describes this situation is provided by the following equation:

$$J = J_m(C_0/C_s) = J_m/[1 + (J_m\delta/C_sD_g)] \tag{47}$$

where J represents the flux of solute at any time from the device and J_m represents the maximum flux or the flux obtained before diffusional resistances within the central core contribute to the overall performance of the device. Here C_s represents the solubility of the solute in the core matrix, δ is the thickness of the region within the core matrix contiguous to the rate-determining membrane that is depleted of suspended solute, and C_0 is the solute concentration in the core directly adjacent to the membrane. The term $J_m\delta/C_sD_g$ represents the ratio of the membrane permeability divided by the diffusive conductance characterizing the device core. If this ratio is small compared with unity or $C_0 \cong C_s$, the device will deliver solute at a constant rate equal to J_m. Conversely, under conditions corresponding to the development of a significant diffusional resistance, the observed flux will be smaller than the flux corresponding to the specific permeability of the membrane per se.

Desalination membranes and membrane-moderated controlled delivery devices are examples of laminated structures exhibiting additivity of properties. In general, the synthesis of a polymer blend membrane is motivated by the desire to superimpose requisite properties upon the basic transport properties of a base polymer. Alternatively, the preparation of a truly miscible blend permits the continuous alteration of membrane properties as a function of blend composition.

Stannett, Hopfenberg and co-workers [77–80] attempted to superimpose the water-insensitive, creep-stable properties of glassy polystyrene upon the excellent transport properties of cellulose acetate by grafting styrene to cellulose acetate in a variety of configurations. Their studies revealed that, in many cases, the grafting procedures led to the formation of homopolymer, as well as graft copolymer, in the resulting multiphase mixture. Water sorption in the resulting grafts, as well as the steady-state water and salt permeability, were completely consistent with the domain model in which styrene-rich domains were dispersed within a matrix of cellulose acetate. These grafting techniques improved the compaction resistance of cellulose acetate. Although the excellent permselectivity of cellulose acetate was retained, the water sorption and, in turn, permeability to water were reduced. Most of their reported results describe the characterization of dense membrane structures, although some attention was directed toward the grafting to asymmetric membranes, and, additionally, attempts were further made to cast asymmetric structures from the pure graft copolymer [80].

Saltonstall and co-workers [81] have reported the successful preparation of membranes from blends of cellulose triacetate (CTA) and cellulose diacetate (CA). These blend membranes were suitable for the single-pass desalination of seawater. These blend membranes have exhibited rejections in excess to 99.9% to aqueous feeds containing approximately 3.5% dissolved sodium chloride.

Cabasso et al. [82] have extended their earlier work [56], sorption of benzene–cyclohexane mixtures in polymeric alloys of polyphosphonates and cellulose acetate, to include a study of permeation of various six-membered ring structures through these blend membranes. The results of the permeation experiments were consistent with the transport parameters determined in the earlier sorption studies. This agreement confirmed the notion that transport in these blends is controlled by molecular diffusion.

The permeability of these structures increases exponentially with temperature and, moreover, with the benzene concentration in the feed solution. The permeability similarly increases as the polyphosphonate concentration in the blend membrane is increased. Only cursory attempts were reported to synthesize high-flux, asymmetric membranes from this miscible blend system. The results suggest, however, that the blending of this miscible pair of polymers can be used as a technique to tailor systematically the permselectivity of the resulting blend structures.

Shchori and Jagur-Grodzinski [83] have described the permselective properties of blends synthesized from a crown ether copolymer and poly-(vinyl pyrrolidone) (PVP). The 18-crown-six rings were incorporated into an aromatic polyamide backbone.

The 8-membered crown ether ring strongly absorbs sodium salts from aqueous solutions. The introduction of the hydrophilic PVP into the blend structure significantly increased the water permeability of the resulting blends. Although Shchori and Jagur-Grodzinski suggest that the stability of the blend membranes in aqueous media indicates miscibility, it is similarly possible that the apparent stability of these blends in water reflects the dispersion of a PVP-rich phase within a continuous polyamide phase. The permeabilities to water of the blend membranes were approximately one order of magnitude higher than that of the corresponding unmodified aromatic polyamides. Salt rejections in the range of 95–99.5% were reported for these blend structures.

Poly(vinyl pyrrolidone) has been blended with polyurethane to form a blend membrane suitable for dialysis by Masuhara *et al.* [84]. Details of the urethane structure and dialysis rates, approximately twice as high as rates normally achieved with conventional cellophane dialysis membranes, were reported.

Reverse osmosis membranes prepared by blending a hydrophilic–hydrophobic block or graft copolymer with a homopolymer composed of the corresponding hydrophilic moieties in the block or graft copolymer were reported in a 1973 Japanese patent. Poly(vinyl pyrrolidone) or poly(acrylic acid) was used as the hydrophilic homopolymer [85].

B. Charged Membranes Including Polyelectrolyte Complexes

Gregor *et al.* [86–88] described the synthesis and properties of ion-selective blend membranes prepared from a polyelectrolyte component and an uncharged, second polymeric component. These studies were motivated by attempts to superimpose the properties of the ionic polymer upon the moisture and thermal stability of the film-forming, nonionic matrix material. The conventional technique for preparing ion exchange resins, or ion exchange membranes, involves cross-linking of the otherwise water-soluble polymer with suitable monomers, such as divinylbenzene, to insolubilize the ionogenic polymer. The blending was studied by Gregor and co-workers in an attempt to produce high charge-density membranes, which were insolubilized by the two-phase nature of the resulting hydrophilic–hydrophobic polymer blend.

Gregor's early work involved blending of poly(styrene sulfonic acid) and a film-forming copolymer of Dynel (a copolymer of acrylonitrile and vinyl chloride). In the second and third papers of this early series, polycarboxylic acid and quaternary ammonium polymers were used as the ion-selective polyelectrolytes, respectively. The early results suggested that the overall concept was compromised somewhat by the agglomeration of the polyelectrolyte component within the film-forming hydrophobic matrix, effectively blocking the active ionic groups from free exchange with the contiguous feed solution. Some 20 years later Gregor and co-workers presented a series of papers [89–91a], in which the efficacy of polymer blend membranes composed of poly(styrene sulfonic acid) and poly(vinylidene fluoride) as ultrafiltration membranes for a variety of specific applications was evaluated. Gregor, in recent work, describes the use of these membranes for concentration of biological proteins [89], purification of primary sewage effluents [90], and purification of effluents from pulp and paper mills [91]. In all cases, motivation for synthesis and, moreover, analysis for the resulting data depended upon the formation of domain structures of the polyelectrolyte component dispersed within a tough and mechanically stable nonionic film forming component.

In 1974 Caplan and Sollner presented a study [92–94] describing the techniques for synthesis and characterization of collodion-matrix membranes containing polyelectrolytes in the pores of the collodion. Two formation techniques were used. A simple "dissolution method" involved the casting of membranes from a mixed solution of polyelectrolyte and inert polymeric matrix material. The inherent incompatibility of the somewhat arbitrarily selected polymers resulted in phase separation which was sufficient to render the membranes mechanically unstable.

Conversely a "sorption method" involved the absorption of polyelectrolyte from an aqueous solution into highly porous collodion membranes. The properties of the resulting membranes were related to the ionic strength of the contiguous solutions and the extent of coiling or expansion of the polyelectrolyte molecules within the confines of the matrix pores. These composite membranes, involving polyelectrolyte-loaded pores, were effective as ion exchange membranes although the configurational changes that occurred in the polyelectrolyte phase, consequent to variations in ambient ionic strength, were required to explain the ion exchange character of these novel membranes.

The pioneering work of Gregor and Sollner was based upon the notion and, in turn, the reduction to practice of blend membranes characterized by a distinct two-phase structure. Blending was attempted in order to superimpose the transport properties of a polyelectrolyte upon the mechanical stability of a nonionic film-forming polymer.

In marked contrast, Michaels and co-workers synthesized and characterized the complex interaction product of polycations and polyanions forming a miscible, truly single-phase, polymer blend [95–97]. These complexes, termed polysalts, are completely insoluble in contact with low ionic strength aqueous solutions. Films of these novel polymer blends can be conveniently cast from ternary solvent mixtures containing water, a miscible organic solvent, and sufficient salt to shield the ionic groups from polymer–polymer interactions, leading ultimately to coprecipitation.

Michaels [98] subsequently reviewed the properties of dialysis and ultrafiltration membranes synthesized from this unique, miscible blend system.

The stoichiometric or "neutral" complex sorbs approximately 30% water by weight, while polyanion- or polycation-rich complexes may absorb up to 10 times their dry weight of water. The high sorptivities contribute to the high moisture permeabilities characteristic of these polysalt structures. Ultrafiltration membranes, prepared from these novel structures exhibit 15 times the specific water permeability of cellulose acetate of corresponding thickness. Moreover, dialysis membranes synthesized from these materials compare favorably with conventional dialysis membranes prepared from cellophane.

Polyelectrolyte complex membranes for use as ultrafiltration barriers have been synthesized and characterized by Kaneko *et al.* [99–101] in Japanese patents. The complexes are miscible blends and the resulting membranes provided raffinose retentions of 92% in contact with a 0.05% aqueous solution of raffinose.

Hollow-fiber dialysis membranes were prepared from a complicated blend of quaternized acrylonitrile–methylvinylpyridine copolymer and an acronitrile–vinyl acetate copolymer. The preparation and properties of these blends are described in a United States patent by Sayler *et al.* [102].

The preparation of polysalt complexes therefore represents a rather ingeneous limiting case of water-stable, although highly hydrated, polymer blends. The water stability is in fact a consequence of the intimacy of the polymer–polymer interactions resulting in mutual shielding of pendant ionic groups by the complementary polymeric counterions.

V. SUMMARY

Polymer blending has already been established as an ·effective means for constructively altering transport properties of polymeric materials. Blends of nylon and polyethylene under conditions of practical importance exhibit a permeability to water vapor that is in fact lower than the permeability of either homopolymer for as yet unexplained reasons. Similarly, under specific conditions of temperature and hexane activity, blends of polystyrene and poly(phenylene oxide) exhibit solubilities and sorption rates which are significantly lower than those measured for the parent homopolymers. Moreover, the tendency of either of these homopolymers to undergo deleterious morphological changes consequent to solvent contact (solvent crazing or solvent-induced crystallization) is reduced significantly by blending.

Homogeneous blend structures involving cellulose triacetate and cellulose acetate provide among the highest rejections reported to date for the selective permeation of water from brackish and sea water feeds. Moreover, the homogeneous blends of polyanions and polycations forming poly-electrolyte complexes represent a class of novel materials, characterized as miscible blends that are suitable for preparation of structurally asymmetric ultrafiltration and dialysis membranes.

In addition to the reduction to practice of blend structures as barrier resins and selective permeation membranes, the study of diffusion, sorption, and permeation in blend structures provides a valuable means for additional characterization of polymer blends. The distinction between miscible and immiscible structures is a first step in the characterization of the nature of a blend of chemically distinguishable homopolymers. The limiting models, describing series and parallel laminates and the effect of dispersed spheres and platelets on transport properties, have been shown in several cases to describe adequately the transport behavior in two phase polymeric blends. These results were consistent with complementary techniques that in-dependently characterize the geometry and distribution of the two phases. Conversely, there has been relatively little attention directed toward modeling the permeation behavior of miscible blend systems. Specifically, although linearity in a semilogarithmic plot of permeability versus blend composition has been cited empirically as a criterion for miscibility, no convincing theory has been advanced which will support this empirical finding. The use of equilibrium vapor sorption determinations to characterize the thermo-dynamic interactions between miscible polymers has only been studied in a cursory way. Moreover, the apparently miscible PS–PPO blend represents a valuable system for the characterization of both unsteady-state and steady-state transport in miscible polymer blends. Specifically, this glassy polymer blend should obey the models developed for the dual-mode sorption of gases in glassy polymers. The characterization would be carried out on a continuous series of polymer structures rather than on an arbitrarily selected, single polymer.

Experimental determination of sorption and permeation behavior appears to be an extremely powerful tool for characterizing the detailed nature of blends. Thermodynamic and kinetic information are readily available from these experiments, and, moreover, the use of penetrants as probes to infer meaningful information regarding the detailed structure of polymers recently has been reduced to a meaningful practice. It would seem, therefore, that continued study of transport phenomena in blends would be motivated not only by the requirement for synthesis of improved materials but also by the continued and growing interest in the nature and characterization of the polymer solid state.

REFERENCES

1. J. F. Henderson and M. Szwarc, *Macromol. Rev.* **3**, 317 (1968).
2. L. H. Sperling and D. W. Friedman, *J. Polym. Sci. Part A-2* **7**, 425 (1969).
3. C. H. M. Jacques and H. B. Hopfenberg, *Polym. Eng. Sci.* **14**, 441 (1974).
4. T. Graham, *Phil. Mag.* **32**, 401 (1866).
5. S. Von Wroblewski, *Wied. Ann. Phys.* **8**, 29 (1879).
6. H. A. Daynes, *Proc. Roy. Soc. London* **A94**, 286 (1920).
7. R. M. Barrer, *Trans. Faraday Soc.* **39**, 237 (1939).
8. R. F. Baddour, A. S. Michaels, H. J. Bixler, R. P. DeFilippi, and J. A. Barrie, *J. Appl. Polym. Sci.* **8**, 897 (1964).
9. H. B. Hopfenberg and H. L. Frisch, *J. Polym. Sci. Part B* **7**, 404 (1969).
10. V. Stannett and H. B. Hopfenberg, *in* "The Physics of Glassy Polymers" (R. N. Haward, ed.), p. 504. Appl. Sci. Publ., London, 1973.
11. W. R. Vieth, J. M. Howell, and J. H. Hsieh, *J. Membrane Sci.* **1**, 177 (1976).
12. W. R. Vieth, C. S. Frangoulis, and J. A. Rionda, *J. Colloid Interface Sci.* **22**, 454 (1966).
13. W. J. Koros, D. R. Paul, M. Fujii, H. B. Hopfenberg, and V. Stannett, *J. Appl. Polym. Sci.* **21**, 2899 (1977).
14. D. R. Paul, *J. Polym. Sci. Part A-2* **7**, 1811 (1969).
15a. D. R. Paul and W. J. Koros, *J. Polym. Sci. Polym. Phys. Ed.* **14**, 675 (1976).
15b. W. J. Koros, D. R. Paul, and A. A. Rocha, *J. Polym. Sci. Polym. Phys. Ed.* **14**, 687 (1976).
16. J. Crank and G. S. Park, *in* "Diffusion in Polymers" (J. Crank and G. S. Park, eds.), p. 1. Academic Press, New York, 1968.
17. T. Alfrey Jr., E. F. Gurnee, and W. G. Lloyd, *J. Polym. Sci. Part C* **12**, 219 (1966).
18. H. B. Hopfenberg, *in* "Controlled Release Polymeric Formulations" (D. R. Paul and F. W. Harris, eds.), Amer. Chem. Soc. Symp. Ser., Vol. 33, p. 26. Amer. Chem. Soc., Washington, D.C., 1976.
19. C. H. M. Jacques, H. B. Hopfenberg, and V. Stannett, "Permeability of Plastic Films and Coatings to Gases, Vapors and Liquids" (H. B. Hopfenberg, ed.), Polym. Sci. and Technol. Ser., Vol. 6, p. 73. Plenum, New York, 1974.
20. A. R. Berens, *Amer. Chem. Soc. Polym. Prepr.* **15**, No. 2, 197 (1974).
21. A. R. Berens, *Amer. Chem. Soc. Polym. Prepr.* **15**, No. 2, 203 (1974).
22. J. C. Maxwell, "Electricity and Magnetism," 3rd ed., Vol. 1. Dover, New York, 1904.
23. L. E. Nielsen, *J. Macromol. Sci. (Chem.)* **A15**, 929 (1967).
24. A. S. Michaels and H. J. Bixler, *J. Polym. Sci.* **50**, 413 (1961).
25. R. M. Barrer, J. A. Barrie, and M. G. Rogers, *J. Polymer Sci. Part A* **1**, 2565 (1963).
26. R. M. Barrer, *in* "Diffusion in Polymers" (J. Crank and G. S. Park, eds.), p. 165. Academic Press, New York, 1968.
27. D. R. Paul and D. R. Kemp, *J. Polym. Sci. Part C* **41**, 79 (1973).
28. L. M. Robeson, A. Noshay, M. Matzner, and C. N. Merriam, *Die Angew. Makromol. Chem.* **29/30**, 47 (1973).
29. W. I. Higuchi and T. Higuchi, *J. Amer. Pharm. Assoc. Sci. Ed.* **49**, 598 (1960).
30. B. G. Ranby, *J. Polym. Sci. Part C* **51**, p. 89 (1975).
31. A. E. Barnabeo, W. S. Creasy, and L. M. Robeson, *J. Polym. Sci.* **13**, 1979 (1975).
32. H. Storstrom and B. Ranby, *in* "Multicomponent Polymer Systems" (N. A. J. Platzer, ed.), Adv. in Chem. Ser. Vol. 99, p. 107. Amer. Chem. Soc., Washington, D.C., 1971.

33. Y. J. Shur and B. Rånby, *J. Appl. Polym. Sci.* **19**, 1337 (1975).
34. Y. J. Shur and B. Rånby, *J. Appl. Polym. Sci.* **19**, 2143 (1975).
35. Y. J. Shur and B. Rånby, *J. Appl. Polym. Sci.* **20**, 3105 (1976).
36. Y. J. Shur and B. Rånby, *J. Appl. Polym. Sci.* **20**, 3121 (1976).
37. Y. J. Shur and B. Rånby, Gas permeation in polymer blends, Part V, *J. Appl. Polym. Sci.* (in press).
38. Y. J. Shur, J. F. Rabek, and B. Rånby, Selective permeation of gases through polymer blends, *Int. Symp. Macromol. IUPAC, 24th, Jerusalem, July 13–18* (1975).
39. Y. Ito, *Kogyo Kagaku Zasshi* **63**, 2016 (1960).
40. J. Barbier, *Rubber Chem. Technol* **28**, 814 (1955).
41. E. T. Pieski, *in* "Polythene" (A. Renfrew and P. Morgan, ed.), 2nd ed., p. 379. Wiley (Interscience), New York, 1960.
42. R. W. Moncrieff, "Man-Made Fibers," 6th ed., p. 606. Wiley, New York, 1975.
43. R. B. Mesrobian and C. J. Ammondson, U.S. Patent No. 3,093,255 (June 11, 1963).
44. M. I. Kohan, *in* "Nylon Plastic" (M. I. Kohan, ed.), p. 418. Wiley, New York, 1973.
45. K. Komatsu and I. Masahiro, *Plast. Age (Jpn.)* **19**, 95 (1973) [*Chem. Abstr.* **81**, 92550j (1974)].
46. A. E. Barnabeo, W. S. Creasy, and L. M. Robeson, *J. Polym. Sci. Part A-2* **13**, 1979 (1975).
47. C. M. Peterson, *J. Appl. Polym. Sci.* **12**, 2649 (1968).
48. K. D. Ziegel, *J. Macromol. Sci. Phys.* **B5**(1), 11 (1971).
49. K. Toi, K. Igarashi, and T. Tokuda, *J. Appl. Polym. Sci.* **20**, 703 (1976).
50. H. Odani, K. Taira, N. Nemoto, and M. Kurata, *Bull. Inst. Chem. Res.* **53**, 216 (1975).
51. H. Odani, K. Taira, T. Yamaguchi, N. Nemoto, and M. Kurata, *Bull Inst. Chem. Res.* **53**, 409 (1957).
52. R. L. Stallings, H. B. Hopfenberg, and V. Stannett, *J. Polym. Sci. Part C* **41**, 23 (1973).
53. D. M. Cates and H. J. White Jr., *J. Polym. Sci.* **20**, 181 (1956).
54. D. M. Cates and H. J. White Jr., *J. Polym. Sci.* **20**, 155 (1946).
55. D. M. Cates and H. J. White Jr., *J. Polym. Sci.* **21**, 125 (1956).
56. I. Cabasso, J. Jagur-Grodzinski, and D. Vofsi, *J. Appl. Polym. Sci.* **18**, 2117 (1974).
57. C. H. M. Jacques, H. B. Hopfenberg, and V. Stannett, *Polym. Eng. Sci.* **13**, 81 (1973).
58. C. H. M. Jacques and H. B. Hopfenberg, *Polym. Eng. Sci.* **14**, 449 (1974).
59. H. B. Hopfenberg, V. T. Stannett and G. M. Folk, *Polym. Eng. Sci.* **15**, 261 (1975).
60. E. P. Cizak, U.S. Patent No. 3,383, 435 (May 14, 1968).
61. R. P. Kambour, E. E. Romagosa, and C. L. Gruver, *Macromolecules* **5**, 335 (1972).
62. B. R. Baird, H. B. Hopfenberg, and V. Stannett, *Polym. Eng. Sci.* **11**, 274 (1971).
63. A. R. Shultz and C. R. McCullough, *J. Polym. Sci. Part A-2* **10**, 307 (1972).
64. T. Nishi and T. T. Wang, *Macromolecules* **8**, 909 (1975).
65. R. L. Imken, D. R. Paul, and J. W. Barlow, *Polym. Eng. Sci.* **16**, 593 (1976).
66. G. Allen, G. Gee, and J. P. Nicholson, *Polymer* **2**, 8 (1961).
67. G. L. Slonimskii, *J. Polym. Sci. Part C* **30**, 625 (1958).
68. A. A. Tager, T. I. Scholokhovich, and J. S. Bessonov, *Eur. Polym. J.* **11**, 321 (1975).
69. S. Ichihara and T. Hata, *Kobunshi Kagaku* **26**, 249 (1969).
70. A. A. Tager, T. I. Sholokhovic, I. M. Sharova, L. V. Adamova, and Y. S. Bessonov, *Vysokomol. Soedin. Ser. A.* **17**, 2766 (1975).
71. S. Ichihara, A. Komatsuj and T. Hata, *Polym. J.* **2**, 640 (1971).
72. T. K. Kwei, T. Nishi, and R. F. Roberts, *Macromolecules* **7**, 667 (1974).
73. D. D. Deshpande, D. Patterson, H. P. Schreiber, and C. S. Su, *Macromolecules* **7**, 530 (1974).
74. O. Olabisi, *Macromolecules* **8**, 316 (1975).

75. P. J. Flory, "Principles of Polymer Chemistry," Chapter 8. Cornell Univ. Press, Ithaca, New York, 1953.
76a. S. Loeb and S. Sourirajan, UCLA Dept. of Eng. Rep. 60-60 (1960).
76b. D. R. Paul and F. W. Harris (eds.), "Controlled Release Polymeric Formulations," Amer. Chem. Soc. Symp. Ser., Vol. 33. Amer. Chem. Soc., Washington, D.C., 1976.
77. H. B. Hopfenberg, F. Kimura, P. T. Rigney, and V. T. Stannett, *J. Polym. Sci., Part C* **28**, 243 (1968).
78. H. B. Hopfenberg, V. Stannett, F. Kimura, and P. T. Rigney, *Appl. Polym. Symp., No. 13*, p. 139 (1970).
79. V. T. Stannett and H. B. Hopfenberg, "Cellulose and Cellulose Derivatives," (N. Bikales and L. Segal, eds.), Vol. 5, Chapter 6, p. 907. Wiley, New York, 1971.
80. F. Kimura-Yeh, H. B. Hopfenberg, and V. Stannett, *in* "Reverse Osmosis Membrane Research" (H. K. Lonsdale and H. E. Podall, eds.), p. 177. Plenum, New York, 1972.
81. W. N. King, D. L. Hoernschemeyer, and C. W. Saltonstall Jr., *in* "Reverse Osmosis Membrane Research" (H. K. Lonsdale and H. E. Podall, eds.), p. 131. Plenum, New York, 1972.
82. I. Cabasso, J. Jagur-Grodzinski, and D. Vofsi, *J. Appl. Polym. Sci.* **18**, 2137 (1974).
83. E. Shchori and J. Jagur-Grodzinski, *J. Appl. Polym. Sci.* **20**, 773 (1976).
84. E. Masuhara, N. Nakaboyoshi, Y. Imai, A. Watanabe, and S. Kazem, Japanese Patent No. 88,077 (1973) [*Chem. Abstr.* **80**, 10946g (1974)].
85. H. Kawai, T. Soen, F. Fujimoto, and T. Shiroguchi, Japanese Patent No. 100,780 (1973) [*Chem. Abstr.* **81**, 64803t (1974)].
86. H. P. Gregor, H. Jacobson, R. C. Shair, and D. M. Wetstone, *J. Phys. Chem.* **61**, 141 (1957).
87. H. P. Gregor and D. M. Wetstone, *J. Phys. Chem.* **61**, 147 (1957).
88. D. M. Wetstone and H. P. Gregor, *J. Phys. Chem.* **61**, 151 (1957).
89. C. C. Gryte and H. P. Gregor, *Amer. Chem. Soc. ORPL Prepr.* **35**, 458 (1975).
90. S. Mizrahi, H. J. Hsu, H. P. Gregor, and C. C. Gryte, *Amer. Chem. Soc. ORPL Prepr.* **35**, 468 (1975).
91. S. Mizrahi, H. J. Hsu, C. C. Gryte, and H. P. Gregor, *Amer. Chem. Soc. ORPL Prepr.* **35**, 475 (1975).
91a. C. C. Gryte and H. P. Gregor, *J. Polym. Sci. Polym. Phys. Ed.* **14**, 1839, 1855 (1976).
92. S. R. Caplan and K. Sollner, *J. Colloid Interface Sci.* **46**, 46 (1974).
93. S. R. Caplan and K. Sollner, *J. Colloid Interface Sci.* **46**, 67 (1974).
94. S. R. Caplan and K. Sollner, *J. Colloid Interface Sci.* **46**, 77 (1974).
95. A. S. Michaels and R. G. Miekka, *J. Phys. Chem.* **65**, 1765 (1961).
96. A. S. Michaels, L. Mir, and N. S. Schneider, *J. Phys. Chem.* **69**, 1447 (1965).
97. A. S. Michaels, G. L. Falkenstein, and N. S. Schneider, *J. Phys. Chem.* **69**, 1456 (1965).
98. A. S. Michaels, *Ind. Eng. Chem.* **57**, 32 (1965).
99. S. Kaneko, T. Koichi, J. Negoro, M. Miwa, and E. Tsuchida, Japanese Patent No. 10,232 (1974) [*Chem. Abstr.* **81**, 14466r (1974)].
100. S. Kaneko, K. Tagawa, M. Miwa, J. Negoro, and E. Tsuchida, Japanese Patent No. 10,233 (1974) [*Chem. Abstr.* **81**, 144675 (1974)].
101. M. Miwa, K. Tagawa, S. Kaneko, J. Negoro, and E. Tsuchida, Japanese Patent No. 15,735 (1974) [*Chem. Abstr.* **81**, 13722Y (1974)].
102. I. O. Sayler, J. S. Tapp, and W. E. Weesner, U.S. Patent No. 3,799,356 (March 26, 1974) [*Chem. Abstr.* **81**, 26689d (1974)].

Appendix

Conversion Factors to SI Units

FORCE (newton, N)
$$1\ N = 1\ m \cdot kg/sec^2$$
$$= 10^5\ dyn$$
$$= 0.1020\ kgf$$
$$= 0.2248\ lbf$$

PRESSURE OR STRESS (pascal, Pa)
$$1\ Pa = 1\ N/m^2$$
$$= 10.0\ dyn/cm^2$$
$$1\ kPa = 7.50\ mmHg = 7.50\ Torr$$
$$1\ MPa = 10.00\ bar$$
$$= 10.20\ kg/cm^2$$
$$= 145.0\ psi$$

ENERGY (joule, J)
$$1\ J = N \cdot m$$
$$= 10^7\ erg$$
$$= 0.2387\ cal$$
$$1\ kJ = 0.9471\ Btu$$
$$1\ MJ = 0.2778\ kWh$$

FORCE PER UNIT LENGTH
$$1\ mN/m = 1\ dyn/cm$$
$$1\ kN/m = 5.710\ lbf/in$$

VISCOSITY
$$1\ Pa \cdot sec = 10.00\ P$$
$$1\ mPa \cdot sec = 1.000\ cP$$
$$1\ m^2/sec = 10^4\ stokes$$

Index

A

Acrylic acid, 35
  at room temperature, 37
Acrylic polymers
  literature on, 79–86
  surface tension of, 261
Acrylonitrile-butadiene copolymers, 460
Additives
  interfacial tension and, 272–274
  surface tension of, 256
Adhesion
  chemical, 287–288
  diffusion theory of, 284–287
  fracture theory of, 279–281
  interfacial tension and, 263
  weak boundary layer theory of, 281–282
  wetting-contact theory of, 282–284
Amorphous PPO blends, morphology of, 214–218
Anisotropy, scattering arising from, 431–433
Antonoff's rule, 263–264

B

Benzyl cellulose
  literature on, 62–63
  at room temperature, 36
Binodal curves, molecular weight and, 148
Binary mixtures
  Flory theory for, 119
  lattice fluid and, 125–128
Birefringence, in polymer blends, 433–438
Block copolymers, surface tension of, 255–256, surface tension of, 255–256, 262
Butadiene compatibility pairs, 34–35
Butadiene-co-styrene, at room temperature, 39

Butyl acrylate, at room temperature, 37
Butyl methacrylate, at room temperature, 37
n-Butyl methacrylate, at room temperature, 38

C

Cellulose acetate
  literature on, 54–61
  at room temperature, 36
Cellulose acetate-butyrate, 34
  at room temperature, 36
Cellulose derivatives, literature on, 53–63
Chlorinated isoprene, 35
Chlorinated polyethylene, literature on, 91
Chlorinated polyisoprene, literature on, 86, 91
Chlorinated poly(vinyl chloride), 87
  at room temperature, 38
Chloroprene, at room temperature, 38, 41
Chlorosulfonated ethylene, at room temperature, 38
Chlorosulfonated polyethylene, 87
  literature on, 91–92
Cloud point, molecular weight and, 171
Cloud point curves, 143
  for SAN-PMMA, 177
  for silicone and polyisoprene mixtures, 145
Cocurrent flow, in two fluids, 304–310
Cohesion, interfacial tension and, 263–264
Coiled tube, viscoelastic flow through, 304
Compatible polymers, blends of, 254
Compatibility
  conditional, 34–36
  defined, 2, 16–18, 447
  dynamic mechanical relaxation in, 194–196

"edge" of, 18
experimental data in, 32–41
experimental determination of, 18–20
hydrogen bonding and, 41
literature on, 16–17, 41–44
with mutual solvents, 19
of PPO-polychlorostyrene blends, 218–219
steps necessary to predict, 48–50
Compatibility criteria, in polymer blend compatibility prediction, 132–138
Compatibility-incompatibility transition, for PPO blends, 224–229
Compatibility prediction, 32, 41–52, 132–138
solubility parameter methods in, 135
Complementary dissimilarity, 7
Composite modulus, *see also* Modulus-composition dependence
experimental data in, 374–381
model calculations for, 369–374
temperature dependence of, 369–381
Copolymers
dispersion modes of, 335
random, 253
solubility parameters of, 30
Correlation distance, defined, 414
Correlation function
in poly(methyl methacrylate)-polystyrene system, 418
in poly(methyl methacrylate)-polyurethane system, 420
scattering and, 413–415
Couette flow, 347
Crystalline components, scattering from blends with, 428–431
Crystalline isotactic polystyrene blends, morphology of, 209–212, 214–217
Crystalline PPO-crystalline iPS blends, morphology of, 212–214, 217–218

D

Debye–Bueche analysis, correlation function and, 419–421
Deformable droplets
microrheology of, 318–322
rheological behavior of suspensions of, 318–325

Deformable particles, macrorheology of, 322–325
Dielectric method, for PPO blends, 221–224
Dielectric relaxation, 196
Differential scanning calorimetry, 20, 219–220
Differential thermal analysis, 20, 193
Dilatometry, in polyblend glass transition temperature studies, 196–197
2,6-Dimethyl-1,4-phenylene ether, 35
Dispersions, rheology of, 296–297
DSC, *see* Differential scanning calorimetry
DTA, *see* Differential thermal analysis
Dynamic mechanical relaxation method, in compatibility studies, 194–196, 204–208, 221–222

E

ε-Caprolactone compatibility pairs, 34–35
Elastic response, in modulus-composition dependence, 356–367
Electron spin resonance, in PPO relaxation studies, 204
EPDM rubber, 35
Epichlorohydrin, at room temperature, 38–39
Equation of state theory, 31–32
as Ising or lattice fluid theory, 116, 121–124
Ester, at room temperature, 39
Ethyl acrylate, at room temperature, 37–38
(Ethyl acrylate)-*co*-(methyl metacrylate), at room temperature, 40
Ethylcellulose
literature on, 61–62
at room temperature, 36
Ethylene, at room temperature, 41
Ethylene-*co*-(vinyl acetate), at room temperature, 39
Ethylene glycol, at room temperature, 38
Ethylene oxide, 35
Ethylene-propylene pairs, 35
Ethylene-vinyl acetate copolymers, 460

F

Fick's laws, 449–450
Filled polymers, rheology of, 325–330

Flory equation of state analysis, 6, *see also* Equation of state theory
Flory–Huggins theory, 4, 25–32, 41, 44, 116–121
  compatibility predictions in, 135–137
  molecular surface area corrections in, 131
  polymer blend compatibility prediction in, 132–138
  simple scheme based on, 45–50
  solubility parameters for, 45–48
  thermal expansion coefficients and, 43
Flow behavior, *see* Rheology
Fractional polarity theory, 264–265
Free energy, Flory–Huggins expression for, 142

G

Gas permeation behavior, in polymer blends, 460–467
Geometric mean equation, 265
  interfacial tension and, 274–275
Glass transition, surface tension and, 249–251
Glass transition temperature
  crystallinity and, 200
  dielectric relaxation in, 196
  differences in, 197–198
  limitations of as compatibility criterion, 197–201
  in polymer mixtures, 20
  of PPO-poly-chlorostyrene blends, 219–222
  of PPO-PS blends, 202–209
  size of dispersed phase and, 229–230
Glass transition temperature measurements, 193–197
  phase composition by, 167–168
Good–Girifalco theory, 264
Group molar attraction constants, 46–47
Guinier equation, 416

H

Halogenated hydrocarbon polymers, surface tension of, 261
Harmonic mean equation, 265
  interfacial tension and, 274–275
HDPE, *see* High density polyethylene

Helfand–Roe lattice theories, 267–268
Helfand–Tagami–Sapse mean-field theory, 266–267
Henry's law, 449, 451
Heterogeneous polymeric systems, permeation through, 454–467
Hexyl methacrylate, at room temperature, 38
High-density polyethylene, blends of, 462, 464
Homopolymers
  glass transition temperatures of, 197–198
  literature on, 86–102
Hume–Rothery rules, in miscibility, 9
Hydrocarbon polymers, surface tension of, 260

I

Incompatibility, defined, 2–3, *see also* Compatibility
Infrared dichrosim, in polymer blends, 439–441
Interaction energy parameter, in phase separation behavior, 151
Interfacial chemical bonding, 287–288
Interfacial tension, 263–278
  additives and, 272–274
  estimation of, 274–277
  mean-field and lattice theories in, 275–277
  molecular weight and, 272
  polarity and, 269–271
  tabulation of data in, 277–278
  temperature dependence in, 268–269
  theories of, 263–268
Interpenetrating networks
  formation of, 17
  of phases, 11
iPS, *see* Isotactic polystyrene
Isobutene, at room temperature, 38
Isoprene, at room temperature, 36
Isoprene-butadiene pairs, 34
Isotactic polystyrene, 209–212, 217–218
  crystalline, 209–212, 214–217

L

Langmuir absorption isotherm, 451
Lattice fluid
  equation of state for, 121–125

phase stability and spinodal in, 128–131
Lattice fluid theory
  compatibility prediction and, 132–134,
    137–138
  interfacial tension and, 267–268, 275–277
LCST, *see* Lower critical solution temperature
Light absorption, in polymers, 396–398
Light scattering
  correlation functions in, 419
  in phase separation behavior, 179–180
  in polymers, 398–400
  transparency and, 401–403
Limiting transport mechanisms, 468–474
Low-density polyethylene blends, 462, 464
Lower critical solution temperature, 10, 21
  dynamical mechanical technique and, 195
  increasing pressure and, 130
  phase separation behavior and, 146–151
  in polymer solutions, 115–116
  spinodals and, 129
  systems exhibiting, 147–151
  thermal expansion coefficients and, 147

M

MacLeod's experiment, for various polymers, 246
Mean-field theory, 266–267
  interfacial tension and, 275–277
MEK, *see* Methyl ethyl ketone
Melt temperature depression, kinetics versus thermodynamics in, 231–232
Membranes
  charged, 483–485
  polyelectrolyte complexes and, 483–485
  for polymer blends, 479–485
  uncharged, 479–483
Methacrylic acid, at room temperature, 38
Methyl ethyl ketone, 212
Methyl methacrylate
  compatibility pairs of, 35
  at room temperature, 37
Mie theory, in scattering of fluids, 409–427
Miscibility
  in amorphous phase, 9
  defined, 3
  of metals, 9
  through specific interactions, 7–9

Miscibility criteria
  optical clarity in, 186–188
  for polyblends, 186–193
  transitions in, 188–193
Mixing, free energy of, 143
Mixtures
  close-packed mass density of, 127
  pair interactions in, 125
Modulus-composition dependence, *see also*
  Composite modulus
  elastic response in, 356–367
  limits or bounds in, 365–367
  mechanical coupling models in, 357–359
  self-consistent models in, 359–365
  viscoelastic response in, 367–369
Mold surfaces, properties of, 256–258
Molecular surface-to-volume ratios, in Flory
  theory, 131
Molecular volume ratios, significance of,
  131
Molecular weight, 4–6
  cloud point and, 171–175
  interfacial tension and, 272
  mutual solubility and, 147
  surface tension and, 247–249
Multiphase polymer blends
  mechanical properties of, 353–388
  modulus-composition dependence in,
    356–367
  small deformations in, 354–355
  and temperature dependence composite
    modules, 369–381
  time-temperature dependence and, 381–
    388
Multiphase polymeric systems, permeation
  through, 454–467
Mutual solubility, molecular weight and, 147

N

NBR, *see* Acrylonitrile-butatiene copolymers
Neutron scattering, in polymers, 398–400
Newtonian fluids, cocurrent flow of, 304–
  307
Newtonian suspension media, rheology of
  rigid spherical suspensions in, 310–313
Nitrocellulose
  compatibility pairs involving, 34
  literature on, 56–58

at room temperature, 36
Noryl resin, PPO blends with, 234–238
Nuclear magnetic resonance, 18
    phase composition by, 168–171
    in PPO studies, 204
    pulsed, 162–165
Nucleation, in phase separation behavior,
    152–154

O

Optical behavior
    absorption and, 396–398
    polarization and, 395–396
    refraction and, 396
    scattering in, 398–400
Optical clarity, miscibility and, 186–188
Optical properties, of polymer blends, 393–
    442

P

PAN, *see* Polyacrylonitrile
Parachor method, 258–259
PBD, *see* Polybutadiene
PE, *see* Polyethylene
PEMA, *see* Polyethyl methacrylate
Phase boundary, in phase separation be-
    havior, 167–175
Phase connectivity, spinodal decomposition
    and, 175
Phase composition
    by glass transition temperature mea-
        surements, 167–168
    by NMR measurements, 168–171
Phase diagrams
    literature on, 42
    three-component, 24–25
    types of, 20–25
Phase equilibrium, 9–11
    thermodynamic criteria for, 142
    upper and lower critical solution tempera-
        ture behavior in, 10
Phase morphology, 11–13
Phases, interpenetrating networks of, 11
Phase separation behavior
    amorphous mixtures in, 165–167
    interaction parameters in, 165–167
    kinetics of, 154–158, 179–183

and light scattering of polymer solutions,
    161, 179–180
morphology of, 175–179
nucleation and growth in, 154–155
phase boundary in, 167–175
phase diagrams and, 152
phase stability and, 152
pulsed nuclear magnetic resonance in,
    162–165
spinodal decomposition mechanism and,
    152–158
stable region in, 159–161
standard methods in, 161
thermodynamics theory of, 20
upper and lower critical solution tempera-
    tures in, 146–151
Phase transition, surface tension and, 249–
    251
PMMA, *see* Poly(methyl metacrylate)
Poiseuille flow, 301–303, 337
Polarity, interfacial tension and, 269–270
Polarization
    in polymers, 395–396
    Raman, 442
Polyacrylonitrile blends, water sorption in,
    468
Polyamides, surface tension of, 262
Polycarbonates, literature on, 94
Polyblend glass transition temperature
    studies, dilatometry in, 196–197
Polyblends, *see also* Polymer blends
    defined, 2
    glass transition temperature in, 193–197
    miscibility behavior for, 186–193
    solid state transition behavior of, 185–238
    unsolved transition behavior problems in,
        229–234
Polybutadiene, 20
    literature on, 70–71
Poly(butadiene-*co*-acrylonitrile), literature
    on, 100, 104
Poly(butadiene-*co*-styrene), 20
    literature on, 99, 103
Polybutene, literature on, 91
Polychloroprene, literature on, 86, 91
Polydichlorostyrene, literature on, 88
Poly(2,6-dimethyl-1,4-phenylene oxide), 208
    blends of with PS, 190
    differential scanning calorimetry and,
        193–194

solid-state transition behavior of blends
with, 201–218
Polyethyl acrylate, 10
Poly(ε-caprolactone)-polycarbonate, 10
Polyepichlorohydrin, literature on, 94
Polyesters, surface tension of, 262
Polyethers, surface tension of, 261
Polyethylene
chlorosulfonated, 87
high density, 462, 464
literature on, 69–70
Poly(ethylene dioxide), literature on, 89
Polyethylene-polypropylene blends, 462
Polyethylmethacrylate, 10
Polyisobutene, literature on, 86, 91
Polyisoprene
chlorinated, 86
literature on, 63–65
Polymer(s)
adhesion between, 278–288
compatibility of at room temperatures,
34–41
filled, 325–330
MacLeod's exponent for, 246
radiation and, 394–400
solubility parameters for, 48–49
surface tension of, 245–252, 260–263
Polymer blends, *see also* Polyblends
applications of, 13
birefringence in, 433–438
as block-and-graft copolymers, 446
classification of, 447–448
compatibility of, 447
compatibility prediction for, 132–138
defined, 445
dispersion modes for, 330–336
dispersion theory for, 344–345
gas permeation behavior in, 460–467
infrared dichroism in, 439–441
membranes prepared from, 479–485
miscibility in, 9
multiphase, *see* Multiphase polymer
blends
normal stress function for, 342–344
optical behavior of, 393–442
permeability and barrier properties of,
460–467
phase and transition behavior in, 9–10
Raman polarization in, 442
rheology of, 295–349

sorption of vapors and liquids in, 468–479
statistical thermodynamics of, 115–138
terminology of, 2
thermodynamic characterization of by
vapor sorption techniques, 474–479
transparency and scattering in, 401–433
transport phenomena in, 445–486
viscosity of, 336–342
Polymer compatability, *see* Compatibility
Polymer melts, bicomponent extrusional,
309
Polymer mixtures, *see also* Polymer blends,
Polymer–polymer mixtures or materials
dilute solutions of, 29
phase relationship in, 142–146
review of data in literature of, 52–106
Polymer pairs
conditionally compatible, 34–36
room temperature compatibility of, 36–40
Polymer–polymer compatibility, 15–106,
see also Compatibility
Polymer–polymer mixtures or materials
compatibility in, 2, 15–106
miscibility in, 3
phase diagrams for, 26
phase separation behavior in, 141–183
thermodynamics of, 2
Polymer–polymer solutions, light scattering
from, 3
Polymer science, expansion of, 1
Polymer solutions
equations of state theories of, 31–32
Flory–Huggins theory of, 25–31
light scattering from, 179–180
thermodynamics of, 31–32
Poly(methyl methacrylate), 10
correlation function technique for, 418
dispersion mode for, 332–334
optical clarity of, 186
Poly(*m*-methylstyrene), literature on, 88
Polyolefin-nylon blends, 463
Poly(*p*-chlorostyrene), literature on, 88
Polypropene, literature on, 86
Polypropylene, carbonate-filled, 326
Polysiloxanes, block copolymers of, 456–
460
Polystyrene
glass transition temperatures for, 202–
209
literature on, 74–78

Poly(styrene-*co*-acrylonitrile)-poly(methyl mathacrylate) mixtures, 175
Polystyrene-ethylcellulose mixtures, 20
Polystyrene-polybutadiene mixtures, 21, 24
Polystyrene-polyethylene blends, viscosity of, 338–339
Polystyrene-poly(vinyl methyl ether) mixtures, 175
Polysulfones, block copolymers of, 456–460
Polyurethane, literature on, 90, 95–96
Poly(vinyl acetate)
  literature on, 71–74
  optical clarity of, 186
Poly(vinyl butyral), literature on, 87–88, 93
Poly(vinyl chloride)
  literature on, 65–69, 100
  miscibility with, 7–8
Poly(vinyl chloride)-ethylene-vinyl acetate copolymers, 460
Poly(vinyl fluoride), 87
Poly(vinylidene chloride), literature on, 87, 92–93
Poly(vinylidine fluoride), 10
Poly(2-vinylpyridine), literature on, 87
Poly(vinyl pyrrolidone), literature on, 93
PPO, *see* Poly(2,6-dimethyl-1,4-phenylene) oxide
PPO-polychlorostyrene blends
  compatibility-incompatibility transitions for, 224–229
  solid state transition behavior of, 218–229
PPO-polystyrene blends, 234–238
Pressure-volume-temperature properties, 116
Prigogine corresponding state theories, 116
Propyl methacrylate, at room temperature, 37
PS, *see* Polystyrene
PSF, *see* Polysulfones
PSX, *see* Polysiloxanes
Pulsed nuclear magnetic resonance, in phase separation behavior, 162–165, *see also* Nuclear magnetic resonance
PVA, *see* Poly(vinyl acetate)
PVC, *see* Poly(vinyl chloride)

R

Radiation, interaction with polymers, 394–400

Random copolymers, surface tension of, 253, 262
Rayleigh–Gans–Debye scattering, model approach for, 402–409
Rayleigh ratio, 406
Refraction, in polymers, 396
Resins, at room temperature, 38–39
RGD scattering, *see* Rayleigh–Gans–Debye scattering
Rheology
  of filled polymers, 325–330
  of rigid axisymmetric particle suspensions in viscoelastic fluids, 310–318
  of suspensions of deformable particles, 318–325
  two-level approach to, 296–297
  viscosities in, 345
Rigid particles, concentrated dispersions of, 314–318
Room temperature, compatibility of polymers at, 34–41
Rubber, gas transmission behavior of, 462

S

SAXS, *see* Small angle x-ray scattering
Scattering
  from anisotropy, 431–433
  from blends with crystalline components, 428–431
  correlation function in, 413–415
  of light, *see* Light scattering
  Mie theory in, 409–427
  Rayleigh–Gans–Debye, 402–409
  statistical description of, 413–427
  types of, 398–410
Second-order fluid, flow of through tubes, 301–303
Small-angle x-ray scattering, 212–215
Small-molecule liquids, surface tension of, 245–247
Solid polymers, surface tension of, 252
Solid state transition behavior of fluids, 185–238
  differences in techniques used for, 232–234
Solubility parameter method, 48, 258–259
Spinodal(s)
  calculation of, 143
  defined, 128

in quasi-binary systems, 144
Spinodal curves, molecular weights and, 148
Spinodal decomposition
  morphology of mixtures in, 158–159
  phase connectivity and, 175–177
  in phase separation behavior, 152, 155–158
Spinodal inequality
  in compatibility prediction, 132
  liquid–liquid temperature–composition diagram and, 130
Stress function, of polymer blends, 342–344
Styrene, 35
  at room temperature, 37
Styrene-*co*-acrylonitrile, 36
Styrene-*co*-maleic acid, at room temperature, 40
Styrene-isoprene mixtures, 21
Surface tension, 245–263
  of blends of compatible copolymers, 254
  of copolymers, blends, and additives, 252–256
  corresponding state method in, 259–260
  estimation of, 258–260
  mold surfaces and, 256–258
  molecular weight and, 247–249
  parachor method in, 258–259
  of random copolymers, 253
  of solid polymers, 252
  tabulation of data in, 260–263
  temperature dependence and, 245–247

Thermorheologically simple materials,
  time–temperature dependence for, 381–383
THF, *see* Tetrahydrofuran
Three-component mixtures, phase diagrams for, 42
Three-polymer mixtures, literature on, 102
Time–temperature dependence
  for multiphase polymer blends, 381–388
  for thermorheologically simple and complex materials 381–388
TOA, *see* Thermal optical analysis
Transitions
  phase equilibrium and, 9–11
  solid-state behavior of, 185–238
Transparency, light scattering and, 401–402
Transport phenomena
  in membranes prepared from polymer blends, 479–485
  and permeation through multiphase polymeric systems, 454–467
  in polymer blends, 445–486
  vapor and liquid sorption in, 468–479
Turbidity, specific, 411–412

T

Taylor instability, 347
Temperature dependence
  in interfacial tension, 268–269
  in surface tension, 245–247
Tetrahydrofuran, 7
Thermal expansion coefficient
  LCST phenomena and, 147
  molecular weight and, 149
Thermal optical analysis, 20, 193
Thermal pressure coefficient, in phase separation behavior, 150
Thermorheologically complex materials,
  time–temperature dependence for, 383–388

U

Upper critical solution temperature, 21
  increasing of, 130
  phase separation behavior and, 146–151
  systems exhibiting, 146
Urethane, at room temperature, 39

V

Van der Poel–Smith equation, 363
Van der Waals attraction, 287–288
Vinyl acetate, at room temperature, 37
Vinyl butyral, at room temperature, 38
Vinyl chloride, at room temperature, 37, 41
Vinyl chloride compatibility pairs, 35
(Vinyl chloride)-*co*-(vinyl acetate), at room temperature, 39–40
Vinylidene chloride, at room temperature, 38
Vinylidene fluoride, at room temperature, 38
Vinyl methyl ether, 35
Vinyl polymers, surface tension of, 261
Vinyl pyrrolidone, at room temperature, 38

Viscoelastic fluids
  cocurrent flow of, 307–310
  constitutive relationships in rheology of, 298–301
  rheology of, 288–304
  rigid axisymmetric particle suspensions in, 313–314
  rigid spheroid suspensions in, 310–318
  secondary flow phenomena in, 303–304
Viscoelastic response, modulus composition dependence and, 367–369

W

Weak boundary layer theory, adhesion and, 281–282
Wide-angle x-ray scattering, 213–216

X

X-ray scattering
  in polymers, 398–400
  small-angle, 212–215
  wide-angle, 213–216